MOSBY'S ACE the BOARDS™

Specialty
Clinical Sciences

MOSBY'S USMLE step 3 REVIEWS

Joseph L. Donnelly III, M.D.
Senior Resident, Department of Emergency Medicine,
Johns Hopkins Hospital,
Baltimore, Maryland

Randolph B. Lipscher, M.D.
Resident, Department of Emergency Medicine,
Strong Memorial Hospital,
University of Rochester,
Rochester, New York

Specialty Clinical Sciences

Mosby

St. Louis Baltimore Boston Carlsbad Chicago Naples New York Philadelphia Portland
London Madrid Mexico City Singapore Sydney Tokyo Toronto Wiesbaden

Mosby
Dedicated to Publishing Excellence

Printed in the United States of America

Mosby–Year Book, Inc.
11830 Westline Industrial Drive
St. Louis, Missouri 63146

Library of Congress Cataloging-in-Publication Data

Donnelly, Joe, M.D.
Mosby's USMLE step 3 reviews—specialty clinical sciences / Joe Donnelly, Randy Lipscher.
p. cm.—(Mosby's ace the boards)
Includes bibliographical references and index.
1. Medicine—Examinations, questions, etc. 2. Medicine—Outlines, syllabi, etc. I. Lipscher, Randy. II. Title. III. Series: Ace the boards.
[DNLM: 1. Clinical Medicine—outlines. 2. Clinical Medicine—examination questions. WB 18.2D685m 1996]
R834.5.D66 1996
616'.0076—dc20
DNLM/DLC 96-29093

ISBN 0-8151-2734-0 (IBM)
ISBN 0-8151-2756-1 (MAC)
99 00 01 02 03 / 9 8 7 6 5 4 3 2

CONTRIBUTORS

Carolyn D'Ambrosio, M.D.
Fellow, Department of Pulmonology and Critical Care,
Yale University School of Medicine,
New Haven, Connecticut

Michael J. Apostolakos, M.D.
Assistant Professor of Medicine,
University of Rochester School of Medicine;
Director, Medical Intensive Care Unit,
Department of Medicine,
Strong Memorial Hospital,
Rochester, New York

Edward S. Bessman, M.D., F.A.C.E.P.
Assistant Professor,
Department of Emergency Medicine,
The Johns Hopkins University School of Medicine,
Baltimore, Maryland

Dennis Blom, M.D.
Resident, Department of Surgery,
Strong University Hospital,
Rochester, New York

Calvin Chiang, M.D.
Clinical Instructor,
University of Rochester School of Medicine;
Department of Emergency Medicine,
Strong Memorial Hospital,
Rochester, New York

Michael J. Goebel, M.D.
Resident, Department of Orthopedic Surgery,
University of Florida,
Gainesville, Florida

David L. Greenberg, M.D.
Resident, Department of Diagnostic Radiology,
Strong Memorial Hospital,
Rochester, New York

Jennifer Griggs, M.D.
Resident, Department of Internal Medicine,
Hematology/Oncology Units,
Strong Memorial Hospital,
Rochester, New York

Mary-Ellen Guido Dean, M.D.
Fellow, Department of Nephrology,
Strong Memorial Hospital,
Rochester, New York

Karen Chodack Johnston, M.D.
Neurovascular Fellow,
Department of Neurology,
University of Virginia Health Sciences Center,
Charlottesville, Virginia

Albert P. Jones Jr., M.D.
Clinical Associate Professor,
University of Rochester School of Medicine;
Department of Obstetrics and Gynecology,
Strong Memorial Hospital,
Rochester, New York

Alan S. Krasner, M.D.
Assistant Professor of Medicine,
Division of Endocrinology and Metabolism,
Department of Internal Medicine,
The Johns Hopkins University School of Medicine,
Baltimore, Maryland

Ann Marie Lenane, M.D.
Assistant Professor of Emergency Medicine and Pediatrics,
University of Rochester School of Medicine;
Department of Emergency Medicine,
Strong Memorial Hospital,
Rochester, New York

Mary K. Marrocco, M.D.
Senior Instructor,
Department of Psychiatry,
University of Rochester School of Medicine,
Rochester, New York

Joseph Modrak, M.D.
Senior Instructor in Medicine,
Director of Graduate Education,
Sleep Disorder Center of Rochester,
Rochester, New York

Marc S. Orloff, M.D.
Associate Professor of Surgery,
University of Rochester School of Medicine;
Department of Surgery,
Strong Memorial Hospital,
Rochester, New York

Ronald D. Plotnik, M.D.
Assistant Professor of Ophthalmology,
University of Rochester School of Medicine;
Department of Ophthalmology,
Strong Memorial Hospital,
Rochester, New York

Sandra M. Schneider, M.D., F.A.C.P.
Professor, University of Rochester School of Medicine;
Chair, Department of Emergency Medicine,
Strong Memorial Hospital,
Rochester, New York

Philip H. Smith, M.D.
Fellow, Department of Allergy and Immunology,
Strong Memorial Hospital,
Rochester, New York

W. Zoe Stitt, M.D.
Resident, Department of Dermatology,
Strong Memorial Hospital,
Rochester, New York

Ellis Tavin, M.D.
Instructor in Surgery,
Division of Plastic Surgery,
University of Rochester School of Medicine,
Rochester, New York

Peter White, Jr., M.D.
Assistant Professor of Medicine,
Division of Pulmonary and Critical Care Medicine,
Department of Internal Medicine,
The Johns Hopkins University School of Medicine,
Baltimore, Maryland

Daniel J. Williford, M.D.
Associate Professor,
Department of Internal Medicine,
University of Rochester School of Medicine,
Rochester, New York

Sophie M. Worobec, M.D.
Assistant Professor,
Department of Dermatology,
University of Rochester School of Medicine,
Rochester, New York

Preface

This book is a powerful tool designed to permit you to review for the USMLE quickly, efficiently, and comprehensively. It was written by resident physicians who recently took—and 'aced'—the Step 3 Exam. This review book is **designed for the busy resident physician** who wants to **ace the last** of the USMLE exams.

You may read all the written text and obtain a full understanding of a particular subject, or **read only the bold,** which will give you a **condensed (abstract) version** of the subject. **Reading the full text** takes approximately **5 minutes per chapter. Reading the bold** takes approximately **20 seconds per chapter.**

If you want more information after reading the bold text, you may simply refer to the unbolded portions. We suggest that you try reading the bold highlights of any chapter to determine how easily this new format makes information accessible.

The material reflects the scope and degree of knowledge necessary to pass the Step 3 exam. The presentation format is consistent and includes

- *basic science*
- *history and physical examination*
- *laboratory and diagnostic studies*
- *formulating the diagnosis* and
- *managing the patient.*

Marginal icons point out each area so you can find exactly what you're looking for, without wasting time.

USMLE-style questions are included at the end of each chapter. Fully explained answers (even rationales for why incorrect choices are wrong) are provided on perforated pages at the end of the book. You can tear out the answer pages and place them right next to the questions to check your work. **Hundreds of additional questions** are included on the free diskette. These questions are categorized first by specialty and then by the USMLE categorization scheme to identify weak areas for further study. The dermatology questions contain full color pictures to test your knowledge of skin disorders.

We designed this textbook to **include only** the **subjects** that the USMLE testing authority indicated would be **presented on** the **actual test. We included only** the **most relevant** aspects of that **material.** Lastly, and perhaps most importantly, **we bold-faced** the most **essential information so** that **you could review** this **entire book quickly** (within several hours).

We hope you find the review text and practice questions in *Ace Specialty Clinical Sciences* helpful, and wish you the best of luck on the USMLE and in your future career.

Joseph L. Donnelly III, M.D.
Randolph B. Lipscher, M.D.

Acknowledgments

We would like to thank Irina Gindentuller, M.D. for carefully reviewing the manuscript. We also appreciate the efforts of Betty Valles for typing the manuscript. Emma Underdown and Christy Wells of Mosby have been instrumental in making this book happen, and Anne Gassett of Graphic World Publishing Services has made the production process as smooth as possible.

To my wife, Heather, and children, Joseph IV and Allison
JLD

To my strong and wise uncle, Arthur A. Lipscher
RBL

Test-Taking Strategies

Suzanne F. Kiewit, M.Ed., PhD

To perform well on the USMLEs, it is imperative that you begin with a **plan.** Preparation time is at a premium, so you will want to be as efficient and effective as possible by planning well.

MONTHS AHEAD OF THE EXAM

- Sit down with a blank calendar and block in your commitments: classes, final exams, scheduled events.
- Include time for activities of daily life: eating, sleeping, exercising, socializing, banking, maintaining your home, and so forth.
- The remaining time is available for study/review.
- Determine an orderly approach to the material you need to cover that fits your particular set of needs (e.g., subject-by-subject approach, systems approach, pathologic state approach).
- Assign the remaining time to content areas. This is done in various ways: material covered freshman year first, easiest first, least comfortable material first, detailed subjects last, whatever. Your plan should reflect your goal: to maximize your score.
- Establish a warm-up, which may consist of breaking the tension in major muscle groups (neck rolls, shoulder rolls, etc.), a quick visualization of you performing successfully, or a brief meditation. Practicing this warm-up routine before each of your study sessions will make it a familiar activity that helps you learn effectively, as well as take exams effectively.
- Designate time at the end of your study period for panoramic review. Depending on your needs, that might be a week or just several days before the exam.
- Plan for feedback on your efforts. Schedule time for answering questions on the material you are reviewing and for taking at least one mock comprehensive exam.
- Do the comprehensive exam midway through your study period so that you can refine your efforts to reflect the degree of your performance.

DAYS AHEAD OF THE EXAM

- Divide each day into thirds: morning block, afternoon block, and evening block.
- Consider the time of day that is most productive for you and do the most difficult or least favorite material at that time.
- Assign more blocks of study to those areas requiring the most review to reach a comfortable knowledge level.
- A popular way to use blocks is to pair subjects or materials. For instance, pair strong content with weaker content so that you are not always in the position of not knowing material (which would invite negative feelings or ineffectiveness). Or pair a conceptual subject with a detail subject, such as physiology with anatomy, so that you are not always doing the same kind of thinking (this invites positive effort).
- Use your most productive blocks of time for actual study/review. Use the nonpeak times for reinforcement of material covered or feedback by answering questions on material that you have covered.

Planning the Blocks

- Once you determine time allocations for each content area and an orderly approach that fits your needs and goals, you want to specify what you plan to do during each block.
- Be specific as to content area, material to study, and task; for example, MICRO: review chart on viruses; PHYS: answer questions on renal; and so forth.
- Each study block will last approximately 3 to 4 hours. To be most efficient and effective, plan to take a 5- or 10-minute break every hour. If you are having difficulty getting into the study mode, plan to study for 25 minutes, then take a 5-minute break. Reserve longer breaks for switches between subjects. Get up and move around on breaks.

BEFORE THE EXAM

Knowing about the USMLEs helps demystify them. In general, the USMLE exams are a 2-day, four-book examination. Approximately 3 hours are allotted per exam book. Each day you will complete one book of about 200 items in the morning and another book in the afternoon.

In each exam book, questions are organized by question type, not content. Specific directions precede each set of questions. Only two question formats are used: one-best-answer multiple choice items (which typically come first on the exam) and matching items (toward the end of the exam). Students have reported that one-best-answer items make up the bulk of the exam (70% to 75%) and negatively stemmed items make up only 10% to 15% of the questions (Bushan, Le, and Amin, 1995). Matching sets, which make up about 15% to 20% of the items, may include short leading lists or long leading lists of up to 26 items from which to choose.

From year to year there may be variations in the organization and presentation of both content and item formats. It would be wise to read the National Board of Medical Examiners' *General Instructions* booklet, which you will receive when you register. This booklet contains descriptions of content, item format, and even a set of practice items. Be certain you read this booklet and familiarize yourself with the questions.

You can further maximize exam performance by taking control. Adults tend to perform better when they feel that they have a measure of control. For the USMLE it is easy to feel out of control. You are told what time to arrive, where to go, what writing instrument to use, when to break the seal, and so on. You want to assume control of as many aspects as possible to maximize your performance:

STUDY Follow the sage advice of planning your work and working your plan to maintain satisfactory preparation with regard to study.

SLEEP Get a good night's rest. Sleep needs vary, but 6 hours is usually minimum. Try to get the appropriate amount of sleep that you require.

NUTRITION Maintain proper nutrition during both study time and exam time. Eat breakfast. Choose foods that help keep you on an even energy level. Eat light lunches on exam days. If you have a favorite food and can take it with you, treat yourself.

MEDICATION It may be cold, flu, or allergy season. Take no medications that may make you drowsy.

CLOTHING Heating and cooling systems are rarely balanced enough to suit everyone. Wear articles of clothing that can be added or removed as necessary. Strive for personal comfort.

READINESS Develop physical, mental, emotional, and psychologic readiness for the exam. Keep your thoughts about the exam and your preparation efforts running positively. You must believe that *you can do this!*

ARRIVAL Plan to arrive as close to the designated time as possible and still allow yourself sufficient time to check in. Keep to yourself so that other people's anxieties will not affect you. Take care of personal needs. Find your seat.

ACCLIMATION Settle in and get comfortable. Take several deep breaths. . . RELAX. A relaxed mind thinks better than a tense one—it's that old "fight-or-flight" syndrome. Do your warm-up routine to help you relax.

ATTENTION Pay attention to the proctor. Complete all identification material as required. Read all instructions carefully. Ask for clarification as needed. Do not open your booklet until told to do so. After you are told to break the seal, quickly glance through the whole test to see how it is set up and how questions are organized. Again, you want to take control of the situation. A quick review eliminates surprises and allows you to develop a plan.

DURING THE EXAM

Plan Your Approach

There are numerous approaches to answering questions. Answer questions in the order that appeals to you. Doing the easier ones first may give a psychologic boost; however, the ones you skip may stay on your mind and cloud your thinking.

Another approach is to answer each question in sequence. Start with the first one in the section with which you begin and fill in an answer for each question. Do not leave any blanks! The theory behind this is that if you spend any time at all on an item, you should mark your best response at that time and go on. If you are not certain of your choice, mark "R" in the test booklet for review and return to it later as you have time.

Some students plan to do the matching items first. Matching items are the last set of questions in the booklets. If you prefer matching items, this is a reasonable plan because it helps you get started with items about which you feel confident. It is also reasonable because matching items are not good items on which to guess if you run short of time. **You** must decide the order in which you want to do the questions.

Complete the Bubble Sheet or Answer Card Carefully

There are two schools of thought on this matter. One is to fill in the bubble sheet **item by item** as you go. This method minimizes transcription errors. The other method is **block transfer.** Complete a logical chunk of questions in the test booklet (one or two pages) and then transfer responses to the bubble sheet. Be sure that the last question number on the page is the last numeral you blacken. This method saves time and offers a mini-mental break at the end of each block. Such minibreaks help decrease fatigue during a long exam. Choose the method that will work for you and *practice* it as you take prep questions.

Budget Your Time

If only the allotted time and the number of questions were considered, you would have approximately 54 seconds per question. Obviously, some questions may go more quickly and balance out the ones that take longer. To keep track, you need a pacing strategy. A good strategy is to establish checkpoints at 30-minute intervals. When you overview the booklet, circle the numerals corresponding to where you should be at 30 minutes, 60, 90, 120, and so on. For example, if you have 200 questions on a 3-hour exam, you should be at question number 33 at 30 minutes. As you complete the exam, check your time at the circled items. This technique keeps you from watching the clock too much, yet permits multiple opportunities to adjust your pacing.

If you find yourself spending too much time on any one question, select your best choice at that time, mark an answer, and "R" it for later review. The point is to keep going. Laboring too long on one question limits you from responding to other items you may know well. Remember, controlling your time helps you maximize your points.

ANSWERING THE QUESTIONS

- **Read and *understand* the stems and alternatives.**
  The most frequent error made on exams is misreading or misinterpreting the various aspects of a question. The **stem** is the introductory question or statement. The **alternatives** are the options from which you select the one best response. To encourage reading and understanding, use a process.
- **Follow a process to answer questions.**
  1. Quickly read the stem.
  2. Quickly read the options. (Combined, the first two steps create a preview of the item.)
  3. Carefully read, underline, and mark the stem in a timely fashion.
     - Selectively underline key words and phrases.
     - Pay attention to nouns, verbs, and modifiers.
     - Circle age and gender.
     - Note data in telescopic form (e.g., ↑ BP).
     - Graphically represent material if it helps you to understand (e.g., diagram the renal tubule to answer a question about reabsorption).
  4. Carefully read each alternative. Mark as appropriate.
- **Consider each alternative as one in a series of true, false, or not sure (?) statements.**
  Read each alternative. Rather than slashing out the ones you eliminate, work with each one and designate it as **true, false,** or varying degrees of **true/false/?.** This marking strategy requires you to make judicious decisions about alternatives relative to the stem. It also provides a record of your original thinking, which will save you rethinking time if you need to reconsider a question. Practice this strategy on preparation questions so it becomes second nature.
- **Avoid premature closure.**
  Sometimes you may read a question and anticipate a response. Such a reaction helps focus your attention. However, be sure to read *all* the options so that you are selecting the *best* response. In one-best-answer multiple choice questions, there is one *best* and several *likely* responses. Avoid being misled; consider all the alternatives.
- **Be leery of negative stems.**
  Negative stems require shifting to a negative thinking mode to determine which alternatives are not correct. You can avoid this shift by using this strategy:
  - Circle such words as *except, least, false, incorrect, not true* to raise your awareness of them.
  - Cross out the negative and read the stem as though it were a positive.
  - Mark each option as T/F/?. The F option will then be the appropriate choice.
- **Keep your original answers.**
  To change or not to change answers is a difficult decision. The answer depends on a person's previous history. If you are the kind of student who, if you change answers, changes them from wrong to right, then selectively changing answers may be worthwhile. If, on the other hand, your past experience has been to change right answers to wrong answers, selectively changing answers is probably not a good idea. Good performers change answers, but only if they have reason, such as acquired insight or discovery of misreading or misinterpretation.

- **Maintain an even emotional keel.**
  If a question upsets you, calm yourself. Take several deep breaths. Tell yourself, "I can do this!" Give yourself a mental or physical break. Pay special attention to the next two or three questions after a bout of emotional uneasiness. It is possible to miss items when attention is diffused.

THINKING THROUGH QUESTIONS

- **Use logical reasoning and sound thinking.**
  - Read the item carefully. After careful reading, ask "What is this question really asking?" Restate it so that you know what is being asked.
  - Engage in a mental dialogue with the question. Talk to yourself about what you do know. Always start with what **you** know. Verbalize your thinking.
  - If a diagram or graphic representation is included, orient yourself to it **first** so that the options do not lead your thinking.
- **Use information found within the questions themselves to help you answer others.**
  There will not be "gimmes" on a nationally standardized exam. However, there may be items or graphics that trigger remembrances.
- **Create a diagram, chart, map, or graphic representation of given information.**
  Material that is visually presented usually helps clarify thinking. Use selective, quick sketching as warranted.
- **Reason through information like a detective.**
  - Sift through the details (preview).
  - Determine the relevant information (selectively mark).
  - Put the clues together as in solving a puzzle (reason).
- **Read carefully and note key descriptors.**
  - Note words such as *chronic, acute, greater than, less than, adult, child.*
  - Attend to prefixes such as hyper-, hypo-, non-, un-, pre-, post-.
- **Analyze base words and affixes.**
  Studying a question at the word level may help you remember salient information. Look for base words or related words. Determine Latin or Greek word parts and use their meanings to assist you.
- **Consider similar options equally.**
  If you mark one alternative as "false" for a particular reason and another option is qualified for the same reason, it's probably "false" as well.
- **Trust the questions.**
  The questions are designed to determine if you have a working knowledge of the material. They are not written to trick you. You need to believe that your medical school curriculum and your study efforts prepared you for most of the questions.
- **Meet the challenge of clinical vignettes.**
  Longer, vignette items challenge you to discern the relevant from the irrelevant material. In doing so, you are given multiple clues to consider. To effectively handle the vignette item, follow this strategy:
  - Scan the stem and read the first several lines.
  - Skip to the end of the stem and read the last several lines.
  - Check the alternatives to narrow your focus.
  - Now that you know what the question is about, go back to the stem; read and mark what's important to your informed decision making.
  - Make good T/F/? decisions.
- **Reread your underlines and markings when you are down to two choices, at 50/50.**
  By the time you work through a stem and numerous alternatives, it is easy to lose the gist of the question. Checking your focus by rereading only the underlines ensures that you are answering the question being posed.

ANSWERING MATCHING ITEM SETS

Matching items are used to measure your ability to distinguish among closely related items. They require knowledge of specific sets of information. As you study, be alert to potential material that could be tested in this way.

Matching items can be formatted in two ways. **Short leading list matching** items include a set of five, lettered options followed by a lead-in statement and then several numbered stems. **Long leading list matching** items include a set of up to 26 lettered options, followed by a lead-in statement and then several numbered stems.

To efficiently deal with a short leading list item, consider it as an upside-down multiple choice item with the same repeated options. To handle it effectively, do the following:

- Scan the list; determine the topic.
- Read the lead-in statement; determine the focus.
- Quickly read the stem; then read and mark key words.
- In the left margin, create a grid with A, B, C, D, E at the top.

- Make good T/F/? decisions about each stem, marking them in the grid. In this way you can see the pattern of your responses. Similarly, a grid with the item numbers can be drawn beside the leading list and responses marked there.

Handling long leading list matching items effectively requires some modifications in the process. It is not efficient to make T/F/? decisions about each option, so follow this strategy:

- Scan the list; determine the topic.
- Read the lead-in; determine the focus.
- Read a stem and generate your own response.
- Narrow the focus. Put a check mark by those related options in the long list.
- Read and mark specifics in the stem to differentiate among those alternatives you marked.
- Make good T/F/? decisions.

For each stem, mark the narrowed-list options with a different symbol (star, dash, etc.). Items are listed in logical order, alphabetically or numerically. When looking for an option such as "xanthinuria," do not start at the beginning of the list. Looking in the appropriate place saves valuable seconds.

TEST WISENESS

How a question is worded can often influence your response to it. Most clues about "test psychology" are a function of the way in which a question is worded—test constructors cannot rename body parts, drugs, diseases, and so forth. Being aware of the psychology behind the wording can often help you answer the test question.

Using techniques of test psychology to arrive at a correct answer has limited value on standardized exams because those who construct the exams are well aware of the use of these techniques. Nonetheless, being wise to these techniques of test psychology may add another point or two to your score, and they can also enhance your sense of control. Knowing these techniques provides additional strategies to employ should the question temporarily stump you.

The best way to take any exam is to be totally prepared with a strong knowledge base and personal test confidence. The following techniques should be used only if you have exhausted your knowledge base, eliminated all distractors, and cannot come up with the answer even with logical thinking and sound reasoning. Such techniques are **not** a substitute for knowledge, nor are they foolproof.

- **Identify common ideas or themes within the options and between the stem and options.**
  - Circle repeated words in the options.
  - Select the option with the most repeated words or phrases.
  - Circle words repeated in both stem and options.
  - Select the option that contains key words or related words from the stem. This is a stem/option repetition.
- **Beware of words that narrow the focus or are too extreme because they tend to be incorrect.**
  Circle such words as *all, always, every, exclusively, never, no, not, none.*
- **Options that are look-alikes are good candidates for exclusion.**
- **Note qualifiers that broaden the focus because they may be correct.**
  Circle words such as *generally, probably, most, often, some, usually.*
- **Identify antonyms or two opposing statements as potentially correct options.**
  Test constructors may use pairs of opposites, so this tip may lose its effectiveness.
- **Select the most familiar-looking option.**
  Always go from what you know. Alternatives with unknown terms may be likely distractors.
- **Select the longest, most inclusive answer.**
  This would include "All of the above" as a strong potential response.
- **In numerical items, knock out the high and low alternatives and select one in the middle that seems most plausible.**
- **In negatively stemmed questions, categorize responses; the one that falls out of the category is a likely candidate.**
- **Mark the same alternative consistently throughout the test if you have no best guess and cannot eliminate distractors.**
  Before the test, decide which letter (A, B, C, D, E) will be your choice. In this way, if you have given a question your best effort and cannot decide, mark your favorite response and move to questions that cover more comfortable material.

AFTER THE EXAM

- **Between sessions and overnight:**
  - Take a well-deserved break. Eat nutritionally.
  - If you feel the urge to study, study material that is comfortable, from a source with which you are familiar (e.g., personally developed study cards or your annotated review book).
  - If you discovered a recurring "theme," you might desire to consult that set of information.

- Do something pleasurable. Relax. Get a good night's rest.
- **After the final booklet:**
  - Recognize that this exam is a measure of what you know on a given day for a given set of information at a given point in time. Keep a reasonable perspective.
  - **Celebrate!**

References

Bushan V, Le T, Amin C: *First aid for the USMLE Step 1*, ed 5, Norwalk, Conn, 1995, Appleton & Lange.

Suzanne Kiewit, M.Ed., PhD, Learning Specialist,
Office of Academic Development and Advising,
Northeastern Ohio Universities,
College of Medicine

Contents

MOSBY'S ACE the BOARDS™

Specialty
Clinical Sciences

Chapter 1

Neurology

Alzheimer's Disease

■ Basic Science **Alzheimer's disease** is the **most common form of dementia.** The etiology of Alzheimer's disease remains unclear, but there is a suggestion that genetics may play a role. Some familial forms of Alzheimer's disease exist that have autosomal dominant inheritance, and abnormalities have been noted on chromosomes 14, 19, and 21. There is a known association between Alzheimer's disease and Down syndrome (trisomy 21).

In patients with Alzheimer's disease, there is a **loss of choline acetyltransferase**, which is **important for the synthesis of acetylcholine.**

The pathological hallmarks of Alzheimer's disease are **neuritic plaques** (containing amyloid proteins) and **neurofibrillary tangles** (paired helical filaments containing tau proteins) in the cortex and hippocampus. There is often **gross atrophy of the brain,** mostly in the temporoparietal regions and frontal poles.

■ History and Physical Examination The risk for Alzheimer's disease increases with **advancing age.** Patients present with a decline in cognition and general functioning. They have **abnormalities** in at least two of the following cognitive spheres: **memory, language, spatial orientation, personality and social skills, abstract reasoning, and judgment. Impairment of memory is usually the first symptom.** As the disease progresses, patients may become apathetic and may develop behavioral disturbances resulting in **restlessness and wandering.** Patients eventually become totally **unable to care for themselves.** They are **unable to speak or walk,** and they **lose basic biological functions** such as sphincter control. Patients may also develop seizures.

On neurologic examination, mental status testing reveals cognitive impairment in two or more spheres. **Disturbances of affect and behavior** may also be noted. The **remainder of the neurologic examination** should be essentially **normal,** except in the later stages. **Focal findings early in the course should alert one to other diagnoses, such as multi-infarct dementia.**

■ Laboratory and Diagnostic Studies Brain imaging sometimes shows atrophy, although the presence of atrophy does not necessarily mean that someone has Alzheimer's disease. Diagnostic **imaging and laboratory tests are generally performed to rule out reversible causes of dementia** (discussed later). Sometimes formal neuropsychological testing is performed.

■ **Formulating the Diagnosis** The diagnosis of Alzheimer's disease requires the **clinical diagnosis** of a **dementia syndrome without another explanation** and a postmortem confirmation of typical pathological findings.

The first step is to determine that the patient is actually demented. Demented patients have acquired, fixed deficits in at least two cognitive spheres. A **"pseudodementia"** should be ruled out. This term refers to the **transient and treatable** cognitive deficits related to **depression. Confusional states,** such as those **secondary to drugs or infections, should also be ruled out.**

If it is determined that the patient is demented, then **reversible causes should be ruled out.** Laboratory tests should include **routine studies** (complete blood count and full chemistry profile) as well as **RPR (syphilis),** B_{12}, **folate, and thyroid function tests.** HIV testing should also be considered in some circumstances. A computerized tomography **(CT)** scan **of the brain** should be considered to rule out a mass lesion, normal pressure hydrocephalus, or subdural hematoma.

■ **Evaluating the Severity of Problems** Alzheimer's disease is a progressive **debilitating disease** that ultimately leads to a **loss of ability to care for oneself** and a **loss of many basic biological functions. Patients need observation because poor judgment predisposes them to accidents.** They may be unable to eat, so alternative forms of feeding may need to be considered. They are at risk for aspiration pneumonia so appropriate precautions should be taken.

■ **Managing the Patient** The only agent currently available for treating cognitive deficits in Alzheimer's disease is **tacrine hydrochloride.** Tacrine is a centrally active, noncompetitive reversible **cholinesterase inhibitor.** This agent is thought to enhance cholinergic function. Other treatments are directed at target symptoms and may include psychotropic medications for the variety of psychiatric and behavioral disturbances that patients often experience. Medications that can worsen mental status should be avoided, and infections should be treated promptly. Patients need extensive supportive care and assistance with activities of daily living.

Brain Abscess

■ **Basic Science** Brain abscesses may develop from multiple causes. These include the following:

- **Direct extension** from an infection in the middle ear or paranasal sinuses
- **Blood-borne spread** from a distant site
- **Direct inoculation** from a penetrating wound

Abscesses caused by direct extension of an ear or sinus infection are usually located in the frontal or temporal lobes near the infected source.

Abscesses caused by **blood-borne infection** may occur anywhere in the brain and are generally **related to bacterial endocarditis or pulmonary infections.**

Congenital heart disease and pulmonary arteriovenous

fistulas predispose one **to abscess formation,** since infected material can be shunted to the brain, bypassing the usual pulmonary filtration.

- **Multiple bacterial species may be found in a single abscess;** these may include common bacterial species such as *Staphylococcus aureus, Streptococcus, Enterobacteriaceae,* and anaerobes.
- The **most common causes of meningitis** (pneumococcus, meningococcus, and *Hemophilus influenzae)* are **rarely involved in abscess formation.**
- **Infants** commonly **suffer from gram-negative abscesses.**
- **Immunocompromised individuals suffer from fungal and parasitic abscesses** such as *Toxoplasma, Entamoeba, Aspergillus,* and *Candida,* besides the usual causes of an abscess.

Regardless of the cause of the abscess, the **pathological process is** the same. First, there is **inflammation** with **destruction** of **brain tissue.** The purulent material is then enclosed in a thick capsule of granulation tissue. **Surrounding regions become edematous** and produce a **mass effect** that causes damage to the adjacent brain tissue.

History and Physical Examination

Clinical symptoms of brain abscess may be attributed either to **local destructive effects of the abscess or to a mass effect in the brain.** Mass effects produce increased intracranial pressure.

Depending on the location of the abscess, local results may include **hemiparesis** (muscular weakness on one side of the body), **aphasia** (defective language function), **personality change, seizure, visual disturbance, or ataxia** (an impaired ability to coordinate movement).

Increased intracranial pressure can cause **papilledema, headache, nausea, vomiting, and cranial nerve palsies.** More severe increases in pressure may lead to a Cushing response, with bradycardia and hypertension. **Fever, meningeal signs, and other systemic manifestations are rare** unless they are related to the septic source of the abscess.

Laboratory and Diagnostic Studies

- **CT or magnetic resonance imaging (MRI) with intravenous contrast will show a ring-enhancing mass,** unless the abscess is not yet encapsulated. There may also be extensive surrounding edema apparent.
- CBC and other **blood tests will usually be normal,** unless other processes are also at play.
- **Spinal fluid is almost always sterile.**
- **Lumbar puncture should be avoided in patients with signs of mass effect,** since this may precipitate herniation. Therefore **imaging should always be done before lumbar puncture, especially if there are focal neurologic signs.**
- **Biopsy is often necessary** to rule out tumor and to culture the offending agent.

- **Formulating the Diagnosis** The diagnosis of brain abscess is made primarily through imaging studies, with or without biopsy.
- **Evaluating the Severity of Problems** **Signs of increased intracranial pressure** demonstrated either by physical examination or by CT scan are alarming since **brain herniation** may follow and rapidly cause **death.**
- **Managing the Patient** Often with **4 to 6 weeks of appropriate antibiotic therapy,** multiple abscesses will resolve fully. A physician's choice of effective antibiotics is guided by two factors:

  - Identification of the organism (biopsy and culture result)
  - Biologic properties of the antibiotic (for example, whether the antibiotic can penetrate the blood-brain barrier)

  Response to treatment should be monitored with repeated imaging studies. An anticonvulsant (e.g., phenytoin) can be used **if seizures occur.** The adjunctive use of corticosteroids to reduce edema remains controversial. **Surgical management** becomes necessary **when the abscess does not respond to appropriate antibiotics,** or when there is acute deterioration from increased intracranial pressure.

Brain Death

- **Basic Science** Death is defined as the **total cessation of life.** Death is determined by fulfilling certain types of defined criteria.

  - The first criterion that can be used to establish death is the **irreversible cessation of circulatory and respiratory function.**
  - The second criterion that can be used to determine death is the **irreversible cessation of all brain function,** including the brain stem (brain death).

  Brain death is usually caused by severe **head trauma, subarachnoid hemorrhage,** or severe **hypoxic or ischemic brain injury.** Less frequently, severe metabolic disease or intoxication causes irreversible loss of brain function. This discussion pertains to **adults** and children greater than 5 years old. A prolonged observation period and a very individualized evaluation should occur in children under age 5 before making the diagnosis of brain death.
- **History and Physical Examination** Establishing brain death mandates that the physician identify a **cause of coma** that is sufficient to account for the loss of brain function (e.g., status post a motor vehicle accident). If such history is not available, then **reversible causes** of severe coma must be considered and ruled out. **Hypothermia, drug intoxication, neuromuscular blockade,** circulatory failure **(shock),** and **major metabolic derangements** may all produce reversible coma.

  In order for a person to be declared brain dead the physical diagnosis must demonstrate **coma, absence of brain stem reflexes, and apnea.**

  The systolic **blood pressure must be greater than 90 mm Hg,** and a **core body temperature $\geq 32°$ C so that there is no**

dispute about potentially reversible causes of coma or apnea. Furthermore, the physical should demonstrate **no cerebral motor responses** such as decorticate posturing. Spinal reflexes, however, may be preserved. **All brain stem reflexes** (i.e., pupillary light response, oculocephalic reflex, oculovestibular reflex, cold caloric test, corneal reflex, gag reflex, cough reflex, and jaw jerk) **must be absent.**

A repeat clinical evaluation, usually between 6 and 24 hours, is recommended.

- **Laboratory and Diagnostic Studies** An **apnea test** must be performed and **show no evidence of respiratory function.** Blood work should be obtained to **rule out metabolic derangements** (such as severe electrolyte disturbance) and hepatic and uremic disease. Screening for **drug intoxication** should be done.
- **Formulating the Diagnosis** **Reversible causes of coma should always be considered.** However, if reversible coma is ruled out and the three criteria (coma, absence of brain stem reflexes, apnea) of brain death are met, the patient may be pronounced dead.
- **Evaluating the Severity of Problems** Disorders that may mimic brain death are generally life-threatening as well and mandate immediate treatment.
- **Managing the Patient** Until the diagnosis of death is made, the treatment is supportive. Once the patient is declared dead by brain criteria the family should be notified of the death. Issues regarding **organ donation** may then be addressed.

Cerebral Vascular Disease

- **Basic Science** The brain receives blood from two different vascular systems, including the carotid and the vertebral. **Stroke** is defined as the **acute onset of neurologic dysfunction** secondary to cerebral ischemia, occurring in a **vascular territory,** lasting longer than **24 hours.** It is the number one cause of morbidity and the third most common cause of death in the United States, following heart disease and cancer. The two major categories of stroke are **hemorrhagic** and **ischemic.**

  Hemorrhages can be **intraparenchymal,** subarachnoid, subdural, or epidural. Intraparenchymal hemorrhagic stroke will be discussed here.

  - **Intraparenchymal hemorrhage** Intraparenchymal hemorrhage generally occurs in two patient populations: hypertensive patients and the elderly.

    - Hypertensive stroke is related to the rupture of **hypertensive microaneurysms.** The most common locations for hypertensive stroke are in the **basal ganglia, thalamus, pons,** and **cerebellum.**
    - In the elderly, intraparenchymal hemorrhage is often caused by **amyloid protein deposits** in the **microvasculature** leading to local areas of increased pressure and decreased vessel wall stability.

Less common causes of intraparenchymal hemorrhages are vascular malformations, trauma, coagulopathies, and ethanol abuse.

- **Ischemic strokes** Ischemic strokes are most commonly the result of **thrombosis, emboli,** or **lacunar infarcts.**

  Atherosclerotic disease is the principle pathologic entity in the formation of thrombus and emboli. Atherosclerotic plaques may rupture, causing platelet adherence to vessel walls. This process leads to the formation of a thrombus. A **thrombus** may grow so large that it **occludes the lumen of an artery,** denying blood supply to distal brain tissue. The thrombus may also fracture, sending emboli distally until they occlude a small vessel. Patients at greatest risk for **embolic events** are those with atrial fibrillation, cardiac thrombus, (secondary to myocardial infarction or cardiomyopathy), or bacterial endocarditis.

  Lacunar strokes result from progressive thickening of small vessels that result in **occlusion of penetrating arteries** (lipohyalinosis). **Chronic hypertension** and **diabetes mellitus** are the most important risk factors for lacunes. Lacunar strokes occur primarily in the basal ganglia, pons, and subcortical white matter.

  Ischemic infarcts, especially embolic events, can undergo hemorrhagic transformation.

■ **History and Physical Examination** The major **risk factors** for stroke include **hypertension, diabetes mellitus, smoking, atrial fibrillation, previous stroke,** valvular disease, coronary artery disease, coagulopathy, endocarditis, and dilated cardiomyopathy, and antiphospholipid antibodies.

Stroke usually presents as **unilateral neurologic dysfunction.** The symptoms usually present acutely though may progress over hours to days. **Cerebral edema** may develop, leading to **clouding of consciousness** or a potentially **fatal herniation** of brain tissue.

Specific neurologic symptoms are attributable to the area of brain infarcted. The following is a partial list of these symptoms by the vasculature affected:

- **Middle cerebral artery** Occlusion of this artery results in **contralateral paralysis, contralateral loss of sensation,** and **homonymous hemianopia.** When the dominant hemisphere is involved, **aphasia** may occur. When the nondominant hemisphere is involved, **sensory neglect, spatial disorientation,** and apraxia (loss of purposeful movement in the absence of paralysis) may occur.

- **Posterior cerebral artery** Occlusion results in contralateral homonymous hemianopia, contralateral sensory loss, and thalamic pain.

- **Vertebrobasilar system** Occlusion results in **cranial nerve abnormalities and sensory, motor, and cerebellar symptoms.** Ophthalmoplegia and pupillary abnormalities may occur. **Stroke in the vertebrobasilar system may be fatal.**

  The **physical examination** should focus on the following:

  - Establish **evidence of atherosclerosis or cardiac conditions** (e.g., detecting a third heart sound or carotid artery bruit).

- Perform mental status testing. **Decreased level of consciousness** is a common finding on physical examination in patients whose area of infarct is large or involves the brainstem.
- **Localize the area of brain affected.** The constellation of deficits on neurologic testing identifies the involved region.

■ **Laboratory and Diagnostic Studies** Laboratory evaluation distinguishes hemorrhagic from ischemic stroke and helps to determine etiology and severity.

- A **noncontrast head** CT scan can identify a hemorrhage almost immediately. The CT scan of an ischemic stroke is usually negative for 24 hours.
- Routine blood work (complete blood count and blood chemistries) and **coagulation studies** should identify evidence of infection, metabolic abnormalities, or coagulopathies.
- **Cerebral angiography is the gold standard** for evaluation of the cerebral vasculature. Angiography may be useful in detecting an area of vessel narrowing.
- **Carotid ultrasound is done** for presumed carotid atherosclerotic disease of the neck.
- **Transcranial Doppler examination** can visualize intracranial vascular disease.
- **MRI** may also be used for sensitive localization of lesions and noninvasive vascular evaluation.
- **Echocardiography** is helpful in the determination of potential cardiac or aortic sources of emboli.
- **Holter monitoring** can identify arrhythmias.

■ **Formulating the Diagnosis** The etiology and location of the stroke are determined by the history, physical exam, and laboratory evaluation. In patients with a gradual onset of neurologic symptoms (weeks, months) intracranial mass lesions should be ruled out with a contrast-enhanced CT scan.

In cases with a relapsing and remitting course, demyelinating disease should be considered.

■ **Evaluating the Severity of Problems** Patients with hemorrhages, brain stem infarcts, or large hemispheric infarcts require close observation and supportive care. A decline in **mental status, autonomic instability,** or evidence of **herniation** all require **emergent attention. Intubation may be necessary if respiratory function is threatened.**

Approximately 20% of stroke patients die in the hospital. The extent of recovery is dependent on the size of infarct, the patient's age, and health status.

■ **Managing the Patient**

- Acute treatment begins with stabilizing respiratory and circulatory function. **Intravenous normal saline** should be used if necessary. Hypotonic solutions may increase cerebral edema.
- **Unless the patient has a hemorrhage, blood pressure should be allowed to increase as high as a mean arterial**

pressure of **120 to 130.** High blood pressure ensures **cerebral perfusion** to threatened but not yet infarcted tissue.

- **Cerebral edema** is usually at **maximum levels by 72 to 96 hours.** During this period the **patient may require intubation and hyperventilation to reduce elevated intracranial pressure (ICP).**
- **Seizures** are a **common sequela** of embolic strokes. Acute seizures should be treated with benzodiazepines. Chronic seizures should be **treated with anticonvulsants.**
- Precautions against aspiration and deep venous thrombosis are indicated.
- Frequent neurologic evaluations should be done.

Specific therapy depends on the etiology of the stroke. **Anticoagulation** is frequently used in patients with embolic disease, venous thrombosis, critical carotid stenosis, and vascular dissection. In these patients **IV heparin** is given by continuous drip. The activated partial thromboplastin time **(aPTT)** is kept at approximately **two times control.** Long-term anticoagulation with **warfarin** may be necessary, and the prothrombin time **(PT)** international normalized ratio **(INR)** is usually kept **between 2 and 3.** Early **rehabilitation** may maximize recovery.

Secondary stroke prophylaxis includes a reduction of risk factors as well as medical therapy such as **aspirin, ticlopidine,** or **warfarin** depending on the etiology of the event. It is known that **warfarin** reduces the risk of stroke in patients with **atrial fibrillation.**

Patients with greater than **70% stenosis of the ipsilateral carotid artery** should be considered for **carotid endarterectomy.**

Infectious etiologies of stroke, such as **endocarditis or syphilis, require** the appropriate **antibiotic therapy;** bacterial endocarditis may require surgical intervention.

Headache

- **Basic Science** The pathophysiology of headache remains largely unknown. Most recently, **abnormal serotonergic pathways** have been considered as a possible etiology for primary headache syndromes.

  There are two types of headache syndromes: primary and secondary.

  - **Primary headache syndromes** include **migraine** (vascular headache) and **tension-type** (muscle contraction headache).
  - **Secondary headache syndromes** are caused by pathologic entities. **Masses, subarachnoid bleeds, and giant cell arteritis** may all cause secondary headaches.

- **History and Physical Examination** A complete and accurate history is the key to the correct diagnosis of the etiology of headache. The physical examination should include a complete neurologic examination.

  - **Primary headache syndromes** The **neurologic examination** is **normal** in primary headache syndromes.
    - ***Tension-type headaches*** Tension-type headaches are usu-

ally **bilateral and pressure or "bandlike."** They may be related to neck pain. Tension headaches can last hours or days and are not associated with autonomic symptoms (tachycardia, diaphoresis). Tension headaches do not generally limit activity.

— ***Migraine headaches*** **Migraine** headaches are commonly **familial** and are sometimes associated with a **history of motion sickness.** Migraine without **aura** (formerly "common") and migraine with aura (formerly "classic") are the most frequently encountered types. The pain is often **unilateral and throbbing;** however, a headache that always occurs on the same side should raise suspicion for a secondary etiology. Migraine may be **triggered** by **specific foods or smells.** The aura typically consists of **scintillating scotomata** (or flashes of light) that precede the headache by minutes and remit before the headache does. The headaches usually last hours. Headaches are associated with **photophobia and phonophobia.** The headaches cause the patient to experience **nausea and/or vomiting.** The patient generally curtails all activity. Episodes are **relieved by sleep.**

- **Secondary headache syndromes** Pathologic entities cause **secondary headaches.** Masses, subarachnoid bleeds, and giant cell arthritis may all cause secondary headaches.

— ***Mass lesions*** **Mass lesions** typically cause headaches that are **worse on awakening. Morning vomiting** is a characteristic sign. The location of the pain does not always correlate with the location of the mass. The headaches may at first be intermittent but eventually become more constant. Associated physical findings include **papilledema; focal motor, sensory, or reflex abnormalities;** and upper motor neuron signs **(spasticity, hyperreflexia, Babinski sign).**

— ***Subarachnoid hemorrhage*** **Subarachnoid, or intracranial parenchymal, hemorrhage** may cause a sudden, severe **"thunderclap"** headache with vomiting, instantaneous **loss of consciousness, or seizure.** Signs include **focal neurologic abnormalities, fundus hemorrhage, and meningismus** (that is, signs of meningeal irritation, such as neck stiffness).

— ***Temporal arteritis*** **Temporal arteritis** occurs primarily in patients **over 60** years old and is associated with **vision loss, ocular motility abnormalities, jaw claudication, weight loss, and general malaise.** The temporal artery may be tender and firm.

■ **Laboratory and Diagnostic Studies** There are no laboratory or diagnostic studies to support or refute the existence of primary headache syndromes. The laboratory workup of secondary headache syndromes is discussed in the following sections.

■ **Formulating the Diagnosis**

- **Primary headache syndromes** Migraine and tension headaches **demonstrate no associated focal neurologic abnor-**

malities. Migraine and tension headaches may be diagnosed based on their historical characteristics.

- **Secondary headache syndromes**

  • **Mass lesions** (such as tumor, abscess, and intracranial hemorrhage) **present with focal neurologic signs. Head CT scan or MRI is abnormal in these patients.** CT scan or MRI should be obtained in any patient with headaches and an abnormal neurologic examination.

  • **Subarachnoid hemorrhage** has a characteristic onset and should be considered in any patient that complains of a sudden severe headache. **Noncontrast head CT scan** should be performed, followed by **lumbar puncture** if the CT scan is negative.

  • **Arteriovenous malformations** should be considered in patients who have **consistently unilateral headaches and seizures.** A **bruit** may be heard over the malformation. **Head CT scan or MRI with contrast** is diagnostic.

  • New onset headaches in **elderly patients** suggest a secondary headache syndrome. Ruptured aneurysms, tumors, and **giant cell arteritis** all become increasingly prevalent with increasing age. In giant cell arteritis, the **sedimentation rate is elevated.** Diagnosis is confirmed by the presence of multinucleated **giant cells on temporal artery biopsy.**

  Headaches are a common complaint of patients with depression. Other causes of headache include subdural or epidural hematoma, meningitis, sinusitis, glaucoma, systemic hypertension, carbon monoxide intoxication, hypothyroidism, and caffeine withdrawal.

■ Evaluating the Severity of Problems

• Headache can be a sign of intracerebral mass lesion. The major complications of **mass lesions** are **herniation and death.**

• **Aneurysmal subarachnoid hemorrhage** often causes **death** before medical care arrives, making the identification of prior "sentinel" bleeding imperative.

• **Giant cell arteritis** can produce **blindness** by restricting blood flow in the ophthalmic and posterior ciliary arteries.

■ Managing the Patient

- **Primary headache syndromes Tension headaches are treated with aspirin, acetaminophen,** or other analgesic therapy. Chronic tension headaches can be treated with a combination of medication, physical therapy and psychotherapy, such as cervical support, muscle relaxation training, and management of underlying stressors.

  The treatment of migraine headaches includes the following:

  — ***Prophylaxis*** **Propranolol** 20 to 40 mg tid is useful in prophylaxis against migraine headaches in about 50% of patients. A calcium channel blocker (such as **verapamil**) may also prove effective.

— ***Treatment of attack*** Mild attacks may be treated with aspirin or codeine. The serotonin agonist **sumatriptan** has proven highly successful in ablating migraine headaches (70% relief of symptoms) and is now considered the therapy of choice for migraine headache. If sumatriptan proves ineffective, **ergotamines** may be tried.

- **Secondary headache syndromes**

  - Patients with suspected **mass lesions or subarachnoid hemorrhage** should be immediately referred to **neurosurgery** for evaluation. **Hyperventilation and corticosteroids** may prove effective in reducing edema and mass effect due to tumors or abscesses.

  - Treat **temporal arteritis** immediately with **high-dose corticosteroids (for example, prednisone 60 mg per day)** to prevent blindness. Further treatment involves 2 to 4 weeks of high-dose steroids.

Meningitis

■ Basic Science Meningitis is an inflammatory reaction of the meninges and subarachnoid space surrounding brain tissue. The major types of meningitis include bacterial, viral, fungal, and chemical (noninfectious).

- **Bacterial meningitis** In **bacterial meningitis** the cerebrospinal fluid (CSF) may become infected by the following:

  - Blood-borne pathogens (such as systemic infection or respiratory infection)
  - Local extension from adjacent infected tissue (for example, sinusitis, mastoiditis, otitis)
  - Direct invasion of pathogens (postsurgical or trauma)

  Infecting organisms vary with the age of the patient and the mode of entry.

  - **In neonates, group B or D streptococci, *E. coli,* and *Listeria monocytogenes* are the major pathogens.** Infection is blood borne, usually beginning with a respiratory infection.

  - **Infants and young children** (1 month to 7 years) **become infected with *H. influenzae,* pneumococci, and the meningococcus.**

  - Inoculation from a blood-borne infection (such as pneumonia) as well as extension of a localized infection (such as otitis media) are the main pathologic processes of bacterial meningitis in this age group.

  - **Older children and adults to age 50** predominantly become infected with **pneumococcus and meningococcus.** The source is typically blood borne.

  - In the **elderly and those with chronic diseases, pneumococci, *E. coli, Hemophilus, Pseudomonas,* and *L. monocytogenes* species** are the predominant organisms.

- The postneurosurgical patient has different pathogens because of the exposure of CSF to skin.

- **Viral meningitis** In **viral meningitis the Enteroviruses (coxsackie, echo, and polio) constitute 40% of all cases.**
- **Fungal meningitis** In **fungal meningitis the host** is usually **immunocompromised secondary to chemotherapy, chronic steroid use, or HIV. The patient presents with chronic complaints.** Etiologic agents include ***Cryptococcus, Coccidioides*** (prevalent in the southwestern United States), ***Candida, Blastomyces,* and *Histoplasma.***
- **Chemical meningitis Chemical meningitis** is an inflammatory reaction in the **absence of an infectious organism.** The reaction **can be the result of** either (1) a parameningeal process **(such as a tumor or abscess) producing irritation** or (2) the presence of a foreign substance (such as air, **medications,** or blood) in the CSF.

■ History and Physical Examination

- **Bacterial meningitis** In **bacterial meningitis the prodrome period is short.** The patient develops the **classic triad of headache, fever, and meningismus. Immunocompromised patients or** those with immature immune systems **(newborns) may not develop these symptoms.** Other symptoms may include photophobia, painful eye movements (from stretching the meninges), altered sensorium, seizures, and rash.

  Physical examination may reveal the source of infection (for example, otitis or pneumonia). Unique signs of meningeal infection include Kernig's sign (neck pain with knee extension while the hip is flexed) and Brudzinski's sign (leg flexion after passive neck flexion). Focal cranial nerve palsies may be present. The fundoscopic examination may show papilledema (a sign of increased intracranial pressure).
- **Viral meningitis** In **viral meningitis the prodrome may be more nonspecific (viral upper respiratory tract infection), but the presenting signs can be identical to bacterial meningitis.** Focal cranial neuropathies are uncommon because of the absence of a purulent exudate in the CSF.
- **Fungal meningitis** In **fungal meningitis** the infection is more **indolent,** and the **presenting signs may be more subtle.** The triad of headache, fever, and meningismus may not be present, and it is not uncommon for patients to present only with headache and a focal cranial neuropathy.
- **Noninfectious meningitis** In **noninfectious meningitis** patients may present with **headache, fever, and meningismus. Medications of note that can precipitate aseptic meningitis include sulfa-containing agents (such as trimethoprim/sulfamethoxazole) and nonsteroidal antiinflammatory agents (such as ibuprofen).**

■ Laboratory and Diagnostic Studies **Once meningitis is suspected, a lumbar puncture must be performed. This should be done after CT scan if mass lesion is suspected.** CSF should be sent for cell count, protein, glucose, and bacteriologic studies (Gram stain and culture). Viral or fungal cultures may be sent if nonbacterial meningitis is suspected.

- In bacterial meningitis CSF cell count and differential generally show a polymorphonuclear leukocytosis. The CSF usually demonstrates an **elevated protein level** because of increased endothelial cell wall permeability. **Glucose levels are usually low** secondary to abnormal transport across the blood-brain barrier and increased cellular metabolism.
- In viral meningitis, the CSF protein level is usually high due to an increased number of viral proteins. CSF blood glucose is unchanged. **The CSF cell differential may show a lymphocytic or monocytic leukocytosis.**
- **In fungal meningitis** CSF blood glucose is usually normal. **The CSF cell differential may show a lymphocytic or monocytic leukocytosis.**
- **In aseptic meningitis** the CSF can have a pleocytosis (increase in cellularity) and an increased protein content; the glucose is usually normal, and organisms are not identified by culture.

■ Formulating the Diagnosis **The combination of signs and symptoms just mentioned suggests the diagnosis of meningitis. Lumbar puncture confirms the diagnosis.** Other important studies include chest x-ray, blood counts, urinalysis, and culture studies to identify the source of infection. Important processes to exclude that may mimic meningitis include brain abscess, epidural abscess, postictal state, and stroke. The postictal state of a seizure and the poststroke period may both have a mild CSF pleocytosis and can be distinguished by other signs.

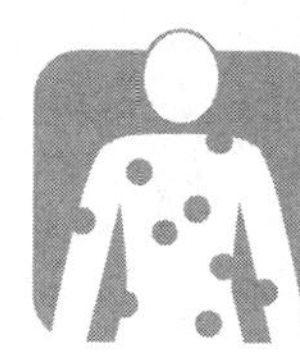

■ Evaluating the Severity of Problems Without treatment of bacterial meningitis the patient will develop brain edema from the bacterial toxins and the associated inflammatory response. Intracranial pressure will increase and blood flow will decrease. Focal infarction of brain tissue may occur.

Patients may die of bacterial meningitis. Other complications of bacterial meningitis include seizures, mental retardation, focal neurologic deficits, and chronic hydrocephalus. **Hearing loss in children with *H. influenza* meningitis is not uncommon.**

Immunizations (*H. influenza,* measles, mumps, and polio) are available for a variety of meningitis-causing organisms.

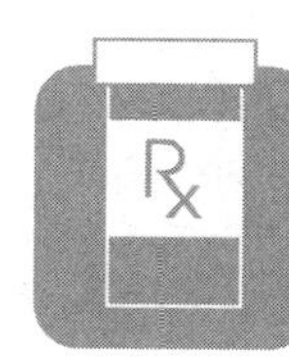

■ Managing the Patient Supportive measures (airway protection, seizure precautions, ventriculostomy for increased intracranial pressure, and electrolyte management) should be begun immediately.

- **Bacterial meningitis** Major complications can be prevented with the early initiation of adequate antibiotic therapy. **If bacterial meningitis is suspected, empiric coverage with broad-spectrum antibiotics should be started immediately. As culture results become available, coverage may be narrowed.**
  - **Ceftriaxone and ampicillin will cover nearly all cases of bacterial meningitis.**

- **Vancomycin and ceftazidime are useful for postsurgical or trauma cases of bacterial meningitis.**
- Dexamethasone given acutely will attenuate the inflammatory response. Dexamethasone has been shown in children to reduce the risk of sensorineural hearing loss and in adults to reduce mortality (limited to adults with significant dysfunction and not aseptic meningitis).

- **Viral meningitis** Viral meningitis is usually a self-limited process and does not produce the complications seen in bacterial meningitis. If herpesvirus is suspected, then acyclovir should be started.
- **Fungal meningitis** Fungal processes should be treated with the appropriate antifungal agent.

Multiple Sclerosis

■ **Basic Science** Multiple sclerosis (MS) is a **chronic disease** of uncertain etiology characterized by scattered **areas of demyelination in the white matter** tracts of the central nervous system. Evidence suggests that multiple sclerosis is an **autoimmune** disease, with **myelin as the predominant protein under attack.** The myelin sheath of peripheral nerves is antigenically different from myelin in central tracts and thus is not affected. Deficits in multiple sclerosis are related to impaired neuronal transmission of signals through the demyelinated axons.

■ **History and Physical Examination** The incidence of multiple sclerosis in the general population is 0.1%. However, those with a close relative have a higher incidence of contracting multiple sclerosis (2% to 5%). Multiple sclerosis most often **affects young adults.** Most people become symptomatic between the ages of 15 and 50 years old. **Whites** have a much higher incidence of multiple sclerosis than blacks or Asians. **Females** are affected twice as often as males.

Although any area in the central nervous system may be attacked, certain locations are preferentially involved, leading to particularly common symptoms of multiple sclerosis. These include the following:

- **Acute decrease in visual acuity in one eye** (optic neuritis) from optic nerve lesions
- **Diplopia and internuclear ophthalmoplegia** from brain stem lesions
- **Ataxia** from cerebellar lesions
- **Spasticity and weakness** from upper motor neuron lesions
- **Urinary frequency and urgency** from autonomic lesions
- **Sensory disturbances** from lesions in the sensory tracts

Later in the course of multiple sclerosis, **partial paralysis and dementia** become more common due to accumulation of white matter damage in the spinal cord and cerebral hemispheres. **Seizures** may occur in up to 5% of multiple sclerosis patients, usually coinciding with active lesions near the cerebral cortex. **Fatigue and depression** are common throughout the disease course but do not clearly relate to nervous system lesions.

The course of the disease is quite variable but can be grossly divided into two distinct presentations.

- The most common presentation is the **relapsing-remitting** pattern of multiple sclerosis, which consists of clear attacks of MS symptoms interspersed with symptom-free periods.
- Multiple sclerosis can also present as a **chronic progressive** disease, where symptoms steadily worsen without disease-free periods. Frequently, patients will present with a relapsing-remitting pattern and then develop a more chronic progressive phase.

Laboratory and Diagnostic Studies

- **MRIs** of the brain and spinal cord can confirm the existence of white-matter lesions in multiple areas.
- **Evoked potentials** of the visual, auditory, and somatosensory systems are also useful in establishing multiple lesions.
- **Cerebrospinal fluid gamma globulin elevation or oligoclonal banding** can confirm the presence of an immune-mediated process.

Formulating the Diagnosis

Traditionally, the diagnosis of multiple sclerosis is made clinically, based on recurrent attacks of deficits related to more than one area of the central nervous system. A diagnosis can usually be made based on the history and neurological examination.

Evaluating the Severity of Problems

Approximately one-third of those with multiple sclerosis will develop severe disability. At the onset of symptoms there is no way to accurately predict the frequency of attacks or subsequent disability that will occur. Life span is reduced only in those with the most extreme deficits, such as **dysphagia, respiratory compromise, or urinary retention,** because of the **increased risk of overwhelming infections.**

Managing the Patient

- **Acute attacks of symptoms can be treated with high-dose intravenous steroids** (for example, methylprednisolone 1000 mg per day for 3 days) to reduce symptoms more rapidly and completely than if untreated.
- **β-interferon can be used chronically to reduce the risk of developing attacks in relapsing-remitting patients.**
- **Spasticity** is treated with **physical therapy, baclofen** (a centrally acting muscle relaxant), and **benzodiazepines.**
- **Urinary frequency and urgency** are reduced by **anticholinergics** (such as oxybutynin). **Intermittent catheterization** may be necessary if the bladder does not empty completely. Chronic catheterization should be avoided due to the risk of infection.
- Fatigue usually responds to **brief naps, amantadine, or pemoline.**
- Depression is treated with standard **antidepressants.**
- Seizures and spasms are best treated with **anticonvulsants** (such as carbamazepine).

Muscular Dystrophy

- **Basic Science** **Muscular dystrophy (MD) is a group of genetically transmitted diseases characterized by progressive symmetric atrophy of groups of skeletal muscles without evidence of involvement or degeneration of neural tissue. In all forms of MD there is gradual loss of strength with increasing disability and deformity, although each type differs in the group of muscles affected, the age of onset, rate of progression, and mode of genetic inheritance.** The five major muscular dystrophies are Duchenne's muscular dystrophy, Becker's muscular dystrophy, myotonic dystrophy, fascioscapulohumeral muscular dystrophy, and limb-girdle muscular dystrophy.

  Duchenne's muscular dystrophy and Becker's muscular dystrophy are X-linked disorders localized to a gene producing the protein dystrophin.

- **History and Physical Examination** **The muscular dystrophies are characterized by progressive weakness and wasting of skeletal muscles.** Their age of onset varies among the specific disorders but as a group can present at any age. In the evaluation for muscular dystrophy it is **important to identify a family history, the rate of progression (typically very slow), and whether systems other than skeletal muscle are involved.**

- **Laboratory and Diagnostic Studies** **Electromyography, muscle biopsy, and immunoblotting** are important diagnostic tests for establishing the diagnosis of a specific muscular dystrophy.

- **Formulating the Diagnosis** The key features needed to identify the presence of a specific dystrophy hinge on clinical presentation (age of onset), serum creatine kinase, neurophysiologic studies, and muscle biopsy.

- **Evaluating the Severity of Problems** **The major concern in all of these diseases includes respiratory insufficiency, cardiac rhythm disturbances, and orthopedic complications.**

- **Managing the Patient** **Treatment of these diseases is mostly symptomatic with the goal of improving ambulation with assistive devices.** Other therapy includes **placement of a cardiac pacemaker** (when indicated) **and providing respiratory assistive devices** (from simple nasal BiPAP to tracheostomy and mechanical ventilation), if necessary. Myotonia (a condition in which a muscle or a group of muscles will not relax) has been treated with anticonvulsants (such as phenytoin or carbamazepine).

Parkinson's Disease

- **Basic Science** Parkinson's disease (PD) is a common neurological disorder of movement caused by abnormalities in the basal ganglia. There is a **loss of dopamine-containing neurons in the substantia nigra.** Other neurotransmitter systems are involved as well, but to a lesser extent. Neuronal loss is progressive, and its cause has not yet been clearly established.

  The pathological hallmark of PD is the **Lewy body,** an intraneuronal inclusion found mostly in the dopaminergic cells of the substantia nigra.

■ **History and Physical Examination** Parkinson's disease usually affects older adults. The four cardinal features of PD are **resting tremor, bradykinesia** (slowness of movement), **rigidity,** and **postural instability.** The most characteristic initial symptom is a resting tremor that is usually unilateral at onset. The tremor can involve the upper extremities, lower extremities, and lower face and jaw. Patients may have difficulty initiating movements and may notice slowed voluntary movements. They have difficulty walking, and their gait is stooped and shuffling. They may also experience freezing (a sudden arrest in movement, particularly when walking through a doorway). Postural instability may result in frequent falls, particularly as the disease progresses. In addition, patients may notice that their handwriting becomes very small (micrographia).

General examination may reveal evidence of seborrheic dermatitis. On neurological examination, mental status testing may reveal dementia or depression. Cranial nerve deficits may include decreased upgaze, infrequent blinking, decreased facial expression (known as **masked facies), soft voice,** excessive drooling, and perhaps a tremor of the lower face or chin. Motor examination usually reveals a **resting tremor** that is often described as **"pill-rolling"** and has an average frequency of 4 Hz. Patients may exhibit a **slowness of movement (bradykinesia).** Passive manipulation of the extremities reveals **rigidity,** often with a **"cogwheeling"** quality (a ratchet-like movement). There is no weakness. Sensory, cerebellar, and reflex exams are generally unremarkable. Gait, as described earlier, is remarkable for **stooped posture, shuffling movements of the feet, and decreased arm swing.** Patients often have difficulty initiating movement, but then steps become **increasingly rapid** and they find it hard to stop **(festination).** Lastly, patients have **postural instability** and may fall if not caught when pulled at the shoulders from behind.

■ **Laboratory and Diagnostic Studies** Although MRI may show decreased signal in the substantia nigra, imaging studies and blood tests are used mainly to rule out other diagnoses. The diagnosis of PD remains a clinical one.

■ **Formulating the Diagnosis** The diagnosis of Parkinson's disease is not difficult when an accurate history and physical reveals the cardinal features. There are **many other causes of secondary parkinsonism** that should be considered whenever there are atypical features or when there is a young age at onset. These causes **include** effects of **medications (such as antipsychotic agents and metoclopramide),** exposure to **toxins (such as carbon monoxide and manganese),** metabolic disturbances (**such as Wilson's disease,** a treatable disorder of copper metabolism), other neurodegenerative diseases, and multiple cerebral infarcts.

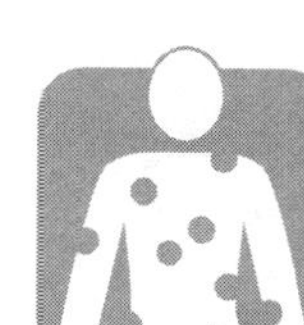

■ **Evaluating the Severity of Problems** The course of PD is variable but is **generally progressive.** Patients can become quite debilitated as the disease progresses. Patients may develop **swallowing problems,** and alternative methods of feeding may need to be considered. Patients are at risk for aspiration pneumonia. Because of gait difficulties, patients are at **risk for falls.** They may become entirely unable to walk or perform the activities of daily living and need extensive supportive care to prevent other complications of **immobility,** such as bedsores.

■ **Managing the Patient** Fortunately, many medications are available to provide symptomatic treatment of PD. The mainstay of treatment is the dopamine precursor **levodopa,** which, unlike dopamine, can cross the blood-brain-barrier. It is generally given with a **decarboxylase inhibitor,** such as carbidopa, which prevents breakdown to dopamine in the periphery. The use of carbidopa results in fewer side effects (such as nausea) and allows for smaller doses of levodopa to be given. Patients who have been on levodopa for a while may develop fluctuations between a rigid state and a nonrigid state (the **"on-off" phenomenon**) and can develop dyskinesias **(involuntary movements)** generally during the "on" phase. This medication can also cause psychosis in some patients, especially when given in high doses. Other medications used to treat PD include **dopamine agonists** (such as bromocriptine and pergolide). **Selegiline** is a specific **MAO-B inhibitor** used to prevent disease progression and to afford symptomatic relief. **Anticholinergic medications** and **amantadine** are sometimes used to help with tremor and bradykinesia. Surgical treatments are generally a last resort. Transplantation of adrenal medullary or fetal midbrain tissue remains experimental.

Seizures

■ **Basic Science** A seizure is an episode of neurologic dysfunction caused by abnormal electric activity of neurons. **Epilepsy is a clinical syndrome characterized by recurrent, unprovoked seizure activity.** The etiology for seizures is classified as either primary (that is, idiopathic or unknown) or secondary. Some causes of secondary seizures include trauma, infection, vascular lesions, anoxia, electrolyte disorders, toxins, and drugs (such as lidocaine). **The etiology of most seizures is idiopathic.**

Seizure types are broadly classified into generalized and partial (focal).

- **Generalized seizures involve loss of consciousness due to abnormal activation of the entire cerebral cortex.** The nomenclature depends on the type of movement noted, such as tonic, clonic, tonic-clonic, atonic, or myoclonic. Apnea, urinary/stool incontinence, rolling of the eyes, and postictal confusion are associated manifestations. Absence seizures are usually characterized by a brief episode of staring and behavioral arrest without associated postictal confusion.
- **Partial seizures stem from activation of localized regions of the cerebral cortex; consciousness may not be affected.** Therefore these seizures may be motor (unilateral tonic-clonic); sensory (paresthesias, numbness); autonomic; or associated with psychic, affective, or cognitive symptoms. Focal seizures may secondarily generalize. The classic complex partial seizure presents with altered consciousness and automatisms such as lip smacking and is followed by transient postictal state.

Febrile seizures are seen in 18-month-old children with temperatures greater than 38° C, and lasting less than 15 minutes. They are familial and are not associated with intracranial lesions or long-term morbidity.

Status epilepticus is a medical emergency in which seizures occur continuously for 20 minutes or longer, or repetitive seizures occur without a fully lucid interval. Mortality may be as high as 20%, and the patient may die of prolonged anoxia, trauma, or cardiovascular collapse. Specific measures to support the patient and initiate anticonvulsant treatment must be undertaken.

Alcoholics may experience nonfocal motor seizures during periods of withdrawal. If untreated, these seizures may progress to delirium tremens.

■ History and Physical Examination **A comprehensive evaluation of seizures begins with a detailed history surrounding the event.** Both the patient and witnesses should be questioned regarding the onset and length of seizure activity, loss of consciousness, and postictal confusion. The presence or absence of aura, localized or generalized motor activity, trauma, precipitating events, or illnesses should all be addressed. In patients with a known seizure disorder, questions regarding compliance with medical regimen, associated alcohol or drug use, sleep deprivation, and intercurrent illness need to be discussed. **The neurologic exam should be thorough, looking for both changes in mentation and focal deficits.**

■ Laboratory and Diagnostic Studies Laboratory studies should be done on an individual basis. **Patients with an initial seizure may require a more extensive workup including serum electrolytes, glucose, BUN, creatinine, calcium, magnesium, toxicology screens, and a complete blood count to search for other etiologies. Anticonvulsant levels may be helpful in detecting compliance with treatment, but should be interpreted cautiously** since therapeutic levels vary greatly between patients.

Radiologic studies include a CT scan with or without contrast to rule out structural causes. In the search for structural lesions, MRI is more sensitive than CT, though it may not be available emergently.

EEGs may help distinguish generalized from focal seizures.

■ Formulating the Diagnosis Seizure disorders must be distinguished from other disorders that present similarly. The differential diagnosis includes syncope, narcolepsy/cataplexy, transient ischemic attacks (TIAs), movement disorders, psychogenic seizures, and malingering. Careful attention needs to be given to clinical features that distinguish seizures from other ailments.

■ Evaluating the Severity of Problems Seizures may be the manifestation of treatable disease, such as brain tumors, infections, and arteriovenous malformations. As such, it is important to aim treatment at the cause when it is identifiable. Patients who are evaluated in the postictal state are not able to give a coherent history and may have transient focal deficits (Todd's paralysis); furthermore, during the seizure they may have suffered an occult injury requiring emergent treatment (such as subdural hematoma).

Table 1.1 *Use of Specific Anticonvulsant Drugs in Different Seizure Types*

	Indications					
Drug	Partial (focal)	Absence	Tonic-Clonic	Usual dose (mg/kg per day)	Half-life (hr)	Side Effects
Carbamazepine	+		+	10-40	10-30	Headache, drowsiness, dizziness, diplopia, blood dyscrasia, hepatotoxicity, arrhythmia
Clonazepam		+	+	0.05-0.10	24-36	Drowsiness, ataxia, secretions, hypotonia, behavior problems
Ethosuximide		+		20-40	24-42	Gastrointestinal distress, rash, drowsiness, dizziness, systemic lupus erythematosus, blood dyscrasia
Phenobarbital	+		+	2-8	48-100	Drowsiness, rash, ataxia, behavioral and cognitive problems
Phenytoin	+		+	4-8	6-30	Drowsiness, gum hyperplasia, rash, anemia, ataxia, hirsutism, folate deficiency
Primidone	+		+	12-25	6-12	Drowsiness, dizziness, rash, anemia, ataxia, diplopia
Valproic acid	+	+	+	10-60	6-18	Gastrointestinal distress, hepatitis, alopecia, drowsiness, ataxia, tremors, pancreatitis, thrombocytopenia

From Johnson RT, Griffen JW: *Current Therapy in Neurologic Disease,* ed 4, St Louis, 1993, Mosby.

■ **Managing the Patient** **Treatment is individualized** to each patient. Certain seizures require more intense management than others (such as status epilepticus). **General measures are taken to protect the patient against injury** (such as padded bedrails), **to stabilize the airway** (oral/nasal airway or intubation), **and to ensure adequate ventilation and circulation** (such as large bore intravenous lines). Definitive control of seizures is geared toward treating the underlying cause (for example, correction of metabolic abnormalities such as hyponatremia). The choice of anticonvulsant depends on the type, frequency, and intensity of the seizure (Table 1.1). Maintenance anticonvulsants should be chosen depending on the most likely etiology. Phenytoin and carbamezipine are most frequently used anticonvulsants for outpatient maintenance therapy.

Management of status epilepticus begins with the establishment of the airway; thiamine and dextrose are given. Initial therapy of active seizures begin with benzodiazepines (i.e., loraxepam). Once the acute seizure has abated, phenytoin should be administered followed by phenobarbitol if necessary. If these measures are not effective, more drastic therapies such as barbiturate coma or general anesthesia are used.

Subarachnoid Hemorrhage

■ **Basic Science** **Subarachnoid hemorrhage** usually results from either **trauma** or a **ruptured intracranial aneurysm.**

• Traumatic subarachnoid hemorrhage usually occurs as result of **major trauma** and is often devastating to the patient.

• Subarachnoid hemorrhage from **aneurysm rupture** is associated with a **death rate of 35% with the first bleed;** an additional 15% of patients die in the ensuing weeks from a second rupture of the aneurysm. Peak incidence of aneurysm rupture is between the ages of 25 and 50, and though rupture may occur at any age, it is very rare in children.

Aneurysms are most commonly **found at bifurcations in the circle of Willis,** where the muscular coat of the cerebral artery is not well developed.

■ History and Physical Examination

- **Traumatic hemorrhage** Patients with traumatic subarachnoid hemorrhage **must be screened for other life threatening injuries.**
- **Aneurysmal hemorrhage Aneurysms are usually asymptomatic until they rupture.** There is often a history of **exertion** at the moment symptoms begin. After rupture has occurred, headaches are often described as the **"worst headache" in the patient's life, a "thunderclap" headache.** The patient may develop **syncope** or appears **obtunded** and **confused.** Alterations in pulse and respiratory effort may occur. **Nausea, vomiting, and meningismus** may also be present as blood entering the CSF irritates the meninges. Twenty-five percent of patients manifest **focal neurological deficits.** The patient may lapse into **coma.**

■ Laboratory and Diagnostic Studies

• **Blood seen on noncontrast CT scan** usually establishes the diagnosis of subarachnoid hemorrhage.

• If the CT scan is negative, lumbar puncture is performed. **Bloody CSF or xanthochromic supernatant** (present after 6 hours) establishes the diagnosis of subarachnoid hemorrhage.

• **Cerebral angiography** is used to identify the cause of the hemorrhage (aneurysm, AVM) as well as asymptomatic vascular abnormalities (unruptured aneurysms).

■ Formulating the Diagnosis **The history, physical, and CT scan (or lumbar puncture) usually establish the diagnosis of subarachnoid hemorrhage.** Cerebral angiography can differentiate aneurysmal subarachnoid bleed from tumor, AV malformation, or subdural hemorrhage.

Occasionally subarachnoid hemorrhage will mimic the symptoms of a myocardial infarction (i.e., ECG changes).

■ Evaluating the Severity of Problems Subarachnoid hemorrhage is an extremely serious disorder, with over 50% mortality within the first 2 weeks of presentation. If subarachnoid bleed is suspected, **neurosurgical consultation should be sought immediately.**

■ Managing the Patient Once hemorrhagic cerebral aneurysm has been diagnosed, the patient's activities should be restricted to **bed rest.** The patient should be placed on agents such as **docusate** that **pre-**

vent constipation and straining while defecating (straining causes increased intracranial pressure). **Narcotics may be necessary for headache control.**

The major serious complications after SAH include ***rebleed, vasospasm,*** and ***hydrocephalus.*** The risk of rebleed is greatest in the first 24 hours after the ictus; early surgery is preventive. Symptomatic vasospasm is usually maximum at 5–14 days and is often treated with oral nimodipine and hypertensive/hypervolemic therapy to prevent ischemic infarction. Acute or delayed hydrocephalus can also occur after SAH and is managed with ventriculostomy and/or chronic CSF diversion (e.g., ventriculoperitoneal shunt) as needed. Other complications that may occur include hyponatremia and seizure.

Transient Ischemic Attacks

■ **Basic Science** The brain receives blood from two different vascular beds, the carotid vascular system and the vertebral vascular system. A transient ischemic attack (TIA) is defined as a brief episode of neurologic symptoms referable to a defined vascular bed that lasts less than 24 hours; however, most TIAs last less than 1 hour.

The most common source of a TIA is an embolus. These small pieces of coagulated blood most commonly emanate from either a thrombus in the heart or a plaque in an atherosclerotic blood vessel. TIAs may also be caused by a decreased amount of blood flow distal to a stenotic lesion in an artery.

■ **History and Physical Examination**

- First, it is critical to determine whether **similar episodes** have occurred in the past.
- TIAs usually have a **rapid onset and resolution and last less than 1 hour.** Associated symptoms may include **palpitations, chest pain, seizure-like movements, and headache.**
- **Carotid-territory symptoms** include **one-sided weakness or numbness** of the face, arm, or leg; **amaurosis fugax** (transient blindness in one eye); and **dysphasia** if the dominant hemisphere is involved.
- **Vertebrobasilar symptoms** include **bilateral or shifting weakness and numbness, bilateral blindness, and homonymous hemianopia.** Vertigo, diplopia, dysarthria, dysphagia, and ataxia also commonly occur with vertebrobasilar TIAs.

Physical examination is usually **normal** unless performed during a TIA, but evidence of neurologic disease such as a pronator drift, mild hyperreflexia, a Babinski sign, mild dysphasia, or cortical sensory findings can confirm the history and suggest past significant ischemic episodes.

Special attention to the **cardiac and vascular examination** is required. An **irregularly irregular pulse or a murmur** may suggest a **cardioembolic** source. **Carotid bruits** correlate poorly with a significant underlying carotid stenosis but along with subclavian, renal, or femoral artery bruits suggest diffuse large vessel **atherosclerotic disease.**

■ **Laboratory and Diagnostic Studies** The initial evaluation of a patient with a TIA should include routine **blood chemistries, complete blood count, coagulation studies, and an ECG. Head CT scan** is helpful for assessing previous infarcts and excluding a hemorrhage or mass lesion.

- Carotid territory TIAs should be evaluated first with a **carotid ultrasound,** and if stenosis is estimated to be greater than 70% of the arterial diameter then a carotid **angiogram** should be ordered.
- Vertebrobasilar TIAs are evaluated first with **transcranial Doppler** and subsequently by angiogram if vessel pathology (such as dissection) is suspected. Patients with cardioembolic risk factors should have transthoracic or transesophageal **echocardiography,** the latter being more invasive and expensive but providing more information about potential cardiac sources of emboli.

■ **Formulating the Diagnosis** Although transient neurologic symptoms (as just described) suggest a TIA, other common disorders such as seizures and migraine should be considered.

■ **Evaluating the Severity of Problems** The overall risk of stroke in a patient with a TIA is approximately 5% in the first month, 15% in 1 year, and 25% in 5 years. Patients with greater than 70% ipsilateral carotid stenosis are at greater risk of stroke and should be treated more urgently.

■ **Managing the Patient**

- The main therapeutic interventions for patients with TIAs are **antiplatelet agents, anticoagulation, and carotid endarterectomy.**
- Patients with **>70% ipsilateral carotid stenosis** and TIAs should have carotid **endarterectomy** followed by antiplatelet therapy and control of risk factors. For patients with 30% to 70% carotid stenosis, it is currently unknown whether medical or surgical management offers more benefit.
- **All patients should receive aspirin in doses from 325 to 1300 mg per day** or ticlopidine (if they fail aspirin or are aspirin intolerant); management of vascular risk factors such as hypertension and smoking is also critical.
- **Anticoagulation** with heparin followed by warfarin is indicated for patients with a well-defined **cardioembolic** source including **atrial fibrillation and/or atrial or ventricular thrombus.**
- Therapy for vertebrobasilar TIAs is not standardized, but initial treatment with an antiplatelet agent is usual.

Multiple Choice Review Questions

A 25-year-old, previously healthy black woman was brought to the emergency room after being struck by a drunk driver. The other car, which was estimated to be going in excess of 50 MPH, hit her broadside. She was the unbelted driver and was found unconscious at the scene with a broken driver side window. She had agonal respirations at the scene and was intubated.

Her examination in the emergency room found her blood pressure at 125/75, she was not breathing over the ventilator rate of 12. On neurologic exam she is unresponsive to all stimuli. Her pupils are fixed and dilated at 6mm and she has no oculocephalic reflex or caloric reflex bilaterally. Jaw jerk, corneal reflexes, and gag reflex are absent.

1. Which of the following is critical to determining whether she is dead by brain criteria?

a. Body temperature
b. Heart rate
c. ECG
d. C-x-ray
e. Head CT

2. Which of the following clinical findings, if present, can still be consistent with the diagnosis of death by brain criteria?

a. Decorticate posturing
b. Pupillary light response
c. Cough reflex
d. Jaw jerk
e. Spinal reflexes

3. If this patient has a seizure in the emergency room while you were examining her, which of the following would you do next?

a. Get an emergent EEG to document death by brain criteria
b. Do an apnea test immediately to document death by brain criteria
c. Get an angiogram to document death by brain criteria
d. Tell the family she is dead by brain criteria and no further tests need to be done
e. Observe and support her because she has not met the criteria for brain death

A 30-year-old woman complained of a severe, throbbing, right-sided headache that lasted 6 hours. There were no preceding symptoms. The headache began gradually, became maximal over 30 minutes, and was associated with vomiting and diarrhea. The headache remitted after sleeping for several hours. She has had no similar headaches in the past, but does recall having occasional bifrontal throbbing headaches that caused her to miss school in her teenage years.

4. What is the most likely diagnosis in this patient?

a. Migraine with aura
b. Migraine without aura
c. Cluster headache
d. Subarachnoid hemorrhage
e. Giant cell arteritis

5. Neurologic examination in this patient would likely reveal which of the following?

a. Left ptosis and miosis
b. Left Babinski sign
c. Meningismus
d. Papilledema
e. Normal examination

6. Appropriate first line abortive treatment for this patient's headaches may include which of the following?

a. Tricyclic antidepressants
b. Beta-blockers
c. Calcium-channel blockers
d. Nonsteroidal antiinflammatory agents
e. Narcotic analgesics

Chapter 2

Ophthalmology

ANATOMY OF THE GLOBE AND ORBIT

The orbital walls are composed of seven bones: ethmoid, frontal, lacrimal, maxillary, palatine, sphenoid, and zygomatic. The orbit is surrounded on three sides by sinuses: the frontal sinus above, the ethmoid and sphenoid sinuses medially, and the maxillary sinus below.

Orbital fractures are most likely to involve the floor, medial wall, and roof. The lateral wall of the orbit is the thickest and strongest of the orbital walls. **The medial orbital wall is paper-thin, as reflected in its name, lamina papyracea; it and the inferior wall are most prone to fracture.**

Following ethmoid fracture, air may enter the orbit when the patient blows his or her nose, causing crepitus on palpating the orbit. Among children, infections of the ethmoid sinuses may extend through the lamina papyracea to cause orbital cellulitis and exophthalmos. **Orbital floor fractures are commonly known as "blowout" fractures. Clinical features of blowout fractures may include diplopia, enophthalmos, hypesthesia in the distribution of the infraorbital nerve, entrapment of orbital tissues or extraocular muscles with restricted upgaze, and radiographic evidence of fluid level or cloudiness of the maxillary sinus.** Orbital roof fractures may be especially serious. Because of the potential involvement of the dura, brain, and frontal sinus, clinical evaluation needs to also establish the possible presence of cerebrospinal fluid rhinorrhea or pneumocephalus (i.e., air in the intracranial cavity). **Usually, orbital computed tomography (CT) scanning, including coronal images, is most appropriate.**

The optic foramen, located at the apex of the orbit, transmits the optic nerve, the ophthalmic artery (which gives rise to the central retinal artery), **and sympathetic fibers from the carotid plexus** (Fig. 2.1). **The superior orbital fissure lies adjacent to the optic foramen and transmits the sympathetic nerve plexus, all other orbital nerves** (cranial nerve [CN] III, IV, VI, and the ophthalmic branch of CN V), **and the superior ophthalmic vein.**

The transparent cornea occupies the center of the anterior pole of the globe and performs most of the refraction of light entering the eye. Abnormalities of the cornea may compromise its transparency, involve loss of corneal tissue, result in neovascularization of the normally avascular cornea, or result in corneal swelling.

In contrast to the transparent cornea, the sclera is opaque and white. It is thicker than most of the cornea and provides insertions to the extraocular muscles. Inflammation of the sclera (or scleritis) may affect individuals with rheumatoid arthritis or other connective tissue diseases.

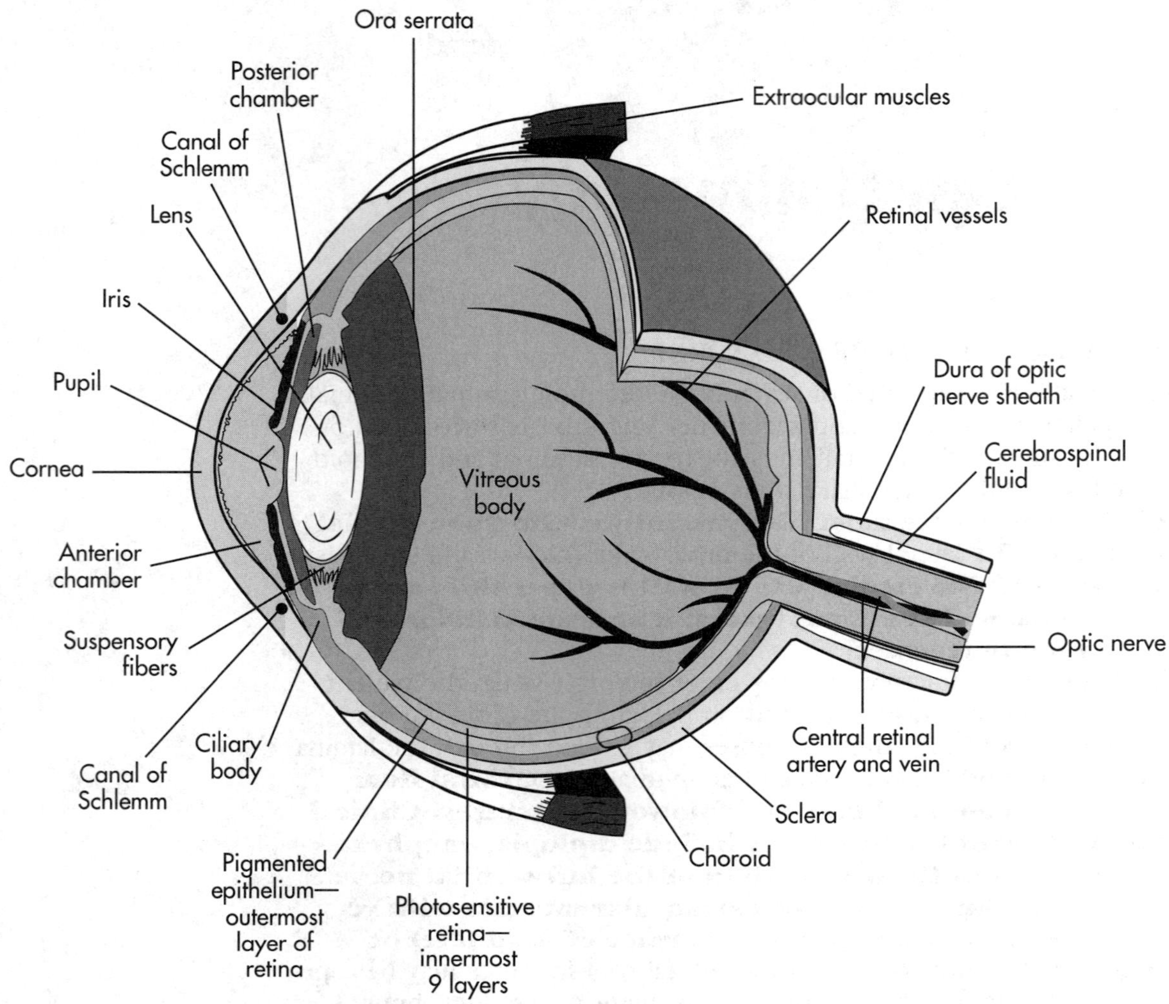

Fig. 2.1 General anatomy of the eyeball, including its tunics and chambers. *(From Burns ER, Cave MD:* Mosby's Ace the Boards: Histology, *St Louis, 1996, Mosby.)*

The limbus, which is the transition zone located between the peripheral cornea and the anterior sclera, is gray and translucent. This is an important area anatomically since many surgical procedures, including cataract extraction, involve entering the eye at this location. The inner lids and sclera are covered by the normally clear conjunctiva. The conjunctiva may become erythematous or edematous or show inflammatory changes in a variety of conditions.

The interior of the eye consists of three chambers. The anterior chamber is the space between the clear cornea and colored iris. This region is filled with aqueous humor produced by the eye. In glaucoma, the pressure in this space is elevated, usually due to poor outflow of this fluid from the eye. **Blood in the anterior chamber is called a *hyphema*** and may occur following ocular trauma or surgery. **Ocular infections or inflammation may result in pus in the anterior chamber, known as a *hypopyon*.**

The posterior chamber is a small space between the iris and the lens immediately behind the dark pupillary space. Aqueous fluid is produced here, and it usually flows forward through the

pupil, into the anterior chamber, and out the chamber *angle,* a space in the eye just interior to the limbus.

The third space is the vitreous cavity, which is behind the lens and occupies the main volume of the eye. This chamber contains vitreous humor, a thick gel-like substance. Bleeding, infection, and inflammation may affect the vitreous. Surgical removal of the vitreous may become necessary in such conditions as severe diabetic retinopathy or infections, or for retinal detachment repair.

Eye movements are controlled by six extraocular muscles: four rectus muscles and two oblique muscles. These muscles insert at varying distances posterior to the limbus and move the eyes in the various directions of gaze. Except for the lateral rectus, which is innervated by the abducens nerve (CN VI), and the superior oblique, which is innervated by the trochlear nerve (CN IV), the remaining extraocular muscles are innervated by the oculomotor nerve (CN III). Complete CN III paralysis results in an inability to move the eye up, down, or in; ptosis (lid droop due to lid elevator paralysis); and a dilated pupil. Palsies of any of the muscles may result in diplopia or double vision. Common causes include strabismus as a child, diabetes, and hypertension. Imaging may be warranted if the cause cannot be readily determined, or if it is long-standing. See Fig. 2.1 for basic eye anatomy.

Cataracts

■ Basic Science **A cataract is an opacity of the lens of the eye that interferes with vision. Cataracts are most common in the elderly** but are also seen in younger individuals. **There are three main types of cataracts:**

- **Congenital or juvenile onset**
- **Secondary**
- **Age related**

Congenital cataracts are generally associated with systemic or ocular diseases, inherited traits, or exposure, but may be of undetermined cause. Associated conditions include galactosemia, hypoglycemia, and intrauterine infections.

Secondary cataracts are directly related to acquired systemic disorders, acquired ocular disease, physical trauma, medications, or local chemical agents. Examples include hypocalcemia, intraocular neoplasms, radiation, and steroids.

Age-related cataracts are the most common form and are caused by lens opacification as its chemical composition changes. Water content decreases, soluble protein becomes more insoluble, and calcium content increases. The lens becomes progressively harder as it opacifies. **About 90% of all cataracts are age related and begin development after 40 years of age.**

■ History and Physical Examination The etiology of congenital and infantile cataracts may often be determined by seeking a prenatal history of maternal infections such as rubella or syphilis, a family history of chromosomal abnormalities (e.g., trisomy 21 or 13), or a family history of cataracts.

In older patients, it is important to determine any secondary causes of cataracts. These include trauma, drug exposure such as chronic steroid administration, dermatologic diseases (e.g., atopic dermatitis), metabolic disorders (e.g., diabetes mellitus or hypothyroidism), and other systemic syndromes or associated ocular abnormalities.

Examination of a patient suspected of having a cataract generally includes visual acuity testing at distance and near, both in lighted conditions and dark. With the use of a slit lamp biomicroscope, one may determine the location and density of most opacities within the lens. A simple way to examine the lens is with a direct ophthalmoscope. This instrument will reveal dark opacities in the normal red reflex. Fundus evaluation by indirect and direct ophthalmoscopy is necessary to evaluate the macula, optic nerve, and retinal vessels. Posterior segment pathology may limit visual rehabilitation after an otherwise uneventful cataract extraction.

■ Laboratory and Diagnostic Studies Special tests estimating potential visual acuity are often helpful in assessing the contribution of the cataract to the patient's visual loss.

Laboratory evaluation assists in determining inborn errors of metabolism and metabolic diseases that may result in cataracts.

■ Formulating the Diagnosis **The occurrence of an intralenticular opacity associated with a decrease in visual acuity in an eye suggests the diagnosis of cataract.** It is confirmed with visual acuity testing, slit lamp examination, funduscopic examination, and potential acuity testing. **Cataract may be confused with corneal opacity, anterior chamber opacity, and intraocular masses, including retinoblastoma.**

■ Evaluating the Severity of Problems **A cataract is clinically important if it causes a significant decline in visual function. Patients often complain of decreased visual acuity but may also experience increased glare, double vision, and decreased contrast sensitivity. Symptoms may worsen in either especially bright or dim light. Often they will complain of difficulty driving at night because of increased glare with oncoming headlights.** Patients often have severe compromise of their activities of daily living such as reading, driving, and working.

■ Managing the Patient **No medical treatment has been proven to prevent or reverse cataract development,** although many are under investigation. Pupillary dilation may improve vision by allowing light to pass peripheral to the lens opacity. Spectacle correction may correct refractive errors induced by the cataract. Light absorption filters are useful for blocking those frequencies of light that create glare. These measures are useful for patients when operative management is neither needed nor desired by the patient. Low vision aids including handheld monoculars and magnifying lenses may be useful for patients in whom surgery is contraindicated.

The decision for surgical management is based primarily on whether reduced vision significantly interferes with the patient's activities. Other factors that must be assessed are

the health of both eyes, the potential for vision, and the general health of the patient.

For congenital cataracts, removal should be considered as soon as possible after birth, preferably before 3 months of age. A clear optical pathway is important during the early stages of visual pathway development to prevent amblyopia or "lazy eye."

Cataract surgery is usually performed using ultrasound power (phacoemulsification) to break up the lens prior to removal. An implant is almost always placed within the eye at the time of surgery for optical rehabilitation; that is, the power of the normal crystalline lens needs to be replaced at the time of surgery to avoid the need for thick "cataract" glasses. The implant is a plastic, acrylic, or silicone lens that acts to refract light once it enters the eye. Visual rehabilitation is usually rapid. Many patients gain their best corrected vision within the first few weeks after surgery.

CONJUNCTIVITIS

■ Basic Science **Conjunctivitis refers to any disorder causing inflammation of the conjunctiva, usually manifested as a "red eye."** There are many primary causes of conjunctivitis, including infections and allergy. Secondary causes of conjunctival inflammation include lid disease, corneal disease, orbital disease, or intraocular inflammation such as iritis. **It is important to differentiate primary conjunctivitis from that due to adjacent ocular disease.**

The conjunctiva may be primarily infected with viral entities (e.g., adenovirus) or with bacteria (e.g., *Staphylococci* and *Streptococci*). Most cases of bacterial conjunctivitis are confined to the conjunctiva, although there may be extension to other ocular structures. Ocular immune responses to infection, especially to viral antigens, may worsen ocular damage.

Certain bacteria are considered normal flora on the conjunctiva. These have been implicated as the etiology of infections in some patients due to compromised local immune function. In others, exposure to contagious individuals or fomites may spread bacteria or virus particles. Adenovirus can live on dry surfaces for many hours and thus is highly contagious.

■ History and Physical Examination It is imperative to establish a history of the present illness. **One should explore the course of development, the presence of a discharge (bacterial infections produce more than viral), pain, recent upper respiratory infection (suggesting viral conjunctivitis), photophobia (suggesting iritis), foreign body sensation, itching (suggesting allergic conjunctivitis), burning, associated fever or rash, or decreased visual acuity.** Elements of the past ocular history may be relevant, including glaucoma and contact lens wear. **A history of atopic disease may be important in the diagnosis of allergic conjunctivitis. A history of herpes simplex virus labialis may be significant.**

Conjunctivitis usually does not decrease visual acuity. The pupils should be equal, round, and reactive to the light.

Glaucoma or iritis may cause abnormal pupils. The periorbital skin should be inspected for lesions such as herpes simplex or zoster and for atopic or allergic changes. The conjunctiva should be grossly inspected for injection (redness and chemosis (swelling). Gross purulence is usually present with bacterial conjunctivitis and sometimes with viral. Gonococcal conjunctivitis usually produces copious purulence. Adenovirus usually causes a watery discharge and swollen preauricular lymph nodes.

The corneal surface and its light reflex should be inspected for whitish opacities (infectious infiltrates), foreign bodies, or decreased sheen and luster. Fluorescein dye may be instilled and observed with a cobalt blue light filter. Any dye uptake by the cornea suggests corneal pathology. **Any conjunctivitis in a contact lens patient should raise immediate suspicion, as corneal infections are increased in frequency in this population.**

Allergic conjunctivitis is often associated with conjunctival swelling and redness, lid swelling and erythema, and significant itching.

Another less common cause of conjunctivitis is *Chlamydia* infection. This is usually sexually transmitted from the hands, and a history of urethral or vaginal discharge should be elicited.

To rule out other ocular diseases such as immune reactions, iritis, scleral or episcleral inflammation (scleritis, episcleritis), keratitis, or other conditions, an examination at a slit lamp is usually needed.

■ Laboratory and Diagnostic Studies **Bacterial cultures are helpful if the patient does not respond to initial treatment.**

■ Formulating the Diagnosis **In the primary care setting, the history and physical usually suffice to make the diagnosis.** Any suspicious history, ocular signs, or past medical history should prompt an ophthalmologic referral.

■ Evaluating the Severity of Problems The following symptoms and signs are especially worrisome and suggest more severe ocular disease: ocular pain, photophobia (light sensitivity), decreased vision, proptosis (eye bulging out due to posterior orbital pressure), abnormal pupils, a cloudy or otherwise abnormal cornea, a history of contact lens wear (especially extended wear), history of trauma or toxic exposure, significant chemosis, history of recent ocular surgery, and others.

■ Managing the Patient **If bacterial conjunctivitis is highly suspected, topical antibiotics can be started. Seven to ten days of treatment is usually sufficient. Never use an antibiotic-steroid combination, or steroids alone, as this may lead to significant complications,** including the potentiation of herpes simplex virus as well as worsened bacterial and viral infections.

Viral conjunctivitis usually runs its course but may lead to serious sequelae. Treatment usually consists of keeping the eye clear of accumulating discharge, cool compresses, and contagious precautions. Topical antibiotics may be added for possible bacterial infection if the diagnosis is in doubt. Viral conjunctivitis can be contagious for approximately 2 weeks from the onset of symptoms. Patients should be advised to use prudent hand washing and to avoid sharing of towels or pillows. **Due to the extremely infectious, often epidemic nature of viral adenovirus conjunctivitis, health care providers and others, including day care workers, must often be secluded.**

Gonococcal conjunctivitis requires systemic antibiotics. Topical antibiotics are often helpful as well. Treatment for *Chlamydia* is also prudent given the significant rate of dual infection in this group. *Chlamydia* is often treated with oral doxycycline for 2 to 3 weeks. It is important to treat the sexual partners of these patients.

Allergic conjunctivitis can be treated with cool compresses and systemic antihistamines. Several topical products exist (e.g., antihistamines, mast cell stabilizers, and nonsteroidal antiinflamatory drugs) and may provide symptomatic relief.

Diabetic Retinopathy

- Basic Science **Diabetic retinopathy is a significant cause of blindness in the United States. Its prevalence within the diabetic population increases with the duration of the disease. Within 7 years of diagnosis, more than half of the patients have some degree of retinopathy. This increases to over 90% for those with diabetes of many years' duration.**

  Diabetic retinopathy is a microvascular disease of the retina. Sustained levels of hyperglycemia cause ongoing vessel damage with basement membrane thickening, endothelial cell loss, and, finally, capillary nonperfusion. Intraretinal capillary closure leads to retinal ischemia and the release of a vasoproliferative factor. Neovascularization may occur on the optic nerve, retina, or iris.

- History and Physical Examination Examination of the retina with a direct ophthalmoscope shows changes that may be categorized in four stages.
  - **Background diabetic retinopathy (BDR)** Impaired autoregulation of the retinal vasculature and alterations in the retinal blood flow lead to **microaneurysms, dot- and flame-shaped hemorrhages, cotton wool spots, and hard exudates.** Patients without retinopathy should be seen by their ophthalmologist at least yearly. Once background retinopathy is detected, patients should be followed at least every 6 to 12 months, or even more often, depending on disease severity.
  - **Maculopathy/preproliferative stage** The central retina area is called the *macula* and is responsible for our precise central vision. **Edema within the macula may lead to central visual loss. Retinal thickening shows characteristic signs including the accumulation of a yellow lipid called *hard exudate.*** This may take the form of a circinate ring in the macula.

    Focal areas of edema may be treated with laser therapy. Diffuse edema has a poorer visual prognosis. Retinal ischemia may also be prominent resulting in significant visual loss. These patients need close ophthalmologic follow-up.

- **Proliferative stage New blood vessels may form in this stage due to extensive ischemia, causing the production of a vasoproliferative factor. These abnormal vessels usually grow from the venous circulation. They are friable and tend to bleed easily. This may lead to hemorrhage within the retina or vitreous cavity, resulting in profound visual loss.**

  This stage generally occurs in diabetics who already manifest other end organ damage. These patients need urgent panretinal laser treatment to destroy hypoxic retina. Laser therapy will stop the production of the angiogenic factor, thus causing regression of neovascularization.

- **Cicatricial stage In this end stage, fibrous traction bands form between the retina and the vitreous face. These bands may eventually contract and may cause tractional retinal detachments.** Surgery is often indicated to remove scar tissue and reattach the retina if needed. The visual potential of these patients is often low.

■ Laboratory and Diagnostic Studies **Fluorescein angiography is often essential to define areas of leakage, aneurysm formation, ischemia, and neovascularization. Serum glucose monitoring is important in defining blood sugar control.**

■ Formulating the Diagnosis **Characteristic retina changes in the setting of diabetes mellitus are diagnostic.**

■ Evaluating the Severity of Problems Blindness is the end result of untreated diabetic retinopathy.

■ Managing the Patient

- **Metabolic control It is generally agreed that tight serum glucose control may retard the onset or progression of retinopathy.** Although some controversy exists, most ophthalmologists advocate as normal a serum glucose as possible.

- **Laser photocoagulation Laser therapy may significantly improve vision in some patients and prevent significant visual loss in others. There are two main groups of patients who benefit. The first group includes those with macular edema. The second group includes those with neovascularization.**

  Appropriate retinal examinations and adequate follow-up are extremely important.

Eyelid and Lacrimal System Pathology

Anatomically, the lids consist of an outer thin layer of skin, a central collagenous plate (tarsus), and the inner conjunctiva. The skin is very thin and lacks subcutaneous fat. Fluid can easily accumulate beneath it. The tarsal plate is firm and rubbery and houses sebaceous glands that secrete lipid into the tear film. The conjunctival layer is in contact with the ocular surface and spreads tears over the eye with each blink. Muscular layers within the lid are responsible for opening and closing the lids.

A wide spectrum of pathologic processes occur on or within the eyelids and can be categorized as infectious (bacterial and viral), degenerative, or neoplastic.

Eyelid Inflammation and Infections

- **Blepharitis The lipid glands within the lid may become plugged. This results in a deficiency of lipid in the tear film that normally lubricates and protects the ocular surface. This condition is known as *blepharitis*.** Patients may complain of eye and lid discomfort including burning and itching, as well as a "gritty" sensation. **Signs include eyelid erythema and edema, crusting or scaling of the eyelid margins, and a red eye. The glands may become secondarily infected, primarily with *Staphylococci,* leading to bacterial toxin production with associated ocular inflammation.** Blepharitis may commonly accompany acne rosacea or acne vulgaris. **Treatment requires patient education as to the chronicity of the problem and eyelid hygiene (warm washcloth compresses followed by gentle scrubbing). An application of an antibiotic-corticosteroid ointment may reduce the infectious and inflammatory components of the disorder,** but this requires closer monitoring. If these measures are unsuccessful, oral doxycycline may be useful.

- **Chalazion This very common, focal inflammation of the lids results from an obstruction of the meibomian glands. The lipid secretions break out of the glands into the surrounding tarsus, muscle, and soft tissues. An acute inflammatory response ensues, along with pain and erythema of the skin and lid.**

  In the acute inflammatory phase, the treatment consists of warm compresses, which soften the sebum and facilitate its passage out of the gland. This may be combined with a topical antibiotic and appropriate lid hygiene. Frequently the lesions become chronic cysts that require incision and curettage to debulk the granulomatous mass. Spontaneous resorption may occur over weeks to months. One should suspect a sebaceous carcinoma whenever a chalazion is unusually persistent or recurrent.

- **Hordeolum (stye) Hordeola are acute bacterial infections involving the meibomian glands or the lash (cilia) follicle.** The onset can be rapid, and an inflammatory mass forms. These may vary in size but may be so big as to distort the vision due to pressure on the globe due to astigmatism. ***Staphylococci* are the usual offending bacteria. A hordeolum usually drains spontaneously, although its resolution can be hastened by the application of warm compresses, antibiotic ointment, and occasionally oral antibiotics.**

- **Herpes simplex Similar in appearance to herpes zoster, simplex lesions are confined to the epidermis.** In comparison to zoster, simplex lesions usually do not scar, but recurrences are relatively common. **For the lid skin, local treatment is rarely needed. If vesicles are present on the eyelid mar-**

gin, the eye may become involved, and thus topical treatment of the eye and lid with an antiviral may be warranted. Oral acyclovir may also be used. Timely ophthalmic referral should be made for any ocular signs or symptoms.

- **Preseptal/orbital cellulitis Cellulitis of the eyelid may occur after seemingly minor trauma or because of a "pimple." The most common offending microbe is *Staphylococcus aureus.* The orbital septum acts as a relative barrier to the spread of the infection posteriorly into the orbit. Preseptal cellulitis involves only anterior structures, does not extend into the orbit, and has no associated ocular motility or vision abnormalities. Appropriate oral antibiotics, outpatient treatment, and close follow-up are needed. Orbital cellulitis with proptosis and motility disturbances requires urgent IV antibiotics and orbital imaging to rule out abscess formation. *Hemophilus influenzae* is a particular culprit when preseptal cellulitis occurs in a young child with coexistent upper respiratory infection. The eyelid frequently has a violaceous hue.**
- **Dacryocystitis** Tears drain from the eye through a small opening in the medial upper and lower lid. This drains through the lacrimal sac (medial to the eye) and the nasolacrimal duct to the inner nose. **Dacryocystitis, or infection of the lacrimal sac, is a common acute or chronic disease that typically occurs in infants or in persons over the age of 40. In acute dacryocystitis the usual infectious organism is *Staphylococcus aureus* or, less frequently, β-hemolytic *Streptococcus.* In chronic dacryocystitis, *Streptococcus pneumoniae* is the predominant agent.** In cases of fungal infection, obstruction of the nasolacrimal duct by a dacryolith sometimes occurs. **In the acute form, inflammation, pain, swelling, and tenderness are present in the tear sac region. A mucopu rulent discharge may be expressed from the puncta by applying pressure to the lacrimal sac. Infantile dacryocystitis may result when the nasolacrimal duct fails to open spontaneously before birth or during the first month of life. Usually, there is a membranous blockage of the one of the valves of the nasolacrimal drainage system near the distal portion of the lacrimal sac beneath the inferior turbinate of the nose. Most obstructions open spontaneously within 4 to 6 weeks after birth. Massage of the lacrimal sac and application of topical antibiotics may be performed. If this conservative management fails, a nasal lacrimal duct probing should be performed.**

Degenerative Changes

- **Entropion Entropion is characterized as a tendency for the eyelid margin to rotate inward, with the lashes rubbing against the globe.** The lower lid is most frequently involved. This seldom occurs in patients under 40 years of age. Senile entropion is a degenerative change resulting from tissue atrophy and detachment of fascial attachments to the lid. Cicatricial entropion results from scarring of the lid conjunctiva with resultant inversion of the lid margin.

Entropion should be distinguished from trichiasis in which the eyelid margin is properly positioned, but eye irritation occurs from the misdirection of the eyelashes rubbing against the cornea.

- **Ectropion Ectropion is a tendency for the eyelid margin to rotate anteriorly, away from the globe. Like entropion, it is a degenerative change and is associated with horizontal lid laxity. Exposure of the eye due to inadequate coverage may cause ocular complications. Paralytic ectropion is a common disorder in patients suffering from facial nerve pathology such as Bell's palsy. These patients often cannot close the eye fully (lagophthalmos).**
- **Dermatochalasis/blepharochalasis Dermatochalasis refers to the redundant, baggy eyelid skin that occurs with age. Blepharochalasis refers to a condition of thinned, stretched eyelid skin that results from repeated bouts of eyelid edema of no apparent cause.**
- **Ptosis Ptosis is a drooping of the upper lid, occurring in two distinct forms. Congenital ptosis is characterized by a mild to severe degree of drooping, poor lid elevation function, and a poorly developed eyelid crease. It will be present at birth. Acquired ptosis is characterized by gradual onset and may be due to previous eye surgery, trauma, or as an aging change. Other conditions associated with ptosis include Horner's syndrome, third nerve palsy, and myasthenia gravis.**
- **Neoplastic growths The most common malignant tumor of the lid is basal cell carcinoma,** accounting for 90% of malignant eyelid neoplasms. **Basal cell carcinomas are more commonly located on the lower eyelid or in the medial canthal region.** Epithelial tumors such as seborrheic keratoses and squamous papillomas, as well as sudoriferous cysts, inclusion cysts, and cutaneous horns, are also quite common. Xanthelasmas are frequently seen lesions that characteristically localize to the nasal portion of both the upper and lower lids. They may occur spontaneously but are often associated with elevated serum cholesterol. These lesions are composed of lipid-filled histiocytes and are typically plaquelike, yellow, and soft. **Sebaceous carcinomas usually arise from the meibomian glands and are the second most common eyelid malignancy. The upper lid is the usual site of origin.** The tumor has a potential for local invasion and metastasis. Frequently they appear as a chalazion or as blepharoconjunctivitis. The diagnosis may be difficult or missed entirely for many months after onset.

Glaucoma

■ Basic Science **Glaucoma is an ocular disease characterized by elevated intraocular pressure (IOP).** The effect of glaucoma on the optic nerve is manifested as "cupping," an atrophy of the optic nerve head that leads to characteristic visual field loss. **In glaucoma, the *cup-to-disc ratio* is increased;** that is, the diameter of the cup takes up a disproportionate amount of the cross section of the optic disc. **This finding suggests significant optic nerve damage.** The damage may be due to a direct effect of elevated pressure on the nerve, disruption of axoplasmic flow, and/or vascular occlusion due to hemodynamic effects.

Physiologically, IOP is determined by the rate of aqueous humor fluid production and outflow. The most common cause of increased IOP is secondary to outflow obstruction.

Normal IOP is between 11 and 21 mm Hg. People with increased IOP are more likely to suffer from glaucomatous optic nerve damage, but increased IOP is not the sole factor. Some people with increased IOP do not develop optic nerve damage and therefore do not have glaucoma, whereas others develop optic nerve damage despite normal to low IOP and therefore do have glaucoma.

■ History and Physical Examination **Clinically, glaucoma may present as acute or chronic varieties. Acute glaucoma usually involves closure of the anterior chamber angle. Closure occurs when the iris is pushed forward against the cornea, blocking the normal fluid outflow path. The most common causes of angle closure glaucoma are primary angle closure (an anatomic variant that is prone to closure), iris neovascularization, and others.**

Patients suffering an acute angle closure attack present with the sudden onset of pain, blurred vision, and photophobia. They may also complain of seeing halos around light sources that result from cornea swelling or edema. Upon examination, the eye may appear red, the cornea appears gray or hazy, and the anterior chamber will be shallow. IOP is usually very high (e.g., 40 to 80 mm Hg). One or both eyes may be involved. The patient may have experienced similar but perhaps milder episodes previously, especially while in a darkened area such as a movie theater. Iris dilation in this situation is a contributing factor. **Pharmacologic dilation (e.g., during a routine examination) in predisposed individuals may also cause an attack of angle closure glaucoma.** Predisposed individuals tend to be farsighted.

Chronic glaucoma is far more common and usually is of the open angle variety. The gross anatomy appears normal, but there is resistance to fluid outflow of the eye at a microscopic level. Patients are usually asymptomatic, but a minority may experience a mild, nonspecific discomfort. Patients with this condition usually, but not always, have increased IOP and increased cupping. When the condition is advanced, loss of peripheral vision and eventually central visual acuity may result. Risk factors for chronic open angle glaucoma include a positive family history of glaucoma or diabetes mellitus and a history of severe ocular trauma. Because early signs of this type of glaucoma (increased IOP and cup-to-disc ratio) are nonspecific, the initial diagnosis may be difficult, and these patients may require serial observations to obtain a definitive diagnosis.

- Laboratory and Diagnostic Studies **The IOP must be measured. Optic nerve damage is evaluated by observing increased cupping by direct ophthalmoscopy. Currently, the most sensitive test of functional damage from glaucoma is visual field testing.**
- Formulating the Diagnosis **The differential diagnosis of a red, painful eye includes glaucoma, conjunctivitis, foreign body, corneal abrasion, keratitis, and intraocular inflammation (uveitis). The differential diagnosis of optic nerve damage includes glaucoma; optic neuropathy; optic neuritis; and compressive lesions of the optic nerve such as gliomas, meningiomas, aneurysms, and metastatic tumors.**
- Evaluating the Severity of Problem **Acute angle closure glaucoma is a medical emergency. Prompt relief of increased IOP and correction of the underlying cause must be accomplished.** Permanent loss of vision can result in as little as an hour to several days.
- Managing the Patient **IOP can be lowered by giving topical beta-blockers, oral or topical carbonic anhydrase inhibitors, and oral or IV osmotic agents such as mannitol. Topical pilocarpine (which constricts the pupil) can often be used to break the attack as it pulls the iris away from the chamber angle. In a primary angle closure attack, the definitive treatment is peripheral iridotomy in which a laser is used to make an opening in the iris, relieving the obstruction to flow.**

  In chronic glaucoma, medical treatment is usually the first line of therapy. Treatments include topical beta-blockers, apraclonidine, carbonic anhydrase inhibitors, and indirect or direct parasympathomimetics (e.g., pilocarpine, phospholine iodide). If patients fail or are intolerant of medical treatment, laser treatment of the trabecular meshwork (part of the outflow channel) may be helpful (argon laser trabeculoplasty). Some patients may be helped by glaucoma filtering surgery. This involves creation of a connection from the anterior chamber of the eye to a space further back on the eye, effectively shunting fluid from the anterior eye.

Hypertensive Retinopathy

- Basic Science **Systemic hypertension may cause localized or generalized arteriolar constriction. This may affect vessels in either the retina (observable with an ophthalmoscope) or choroid.** The choroid is an intricate network of smaller vessels beneath the retina.

  Acutely and significantly elevated blood pressure may lead to vascular nonperfusion of the choroidal layer. This may affect the overlying retina with fluid leakage (retinal edema), ischemia, and eventual retinal photoreceptor loss.

The optic nerve head may also become swollen on an ischemic or mechanical basis.

Chronically elevated systemic blood pressure leads to retinal vessel constriction secondary to autoregulation. This leads to vessel leakage of fluid into the retina, thickening of the vessel walls, and formation of aneurysms of retinal vessels.

- History and Physical Examination **The fundus should be examined with a direct ophthalmoscope. Dilation may be needed for an adequate examination. The following signs may be noted:**
  - **Microaneurysms/macroaneurysms** Microaneurysms and macroaneurysms are small, red, dotlike vascular outpouchings in the retina. They are well defined and round and need not be adjacent to observable arterioles.
  - **Cotton wool spots** Inner retinal infarction may cause cotton wool spots, which are white nebulous areas in the retina.
  - **Optic nerve swelling** The nerve head may be swollen; this is demonstrated by the loss of the normally sharp outer rim. The nerve bulges toward the vitreous cavity instead of being flush against the ocular wall.
  - **Retinal vessel changes**
    - ***Narrowing of retinal arterioles*** The absolute diameter of the vessels is decreased.
    - ***Thickening of retinal arteriole walls*** The light reflex off the vessel wall is broader and brighter.
    - ***Arteriovenous crossing changes*** Thickened arterioles may cause compression of veins at points where they cross; this is called *nicking.*
  - **Hemorrhages** Retinal hemorrhages may be dot, blot, or flame shaped.
  - **Hard exudates** Hard exudates are well-defined yellow accumulations of lipid within the retina that are due to chronic edema.
- Laboratory and Diagnostic Studies **Fluorescein angiography may be performed by an ophthalmologist.** Dye is injected into a peripheral vein, and specialized pictures are taken of the retinal vasculature as the fluorescent dye travels within it. **Vessel wall anatomy, patency, and leakage can be well visualized.**
- Formulating the Diagnosis **Characteristic retinal vascular changes in the setting of either acute or chronic systemic hypertension may be diagnostic.**
  - **Evaluating the severity of problems** Hypertensive retinopathy suggests that hypertensive changes have occurred elsewhere within the cardiovascular system, predisposing the patient to an increased risk of stroke and heart disease.
- Managing the Patient **Control of chronic hypertension may slow or halt disease progression. Argon laser photocoagu-**

lation may be performed to treat some retinal changes including fluid leakage, ischemia, and aneurysm formation. If the retina shows signs of acute malignant hypertension (optic nerve head swelling, ischemic/necrotic changes), blood pressure should be brought down urgently but gradually to prevent optic nerve infarction.

Multiple Choice Review Questions

1. A 60-year-old woman relates progressive painless reduction of vision in both eyes over the previous 6 months. She reports that vision seems impaired in bright sunlight and while driving at night. She denies any medical problems. On examination, her visual acuity measures 20/50, and her eyes appear quiet. The most likely diagnosis is which of the following?

a. Corneal abrasion
b. Retinal detachment
c. Cataract
d. Acute angle closure glaucoma
e. Central retinal artery occlusion

2. Of the following preoperative factors, which has the least importance in determining the final visual acuity following cataract removal?

a. Astigmatism
b. Amblyopia or lazy eye
c. Retinal disease
d. Corneal opacity
e. Optic nerve atrophy

3. A 30-year-old woman presents herself with a 3-day history of a unilateral red eye. The vision is decreased and there is significant light sensitivity (photophobia). There is a mucus-type discharge present. She had the "flu" with fever 2 weeks earlier. There is no history of exposure to an individual with a red eye, no history of contact lens wear, and no other associated nonocular findings. Past medical history is positive for herpes simplex virus labialis and hayfever. What is the best course of action?

a. Gentamycin ophthalmic solution q.i.d. for presumed bacterial conjunctivitis
b. Cool compresses and observation for presumed viral conjunctivitis
c. Ophthalmologic referral for presumed acute angle closure glaucoma
d. Ophthalmalogic referral for presumed herpes simplex virus-related ocular disease
e. Observation and topical antihistamines for presumed allergic conjunctivitis

4. A mother of a child in daycare presents herself with a unilateral red eye, 5 days after her son begins topical ocular antibodies for a bilateral red eye. Although the vision is not decreased, there is significant mucopurulent discharge. Her son had resolution of symptoms in 2 days. What is the most likely diagnosis?

a. Bacterial conjunctivitis
b. Viral epidemic conjunctivitis
c. Herpes simplex virus conjunctivitis
d. Fungal conjunctivitis
e. Allergic conjunctivitis

5. A healthy 65-year-old white male presents himself with complaints of bilateral red eye with associated foreign body sensation, burning, and a "heavy feeling" about the eye. The symptoms have been present for approximately 2 months without significant change. The patient notes a small "bump" is present in the lower left lid which was initially painful to touch but which is now smaller, firm, and nontender. The swellings have been recurrent. The patient denies any exudative-type drainage but does have significant tearing in both eyes. What is the most likely diagnosis?

a. Bacterial or viral conjunctivitis
b. Recurrent attacks of acute angle closure glaucoma
c. Herpes simplex keratitis
d. Blepharitis with internal hordeolum
e. Corneal abrasion

6. What would the best management for this patient be?

a. Routine bacterial culture of the eyelids followed by a 7 to 10 day course of oral antibiotics
b. Hot soaks to the eyelids bid with a topical Gram-positive antibiotic ointment applied to both lids once a day
c. Viral culture of the eyelids
d. Herpes simplex virus culture of the eyelids
e. Send the patient home with a topical anesthetic qid with scheduled ophthalmology follow up in 1 week

Chapter 3

Otorhinolaryngology

Allergic Rhinitis

- **Basic Science** Allergic rhinitis is the nasal manifestation of the body's allergic response. It may be triggered by environmental or food allergies. Binding of **IgE** to mast cells triggers the release of **histamine** and **leukotriene** within the nose, causing swelling of the nasal lining and increased production of mucus.

- **History and Physical Examination** **Symptoms** are commonly **seasonal,** but may be **perennial** if the allergens are food, dust, or animal dander. Nasal symptoms of **sneezing, congestion, and watery rhinorrhea** may be accompanied by **itchy, watery eyes,** "postnasal drip," and a scratchy or sore throat. Facial pain may be intermittently present as a result of edematous blockage of sinus ostia. Decreased hearing and clogged ears may result from concurrent serous otitis media.

  Physical examination will reveal **pale, boggy, bluish tinged nasal mucosa and swollen inferior turbinates.** Nasal secretions will be clear or white. Conjunctival injection and prominent pharyngeal lymphoid follicles may also be present.

- **Laboratory and Diagnostic Studies** One may examine the **nasal secretions** for the presence of **eosinophils. Skin testing or radioallergosorbent test (RAST)** testing may be used to determine the causative allergen.

- **Formulating the Diagnosis** The **diagnosis of allergic rhinitis is made clinically** based on the constellation of signs and symptoms. Symptoms of an upper respiratory infection or sinusitis may mimic some of those of allergic rhinitis. Chronicity of symptoms will help differentiate allergic rhinitis from viral upper respiratory infection. Vasomotor rhinitis, a runny nose triggered by cold weather, may present similarly to allergic rhinitis, but without the itching and sneezing. Sinusitis will present with facial pain and pressure and purulent rhinorrhea.

- **Evaluating the Severity of Problems** Allergic rhinitis, although potentially quite annoying, is not a life threatening condition. Allergic edema may predispose a patient to bacterial sinusitis.

- **Managing the Patient** The first line of treatment is **avoidance,** if possible, **of the offending allergen.** Rugs, bed sheets, and heavy draperies are reservoirs for dust mites and should be removed and cleaned frequently. Those patients with pollen allergies should keep their windows closed and rely on air conditioning during the pollen season. Oral **antihistamines,** either the older over the counter forms or the newer prescription, non-sedating types are useful in treating both nasal and ocular symptoms. The physician must be aware of the **anticholinergic side effects of over the counter antihistamines** (drying, increased risk of urinary retention in patients with prostatic hypertrophy, and acute angle closure in narrow angle glaucoma). In addition, several of the **non-sedating antihistamines may produce cardiac arrhythmias** in patients with hepatic dysfunction or those taking macrolide antibiotics or certain antifungal drugs.

Decongestants such as pseudoephedrine or phenylpropanolamine are useful in improving nasal breathing. They should be avoided, however, in patients with hypertension or cardiac disease.

Increasingly **nasal steroid sprays** are becoming a mainstay of therapy for allergic rhinitis. They have the advantages of no appreciable systemic side effects, and minor nasal side effects (i.e. stinging). *Candidal* infection of the nose is a rare complication of intranasal steroid use. Nasal steroids may take 3-10 days to reach a therapeutic intranasal level. They need to be used continuously to be effective, but can be discontinued after allergy season has ended.

Cromolyn sodium nasal spray may be used as an adjunct to intranasal steroids. It has the disadvantage of needing to be used four times a day versus the once or twice a day for the nasal steroids.

Allergen desensitization should be considered in patients whose symptoms are not relieved by the above medications.

Patients with **nasal polyps** or **turbinate hypertrophy** unresponsive to medications **may require surgery** to relieve nasal obstruction.

Epistaxis

- **Basic Science** The most **common site** of epistaxis is located on the anterior, inferior portion of the nasal septum. This area, known as Kiesselbach's plexus, is the site of a confluence of branches of the ethmoidal artery (via the internal carotid) and internal maxillary artery (via the external carotid). Since this area is near the nares, it is subject to trauma and desiccation and erosion of superficial blood vessels. The most frequent cause of epistaxis is the erosion of the vasculature of the anterior nasal septum. Posterior epistaxis is uncommon and occurs secondary to arteriosclerosis.

- **History and Physical Examination** Epistaxis occurs more commonly during the winter season, in the presence of artificial heat and low humidity. Patients with underlying bleeding disorders and those on anticoagulant medication (e.g., warfarin) are at increased risk for epistaxis. Questions should be directed toward detecting the frequency, severity, duration, and side of the nosebleed.

  On physical examination the **physician should localize the site of bleeding.** Examining the anterior nasal cavity and pharynx with a pen light for trickling blood should be done. If the patient is not currently bleeding, it is often best not to disturb a clot. Although **hypertension does not cause epistaxis, it may make control of bleeding more difficult once it has begun.**

- **Laboratory and Diagnostic Studies** A complete blood count is the only test necessary in most cases of epistaxis: however, a protime (PT) and partial thromboplastin time (PTT) may be useful in the presence of a history of bleeding disorder or use of anticoagulant medicine.

- **Formulating the Diagnosis** When there is blood dripping from the nose, the diagnosis is obvious. However, if a patient spits up blood, one might confuse hemoptysis or hematemesis with epistaxis. Essentially, to make the diagnosis of epistaxis, **blood must be observed in, or coming from, the nose.**
- **Evaluating the Severity of Problems** Most cases of epistaxis can be controlled with 10 minutes of **continuous, direct pressure** on the external nasal skin below the nasal bones. If bleeding persists in spite of conservative measures, further intervention is indicated.
- **Managing the Patient** Localizing the site of bleeding requires light, a nasal speculum, and often a suction device. **Bleeding** may be **controlled with topically applied vasoconstrictors (e.g., cocaine or phenylephrine),** oxidized cellulose, electric **cautery,** or chemical cautery with silver nitrate. If these measures are unsuccessful, **anterior nasal packing** should be placed. When bleeding is posterior in origin, an otolaryngologist should be consulted for placement of a **posterior pack,** followed by bilateral anterior packing. Such a patient should be admitted and have continuous pulse oximetry while the packing is in place. Bleeding that continues despite adequate packing may require surgical or angiographic intervention.

  All patients with epistaxis will **benefit from** humidification of the nasal mucosa. This can be accomplished with **saline nasal sprays** and room humidifiers.

Head and Neck Cancer

- **Basic Science** The upper aerodigestive system is lined by squamous epithelium, so the most common form of cancer in the head and neck is squamous cell carcinoma. **Risk factors for squamous cell carcinoma include smoking and drinking alcohol.** The presence of both factors increases risk geometrically. Like many cancers, squamous cell carcinomas do not manifest their presence until they invade surrounding structures. Referred pain to the ear along cranial nerves IX and X may be the first manifestation of an oropharyngeal or hypopharyngeal tumor. The most common site for a cancer is the larynx.
- **History and Physical Examination** **Sores** on the tongue, cheek, palate, or lips **that do not heal** in 2 weeks and sore throats that do not respond to 7 days of antibiotics require investigation. **Difficulty swallowing (dysphagia), pain on swallowing (odynophagia),** and blood-tinged saliva may all be signs of an occult malignancy. **Ear pain** with a normal ear examination demands examination of the base of the tongue and hypopharynx. Hoarseness that lasts more than 6 weeks should prompt examination of the vocal cords. Many of these symptoms may be caused by benign processes, but if ignored may result in a lost opportunity to find an early-stage cancer. A **neck mass in an adult is a metastatic lymph node until proven otherwise.**

  A thorough examination of the tongue, buccal mucosa, floor of the mouth, and palate can be performed with two tongue depressors. The base of the tongue, hypopharynx, and larynx can be examined with a dental mirror.

- **Laboratory and Diagnostic Studies** No blood test exists that can detect head and neck cancer. Plain x-ray examinations are also of no value. **Fiberoptic laryngoscopy** is an office procedure performed under topical anesthesia that can visualize those areas of the mouth and throat that are not visible perorally. Suspicious lesions should be biopsied. **Neck masses** (nonpulsatile) should be **aspirated** with a **fine needle** and sent for cytologic examination. **Computed tomographic (CT) scan** with **intravenous contrast** is the imaging study of choice for delineating the extent of **confirmed cancers.**

- **Formulating the Diagnosis** Many symptoms of head and neck cancers may be caused by benign conditions, but in a patient who smokes or drinks, **a malignancy should be assumed to be present until proven otherwise.**

  The following benign conditions may present with the same symptoms as head and neck cancers: gastroesophageal reflux, vocal cord polyp (or nodule), and temporomandibular joint dysfunction.

- **Evaluating the Severity of Problems** Late stage head and neck cancer may require tracheotomy for upper airway obstruction and nasogastric or gastric tube feedings because of poor oral intake.

  Endoscopy under general anesthesia is usually necessary to evaluate the extent of base of tongue, hypopharyngeal, and laryngeal tumors.

- **Managing the Patient** **Early stage cancers** may be managed by **surgery *or* radiation. Later stage tumors** are better treated by **surgery *and* radiation.** Encourage all patients to **stop smoking and drinking.**

Hearing Loss (Sensorineural)

- **Basic Science** The inner ear is divided into two components, the cochlea and the vestibular apparatus.

  The **cochlea** converts mechanical energy into neural signals. This occurs when the endolymph stimulates the movement of specialized projections on **hair cells.** The hair cells are topographically organized so that sounds of different pitch are recognized by hair cells at different locations within the cochlea. Malfunction or damage to the cochlea can **result in sensorineural hearing loss or tinnitus** (noise in the ear).

- **History and Physical Examination** **Sensorineural hearing loss** may be congenital or acquired. The most common form is **associated with aging.** People with sensorineural hearing loss will often report **hearing people speak but being unable to understand what is being said.** They often have **difficulty hearing in the presence of background noise.**

  Risk factors for sensorineural hearing loss are the presence of autoimmune diseases, collagen vascular disorders, or tertiary syphilis, or therapy with ototoxic drugs (i.e., aminoglycosides, high dose erythromycin, loop diuretics, asplatinum).

  Physical examination of the outer and middle ear (otoscopy) will be **normal** in inner ear disorders.

- **Laboratory and Diagnostic Studies** Many tests have been recommended to evaluate inner ear disorders. All patients should have a behavioral audiogram performed by an audiologist in a soundproof booth. Patients with unexplained sensorineural hearing loss should have erythrocyte sedimentation rate, antinuclear antibody, rheumatoid factor, and VDRL (i.e., a syphilis test) tests.
- **Formulating the Diagnosis** The most common cause of gradual **sensorineural hearing loss** is the idiopathic change of **aging** (i.e., **presbycusis).** Causes of **sudden hearing loss** include **trauma, viral infections** such as **mumps and measles, autoimmune diseases, and cerebrovascular accidents.**
- **Evaluating the Severity of Problems** Sudden hearing loss requires early intervention if hearing is to be successfully salvaged. A tuning fork can be used to perform a Weber test to differentiate conductive from the sensorineural hearing loss. The patient is asked to localize the sound to the right, left, or midline. Sound localizes toward the ear with conductive loss and away from the ear with sensorineural hearing loss.
- **Managing the Patient** **Presbycusis** can be **managed with a hearing aid. Sudden sensorineural hearing loss** (with an unremarkable workup) should be treated with a short course of **high-dose steroids** for presumed viral etiology.

Otitis Externa

- **Basic Science** The ear canal is protected by hair and cerumen. **The bacteria that normally live in the ear canal are** *Pseudomonas aeruginosa* and *Staphylococcus* species. **Trauma** to the ear canal skin allows these bacteria to invade, resulting in **otitis externa.**
- **History and Physical Examination** Otitis externa (swimmer's ear) is usually the result of trauma to the ear canal. The two most common causes are the use of cotton swabs and maceration from water. The patient will complain of severe, worsening pain in the ear, associated with purulent, foul-smelling drainage. As the drainage accumulates, a hearing loss will become apparent.

  Physical examination will demonstrate tenderness induced by pulling on the pinna or pushing on the tragus. The ear canal will be swollen and red. Drainage may obstruct visualization of the eardrum.
- **Laboratory and Diagnostic Studies** No laboratory or diagnostic studies are necessary to make the diagnosis of acute otitis externa.
- **Formulating the Diagnosis** **The diagnosis is made clinically,** based on the history and physical examination.

  Ear pain may also be the result of dental infections, teething, temporomandibular joint disease, otitis media, and head and neck cancers.
- **Evaluating the Severity of Problems** Most outer ear infections are easily treated on an outpatient basis. Occasionally, patients with otitis externa will develop a chondritis or, rarely, a *Pseudomonas* osteomyelitis of the temporal bone requiring intravenous antibiotics.
- **Managing the Patient**

  - The **treatment** of acute **otitis externa** consists of **removal of debris** and pus from the canal and instillation of **antibiotic and cor-**

ticosteroid drops. If the canal is **too swollen to allow the entry of drops,** a **wick should be inserted.**

- Patients should be advised to avoid inserting cotton swabs into their ears and to thoroughly dry the ear canal after swimming and showering to prevent recurrence of otitis externa.

Otitis Media

■ **Basic Science** The middle ear is lined by respiratory-type mucosa. Air enters the middle ear through the eustachian tube. When the eustachian tube is obstructed, the cells lining the middle ear secrete fluid. This fluid may become infected by bacteria from the nasopharynx. The most common **organisms in otitis mediaare *Streptococcus pneumoniae, Hemophilus influenzae, Moraxella catarrhalis,* and group A streptococcus.**

■ **History and Physical Examination** **Otitis media** often **follows a viral** upper respiratory **infection** that has caused edema and obstruction of the eustachian tube. Children with cleft palate or Down syndrome are at increased risk for otitis media. **Infants** and young **children** become febrile, **appear irritable,** and may **tug at their ear.** Older children and adults may notice a **decrease in hearing** secondary to the accumulation of middle ear fluid. When the fluid becomes infected, an **earache** results. If acute otitis media is not treated, tympanic membrane perforation may occur, resulting in drainage of pus from the ear and decreased pain.

Physical examination reveals an **opaque, red, bulging eardrum.** Tympanic membrane landmarks may disappear. The presence or absence of a light reflex is not a useful sign. If the tympanic membrane is perforated, there may be pus in the ear canal. Pinna movement does not cause pain as it does in otitis externa.

■ **Laboratory and Diagnostic Studies** No laboratory or diagnostic studies are necessary to make the diagnosis of acute media.

■ **Formulating the Diagnosis** **The diagnosis is made clinically.** Familiarity with appearance and mobility of the normal tympanic membrane is probably the most important factor in differentiating middle ear infection from other causes of ear pain.

Dental infections, teething, temporomandibular joint disease, and head and neck cancers can all cause the perception of ear pain. A crying child will have pink and injected ear drums, even in the absence of infection.

■ **Evaluating the Severity of Problems** Serious complications of otitis media may be heralded by sudden hearing loss, vertigo, fever, and chills. Acute mastoiditis, facial nerve paralysis, meningitis, and brain abscess may occur.

■ **Managing the Patient** **Acute otitis media** should be treated with a **10-day course of amoxicillin** or trimethoprim-sulfamethoxazole. If fever or pain persist beyond 48 to 72 hours, the patient should be given a broader-spectrum antibiotic.

- Patients with **recurrent otitis media** should be given a trial of low-dose **prophylactic antibiotics.** The **failure of antibiotic pro-**

phylaxis in combination **with** a **conductive hearing loss** (90 days after the most recent episode of otitis media) are indications for the **insertion of tympanostomy tubes.**

Sinusitis

- **Basic Science** There are four sets of paranasal sinuses. **Most** of the **sinuses drain into** the nose through the **middle meatus.** Obstruction of the middle meatus by viral-induced edema, inflammatory polyps, severe nasal septal deviation, or tumors can trigger bacterial sinusitis. Bacteria responsible for sinusitis include *Streptococcus pneumoniae, Hemophilus influenzae, Moraxella catarrhalis,* and group A streptococcus.
- **History and Physical Examination** Symptoms of sinusitis include facial pain and headache, purulent nasal discharge, postnasal drip, and cough. There is **usually a history of antecedent viral upper respiratory infection.**

  On physical examination, patients may have at most a low-grade fever. Inspection of the nose will show swollen, red turbinates with **purulent yellow or green rhinorrhea.** Generally there is tenderness to palpation over the infected sinus. There may be an increase in facial pain when the patient is asked to bend forward at the waist.
- **Laboratory and Diagnostic Studies** X-ray films of the sinuses can help differentiate sinusitis from other causes of facial pain or headache. If the physician is confident of his or her diagnosis based on the history and physical examination alone, then x-ray examinations are not necessary.
- **Formulating the Diagnosis** Sinusitis may be confused with a viral upper respiratory infection or allergic rhinitis. The diagnosis of sinusitis should be considered when **nasal discharge lasts longer than 1 week,** the color of the discharge changes from clear or white to **dark yellow or green,** or **facial pressure or pain,** not normally present in a viral illness, develops.
- **Evaluating the Severity of Problems** Since the sinuses are adjacent to the orbits and the brain, it is possible for infection to spread to these structures. In patients with evidence of acute or recent sinusitis, eyelid swelling, conjunctival edema, proptosis, or lethargy, demand evaluation by specialists. These patients may need a CT scan (the most sensitive imaging study to identify patients with sinusitis).
- **Managing the Patient** Patients with sinusitis should be **treated with oral antibiotics (e.g., ampicillin, trimethoprim-sulfamethoxazole,** or first- or second-generation cephalosporin); additionally, **oral and topical decongestants (e.g., pseudoephedrine and phenylephrine nasal spray) should be prescribed.** Antihistamines should be used only in patients with underlying allergic rhinitis, as they thicken nasal secretions and retard drainage.

  Patients with orbital or intracranial complications will require intravenous antibiotics and (possibly) surgery.

  Patients with recurrent or chronic sinusitis may benefit from a workup to detect underlying structural abnormalities that predisposes them to sinusitis.

Tonsil and Adenoid Inflammation

- **Basic Science** The adenoids and tonsils are lymphatic structures that function as the first line of defense against oral and respiratory infections. These structures may **cause problems** when they become a **repeated site of infection.** The adenoids and tonsils may also hypertrophy to the extent that they obstruct the oral or nasal airways.

  The tonsils are separated from the pharyngeal musculature by a fibrous capsule. If the **infection** spreads from the tonsil **into** this **potential space,** a **peritonsillar** abscess forms.

  The adenoid pad may contribute to otitis media by acting as a reservoir for bacteria to spread to the middle ear via the eustachian tubes.

- **History and Physical Examination** Adenotonsillitis may be of **viral or bacterial etiology.** Patients complain of a **sore throat, fever,** nasal congestion, and **difficulty swallowing.**

  On physical examination, the **tonsils will be enlarged and red and may be covered with an exudate.** Some viral infections cause vesicles to form on the tonsils (e.g., Coxsackie virus). There may be a fine, papular, generalized **rash present in the patient with strep throat.**

  Trismus (inability to open the mouth) and a muffled voice are signs of a peritonsillar abscess. The tonsil on the side of the abscess side is often obscured.

- **Laboratory and Diagnostic Studies** **A throat culture or rapid strep test may be used to diagnose group A β-hemolytic strep throat.**

  A monospot test may be indicated in young patients with tonsillitis and generalized adenopathy, or in those patients who do not respond to antibiotics. A white count with differential may be useful in suggesting infectious mononucleosis when the monospot test is negative (within the first 3 days of infection).

- **Formulating the Diagnosis** **The diagnosis of adenotonsillitis is made clinically.**

- **Evaluating the Severity of Problems** Most patients with tonsillitis or peritonsillar abscess can be **treated as outpatients.** If a patient **cannot drink** because of the pain, he or she should **be admitted for intravenous antibiotics and hydration.** Occasionally, acute adenotonsillar enlargement can impinge on the airway, necessitating placement of a nasopharyngeal airway to relieve the obstruction.

- **Managing the Patient** The **decision to treat adenotonsillitis is usually based on the severity of symptoms.** The antibiotic of choice for bacterial tonsillitis is **penicillin.** Erythromycin is an alternative drug for penicillin-allergic patients.

  Peritonsillar abscesses should be drained by needle aspiration or by incision and drainage. The carotid artery lies in close proximity to the tonsils—therefore, abscess drainage is usually performed by an otolaryngologist.

  Patients with **recurrent tonsillitis** or upper airway obstruction from adenotonsillar hypertrophy will benefit from **adenoidectomy or tonsillectomy, or both.**

Upper Airway Obstruction

■ **Basic Science** The upper airway consists of the nose, nasopharynx, oral cavity, oropharynx, laryngopharynx, and trachea. Obstruction at any of many levels may impede respiration.

The primary function of the larynx is protection of the lung from aspiration. A reflex arc is mediated through the sensory and motor branches of the vagus nerve to adduct the vocal cords during swallowing.

■ **History and Physical Examination** Signs of upper airway obstruction may include a history of **shortness of breath, pain or difficulty swallowing, change in voice,** and the presence of fever.

Physical findings include visible **swelling of upper airway structures, drooling,** and, most importantly, **stridor. Stridor is a loud noise created during** respiration. In upper airway obstruction, stridor is predominantly heard on **inspiration.** Stridor can be differentiated from wheezing (a lower respiratory sign) by its inspiratory nature. Stridor may also produce **loud noises in the neck;** in contrast, wheezing is louder in the chest.

■ **Laboratory and Diagnostic Studies** **AP and lateral neck x-ray** examinations should be ordered to evaluate the upper airway. Fiberoptic laryngoscopy may also be necessary, but should not be attempted in children suspected of having epiglottitis.

Because of potential airway instability, the **patient should not leave the emergency department.** All x-ray films should be taken in the emergency department under physician supervision.

An arterial blood gas measurement is not a helpful diagnostic or prognostic test in upper airway obstruction, as it may be normal until respiratory arrest occurs.

■ **Formulating the Diagnosis** Upper airway obstruction can be secondary to any of the following:

- **Foreign body aspiration**
- **Epiglottitis** (or other infection)
- **Allergic reactions**
- **Trauma** (e.g., fracture of both sides of the mandible, which may result in retrodisplacement of the tongue)
- **Acquired narrowing** (e.g., subglottic stenosis)
- **Congenital narrowing** (e.g., choanal atresia)
- **Neoplasm** (e.g., laryngeal cancer)

■ **Evaluating the Severity of Problems** Upper airway obstruction is a **medical emergency.**

■ **Managing the Patient** First, the site of the problem must be established (i.e., upper versus lower airway).

Upper airway obstruction is a medical emergency. An immediate assessment must be made of the severity of respiratory difficulty and the degree of obstruction. If the problem can be found, it must be treated immediately. **If the patient's airway is in jeopardy, an endotracheal tube must be placed. When an endotracheal tube cannot be placed, a surgical airway (e.g., cricothyroidotomy) must be established.**

Treat the cause:

• *Foreign bodies.* **Foreign objects** should be visualized if possible with **indirect laryngoscopy.** They should be removed **with McGill forceps.** If this is impossible, an ENT should be called to perform flexible bronchoscopy.

• *Epiglottitis.* Because of the airway instability, **endotracheal intubation** should be performed **immediately,** preferably in the operating room by an anesthesiologist with an otolaryngologist in attendance. Intravenous antibiotics should be given.

• *Allergic reactions.* **Allergic reactions, manifested by edema of the lips, tongue, and uvula,** are all signs of progressive airway compromise. The patient should immediately be given **subcutaneous epinephrine** before the workup continues. Additional management includes nebulized epinephrine, intravenous antihistamines, and intravenous steroids. If the patient **deteriorates further, he or she should be intubated; failing intubation, a surgical airway must be established.**

Vertigo

■ **Basic Science** The inner ear **vestibular apparatus** is one part of the body's vestibular system. It is responsible for alerting the brain to changes in head position. Malfunction or damage to the vestibular apparatus can result in **peripheral vertigo.**

Damage to the **eighth cranial nerve** or the **vestibular nuclei and pathways** (which travel through the cerebellum and brain stem) can cause **central vertigo.**

■ **History and Physical Examination** **Peripheral vertigo** may last **minutes, hours, or days.** Patients report the sensation that they (or their environment) are spinning. Patients with peripheral vertigo may be **nauseated, vomit, or fall,** but they will never have syncope or vision or speech difficulties.

Physical examination of the outer and middle **ear** (otoscopy) will be **normal. Peripheral vertigo** may be **initiated** with a **change in head position.** The patient will manifest **horizontal-rotary nystagmus.**

In central vertigo nystagmus may be of any variety, but vertical nystagmus is associated only with central vestibular dysfunction.

Evaluation of vertigo should include **checking the pulse for an irregular heart rhythm and a complete neurologic examination.**

■ **Laboratory and Diagnostic Studies** Many tests have been recommended to evaluate inner ear disorders. All patients should have a behavioral audiogram performed by an audiologist in a soundproof booth. Patients may also benefit from a **magnetic resonance image to rule out brain stem tumor, hemorrhage, or infarct.**

■ **Formulating the Diagnosis** The **most common cause of peripheral vertigo is vestibular neuronitis,** a viral infection of the vestibular nerve. Patients report the sudden onset of unremitting vertigo that lasts for several days before symptoms begin to spontaneously relent.

Ménière's disease is a clinical diagnosis. Patients with Ménière's disease have episodic **bouts of vertigo, hearing loss, and a sensation of a clogged ear** that last from **1 to 24 hours.** Tinnitus may be constant or intermittent. Ménière's disease usually affects only one ear at a time. As the vertigo is recurrent the diagnosis of Ménière's cannot be made on first presentation.

Basilar artery insufficiency, orthostatic hypotension, and cardiac arrhythmias can also cause the sensation of vertigo or lightheadedness. Diabetic peripheral neuropathy may give a patient the feeling of unsteadiness.

- **Evaluating the Severity of Problems** **Peripheral vertigo** must be **differentiated from central vertigo.** Untreated peripheral vertigo will not result in permanent injury; however, **central vertigo may be life-threatening** and needs to be identified immediately.

- **Managing the Patient** **Symptoms of peripheral vertigo** may be **relieved** with the following:
  - ***Benzodiazepines*** for vestibular sedation. This is the fastest, most effective treatment for acute vertigo.
  - ***Antihistamines*** (e.g., meclizine or diphenhydramine) for vestibular system sedation
  - ***Anticholinergics*** (e.g., scopolamine or atropine) for control of nausea

  Reduction of sodium intake and administration of a diuretic may control the vertigo of Ménière's disease.

  Some patients with unrelenting vertigo may require vestibular nerve section or other ablative procedures directed towards the malfunctioning vestibular aparatus.

Multiple Choice Review Questions

1. A 15-year-old boy comes to the emergency room complaining of a sore throat. He has a temperature of 102° F. Which of the following signs and symptoms does NOT help distinguish a peritonsillar abscess from tonsillitis?

a. Trismus
b. Sore throat
c. Bulging of the anterior tonsillar pillar
d. Muffled voice
e. Difficulty visualizing one tonsil

2. The boy in the previous question did not have any of the signs of a peritonsillar abscess. You diagnosed him with his first episode of bacterial tonsillitis and treated him with oral penicillin and analgesics, but 5 days later he returns to your office unimproved. Physical examination findings are unchanged, that is, the tonsils are red, swollen, and covered with white exudates. What do you do next?

a. Obtain a throat culture for *Streptococcus*
b. Obtain a monospot test
c. Treat with intravenous penicillin
d. Treat with oral clindamycin
e. Obtain a tonsil biopsy

3. A 30-year-old woman comes to your office complaining of a green nasal discharge. Which of the following is most useful in making the diagnosis of acute sinusitis?

a. History and physical examination
b. Plain sinus x-rays
c. CT scan of the sinuses in the coronal plane
d. CT scan of the sinuses in the axial plane
e. Swab culture of the nasal discharge

4. You treat the previous patient with the appropriate antibiotics, but forget to give decongestants. The following week her ophthalmologist calls you to say he has the patient in his office and is concerned about an orbital cellulitis or abscess. The presence of which of the following signs has the ophthalmologist worried?

a. Excessive tearing
b. Clear nasal drainage with occasional flecks of blood
c. Conjunctival edema and proptosis
d. Conjunctival erythema and purulent drainage on the lashes
e. Enophthalmos

5. You are called to see a nursing home patient brought in with respiratory distress. He is somnolent and stridorous. The first step in management of his airway is which of the following?

a. Endotracheal intubation
b. Cricothyroidotomy
c. Tracheotomy
d. Inspection of the mouth and jaw thrust
e. Insertion of a nasopharyngeal airway

Chapter 4

Pulmonology

ACUTE RESPIRATORY FAILURE

■ Basic Science **Acute respiratory failure occurs when the lungs are unable to maintain normal levels of oxygen and carbon dioxide** in the systemic arterial blood.

Hypoxemic respiratory failure occurs when arterial oxygen tension falls below 60 mm Hg while breathing room air; arterial CO_2 tension is normal or reduced. **Hypercapnic respiratory failure occurs when arterial CO_2 tension rises above 45 mm Hg, associated with a respiratory acidosis (pH <7.35);** arterial oxygen tension is secondarily reduced.

The prototypic example of acute hypoxemic respiratory failure is the adult respiratory distress syndrome (ARDS). Many clinical disorders lead to ARDS, but the most common are sepsis, major trauma, and aspiration of gastric contents. The mechanisms of injury are poorly understood but probably involve both cellular and humoral mediators.

With appropriate support, few patients die of respiratory insufficiency; however, the mortality remains at 60% to 70%. One third of the deaths occur within 72 hours as a direct result of the underlying illness. Most of the remaining deaths occur within 2 weeks from infection and multiple organ failure. The prognosis for survivors appears good. Few are symptomatic, and residual pulmonary function abnormalities, if present, are usually mild.

Hypercapnic respiratory failure often occurs secondary to obstructive lung diseases (e.g., COPD, asthma), **upper respiratory obstruction** (e.g., foreign body aspiration, bacterial epiglottitis, and inhalation injury), and **chest wall diseases that increase the stiffness of the chest** (e.g., kyphoscoliosis). A sudden increase in chest wall compliance can also disrupt normal ventilation. For example, multiple rib fractures may cause part of the chest wall to become unstable ("flail chest"). The paradoxic respiratory motion of the unstable section of chest wall and severe pain associated with respiration can cause hypoventilation and atelectasis.

A variety of disorders result in hypercapnic respiratory failure by interfering with the function of respiratory muscles and nerves (e.g., myasthenia gravis, muscular dystrophy, and bilateral phrenic nerve disruption).

Hypercapnic respiratory failure frequently complicates disorders that interfere with brain stem respiratory center function or efferent impulse transmission in the cervical spinal cord. **Overdose of CNS depressant drugs is the most common cause of respiratory center suppression.** Trauma, hemorrhage, or stroke that involves the respiratory center or cervical spinal cord (third through fifth cervical nerve roots) are other causes.

■ History and Physical Examination **Typically, patients with ARDS develop respiratory distress within the first 72 hours of the inciting event with pink, frothy sputum, and diffuse crackles upon examination of the chest.**

The clinical features of patients with acute hypercapnic respiratory failure will, in part, depend on the specific underlying cause. Acute hypercapnia can produce dramatic changes in central nervous system function, including obtundation and coma. Patients with hypercapnic respiratory failure due to central nervous system disorders or neuromuscular disease may not develop respiratory distress. Patients with intrinsic lung, airway, or chest wall disorders will generally complain of dyspnea and exhibit tachypnea and use of accessory respiratory muscles. Patients with severe exacerbations of COPD or asthma typically present with a recent history of worsening dyspnea, confusion, and agitation. They are tachypneic with diffuse wheezing and decreased expiratory flow rates. Patients with acute upper airway obstruction present with severe dyspnea and tachypnea. If the airway obstruction is extrathoracic (above the vocal cords), stridorous breath sounds are heard on inspiration, whereas obstruction in the lower trachea generates wheezing. Patients with respiratory muscle weakness present with variable complaints of dyspnea and difficulty coughing. Worrisome signs include ineffective cough, difficulty in handling secretions and swallowing, and increasing respiratory rate.

■ Laboratory and Diagnostic Studies **All patients with acute respiratory failure should undergo arterial blood gas (ABG) measurement. Additionally, a chest x-ray should be obtained.**

■ Formulating the Diagnosis **The differential diagnosis of acute respiratory failure includes disorders that affect the parenchyma of the lung, such as pulmonary edema or obstructive airway disease, or problems that decrease ventilation in a normal lung, such as upper airway obstruction or diminished respiratory drive or muscle strength.**

ARDS is defined by the combination of severe dyspnea, diffuse alveolar infiltration on the chest radiograph, marked hypoxemia, and the absence of left ventricular failure. Pulmonary edema due to left ventricular failure must be ruled out. Sometimes this can be accomplished by history, physical examination, ECG, and echocardiography. Often the distinction between ARDS and left ventricular failure will not be clear, and pulmonary artery catheterization will be required.

In considering the differential diagnosis of hypercapnic respiratory failure, it is useful to categorize disorders according to the component of the respiratory system affected. Obstructive lung diseases such as COPD and asthma cause hypercapnic respiratory failure primarily by increasing pulmonary dead space and precipitating respiratory muscle fatigue. In patients with COPD, acute respiratory failure frequently occurs in the setting of chronic respiratory failure and may be precipitated by respiratory tract infections, congestive heart failure, pulmonary emboli, or pneumothorax. Upper airway obstruction causes hypercapnia by decreasing total ventilation.

- **Evaluating the Severity of Problems** In hypercapnic respiratory failure, the severity of disease is reflected by the degree of hypercapnia detected by the ABG. Similarly, the severity of hypoxemic failure is gauged by the magnitude of hypoxemia on the ABG and indirectly through pulse oximetry.

- **Managing the Patient** The physician should consider the possibility of upper airway obstruction in every patient with acute respiratory failure. **To manage acute respiratory failure successfully, the physician must accomplish three goals: (1) establish an airway, (2) administer an appropriate amount of oxygen, and (3) maintain adequate alveolar ventilation.** Furthermore, the underlying cause of respiratory failure should be identified and treated.

  Most patients with severe respiratory failure of either type will require mechanical ventilation. With regard to ventilator settings, the degree of hypoxemia may be largely controlled with increasing concentrations of oxygen and positive end-expiratory pressure. Hypercapnia is treated with increasing respiratory rate and tidal volume.

ASTHMA

- Basic Science **Asthma is a disease characterized by hyperresponsiveness of airways to irritants, causing bronchoconstriction. Pathologic changes include the following:**

  - Infiltration of airway mucosa with eosinophils
  - Airway edema
  - Overproduction of airway mucus
  - Inflammatory cell-mediated injury to airway epithelium

  Because of these changes, alveoli become unable to exchange gas with the outside air. The alveolar oxygen content becomes depleted. Thus venous blood returns to the left heart with much less oxygen than normal (decrease in Po_2).

  Between 2% and 6% of the U.S. population have asthma. **Asthma may present at any age and is loosely classified as either allergic or nonallergic. Allergic asthma develops in childhood.** Many asthmatic patients have allergic conditions such as hay fever or eczema. Common allergens include pollen, house dust, drugs, feathers, and animal fur. The airways of allergic asthmatics contain mast cells with IgE directed against specific antigens. Contact with these antigens results in release of histamine, bradykinin, and leukotrienes. Continued exposure to the antigen leads to an influx of eosinophils and lymphocytes that cause damage to the airway epithelium.

  Nonallergic asthma occurs in older individuals with normal IgE levels. Precipitants include exercise, drugs (e.g., aspirin), chemicals, gastroesophageal reflux, atmospheric pollution, chronic sinusitis, and viral infection. Although asthma may improve during adolescence and with age, the condition may recur.

- History and Physical Examination **Most attacks begin with shortness of breath; sudden wheezing; and a hacking, nonproduc-**

tive cough. The patient may give a history of an upper respiratory infection. Patients should be asked about the severity of their asthma, known precipitants, previous hospitalizations (ICU admissions and intubations), and medication compliance.

In an acute attack the patient appears apprehensive and may speak in a panting manner. The patient may bend over to ease breathing. Wheezing, tachycardia, or the use of accessory respiratory muscles may be found.

- Laboratory and Diagnostic Studies During an acute asthma attack, an arterial blood gas should be obtained. **A rising (or "normal") P_{CO_2} with dyspnea represents a severe exacerbation and impending respiratory failure.** The patient should be placed on a pulse oximeter to monitor changes in arterial oxygenation.

  A chest x-ray should be ordered for patients whose history and physical exam suggest pneumonia or pneumothorax. Hyperinflation is typically seen because of air trapping.

  Pulmonary function tests should be obtained during the workup of a chronic asthmatic. Spirometry will reveal an obstructive pattern, with an FEV_1/FVC ratio less than 70%. Skin tests should be done, as they may help identify a precipitant.

- Formulating the Diagnosis **The diagnosis is based on the prior history and the typical symptoms of asthma (cough, dyspnea, and wheezing).**

- Evaluating the Severity of Problems Asthmatic episodes vary greatly in frequency, duration, and the degree of symptoms, ranging from occasional periods of wheezing, mild coughing, and slight dyspnea to severe attacks that can lead to total airway obstruction and respiratory failure.

  Features of a severe attack include a heart rate >120/min, tachypnea (> 30/min), pulsus paradoxus (i.e., a >20 mm Hg fall in systolic blood pressure during inspiration), cyanosis, reduced or absent air entry on chest examination, and a peak flow rate less than 120 ml/sec. Patients with these signs should be hospitalized.

  Intubation and mechanical ventilation may be necessary if the patient becomes fatigued or severely hypoxic.

- Managing the Patient **The management of an acute asthma attack includes high-flow O_2 and nebulized albuterol.** Albuterol nebs may be administered continuously if necessary. **In a severe asthma attack, intravenous corticosteroids (e.g., methylprednisolone) should be given.** Subcutaneous epinephrine and aminophylline can be given during severe exacerbations if there are no contraindications.

  The treatment of chronic asthma involves drug therapy and identification and avoidance of asthma triggers. Three main types of agents are used in the long-term treatment of asthma: bronchodilators, corticosteroids, and other anti-inflammatory agents.

  - **Albuterol (β_2-agonist) Albuterol acts on airway smooth muscle, causing bronchodilation.** It is given as a metered dose inhaler (MDI) or nebulizer. Albuterol often provides rapid relief of symptoms. Side effects include tachycardia and irritability. Salmeterol is a newer β-agonist with a longer half-life. It is used for maintenance therapy and never during an acute attack.

- **Corticosteroids Corticosteroids decrease the activity of proinflammatory cells;** they are given as MDIs, IV, or orally. Corticosteroid MDIs should be used initially. Oral candidiasis is usually the only side effect. Oral or parenteral corticosteroids are usually reserved for acute exacerbations or for those few individuals not controlled by MDIs.
- **Cromolyn sodium Cromolyn sodium is a mast cell stabilizer administered via an MDI; it is a particularly useful antiinflammatory agent in younger patients.**
- **Methylxanthines The methylxanthines include theophylline (given by mouth) and aminophylline (given intravenously). They are used as second-line agents and are not as effective as β-agonists in causing bronchodilation;** additionally, methylxanthines have many side effects, including nausea, vomiting, and cardiac arrhythmias. **These drugs also have pharmacokinetic interactions with warfarin and anticonvulsants; the chronic use of theophylline requires therapeutic (i.e., drug level) monitoring.**

A peak flow meter should be given to asthmatics. This allows an objective measure to evaluate the severity of symptoms.

Chronic Obstructive Pulmonary Disease

■ **Basic Science** The term *chronic obstructive pulmonary disease* (COPD) refers to a spectrum of chronic respiratory diseases characterized by cough, sputum production, dyspnea, expiratory airflow limitation, and impaired gas exchange.

Chronic bronchitis is defined as **presence of a productive cough for at least 3 months of the year for 2 successive years** in a patient in whom other causes of chronic cough, such as infection and carcinoma, have been excluded.

Emphysema is defined by pathologic criteria as destruction of alveolar septa causing permanent distal airspace enlargement.

Chronic bronchitis and emphysema are **most often associated with cigarette smoking.** Less common causes of chronic expiratory airflow obstruction include bronchiectasis, cystic fibrosis, allergic bronchopulmonary aspergillosis, and bronchiolitis obliterans.

■ **History and Physical Examination** Patients typically have a chronic productive cough followed by slowly progressive dyspnea. They may present to the physician with a **long history of cigarette smoking, cough, dyspnea, expectoration, and/or wheezing, usually in the sixth or seventh decade.** Hemoptysis may occur in chronic bronchitis but should always raise the possibility of lung cancer in a person with a smoking history.

Physical examination may reveal little abnormality except

wheezing. In moderate to severe obstructive lung disease, evidence of hyperinflation may be present with enlarged anteroposterior diameter of the thorax **("barrel chest"),** decreased diaphragmatic motion, or **use of accessory muscles** during breathing. Signs of pulmonary hypertension and right ventricular hypertrophy may be present in end-stage COPD.

- Laboratory and Diagnostic Studies The presence of COPD is often evident from the history and physical findings, but the **quantitation of the disease and evaluation of its response to therapy are based on pulmonary function tests, chest radiography, and arterial blood gases.**

Chest radiographs often show **low, flattened diaphragms.** In severe emphysema, the lung fields may be hyperlucent, with diminished vascular markings and bullae.

Pulmonary function testing will reveal that the **FEV_1 and the FEV_1/FVC ratio fall** progressively **as the severity of COPD increases. Total lung capacity (TLC), functional residual capacity (FRC), and residual volume (RV) may be increased.** The diffusing capacity for carbon monoxide (DLCO) is reduced in patients with moderate to severe emphysema.

Arterial blood gases reveal worsening hypoxemia and hypercarbia as the severity of the disease progresses.

With an exacerbation of COPD, sputum usually becomes more purulent with an influx of neutrophils. The **most frequent pathogens** cultured are a **mixture of** organisms, including ***Streptococcus pneumoniae, Hemophilus influenzae,*** **and** ***Moraxella catarrhalis.*** However, cultures are rarely necessary before beginning antimicrobial therapy in outpatients. In hospitalized patients, Gram's stain and culture may reveal gram-negative or, rarely, a staphylococcal infection.

- Formulating the Diagnosis Although the history and physical examination suggest the possibility of COPD, a **chest radiograph** excludes other diagnoses (e.g., tuberculosis, lung cancer) and may reveal the findings of emphysema or a complicating pneumonia. **Forced expiratory spirometry and arterial blood gas measurements** quantify the degree of airway obstruction and the presence and severity of hypoxia and hypercapnia. The response of FEV_1 to a bronchodilator may help to differentiate COPD from asthma and may suggest that airflow obstruction will respond to corticosteroids. The prognosis of patients with mild airway obstruction is favorable; it is poor in those with severe obstruction, especially if accompanied by hypercapnia. In patients presenting with values of FEV_1 <0.75 L, the mortality rate at 1 year is approximately 30%.

- Evaluating the Severity of Problems Although infection is a common cause of COPD exacerbation, pneumonia, pulmonary emboli, pulmonary edema, pneumothorax, and other causes of dyspnea should be suspected. **Acute respiratory failure in COPD is defined as an arterial P_{CO_2} >50 mm Hg, or an arterial P_{O_2} <50 mm Hg, in association with recent clinical worsening. The first goal of treatment is to improve hypoxemia and prevent tissue hypoxia by administering low, controlled concentrations of oxygen to raise the P_{O_2} to 55 to 60 mm Hg, which will**

prevent tissue hypoxia but not abolish hypoxic ventilatory drive. The next step in management is to reverse airway obstruction with inhaled β_2-agonists, anticholinergics, and/or aminophylline. Adequate hydration should be maintained, and empiric antibiotics may be started. A course of steroids is usually given and then tapered quickly as the clinical status permits. Most patients can be managed conservatively; however, worsening hypoxemia and hypercapnia in association with clinical deterioration mandates mechanical ventilation.

■ **Managing the Patient** **Cigarette smoking is the most important risk factor,** and patients should be urged to quit. Occupational and environmental irritants should also be evaluated. **Influenza vaccine** should be given annually; the 23-strain **pneumococcal vaccine** should be given once.

Ambulatory therapy consists of symptomatic treatment directed against the reversible elements of airways obstruction (e.g., secretions, bronchospasm, and inflammation) and long-term oxygen therapy, which improves functioning of the patient without affecting lung physiology.

The mainstays of symptomatic therapy are β_2-agonists, anticholinergic agents, and theophylline.

- **Sympathomimetics** β_2-adrenergic agonists such as metaproterenol, albuterol, and terbutaline are effective bronchodilators. They relieve smooth muscle spasm and enhance mucociliary clearance.
- **Anticholinergics** Unlike atropine, inhaled ipratropium bromide is not systemically absorbed and is free of anticholinergic side effects. In many studies, ipratropium has had a greater bronchodilating effect in COPD than the β_2-aerosols.
- **Theophylline** The mode of action of theophylline is poorly understood. Toxicity is poorly related to its blood level. More serious toxicity, such as arrhythmias and seizures, occurs at blood levels greater than 20 μg/ml.
- **Corticosteroids** About 15% of patients show objective ventilatory improvement after corticosteroid administration. A trial of oral corticosteroids should be initiated only after patients have been on an optimum treatment regimen with the agents just described. Inhaled corticosteroids are thought to offer more local control of inflammation without the systemic effects of oral steroids.
- **Antibiotics** An upper respiratory infection is the most common identifiable cause of a COPD exacerbation. Antibiotic-treated exacerbations are briefer and less likely to have serious consequences. Routine cultures are not indicated before instituting therapy.
- **Home oxygen** Long-term oxygen therapy prolongs life in hypoxemic COPD patients. Indications include a resting room-air arterial Po_2 of ≤55 mm Hg, oxygen saturation of <89%, or a Po_2 between 56 and 59 mm Hg in the presence of erythrocytosis or cor pulmonale.

Interstitial Lung Diseases

- Basic Science **Interstitial lung diseases (ILDs) have a variety of causes (known and unknown) and a single result (inflammation and fibrosis of the interalveolar septum). The initiating event is an injury to the interalveolar septum. Excessive repair leads to progressive deposition of extracellular matrix proteins and distortion of normal alveolar architecture.** The hallmark of active ILD is "alveolitis," that is, infiltration of the alveolar space with inflammatory cells. The unremitting inflammation appears to be the prerequisite for the development of fibrosis. **The known causes of ILDs include infections and exposures (to dusts, fumes, radiation, and drugs); however, the cause of ILDs in most patients is unknown. Patients in whom the cause of ILD is unknown most often receive the diagnosis of idiopathic pulmonary fibrosis (IPF).**

- History and Physical Examination **The typical patient with ILD is middle-aged and presents with the insidious onset of dyspnea on exertion and a dry, nonproductive cough.** A careful history may elicit clues to an environmental or occupational exposure that may be the cause of the ILD.

  Physical examination usually reveals bibasilar end-inspiratory crackles (i.e., "velcro rales"). Bibasilar crackles in a middle-aged individual without signs of congestive heart failure (e.g., elevated jugular venous pressure, S_3 gallop) should raise the suspicion of ILD. **Clubbing of the fingers is common** in patients with IPF but rare in other causes of ILD (e.g., sarcoidosis).

- Laboratory and Diagnostic Studies There are no laboratory abnormalities diagnostic of ILD; however, an elevated ESR and hypergammaglobulinemia are common. Patients commonly come to medical attention because of an abnormal chest x-ray. **The most common radiographic abnormalities are a reticular or reticulonodular pattern.** A high-resolution thoracic CT scan also is an important test in evaluating ILD.

  Pulmonary function testing in patients with ILD typically reveals a restrictive pattern (i.e., vital capacity, functional residual capacity, and total lung capacity are reduced). Pressure-volume measurements will reveal a **reduced lung compliance. The DLCO is reduced** as a result of a destruction of alveolar capillary units. The resting **arterial blood gas measurement may reveal a normal or reduced Po_2** and a mild respiratory alkalosis. The arterial Po_2 usually falls with exertion.

- Formulating the Diagnosis **A careful clinical history and physical examination with particular emphasis on possible occupational or environmental exposures and the determination of whether the ILD is occurring in the setting of a systemic illness will allow the majority of cases to be identified.** It is important to consider chronic infectious processes such as mycobacterial and fungal disease. **Following this initial evaluation, the next important issue is to confirm the diagnosis** and determine the level of impairment in order to obtain information on prognosis and decide on therapeutic options. **Fiberoptic bronchos-**

copy with bronchoalveolar lavage and transbronchial biopsies should be the first step in the majority of cases. However, a number of the causes of ILD often cannot be established by bronchoscopy because inadequate lung tissue is obtained. In these cases, traditional open lung biopsy or video-assisted thoracoscopic biopsy is needed.

■ **Evaluating the Severity of Problems** Cardiopulmonary exercise testing quantifies the amount of work the patient can perform and the need for supplemental oxygen during periods of exertion. Signs of pulmonary hypertension and overt right ventricular failure (cor pulmonale) indicate advanced disease.

■ **Managing the Patient** **Management of the patient with ILD is determined by the etiology. The patient with an occupational or environmental exposure should be removed from the offending agent. Smokers should be encouraged to quit.**

Corticosteroids are the mainstay of therapy for IPF, collagen vascular disease-associated ILD, inorganic dust exposure, acute radiation pneumonitis, and drug-induced disease.

The goals of therapy are to suppress the ongoing inflammatory response and prevent further fibrosis. Objective response commonly takes 6 to 12 weeks in most forms of ILD; however, many forms of ILD are not responsive to therapy. Many patients will relapse and require lifelong therapy.

In IPF only 20% to 30% of patients have objective evidence of improvement. Patients with IPF who demonstrate progression of disease on corticosteroids may be candidates for cytotoxic therapy with cyclophosphamide or azathioprine. The treatment of choice for patients who do not respond to therapy is lung transplantation. All patients should be encouraged to maintain an exercise routine to improve oxygen utilization.

Obstructive Sleep Apnea Syndrome

■ **Basic Science** **Obstructive sleep apnea syndrome (OSA) is characterized by repetitive episodes of complete or partial obstruction of the upper airway during sleep resulting in apneas** (complete cessation of air flow for at least 10 seconds) or hypopneas (50% reduction in air flow for more than 10 seconds). The underlying mechanism for these obstructive episodes is a combination of abnormal upper airway anatomy and normal sleep-induced reduction in pharyngeal muscle tone. **These apneas and hypopneas may lead to hypoxemia and hypercapnia, which can cause nocturnal arrhythmias,** particularly bradyarrhythmias. The respiratory events also cause transient (3-5 second) arousals and thus lead to sleep fragmentation, **with symptoms of chronic sleep deprivation.**

The prevalence of obstructive sleep apnea syndrome is approximately 4% for adult males and 2% for adult females. There is an association between obstructive sleep apnea syndrome and hypertension, some of which is due to shared risk factors (obesity, male sex), but OSA also appears to be an independent risk factor for systemic hypertension. **Pulmonary hypertension occurs in 15% to 20% of patients with OSA.**

A subset (5% to 10%) of OSA patients suffers from the obesity-hypoventilation syndrome. This syndrome is characterized by waking hypoventilation ($PaCO_2 > 45$ mm Hg) and severe OSA. If untreated, these patients often develop cor pulmonale.

■ **History and Physical Examination** Symptoms of OSA arise either as a direct manifestation of the obstructive episode or as a sequelae of chronic sleep disruption. **Clinical presentations include frequent awakenings during the night, sometimes with choking or gastroesophageal reflux.** Morning headaches secondary to hypercapnia occur in 10% to 15%. Impotence can occur in males. **Almost all patients with OSA snore,** though most snorers do not have OSA. In addition to snoring, bed partners may report apneic episodes, which often are terminated with a snort and a jerk. Also, patients typically suffer from excessive daytime sleepiness. Increased irritability and an impaired ability to concentrate may arise from chronic sleep fragmentation.

Risk factors for OSA include obesity, anatomic abnormalities (e.g., nasal obstruction, mandibular or maxillary hypoplasia, macroglossia, and tonsillar or adenoid hypertrophy), neuromuscular diseases, hypothyroidism, acromegaly, and Down syndrome.

Physical examination should be directed toward identifying predisposing anatomical abnormalities as well as sequelae of OSA, such as systemic or pulmonary artery hypertension.

■ **Laboratory and Diagnostic Studies** **Laboratory evaluation plays a limited role.** Patients with severe OSA, especially those with obesity hypoventilation syndrome, may be polycythemic and/or have elevated serum bicarbonate. Thyroid function tests are indicated in those with symptoms of hypothyroidism. Arterial blood gases are typically only indicated for patients with severe OSA.

Imaging studies such as cephalometrics, or CT/MRI imaging of the upper airway, sometimes may identify sites of collapse during sleep. Although these approaches have generated research interest, clinical applicability has yet to be defined.

The diagnosis of OSA requires a nocturnal polysomnogram (sleep study), during which respiratory efforts, nasal and oral air flow, and oximetry are measured. Sleep staging is also frequently performed. Overnight oximetry is sometimes used as a screening test. An oximetry study showing oxyhemoglobin desaturation is very suggestive of OSA. The absence of desaturation below 90% does not, however, exclude the diagnosis as many patients can be very symptomatic but not desaturate. Analysis of the variability in oxygen saturation during the night may increase the sensitivity of overnight oximetry significantly, but this is still in the developmental stages.

■ **Formulating the Diagnosis** **Clinical history is the determining factor in deciding whether to work-up a patient for OSA.** Current indications for a sleep study include excessive daytime sleepiness, frequent nocturnal awakenings, observed apneic episodes during sleep, cyclical nocturnal arrhythmias, waking hypoventilation, or pulmonary hypertension without an obvious cause. Snoring, obesity, or hypertension alone without other symptoms of OSA are not indications for a sleep study.

- **Managing the Patient** **All patients with OSA should be advised to avoid alcohol or sedative drugs as these can worsen disease severity.** Some patients have obstructive episodes only while supine; sleep position training to keep them on their side can eliminate these obstructive episodes. In obese patients, weight loss is critical and should be emphasized even if other modes of treatment are anticipated. **Weight loss alone can often be curative. Underlying conditions,** such as hypothyroidism or chronic nasal congestion, **should be treated.** Patients with excessive daytime sleepiness should not engage in dangerous activities (driving), where falling asleep could be lethal until they have been adequately treated.

  The first-line and most effective therapy for OSA is continuous positive airway pressure delivered by a nasal mask (nasal CPAP). Nasal CPAP works by creating a pneumatic splint, which holds the pharynx open during sleep, preventing obstructive respiratory events. The major drawback to nasal CPAP is poor acceptance of the device by patients resulting in compliance rates of only 50% to 70%.

  Several surgical procedures have been used to treat OSA. Currently the most popular procedure is uvulopalatopharyngoplasty (UPPP), which involves resecting the uvula, portions of the soft palate, tonsils and adenoids (if present), as well as the tonsillar pillars. Unfortunately, this procedure is effective in only 40% to 50% of cases. Recent studies suggest that procedures combining UPPP with surgery to advance the base of the tongue forward may be more effective. Tracheostomy, by bypassing the site of obstruction, is generally curative, but this procedure is reserved only for those with severe OSA who fail other treatment modalities. In children, enlarged adenoids are the major cause of OSA, so tonsillectomy and adenoidectomy is indicated as a first line treatment.

  Dental appliances that act to advance the mandible and tongue can be of some benefit, especially in milder cases.

Pleural Effusion

- **Basic Science** The pleural space normally contains less than 10 ml of fluid, which is formed continuously in small amounts. **Excess fluid, known as a pleural effusion, results when the net fluid formation exceeds the clearance capacity of the pleural lymphatics** (300-400 ml/day) **or when a process diminishes or blocks lymphatic flow** (such as a parasite or neoplasm).

- **History and Physical Examination** Although a pleural effusion can be asymptomatic, there may be symptoms referable to the underlying condition (e.g., abdominal pain in pancreatitis, fever and sputum production in pneumonia, or joint pain in SLE or rheumatoid arthritis). Irritation of the pleural surfaces or mediastinal shift can produce a **dry, nonproductive cough.** A process involving the visceral pleura can result in **dull, aching pain,** whereas parietal pleura involvement leads to **sharp, knifelike pain worsened with deep inspiration. Dyspnea** typically reflects a very large or rapidly accumulating effusion.

  Physical examination findings include **decreased breath sounds** by auscultation, **decreased resonance** by percussion, and **de-**

creased tactile fremitus by palpation. **Thoracic expansion is diminished** on the ipsilateral side, and if the effusion is large enough, the **trachea can be shifted** to the contralateral side.

Laboratory and Diagnostic Studies

- **Diagnostic imaging** A **PA and lateral chest radiograph** is the first step in evaluating suspected pleural effusion. The presence of as little as 100 ml can be detected with the lateral decubitus view, which is also useful in confirming whether the fluid is free flowing or loculated (trapped in pockets from adhesions).

  Ultrasound is useful for confirming the free-flowing nature of the fluid and selecting a site for thoracentesis when loculation is suspected.

  CT scanning can be useful in distinguishing small pleural effusions from atelectasis or masses and can be useful in detecting the presence and degree of loculation.

- **Invasive procedures**
  - ***Thoracentesis*** A thoracentesis is **needed in almost every case** of a newly discovered pleural effusion to help with definitive diagnosis and management. Exceptions to the need for diagnostic thoracentesis include effusions secondary to congestive heart failure or pneumonia that are improving with therapy.
  - ***Pleural biopsy*** Pleural biopsy is **indicated if malignancy or granulomatous disease** is **suspected,** or where serial **thoracenteses** have revealed only **nondiagnostic** exudative material.

Formulating the Diagnosis

The **first decision point** depends on the characterization of the fluid as a **transudate or exudate.** Exudates have at least one of the following criteria, whereas transudates will have none: (1) ratio of pleural fluid to serum protein >0.5, (2) ratio of pleural fluid to serum LDH >0.6, (3) pleural fluid LDH >⅔ top-normal serum LDH.

Congestive heart failure is the most common cause of a transudative effusion; exudative effusions are most commonly the result of malignancy. If the fluid is a **transudate, no further workup** is needed, and efforts are focused on treating the underlying condition. **Exudates,** however, require **further investigation.** The diagnostic effort can be focused with data from the following **tests on the pleural fluid:**

- ***LDH.*** Very high LDH, especially without elevated protein, suggests malignancy.
- ***Protein.*** >6.0 g/dl suggests TB or parapneumonic effusion.
- ***Glucose.*** <60 mg/dl suggests TB, rheumatoid arthritis, malignancy, or a parapneumonic process.
- ***White blood cell count.*** This is rather nonspecific; counts >10,000/mm^3 are seen with TB, malignancy, infection, collagen-vascular disease, and pulmonary infarction.
- ***Differential white cell count.*** Lymphocyte predominance suggests TB, sarcoidosis, and malignancy. Neutrophil predominance

suggests bacterial infection, pulmonary embolism, or pancreatitis. Although it is nonspecific, eosinophilia is seen with bleeding and pneumothorax.

- ***Red blood cell count.*** Generally, >100,000 cells/mm^3 are required to define an effusion as grossly bloody. Grossly bloody effusions are seen in trauma, malignancy, TB, and pulmonary embolism.
- ***Amylase.*** This is elevated in pancreatitis, esophageal perforation, and adenocarcinoma.
- ***Triglycerides.*** An elevated triglyceride level (>150 mg/dl) is diagnostic of chylothorax.
- ***pH.*** A parapneumonic effusion with a pH <7.20 suggests a complicated effusion with risk of empyema. In malignancy, a pH <7.30 connotes a poor prognosis.
- ***Gram's stain and culture.*** Empyema is defined if either is positive in a parapneumonic effusion. If TB is suspected, AFB stain should be performed.
- ***Cytology.*** The diagnostic yield for metastatic malignancy is 60% on a single specimen. Three separate specimens will improve the yield to 90%.

■ **Evaluating the Severity of Problems** Respiratory compromise from large effusions requires therapeutic thoracentesis until diagnosis permits definitive management.

■ **Managing the Patient** In **most cases** of effusions, a definitive diagnosis is possible, and **therapy is directed at the underlying disorder.** Parapneumonic and malignant pleural effusions, however, merit further attention.

- **Parapneumonic effusion** Effusions adjacent to pneumonia or abscesses are common. The crucial **management decision,** besides selection of appropriate antibiotics, centers on **whether tube thoracostomy is needed to drain the effusion. Indications for chest tube drainage** are pleural fluid that shows **gross pus, positive Gram's stain, glucose <40 mg/dl, or pH <7.00.** Complicated effusions, which require chest tube drainage, are those that are grossly infected (an empyema) or those likely to develop fibrosis and loculations.
- **Malignant effusion** Pleurodesis (the production of adhesions to obliterate the pleural space) may be performed when pleural metastases exist. Before considering pleurodesis, one needs to **ascertain whether the patient has symptoms** attributable to the effusion and, if so, **demonstrate symptomatic improvement after therapeutic thoracentesis.** If these conditions are met, pleurodesis is done (via tube thoracostomy or video-assisted thoracic surgery) with chemical irritants such as doxycycline or talc, or a chemotherapeutic agent such as bleomycin.

PNEUMOTHORAX

■ **Basic Science** A **pneumothorax** is defined as **gas present in the pleural space.** Pneumothoraces are characterized as **traumatic** or **spontaneous. Spontaneous** pneumothoraces are further subdivided

Box 4.1

CLASSIFICATION OF PNEUMOTHORAX

Spontaneous
- Primary (no underlying lung disease)
- Secondary (associated with known cause of pneumothorax, e.g., cystic fibrosis)

Traumatic
- Penetrating
- Nonpenetrating

Iatrogenic

into **primary,** if there is no known underlying lung pathology, **or secondary,** if related to a known cause of spontaneous pneumothorax such as COPD, cystic fibrosis, or the menses (Box 4.1).

Normally the pressure in the pleural space is negative relative to alveolar pressure and the atmospheric pressure. A pneumothorax develops if a communication is established between either the alveoli or the chest wall, in which case air will enter the pleural space until the intrapleural pressure equals the atmospheric pressure. This increased pressure results in compression of the underlying lung and outward expansion of the hemithorax. If a **ball-valve effect** is present, air will move into the pleural space during inspiration but will be trapped during expiration, creating the situation known as a **tension pneumothorax.**

■ History and Physical Examination The **symptoms** of pneumothorax are usually **sudden** in onset and include **dyspnea** and **sharp chest pain that is worsened with deep inspiration.** The symptoms are typically more severe and less well tolerated in those with secondary rather than primary pneumothorax. A history of antecedent penetrating or blunt trauma is usually obvious in cases of traumatic pneumothorax.

Physical findings in pneumothorax can include **tachypnea,** with **decreased tactile fremitus** and breath sounds on the ipsilateral side. **Percussion resonance is increased** on the side with the pneumothorax. The **trachea may be shifted** to the **opposite side.** When a **tension pneumothorax** is present, these signs are present, in addition to **tachycardia, hypotension, elevated jugular venous pressure, distended external jugular veins, thready peripheral pulses,** and **cyanosis.**

■ Laboratory and Diagnostic Studies A **plain chest radiograph** is usually adequate to confirm the suspected diagnosis of pneumothorax, in which case a visceral-pleural line can be seen. Occasionally, a pneumothorax will be loculated and better visualized on a **lateral radiograph** or **CT scan.**

■ Evaluating the Severity of Problems **Imaging has no role in confirming the presence of tension pneumothorax,** as the condition is life-threatening, and **a delay in diagnosis can be fatal.** In the case of suspected tension pneumothorax, the clinician should proceed immediately to needle or tube thoracostomy.

■ **Managing the Patient** In a case of **suspected tension pneumothorax,** a large-gauge **needle** should be **inserted immediately** into the ipsilateral second intercostal space in the midclavicular line. An **alternative is the insertion of a chest tube** if equipment and skilled personnel are immediately available. In either case, an immediate **rush of air through the needle or chest tube upon insertion confirms the diagnosis.**

In cases of **traumatic or iatrogenic pneumothorax,** where there is greater than **30% to 40% pneumothorax, tube thoracostomy will be required to reexpand the lung.** If the size of the pneumothorax is smaller, observation, with or without supplemental oxygen, is standard therapy.

In **primary pneumothorax,** initial management can include **aspiration of the air with a 16-gauge needle. Recurrence rates are high** (>50% in 2 years), and recurrences should be **managed with tube thoracostomy** and sometimes with open thoracotomy and pleural abrasion.

Most cases of **secondary pneumothorax** should be **managed with tube thoracostomy; chemical pleurodesis** should be considered.

Pulmonary Embolism

■ **Basic Science** Pulmonary emboli are blood clots originating in the venous system that dislodge, traverse the right heart, and occlude portions of the pulmonary vasculature. Rarely, thrombotic material from upper extremity or right cardiac sources or nonthrombotic material such as tumor, intravascular foreign bodies, or bone marrow can manifest as pulmonary emboli. However, the vast majority of **pulmonary emboli comes from venous thrombosis in the proximal veins of the lower extremities.** The **cardinal manifestations** of pulmonary emboli are **hypoxia** and **increased right ventricular afterload.**

Patients with large emboli may present in right ventricular failure and shock or may suffer minimally symptomatic recurrent episodes and present with irreversible pulmonary hypertension. Ten percent die before reaching medical care.

■ **History and Physical Examination** The **predisposing factors** leading to pulmonary emboli are those that predispose to deep venous thrombosis:

- **Stasis**
- **Vascular injury**
- **Enhanced clotting**

Clinical settings include **sedentary or bedridden** patients, patients with **trauma, postoperative patients,** patients with cancer (especially adenocarcinoma), patients who are **pregnant or using oral contraceptives,** and patients with a **previous history of pulmonary embolism or deep venous thrombosis.** Patients complain of the **sudden onset of dyspnea accompanied by pleuritic chest pain.** Unilateral leg pain, warmth, and swelling are helpful diagnostically but are often absent. Cough, either nonproductive or hemoptic, occurs in about half of patients, and 10% to 15% may present with syncope. **On physical exam,** the patient is anxious, **tachypneic,** and **tachycardic,** and may have low-grade fever. If the emboli are large, the blood pressure will be low, and the patient may be cyanotic. There may be jugular venous distension and a loud second heart sound. The lungs may be clear, may demonstrate wheezing from the release of bronchoconstrictive mediators, or may have crackles.

- Laboratory and Diagnostic Studies Routine blood work is generally not helpful. The **chest x-ray may be normal or may show a variety of abnormalities,** including unilateral pleural effusion, an elevated diaphragm or atelectasis from splinting, focal oligemia, or a localized pleural-based infiltrate indicating pulmonary infarction. The **ECG usually shows sinus tachycardia** but may also show right ventricular strain, right bundle branch block, right axis deviation, or a supraventricular arrhythmia; the "classic" **S1Q3T3 pattern is rare. Arterial blood gases usually show hypoxia** (Po_2 <80 on room air) **and a respiratory alkalosis.** The absence of tachypnea, tachycardia, and hypoxia in a patient presenting with symptoms suggestive of pulmonary embolism almost certainly excludes that diagnosis.

  The most important diagnostic study for suspected pulmonary embolism is ventilation-perfusion lung scanning. The finding of multiple segmental perfusion defects without corresponding ventilation defects is highly suggestive of pulmonary embolism and is virtually diagnostic when combined with a typical clinical history. Conversely, the absence of any perfusion defects as large as a single segment makes emboli very unlikely. However, many scans lie between these two extremes and are read as "indeterminate," that is, requiring further study. Further tests may include noninvasive studies of the veins of the legs such as venous plethysmography or duplex scanning. These tests are highly sensitive and specific for proximal deep venous thrombosis. A positive finding both increases the specificity of the indeterminate lung scan and necessitates anticoagulant therapy that treats pulmonary embolism as well as the venous thrombosis. **If the venous study is negative and the concern about emboli remains, pulmonary angiography remains the definitive study.**

- Formulating the Diagnosis **Patients presenting with the signs and symptoms of pulmonary embolism, hypoxia, and a "high probability" lung scan have a greater than 95% probability of having emboli and should be treated as such.** Patients whose history and physical examination are less suggestive of emboli who have normal or low probability scans should not be treated. Other patients presenting with dyspnea, pleuritic chest pain, and hypoxia may have alternative diagnoses such as pneumothorax, pneumonia, or empyema, which become obvious with evaluation. However, many patients with equivocal clinical presentation and hypoxia require further study, occasionally including pulmonary angiogram. Untreated, pulmonary emboli have a 30% mortality, which falls to less than 5% with therapy. Some patients survive multiple unrecognized episodes and present with irreversible pulmonary hypertension and right ventricular failure.

■ **Evaluating the Severity of Problems** Patients in shock and patients who require intubation and mechanical ventilation obviously require management in an intensive care unit. Patients with underlying cardiac or pulmonary disease, those requiring high (>50%) concentrations of inspired oxygen, and those with three or more segmental defects on perfusion scanning are also at increased risk of dying from additional emboli and may benefit from intensive monitoring.

■ **Managing the Patient** Hypoxia must be relieved immediately with **supplemental oxygen.** Shock must also be treated rapidly, first with intravenous isotonic crystalloid and then, if necessary, with the vasopressors dopamine or norepinephrine. Patients with suspected pulmonary emboli should also begin treatment with **intravenous heparin** (bolus and continuous infusion) immediately **while diagnostic studies are obtained.** Unless the patient has a major contraindication to anticoagulation, the risk of additional emboli far exceeds the risk of hemorrhage while awaiting definitive tests.

Anticoagulation with heparin is the mainstay of acute therapy for pulmonary emboli. Heparin activates antithrombin III, allowing endogenous thrombolysis and preventing clot propagation, but does not directly dissolve existing clot. Heparin should be administered by continuous infusion to elevate the activated partial prothrombin time (PTT) to 1.5 to 2 times its control value. **Patients in shock or with refractory hypoxia should be considered for thrombolytic therapy with streptokinase, urokinase, or tPA.** These agents more rapidly reverse the hemodynamic abnormalities, although the ultimate degree of pulmonary clot resolution is similar to that in patients receiving heparin alone. Contraindications to thrombolysis include active bleeding from a noncompressible site, pregnancy, recent surgery, and central nervous system trauma, surgery, bleeding, or infarct in the preceding 2 months. After thrombolytic therapy, patients are restarted on heparin for the usual course of 7 to 10 days. During this time, **patients are converted to oral warfarin.** Oral anticoagulants should be **continued for 3 to 6 months and require close monitoring of the prothrombin time.** They may be continued indefinitely in patients who have had recurrent emboli or have unavoidable risk factors such as hemiplegia. If there is a contraindication to warfarin such as pregnancy, patients may self-inject subcutaneous heparin.

Patients who hemorrhage on anticoagulants, who develop recurrent emboli despite appropriate anticoagulation, or in whom any additional embolization would be fatal should be considered for vena caval interruption. This is performed via transvenous placement of a wire filter in the inferior vena cava, which intercepts clots embolizing from the lower extremities.

The prevention of pulmonary emboli is achieved through the prevention of deep venous thrombosis. Patients with venous insufficiency should avoid sitting for long, continuous periods. Hospitalized bedridden patients, particularly those with fractures, surgery, or venous insufficiency, should receive 5000 units of subcutaneous heparin twice daily or wear sequential compression stockings. The role of low-molecular-weight heparins in the treatment and prevention of deep venous thrombosis remains to be determined.

Sarcoidosis

- Basic Science **Sarcoidosis is a systemic disease characterized by the presence of noncaseating granulomatous inflammation in involved organs.** Although much research has been undertaken to find the cause of sarcoidosis, **its etiology remains unknown.** The CD_4 T-helper lymphocyte and a panel of cytokines, including interleukin-2 and γ-interferon, are thought to contribute to the pathogenesis of the disease. This inflammatory response may be self-limiting and spontaneously regress, or it may go on to fibrosis.

- History and Physical Examination **The onset of sarcoidosis usually occurs between the ages of 20 and 40 years.** It rarely occurs in childhood. In the United States, **it is approximately 14 times more common in blacks** than in the white population. The sex distribution is equal.

  The clinical presentation of patients with sarcoidosis falls into three broad categories: (1) the asymptomatic patient with an abnormal chest x-ray, (2) the patient with pulmonary symptoms, and (3) the patient with extrapulmonary manifestations. Over 90% of patients with sarcoidosis have lung involvement, with dominant symptoms being nonproductive cough and dyspnea on exertion. The patients may have very little symptomatology despite extensive radiographic abnormalities, and their chest examination is often unremarkable. **Any organ system can be involved. The most common skin manifestations are erythema nodosum** (which is a herald of acute sarcoidosis and generally has a better prognosis) and chronic plaques. **Granulomatous anterior uveitis detected by slit-lamp examination is the most common eye finding.** Other disorders seen on physical examination may include lupus pernio (nodular skin lesions around the nares and cheeks), hepatosplenomegaly, lymphadenopathy, arthritis, cranial and peripheral nerve deficits (most commonly the facial nerve), and lacrimal and parotid gland enlargement.

- Laboratory and Diagnostic Studies **Fiberoptic bronchoscopy provides the most common source of biopsy tissue. However, in appropriate patients, biopsies may be obtained from other affected tissues (e.g., the skin).** Pulmonary function tests typically show restrictive lung disease on spirometry and lung volume studies; however, an obstructive pattern can be seen with endobronchial sarcoidosis. The use of angiotensin-converting enzyme (ACE) levels in the diagnosis and management of sarcoidosis is controversial. It is elevated in 60% of patients with sarcoidosis but has a low specificity and little predictive value. Similarly, the clinical usefulness of gallium scans is limited. Laboratory abnormalities common to sarcoidosis include leukopenia, eosinophilia, hypercalcemia, hypercalciuria, and an elevated erythrocyte sedimentation rate.

- Formulating the Diagnosis **The diagnosis of sarcoidosis requires a compatible clinical, radiographic, and laboratory scenario with supporting histopathology demonstrating noncaseating granulomas.** Care must be taken to rule out infectious diseases such as tuberculosis and, in the presence of hilar or peripheral adenopathy, lymphoma. **Given sarcoid's systemic involvement, it may mimic many other disease processes.**

Table 4.1 *Medical Conditions Associated with an Increased Risk of Developing Active Tuberculosis Once Tuberculous Infection Has Occurred*

HIV infection
Diabetes mellitus
Silicosis
Immunosuppressive therapy
End-stage renal disease
Hematologic malignancy
Being 10% below ideal body weight
Following gastrectomy and intestinal bypass

- **Evaluating the Severity of Problems** Clinical symptoms such as progressively worsening dyspnea on exertion, besides abnormal pulmonary function testing and chest x-ray findings, serve as monitors for the severity and progression of pulmonary sarcoidosis. Transaminases and alkaline phosphatase can be used to monitor liver involvement. Other laboratory tests such as BUN, creatinine, calcium, complete blood cell count, and cerebrospinal fluid analysis can be used to monitor the severity of the disease.
- Managing the Patient **Many patients with sarcoidosis do not require therapy since spontaneous remission often occurs. Patients who should be considered for therapy are those with symptomatic pulmonary disease and extrapulmonary sarcoidosis, especially those with ocular, CNS, hepatic, and myocardial involvement. The mainstay of therapy is corticosteroids,** which are usually taken systemically but can be used as a topical application in sarcoid confined to the skin or eyes. Serial clinical evaluations, pulmonary function tests, chest x-rays, and appropriate laboratory tests (e.g., CBC, chemistry panel) are useful to follow response to therapy and relapse.

Tuberculosis

- Basic Science ***Mycobacterium tuberculosis*** **is carried through the air in infectious droplet nuclei produced when infected individuals cough, sneeze, or speak. The pathogenesis of tuberculosis (TB) occurs in two phases: the acquisition of the infection and the subsequent development of the disease. Active tuberculosis may develop immediately following infection, later, or not at all.** Certain patients are predisposed to the development of active TB once infection occurs (Table 4.1).

  If the number of bacilli encountered at the initial infection is sufficient, a localized tuberculous pneumonia results, which generally occurs in the lower lobes. This pneumonia heals spontaneously, leaving only a calcified parenchymal focus that may be accompanied by calcification of a hilar node. During this initial phase of infection, a tuberculous bacillemia occurs, and the organism settles in the lung apices. The subsequent development of disease usually begins there.

Table 4.2 *Classification of a Positive Purified Protein Derivative*

5-mm induration	
	Close contacts of newly identified active tuberculosis
	Persons with radiographic findings suggesting old healed pulmonary TB
	Persons with known or suspected HIV infection
10-mm induration	
	Persons with medical conditions that increase the risk of TB (Table 4.1)
	Foreign-born from high-prevalence countries
	Low-income populations, including high-risk minorities
	Residents of long-term care facilities, including prisons
	Intravenous drug abusers
15-mm induration	
	Classified as positive in all other persons

The incidence of tuberculosis had been steadily decreasing until the mid-1980s but subsequently increased because of the emerging AIDS epidemic.

- History and Physical Examination **Pulmonary tuberculosis should be suspected in patients with a persistent productive cough, fever, chills, malaise, loss of appetite, night sweats, and weight loss.** With extensive disease, **hemoptysis and dyspnea may occur.** Rales may be heard in the area of involvement, along with dullness and bronchial breath sounds if there is lung consolidation.

  Extrapulmonary disease (e.g., pleural, lymphatic, gastrointestinal) occurs in 15% of all patients but in 60% of those with HIV.

- Laboratory and Diagnostic Studies
  - **Sputum Patients should have at least three sputum specimens examined by smear and culture. Cultures appear between 2 and 6 weeks.** If there is no growth at 6 weeks, they are reported as negative. A more rapid method (that relies on detection of labeled $^{14}CO_2$ liberated by mycobacterial metabolism) can detect growth in 7 to 10 days. Identification of the different mycobacterial species requires more sophisticated testing (e.g., gene probes).
  - **Tuberculin skin testing Dermal reactivity to purified protein derivative (PPD) is the hallmark of a cell-mediated response to *M. tuberculosis.*** The test consists of the intradermal injection (Mantoux's method) of 0.1 ml of 5 TU (0.0001 mg) test strength. The reading is done 48 to 72 hours after the injection. **The reaction size is determined by the diameter of induration. What is considered a positive reaction depends on a person's pretest probability of having tuberculosis (Table 4.2).** If the prevalence in a risk group is high, then 5 mm is considered positive. Similarly, if a person has no risk factors, 15 mm is considered positive, thereby decreasing the occurrence of false positives. Moderately high-risk persons are considered positive at 10 mm. Overwhelming tubercu-

losis itself can cause a diminished or absent tuberculin response. Other causes of false negatives include depressed cell-mediated immunity from HIV infection, lymphoma, malnutrition, and sarcoidosis.

The "booster effect" occurs when a person who has been infected in the past loses skin test reactivity over the course of several years. When this occurs, the PPD is falsely negative, but a subsequent test would resume reactivity. A converter is an individual with a newly positive PPD, or a 10 mm increase in response if under 35 years of age, and a 15 mm increase if older than 35.

- **Chest x-ray** Pulmonary tuberculosis nearly always causes detectable abnormalities on the chest film. **In a primary tuberculosis infection the process is generally seen as a middle or lower lung infiltrate,** often associated with ipsilateral adenopathy. **Tuberculosis that develops at a time remote from the original infection (endogenous reactivation) usually causes disease (often with cavitation) in the upper lobes** of one or both lungs. Healing of the tuberculous lesions usually results in development of a fibrotic, calcified scar with shrinkage of the lung parenchyma. In immunocompromised patients, particularly those with AIDS, "atypical" radiographic presentations are the rule rather than the exception. Findings may include diffuse lung involvement and more lower lung disease. Cavitation is less common.

■ **Evaluating the Severity of Problems** Multidrug-resistant TB is hampering efforts to control the spread of this disease. Directly observed therapy is effective in reducing the incidence of treatment failures. Patients who become infected with multidrug-resistant TB require therapy individualized depending on sensitivities.

■ **Managing the Patient**

- **Preventive therapy** The use of chemoprophylaxis for patients with a positive PPD has been shown to decrease the incidence of tuberculosis. **Certain groups within the infected population are at greater risk of developing tuberculosis than others. Persons in these groups should be treated regardless of age (Table 4.3). Currently, 6 months of therapy is required,** except in HIV-infected patients, where therapy is increased to 12 months.

- **Tuberculosis treatment The mainstay of treatment for tuberculosis is prolonged multiple drug therapy to reduce the incidence of resistance and treatment failure. The current recommendation of the American Thoracic Society and Centers for Disease Control is a 6-month regimen of isoniazid and rifampin supplemented by pyrazinamide for the initial 2 months. Ethambutol should be included in the initial regimen while awaiting drug susceptibility studies, if there is significant drug resistance in the community.**

  Isoniazid is the most widely used of the antituberculous agents and should be included in all regimes. Hepatitis is the major toxic effect, increasing with age and alcohol consumption. **The pe-**

Table 4.3 *Candidates for Preventive Therapy*

Skin test positive regardless of age
Close contacts of newly identified active TB
Recent PPD converters
Previously untreated or inadequately treated persons with radiographic findings suggesting old healed pulmonary TB
Persons with medical conditions that increase the risk of TB (Table 4.1)
Intravenous drug abusers
Persons with known or suspected HIV infection (12-month course)
Skin test positive, less than 35 years of age
Foreign-born from high-prevalence countries
Low-income populations, including high-risk minorities
Residents of long-term care facilities, including prisons

ripheral neuropathy associated with isoniazid is prevented or reversed by administration of pyridoxine (vitamin B_6). Adverse reactions to rifampin include hepatitis, gastrointestinal upset, and, rarely, thrombocytopenia. Pyrazinamide can cause dose-dependent liver injury. Hyperuricemia occurs in nearly all patients, but clinical gout is not common. Retrobulbar neuritis is the main adverse effect of ethambutol. Symptoms include blurred vision, central scotomata, and red-green color blindness; this complication is dose related.

Multiple Choice Review Questions

1. A 23-year-old woman presents to the emergency room with 2 days of upper respiratory tract infection (URI)-type symptoms and slowly increasing wheezing. In the emergency room she is extremely dyspneic, with a respiratory rate of 30, pulsus paradoxus of 25 mm Hg, heart rate of 125 beats/min, and a peak flow rate of 100 L/min. All of the following therapies are indicated *except* for:

a. Nebulized beta-2 agonist.
b. Intravenous methylprednisolone.
c. Inhaled corticosteroids.
d. Intravenous aminophylline.
e. Subcutaneous epinephrine.

2. A 27-year-old white man with AIDS presents to the hospital with shortness of breath, fevers, chills, night sweats, and increasing sputum production. A chest x-ray reveals bilateral pneumonia. This pneumonia does not respond over several days to routine intravenous antibiotics. Sputum is obtained for AFB smear and culture. The smears and cultures confirm tuberculosis. All of the following statements regarding this patient and tuberculosis are true *except:*

a. The patient should be started on a 4-drug antituberculosis regimen: isoniazid, rifampin, ethambutol, and pyrazinamide.
b. Patients with AIDS are more likely to have "atypical" radiographic presentations.
c. The incidence of tuberculosis has increased with the increased incidence of HIV disease.
d. It is less likely that this patient has extrapulmonary tuberculosis.
e. The skin test of this patient may be negative.

3. A 25-year-old white woman is found to have bilateral hilar adenopathy on a routine chest x-ray obtained before a minor operation. The patient is asymptomatic. Because of concern about lymphoma, the patient undergoes bronchoscopy with transbronchial biopsy of the lung. The parenchyma reveals noncaseating granulomatous disease consistent with sarcoidosis. Serum chemistries (including BUN, creatinine, calcium, liver function tests), urinalysis, and slit-lamp examination of the eyes are all within normal limits. The patient's pulmonary function tests are normal. Further appropriate care of this patient includes:

a. A CT scan of the chest to better evaluate the adenopathy and make further comparisons.
b. Starting prednisone to treat her pulmonary sarcoidosis and to prevent progression of the disease.
c. Following the patient with serial chest x-rays, clinical examination, and ophthalmologic examination.
d. A gallium scan of the lung to rule out "active" sarcoidosis.
e. Referral of the patient to a neurologist to rule out neurosarcoidosis.

4. A 35-year-old black woman presents with 3 months of nonproductive cough, dyspnea on exertion, and low-grade fevers. She has no other systemic complaints. She works as a secretary and denies any dust or animal exposures. Physical examination is unremarkable. CXR shows bilateral reticular changes, most prominent on the lower lobes. She also has bilateral hilar and right paratracheal adenopathy. PFTs show a restrictive defect with a reduction in DLCO. The most likely diagnosis is:

a. Sarcoidosis.
b. Lymphoma.
c. TB.
d. Hypersensitivity pneumonitis.

5. A 40-year-old, moderately obese man presents with excessive daytime sleepiness. He reports that when he is sedentary, such as watching TV or reading, he falls asleep. These symptoms started 3 years ago, have progressed, and are present no matter how much sleep he gets the night before. He has been told that he is a loud snorer, but does not know if he stops breathing during his sleep. He denies morning headaches or symptoms of right-sided heart failure. Physical examination is remarkable only for moderate obesity. At this point you would:

a. Do no further work-up, as he does not have morning headaches and his sleepiness is not severe.
b. Obtain a nocturnal polysomnogram.
c. Obtain a screening oximetry study to see if he desaturates below 90%.
d. Obtain an echocardiogram to assess for pulmonary hypertension.

Chapter 5

Cardiology

Angina Pectoris

- **Basic Science** Angina pectoris is divided into **three broad categories: stable angina, unstable angina, and variant (or Prinzmetal's) angina.** Stable angina is characterized by episodic chest pain, lasting 5 to 15 minutes, exacerbated by exercise or stress, and relieved by rest or nitroglycerin. Unstable angina is also divided into three general categories: (1) new onset exertional angina; (2) angina with increasing severity, duration, or requiring more nitroglycerin (crescendo angina); and (3) angina at rest (or angina decubitus).

  Finally, variant angina usually occurs at rest, without provocation, and is thought to be secondary to spasm of epicardial coronary arteries.

- **History and Physical Examination** **History is the most useful tool** in formulating a diagnosis of angina. **Anginal pain** frequently is described as **retrosternal, radiating to the neck, jaw, shoulders, and at times down the left arm.** Patients may also complain of nausea, vomiting, lightheadedness, diaphoresis, and shortness of breath.

  Risk factors for angina are the **same as those for coronary artery disease: hypertension, lipid abnormalities, male gender, postmenopausal females, age greater than 55, family history of heart disease before 55, diabetes mellitus, cigarette smoking, and obesity.** Physical examination may yield an anxious patient with a transient S_4 or an apical systolic murmur indicative of mitral regurgitation. Conversely, the physical examination may be unremarkable.

- **Laboratory and Diagnostic Studies** An **ECG** taken during an anginal attack will show **changes 50% of the time,** usually **ST depression;** ST elevation may be seen during episodes of variant angina. Serum **creatine kinase should not be elevated.** Chest x-ray is useful in ruling out other potential causes of chest pain. Other routine lab work is not typically helpful in making the diagnosis of angina.

- **Formulating the Diagnosis** A prior history of angina is critical when diagnosing stable angina. **All new onset angina is unstable angina until proven otherwise.** A diagnosis of unstable angina should be entertained if the chest pain is new, different, progressive, or not responding to nitroglycerin, or if pain is occurring at rest. Variant angina tends to recur at the same time of day and frequently is associated with ST segment elevation; one third of patients with variant angina have no identifiable lesion on coronary angiography.

■ **Evaluating the Severity of Problems** If any of the features of unstable angina are present, there is a 40% incidence of acute infarction and a 17% incidence of death in the next 3 months; thus it is imperative to make the diagnosis. In unstable angina, medical therapy can reduce the risk of acute myocardial infarction to 8% and the risk of early death to 3%.

■ **Managing the Patient** **The most critical part of managing the patient with angina is to recognize and admit those patients with an acute myocardial infarction or unstable angina.** With respect to stable angina, modifiable risk factors should be identified and corrected: smoking stopped, hypertension and diabetes controlled, and blood lipids lowered with diet or medication. The cornerstone of acute therapy is sublingual nitroglycerin (tablet or spray). Long-acting preparations can be used, but with caution to prevent the development of tolerance. Beta-blocking agents have also been shown to be useful in the management of stable angina; additionally, they reduce infarct size and lower mortality in acute myocardial infarction. Beta-blockers are contraindicated in patients with CHF, AV block, variant angina, chronic obstructive lung disease, asthma, and insulin dependent diabetes. Calcium channel blockers are also an effective adjunct in the management of stable and variant angina.

Arrhythmias and Conduction Disorders

■ **Basic Science** A basic understanding of electrocardiography is assumed.

■ **History and Physical Examination** **It is important to know if the patient is symptomatic** because of the arrhythmia or conduction disturbance. An arrhythmia without symptoms may be well tolerated and may not require intervention. Symptoms such as syncope, presyncope, lightheadedness, weakness, shortness of breath, nausea, and chest pain should be sought. Signs such as diaphoresis, tachypnea, hemodynamic instability, or mental status changes demand urgent evaluation.

■ **Laboratory and Diagnostic Studies** The basic evaluation of a patient with an arrhythmia is a thorough analysis of the **12-lead ECG.** Single-lead rhythm strips (e.g., lead II) are helpful in determining the regularity of the rhythm. **Holter monitors** allow continuous recording of a patient's rhythm. **Electrophysiologic studies** are evaluations of the conduction system by surface recordings during a cardiac catheterization; they allow attempts at inducing the arrhythmia in a controlled setting.

Laboratory studies should be limited to serum electrolytes, particularly potassium, calcium, and magnesium, or drug levels of therapeutic agent(s) that the patient is taking.

■ **Formulating the Diagnosis**

- **Sinus bradycardia** (Fig. 5.1) P waves are normal. Rate is less than 60 beats/min. Some athletic patients have normal rates between 40 and 60.

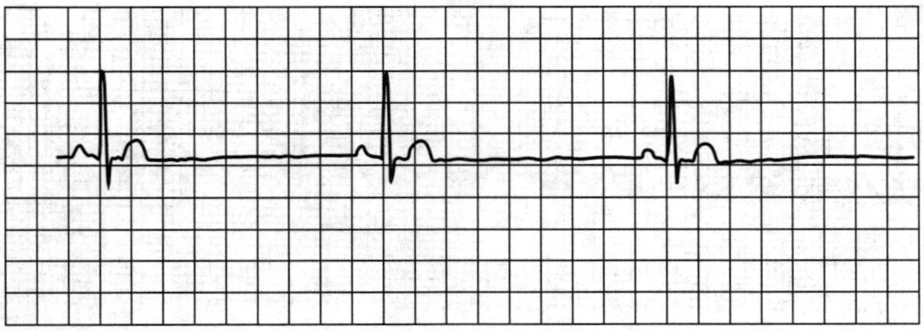

Fig. 5.1 Sinus bradycardia. *(From Cerra FB:* Manual of critical care, *St Louis, 1987, Mosby.)*

- **Sinus tachycardia** (Fig. 5.2; see also Fig. 5.5 *A*) P waves are normal. Rate is greater than 100 beats/min.
- **Premature atrial contractions (PAC)** (Fig. 5.3 *A*) P waves are abnormal in contour and axis. QRS complexes occur early with a short RR interval. QRS is normal.
- **Atrial tachycardia** (Fig. 5.3 *B*) P wave axis and contour are abnormal. Look for associated block (may be variable block) as this is pathognomonic of digitalis toxicity.
- **Multifocal atrial tachycardia (MAT)** (Fig. 5.4) Abnormal P waves are present with at least three different contours or axes. This is commonly associated with chronic lung disease.
- **Supraventricular tachycardia (SVT)** (Fig. 5.5 *C*) P wave is buried in or following the QRS (retrograde P wave). QRS is normal (unless conduction occurs aberrantly because of partial refractoriness or an aberrant pathway). SVT is frequently caused by a reentrant circuit that can produce a fast, regular ventricular response rate (180-250 beats/min).
- **Atrial flutter** (Fig 5.5 *D*) Atrial flutter features sawtoothed "P waves." Often a 2:1 block is present, giving a ventricular rate of 150.
- **Atrial fibrillation (AFib)** (Fig 5.5 *B*) Atrial fibrillation shows an irregular baseline without discernible P waves with an irregular ventricular response. The ventricular response may be quite fast (130 to 180 beats/min), requiring prompt intervention. The pulse is irregularly irregular.
- **Atrioventricular block (AVB)**
  - ***First degree*** (Fig. 5.6) Prolonged PR is greater than 0.20 seconds. This may be normal but is also associated with digitalis toxicity, beta blockers, and calcium channel blockers (diltiazem and verapamil).
  - ***Second degree, mobitz type I*** (Fig. 5.7) PR progressively lengthens until a QRS is dropped (Wenckebach phenomenon).
  - ***Second degree, mobitz type II*** (Fig. 5.8) A QRS is unexpectedly dropped after a normal P wave. This may occur with a regular or irregular block pattern. The ratio of conducted to blocked beats may be irregular, or may be fixed (e.g., 2:1,

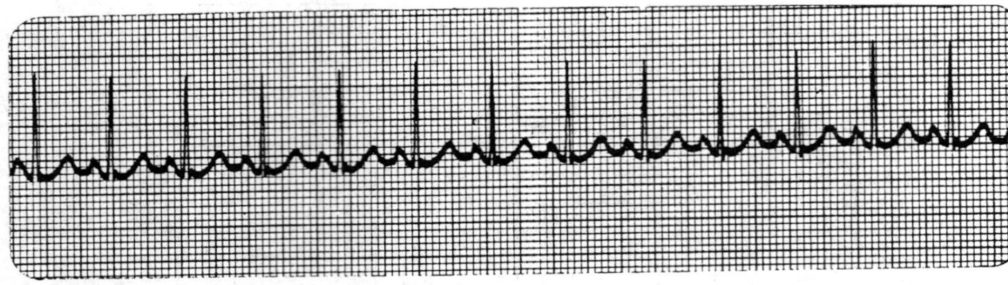

Fig. 5.2 Sinus tachycardia. *(From Goldberger AL, Goldberger E:* Clinical electrocardiography, *ed 5, St Louis, 1994, Mosby.)*

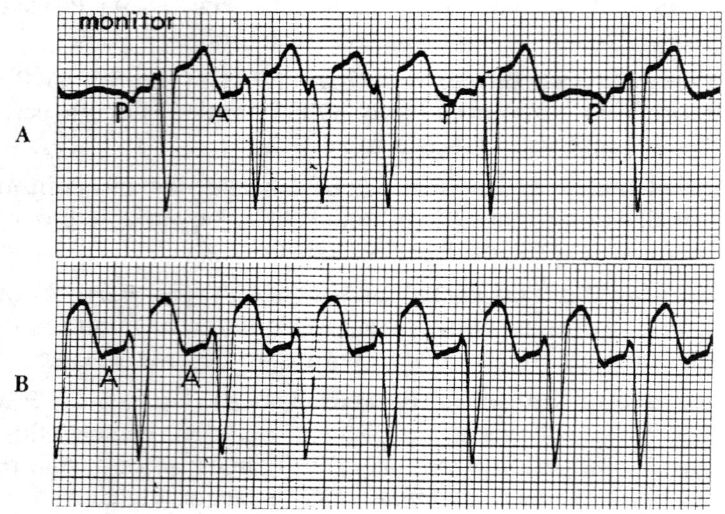

Fig. 5.3 *A* and *B* taken a few seconds apart. *A,* Sinus beat (*P*) followed by three premature atrial contractions *(A)*. *B,* Atrial tachycardia has developed. *(From Goldberger E:* Treatment of cardiac emergencies, *ed 5, St Louis, 1990, Mosby.)*

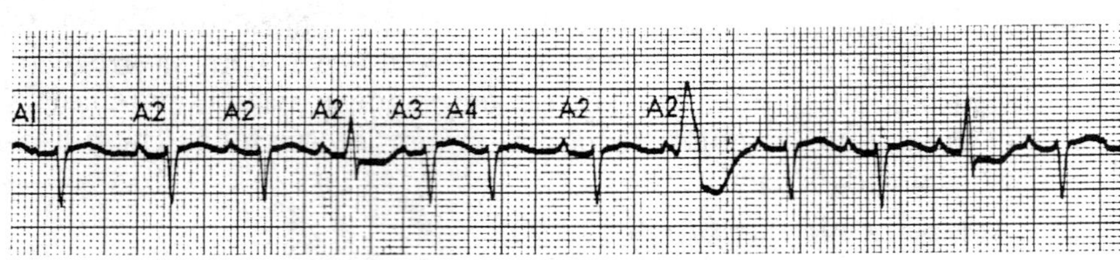

Fig. 5.4 Multifocal atrial tachycardia. *A1, A2, A3,* and *A4* show premature atrial beats from varying foci. Notice that fourth, eighth, and eleventh QRS complexes are aberrant. *(From Goldberger E:* Treatment of cardiac emergencies, *ed 5, St Louis, 1990, Mosby.)*

3:1, etc.). This pattern implies a block below the atrioventricular node and has a propensity to progress to third degree heart block.

— ***Third degree (AV dissociation)*** (Fig. 5.9) P waves are normal and regular but are independent of the QRS complexes (which are also regular but may be wide).

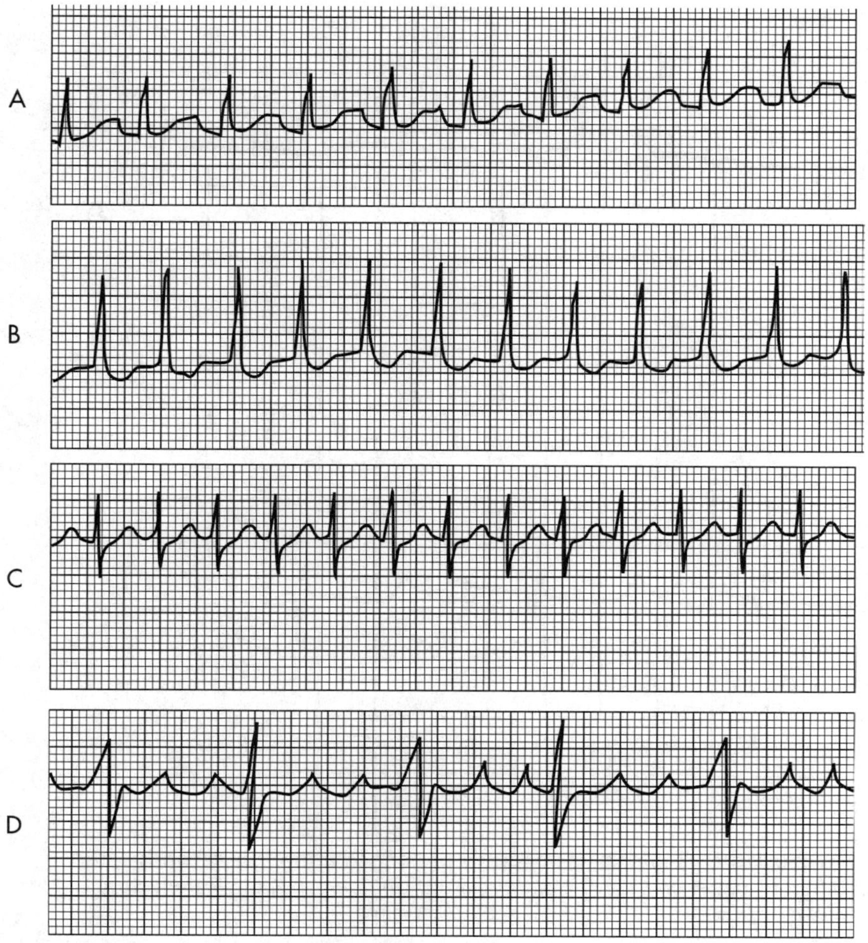

Fig. 5.5 Four look-alike narrow-complex tachycardias. *A,* Sinus tachycardia. *B,* Atrial fibrillation. *C,* Paroxysmal SVT due to AV nodal reentrant tachycardia. *D,* Atrial flutter with 2:1 block. When the ventricular rate is 150 beats/min, these four arrhythmias may be difficult if not impossible to tell apart with the standard ECG. In this example of sinus tachycardia *(A)* the P waves can barely be seen. Notice that the irregularity of the atrial fibrillation *(B)* is very subtle. In PSVT *(C)* the rate is quite regular without evident P waves here. In atrial flutter *(D)* the F waves cannot be seen clearly in this lead. *(From Goldberger AL, Goldberger E:* Clinical electrocardiography, *ed 5, St Louis, 1994, Mosby.)*

- **Right bundle branch block (RBBB)** (Fig. 5.10) Wide QRS (>0.120 msec) is present with a broad S wave in leads I and V_6. An rSR' pattern is in V_1.
- **Left bundle branch block (LBBB)** (Fig. 5.11) Wide QRS (>0.120 msec) is present with a QS wave in leads V_1 and III, and an R wave in leads I and V_6.
- **Premature ventricular contractions (PVC)** (Fig. 5.12) There is no P wave prior to a wide QRS, which appears early (short RR interval). A compensatory pause follows this complex.
- **Ventricular tachycardia (VT)** (Fig. 5.13 on page 85) P waves may or may not be present, but there is AV dissociation with a wide QRS and a rate of greater than 100 beats/min. The patient

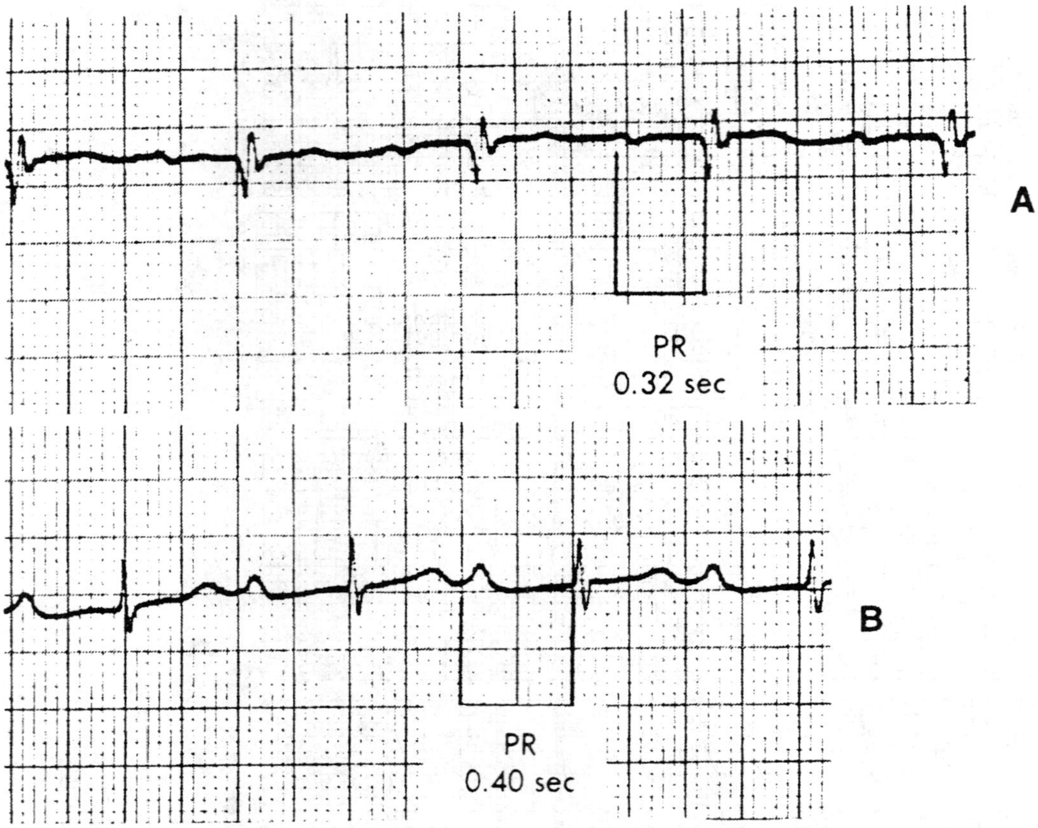

Fig. 5.6 With first-degree AV block the PR interval is uniformly prolonged above 0.2 seconds with each beat. *A* and *B* are from different patients. *(From Goldberger AL, Goldberger E:* Clinical electrocardiography, *ed 5, St Louis, 1994, Mosby.)*

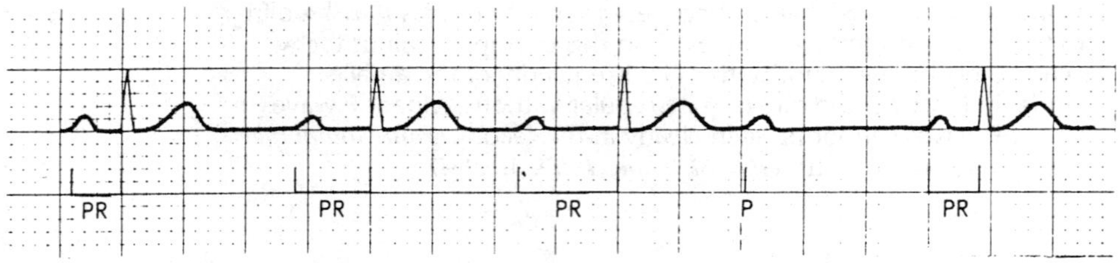

Fig. 5.7 In second-degree AV block (Mobitz type I) the PR interval lengthens progressively with successive beats until one P wave is not conducted at all; then the cycle repeats. *(From Goldberger AL, Goldberger E:* Clinical electrocardiography, *ed 5, St Louis, 1994, Mosby.)*

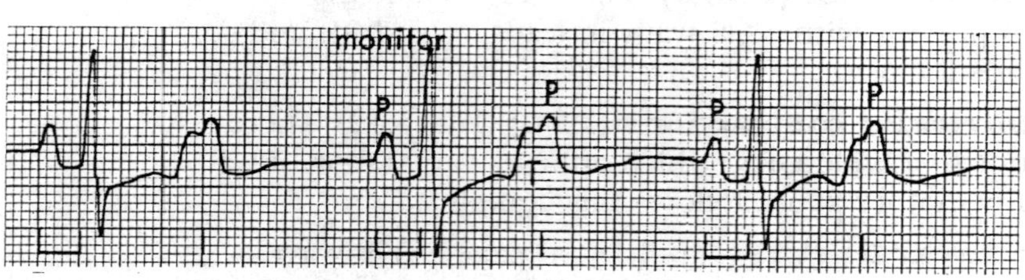

Fig. 5.8 Mobitz type II AV block. Notice that every alternate P wave is blocked. *(From Goldberger E:* Treatment of cardiac emergencies, *ed 5, St Louis, 1990, Mosby.)*

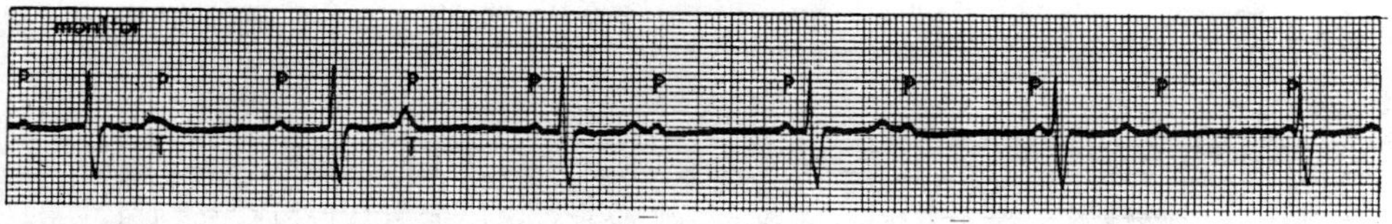

Fig. 5.9 Complete AV block. *(From Goldberger E:* Treatment of cardiac emergencies, *ed 5, St. Louis, 1990, Mosby.)*

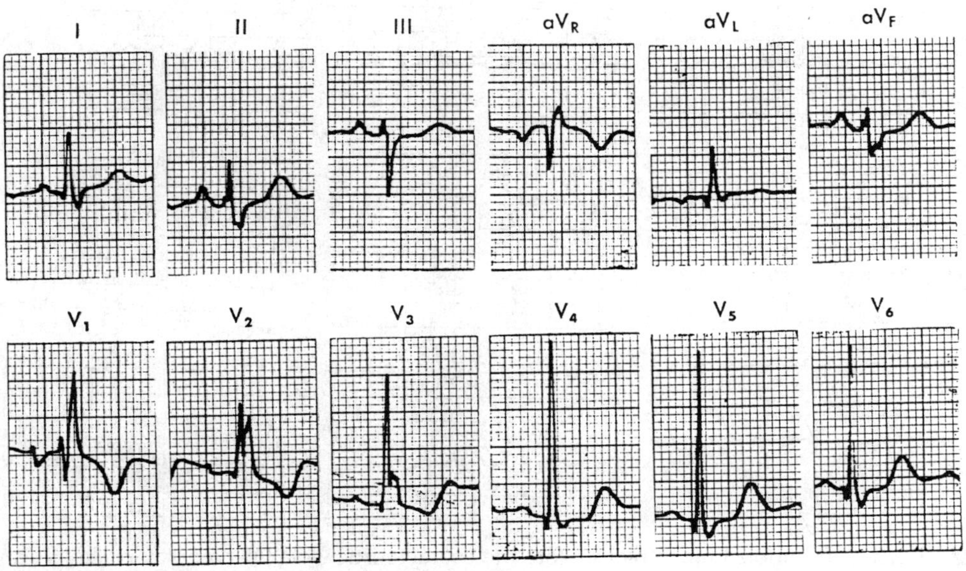

Fig. 5.10 Right bundle branch block. Note the wide rSR′ complex in lead V_1 and the qRS complex in lead V_6. *(From Goldberger AL, Goldberger E:* Clinical electrocardiography, *ed 5, St Louis, 1994, Mosby.)*

will not necessarily be unconscious, as occasionally VT at a slow rate is well olerated.

- **Ventricular fibrillation (VFib)** (Fig. 5.14) There is an undulating baseline without discernible QRS complexes. This rhythm (as opposed to VT) is not compatible with a pulse or consciousness.
- **Asystole** (Fig. 5.15) There are either no QRS complexes or wide, infrequent QRS complexes without a pulse.
- **Pulseless electrical activity (PEA)** The PEA classification recognizes electrical activity without an obtainable pulse or blood pressure.

■ Evaluating the Severity of Problems **As with any potentially life-threatening condition, attention must first be directed toward the ABCs.** If the airway is not patent or maintainable the patient should be intubated. If circulation is inadequate as evidenced by a low blood pressure, delayed capillary refill, or mental status changes, then rapid intervention including pharmacologic agents or cardiac pacing or cardioversion should be employed.

■ Managing the Patient **Asymptomatic arrhythmias require evaluation but may not require therapy unless the potential for adverse or life-threatening events is present.** Patients with a

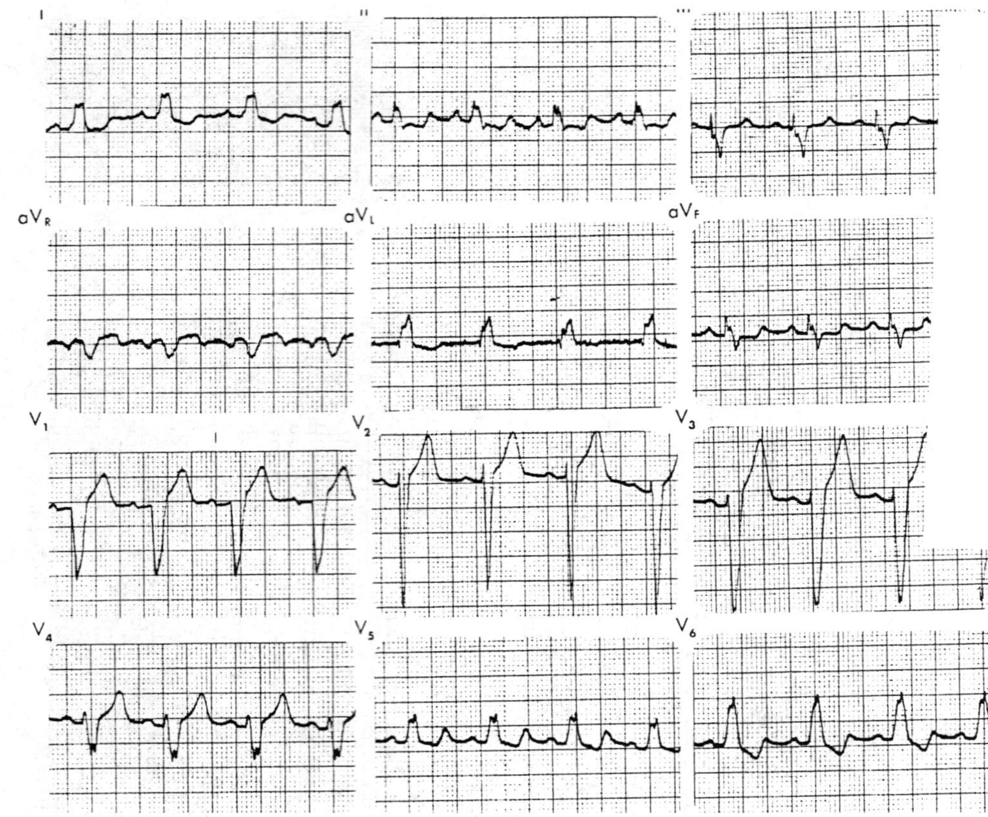

Fig. 5.11 Left bundle branch block. Note the characteristic wide QS complex in lead V_1 and the wide R wave in lead V_6 with slight notching at the peak. *(From Goldberger AL, Goldberger E:* Clinical electrocardiography, *ed 5, St Louis, 1994, Mosby.)*

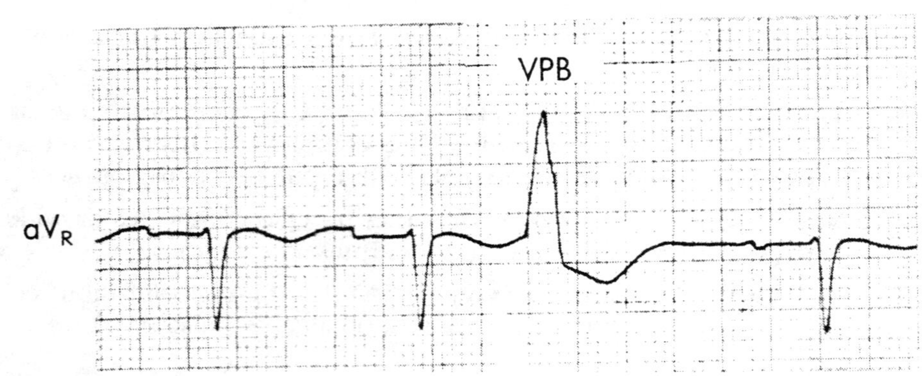

Fig. 5.12 Premature ventricular beat (PVB). *(From Goldberger AL, Goldberger E:* Clinical electrocardiography, *ed 5, St Louis, 1994, Mosby.)*

history of syncope and a normal 12-lead ECG require an evaluation for a potential arrhythmogenic source; with careful follow-up many of these patients can be evaluated by a short inpatient stay with monitoring, or with an outpatient Holter monitor.

Atrial fibrillation is a common arrhythmia. The **first goal is to control the rapid ventricular rate that sometimes accom-**

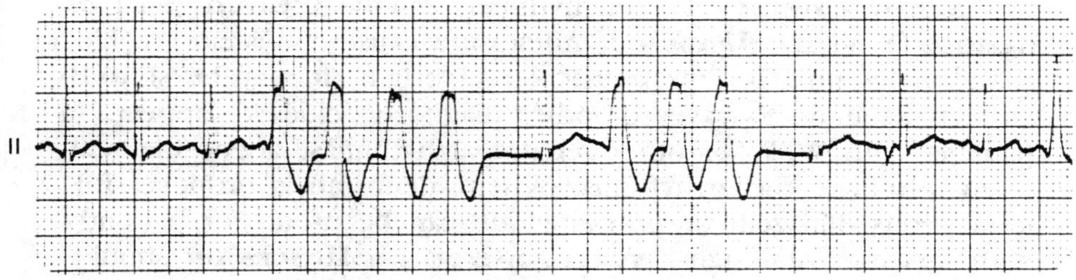

Fig. 5.13 Nonsustained ventricular tachycardia. *(From Goldberger AL, Goldberger E:* Clinical electrocardiography, *ed 5, St Louis, 1994, Mosby.)*

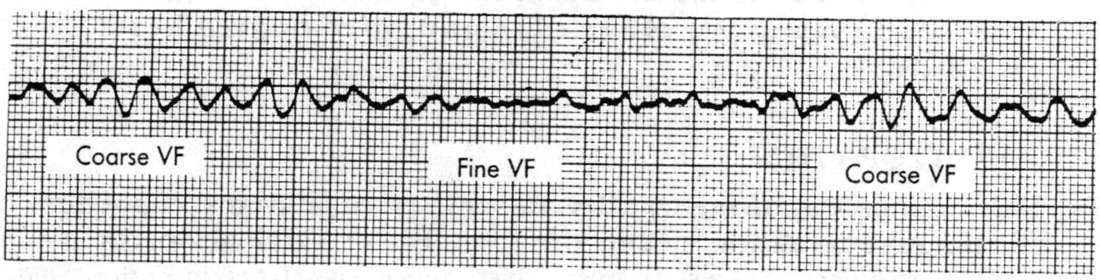

Fig. 5.14 Ventricular fibrillation. *(From Goldberger AL, Goldberger E:* Clinical electrocardiography, *ed 5, St Louis, 1994, Mosby.)*

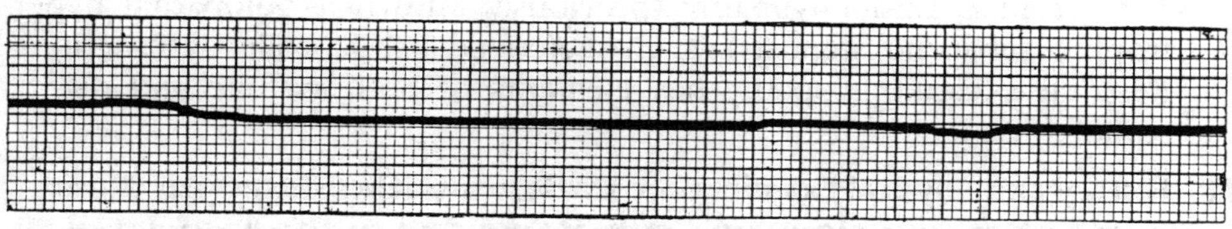

Fig. 5.15 Complete ventricular standstill produces a straight-line pattern during cardiac arrest. *(From Goldberger AL, Goldberger E:* Clinical electrocardiography, *ed 5, St Louis, 1994, Mosby.)*

panies it. Calcium channel blockers, beta blockers, or digoxin are sometimes used for this purpose. **In the absence of an intracardiac thrombus, electrical cardioversion may be attempted. Failing that, the patient should be anticoagulated.**

Symptomatic bradyarrhythmias should be **treated initially with external or internal pacing,** withdrawal of drugs with bradycardic side effects, and evaluation for other potential sources, including myocardial infarction. **Atropine is a temporizing measure,** and the patient nearly always requires a more definitive intervention.

Symptomatic tachyarrhythmias are treated according to the width of the QRS complex. **Wide complex** tachycardias are assumed to be ventricular tachycardia. Initial therapy should be with **lidocaine** if the patient is hemodynamically stable. Synchronous **cardioversion** should be performed **on any patient with unstable vital signs.** If the patient is in extremis, unsynchronized cardioversion should be performed.

Narrow complex tachyarrhythmias can be **treated (and diagnosed) with adenosine.** Adenosine causes transient AV block. If the arrhythmia is supraventricular in origin, it may be slowed enough to allow P waves or other diagnostic clues to appear. In patients with WPW or other reentry arrhythmias, adenosine will often terminate the arrhythmia. When adenosine is unsuccessful, intravenous diltiazem or procainamide can be used. For patients with unstable vital signs, **synchronized cardioversion** should be performed.

- **Indications for emergent pacemaker placement** Certain rhythms may require placement of a transvenous or external pacemaker until a permanent pacemaker is placed. These rhythms are **symptomatic sinus bradycardia or sinus arrest, and second (especially Mobitz II)- and third-degree heart block.**

Cardiomyopathy

■ Basic Science **Cardiomyopathy describes a primary disorder of the myocardium.** Before the diagnosis of cardiomyopathy is made, secondary causes of heart dysfunction, such as infection, ischemia, and collagen disease must be identified or ruled out. **The cardiomyopathies are classified pathophysiologically as dilated, hypertrophic, or restrictive** (Table 5.1). Dilated cardiomyopathy is characterized primarily by systolic dysfunction; hypertrophic cardiomyopathy (previously known as idiopathic hypertrophic subaortic stenosis) is characterized by diastolic dysfunction. Restrictive cardiomyopathy has elements of both systolic and diastolic dysfunction.

■ History and Physical Examination **Patients with dilated cardiomyopathy develop signs and symptoms due to the associated biventricular failure.** Complaints include exertional dyspnea, fatigue, orthopnea, and cough. Physical findings may consist of jugulovenous distension, cardiomegaly, S_3 and/or S_4 gallop, edema, hepatosplenomegaly, ascites, and a mitral regurgitation murmur.

Common symptoms in patients with hypertrophic cardiomyopathy are related to pulmonary congestion (e.g., exertional dyspnea and paroxysmal nocturnal dyspnea). Angina, palpitations, and syncope are other typical complaints. The physical examination is notable for brisk carotid upstrokes, a prominent apical impulse, and an S_4. A systolic ejection murmur is heard in patients who have developed outflow obstruction as a consequence of myocardial hypertrophy.

Patients with restrictive cardiomyopathy present with signs and symptoms much like those in patients with dilated cardiomyopathy, although clinical evidence of cardiomegaly is less dramatic.

■ Laboratory and Diagnostic Studies **All patients suspected of having cardiomyopathy should have a chest x-ray, ECG, and echocardiogram.** Table 5.1 lists common findings of these tests. Non-cardiac tissue biopsy is becoming more routine in the evaluation

Table 5.1 *Laboratory Evaluation of the Cardiomyopathies*

	Chest X-Ray Examination	Electrocardiogram	Echocardiogram
Dilated cardiomyopathy	Moderate to severe cardiomegaly with pulmonary congestion	Sinus tachycardia, arrhythmia, bundle branch block	Dilated cavities with normal wall thickness, MR, TR, and decreased ejection fraction
Hypertrophic cardiomyopathy	Often unremarkable	LVH, exaggerated septal Q waves, arrhythmias	Normal or small LV cavity with asymmetrical hypertrophy, possible outflow obstruction
Restrictive cardiomyopathy	Mild to moderate cardiomegaly	Low voltage, AV conduction abnormalities	Normal to mild dilatation of ventricle, reduced systolic function, pericardial effusion

LVH, Left ventricular hypertrophy; *AV*, atrioventricular; *MR*, mitral regurgitation; *TR*, tricuspid regurgitation; *LV*, left ventricular.

of patients since it often identifies diseases that mimic cardiomyopathy (e.g., amyloidosis, sarcoidosis, hemochromatosis, myocarditis, and unusual infections). Cardiac catheterization may be useful in measuring pressure/volume relationships and in identifying individuals suspected of having coronary artery disease.

- Formulating the Diagnosis **The diagnosis of cardiomyopathy is made by the history and physical examination and confirmed by laboratory studies** as outlined earlier. The primary cause of the cardiomyopathies is idiopathic.

  Conditions frequently associated with dilated cardiomyopathy include alcoholism, doxorubicin treatment, pregnancy, and endocrinopathies. Hypertrophic cardiomyopathy may be seen in families, since a genetic defect has been identified as a risk factor; chronic hypertension may be seen in association with this disease. Restrictive cardiomyopathy is commonly confused with constrictive pericarditis; it may be seen in association with amyloidosis, postradiation therapy, and endomyocardial fibrosis.

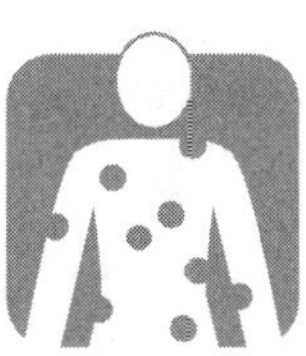

- Evaluating the Severity of Symptoms **Patients with hypertrophic cardiomyopathy are at risk from sudden death secondary to lethal ventricular arrhythmias.** This manifestation is occasionally seen in young athletes during extreme physical exertion.

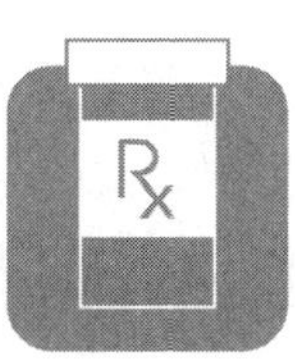

- Managing the Patient Since the cause of the cardiomyopathies is often unknown, **therapy is directed at symptomatic relief.** In patients with dilated cardiomyopathy, this includes eliminating aggravating factors such as alcohol and treating endocrinopathies when present. Further management consists of therapy for congestive heart failure, such as digitalis, diuretics, and vasodilators (see "Congestive Heart Failure," next section, for details). When systolic function is severely impaired, chronic anticoagulation should be considered to limit thromboembolic events.

Symptomatic patients with hypertrophic cardiomyopathy are often responsive to negative inotropes, including beta-blockers, calcium channel blockers, and disopyramide. Because of its antiarrhythmic properties, amiodarone may be useful in preventing sudden cardiac death. Surgical removal of part of the myocardial septum may benefit some patients with outflow obstruction.

The basic tenets of treatment for restrictive cardiomyopathy are similar to those for dilated cardiomyopathy.

Congestive Heart Failure

- **Basic Science** Congestive heart failure (CHF) is the inability of the heart to supply sufficient output to meet metabolic demands. Manifestations of CHF derive from tissue hypoperfusion (forward failure) as well as pulmonary and systemic venous congestion (backward failure). In addition, **several compensatory processes occur that augment cardiac output. The most important are (1) cardiac dilatation and hypertrophy, (2) elevated systemic vascular resistance and myocardial contractility (due to increases in sympathetic tone and circulating catecholamine levels), and (3) increased plasma volume due to activation of the renin-angiotensin vasopressin and aldosterone systems.** Beyond a certain point, each of the compensatory mechanisms will no longer improve cardiac output and instead will exacerbate the underlying cardiac insufficiency. The interplay between the primary pump failure and the secondary compensatory mechanisms determines the course of CHF in a given individual.

- **History and Physical Examination** **The classic symptoms and signs of CHF are due to venous congestion.** A failing left ventricle causes pressures in the pulmonary circuit to rise. Eventually the right ventricle begins to fail, and pressures in the systemic venous system rise.

Interstitial edema typically leads to a dry, hacking **cough.** The patient perceives decreased pulmonary compliance and impaired oxygenation as **dyspnea.** At first, shortness of breath will occur with exertion, but as pump failure worsens, dyspnea will appear with less and less activity. The patient will complain of **fatigue** and eventually will note **dyspnea at rest.**

With a failing heart, the patient may experience dyspnea when recumbent (**orthopnea**). During sleep, fluid that had seeped into the peripheral interstitium will shift back into the vasculature. The subsequent volume overload will precipitate sudden shortness of breath or **paroxysmal nocturnal dyspnea** (PND). The kidneys will attempt to reduce the apparent volume expansion, resulting in **nocturia.**

Pump failure will reduce renal perfusion, activating the renin-angiotensin system and promoting sodium and water retention. The patient will complain of **weight gain** and **edema.**

Obesity, smoking, hypertension, diabetes, coronary artery disease, and a positive family history are the most common clinical conditions associated with CHF.

Physical examination of the chest often reveals **wheezing** in mild

CHF, but the cardinal sign is the presence of **rales.** An **effusion,** typically right sided or bilateral, may be present. The heart, when percussed, will be **enlarged** with the **PMI displaced laterally,** and a heave will often be present. The **S_3 gallop** is the most specific sign of heart failure. Inspection of the neck will reveal **jugular venous distension.** Palpation of the liver will reveal hepatomegaly, and frequently hepatojugular reflux can be elicited. Dependent edema is common, usually noted first at the ankles but in advanced cases will also be evident more proximally. In the bedridden, **presacral edema** may be noted.

- **Laboratory and Diagnostic Studies** The **chest x-ray** is particularly valuable in diagnosing CHF. In addition to cardiomegaly, characteristic patterns of vascular redistribution and engorgement and interstitial edema can be seen. The **ECG** can be helpful in diagnosing underlying disease or rhythm disturbances, but there are no diagnostic changes in CHF.

- **Formulating the Diagnosis** **When the history and physical examination indicate the presence of dyspnea, pulmonary congestion, and edema, the diagnosis of CHF becomes straightforward.** When the clinical picture is incomplete, other diagnoses must be considered. Common noncardiac causes of dyspnea include COPD, asthma, and pneumonia. The differentiation between CHF, COPD, and pneumonia on physical examination alone can be difficult, and these diseases commonly coexist. Peripheral edema may be due to venous incompetence and thrombosis, trauma, renal disease, malignancy, and cirrhosis. Careful attention to details of the history and physical examination coupled with the chest x-ray will enable a correct diagnosis.

- **Evaluating the Severity of Problems** The most malignant form of CHF, **acute pulmonary edema,** is characterized by the abrupt onset of life-threatening congestive symptoms (e.g., pink, frothy sputum), often in association with significant forward failure. Treatment includes vigorous diuresis, supplemental oxygen, and management of hypertension when present. In severe cases, respiratory support via endotracheal intubation and mechanical ventilation will be required. Additionally, **intraaortic balloon counterpulsation** may be employed to reduce afterload, and **left ventricular assist devices** can help support a failing heart.

- **Managing the Patient** The medical management of chronic CHF is directed toward reducing preload and afterload and improving contractility. The mainstay of chronic therapy has traditionally been **diuretics,** which decrease fluid retention and lower preload. Various agents, used singly or in combination, include **thiazides, potassium-sparing diuretics** (e.g., spironolactone, triamterene), and **loop diuretics** (furosemide and bumetanide). Additionally, venodilators, primarily **nitrates,** reduce preload by increasing venous capacitance. Dietary salt restriction is also helpful.

  Afterload is reduced primarily by arterial vasodilatation. The agents used most commonly are nitrates, **alpha-adrenergic blocking agents** (e.g., prazosin), **calcium-entry blocking agents** (particularly amlodipine), and various angiotensin-converting enzyme **(ACE)**

inhibitors (e.g., captopril, enalapril, and others). Increasingly, the ACE inhibitors are supplanting the diuretics as the cornerstone of therapy as they have been shown to improve survival.

Inotropic drugs improve contractility. Currently, **digitalis** preparations are the primary agents chronically used for this purpose. In acute situations dobutamine, milrinone, or amrinone can be used intravenously.

Finally, severe refractory CHF may be amenable to cardiac transplantation in selected cases.

ENDOCARDITIS

- Basic Science **Endocarditis refers to an infective process involving the heart valves or any other part of the inner lining of the heart (i.e., the endocardium).** The characteristic lesion of endocarditis is a vegetation, typically found on a heart valve, although it can occur anywhere on the endocardium. A vegetation represents deposition of fibrin and platelets about the site of infection, often with ulceration and significant tissue damage. Although **most infections are bacterial, caused by staphylococci or streptococci,** other microorganisms such as fungi and *Rickettsia* can result in endocarditis. A spectrum of symptoms is associated with endocarditis and is dependent on the type of infecting microbe, the condition of the heart valves, and the health of the host. A cardinal feature of this disease is that **most patients will have a continuous bacteremia regardless of symptoms.**

  Endocarditis can present in an acute or subacute form, and both forms can be fatal unless treated. **Intravenous drug users and those individuals with heart valves damaged due to disease, surgery, or congenital malformation are predisposed to acquiring this disease.** In general, intravenous drug users develop *Staphylococcus aureus* infection of their tricuspid valve, but multiple organism infections also occur in this population. Infection with *Staphylococcus* sp. usually results in acute endocarditis with rapid tissue destruction and progression of symptoms. In comparison, endocarditis due to *Streptococcus* sp. tends to present subacutely and to involve left-sided heart valves, and usually occurs in patients with prosthetic valves or those with a history of rheumatic valvular disease or mitral valve prolapse.

- History and Physical Examination Individuals with preexisting cardiac lesions are prone to developing subacute endocarditis, particularly after undergoing an invasive procedure involving the oropharynx (e.g., dental work), the dermis, the genitourinary tract (e.g., catheterization), or the gastrointestinal system (e.g., colonoscopy). Intravenous drug users generally acquire their infection from skin flora.

  Manifestations of endocarditis vary, but **almost all patients will have or will complain of a fever.** Arthritis and arthralgias are common, as are weight loss, fatigue, and malaise. **Classically, a murmur is present** or there is a change in the sound of a preexisting murmur. However, intravenous drug users with infection of the tricuspid valve often do not have a murmur. Other physical findings include splenomegaly, petechiae, and clubbing. Found less commonly are

Roth's spots (retinal hemorrhages with a clear center), Janeway's lesions (small nodular hemorrhages found on the palms and soles), splinter hemorrhages (small linear hemorrhages found in the nail bed), Osler's nodes (tender subcutaneous nodules of the finger pulp), and septic emboli.

- **Laboratory and Diagnostic Studies** The essential element to making the diagnosis of endocarditis is the presence of positive blood cultures. **Usually three sets of anaerobic and aerobic blood cultures are obtained** from three different venipuncture sites to improve sensitivity. Endocarditis is often associated with hematuria. Other laboratory findings include an elevated erythrocyte sedimentation rate and a normochromic-normocytic anemia. Chest x-ray often reveals cardiomegaly, and echocardiogram will reveal vegetations in most patients.

- **Formulating the Diagnosis** The diagnosis of endocarditis is dependent on the presence of positive blood cultures, without evidence of an obvious source of infection. Endocarditis may be present with other diseases, including cellulitis or pneumonia. **Fever and positive blood cultures with a history and risk factors consistent with endocarditis are essential to making the diagnosis.**

- **Evaluating the Severity of Problems** **The major complications of endocarditis include damage to the infected heart valves and embolic events.** Damage to the heart valves can result in congestive heart failure. Septic emboli and infarcts can occur in different parts of the body, including the lung, brain, and kidney, as parts of the vegetation break off and are dispersed in the bloodstream.

- **Managing the Patient** **The management of the patient with endocarditis includes hospitalization and intravenous antibiotic therapy after the patient has been appropriately cultured.** Treatment should begin empirically while awaiting culture results. **Combination therapy with** a β-lactamase-resistant penicillin agent and an aminoglycoside (e.g., **oxacillin and gentamicin**) **is an acceptable choice.** When culture and sensitivity results are available, therapy should be adjusted accordingly.

  All patients who have been successfully treated for an episode of endocarditis are at risk for recurrence. Routine **prophylaxis is suggested for many procedures,** including those mentioned previously.

Hypertension

- **Basic Science** Hypertension is a common pathologic condition, occurring in 15% to 30% of all adults, and over 60% of patients over the age of 65 in the United States. It is usually asymptomatic, noted on routine screening. The presence of hypertension increases a patient's risk for other cardiovascular disease.

  More than 90% of hypertension cases are classified as primary, or essential, hypertension. These cases are more commonly found in older patients, in males, and in blacks. Family history, obesity, and a high-sodium diet also contribute to this form of hypertension.

Less than 10% of cases represent secondary hypertension due to renal vascular insufficiency; parenchymal renal disease; pheochromocytoma; mineralocorticoid hormone excess; coarctation of the aorta; or use of oral contraceptives, steroids, or other medications.

■ **History and Physical Examination** Elevated blood pressure is usually noted incidentally. Attention should be directed toward identifying any family history of hypertension and cardiovascular disease. The patient's history of medication use, diet, exercise, and other cardiovascular risk factors should be explored.

Physical examination should focus on evidence of the causes of secondary hypertension and on evidence of **target organ damage** due to hypertension. The five primary target organs affected by hypertension are as follows:

- **Brain** (hemorrhagic and lacunar strokes, encephalopathy)
- **Eye** (fundal hemorrhages, exudates, papilledema)
- **Heart** (left ventricular hypertrophy, CHF, acute ischemia)
- **Kidney** (nephrosclerosis, acute hematuria and proteinuria)
- **Vessels** (aortic dissection, necrotizing vasculitis)

■ **Laboratory and Diagnostic Studies** Initial studies of a newly diagnosed hypertension patient should include hematocrit, electrolytes, glucose, renal function tests, cholesterol level, urinalysis, chest x-ray, and an ECG.

Additional diagnostic studies to evaluate possible causes of secondary hypertension are usually unnecessary. However, they should be undertaken if indicated by the history and physical examination, or if continued failure of antihypertensive therapy causes the initial diagnosis of essential hypertension to be questioned.

■ **Formulating the Diagnosis** The diagnosis of hypertension should be based on multiple measurements over time. Transient mild to moderate hypertension in response to stress or medication is not pathologic. If persistent hypertension is documented without evidence of secondary hypertension on history, physical examination, and routine laboratory evaluation, then essential hypertension may be presumed.

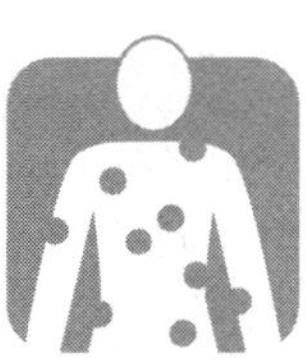

■ **Evaluating the Severity of Problems** The degree of hypertension is classified as follows: (Table 5.2).

Isolated systolic hypertension may also occur with a systolic blood pressure above 160 mm Hg and a diastolic pressure below 90 mm Hg. This is more common in the elderly, in whom vascular elasticity is lost.

Patients presenting acutely with an exacerbation of hypertension may be classified as having a **hypertensive emergency** when acute target organ damage is evident, regardless of the numerical value of the blood pressure. This condition demands immediate treatment, often with parenteral agents. In contrast, patients with a **hypertensive urgency** have no target organ damage but display a dangerously high diastolic blood pressure (i.e., >115 mm Hg), which, if allowed to persist, may be expected to cause target organ damage within 24 to 48

Table 5.2 *Classification of Blood Pressure for Adults Aged 18 Years and Older**

Category	Systolic	Diastolic
Normal†	<130	<85
High normal	130–139	85–89
Hypertension‡		
Stage 1 (mild)	140–159	90–99
Stage 2 (moderate)	160–179	100–109
Stage 3 (severe)	180–209	110–119
Stage 4 (very severe)	>210	>120

From Joint National Committee. The fifth report of the Joint National Committee on Detection, Evaluation, and Treatment of High Blood Pressure (JNC V). *Arch Intern Med* 1993; 153:154–183.

*These definitions apply to adults who are not taking antihypertensive drugs and who are not acutely ill. When systolic and diastolic blood pressures fall into different categories, the higher category should be selected to classify the individual's blood pressure status. Isolated systolic hypertension is defined as a systolic blood pressure of 140 mm Hg or more and a diastolic blood pressure of less than 90 mm Hg and staged appropriately.

†Optimal blood pressure with respect to cardiovascular risk is less than 120 mm Hg systolic and less than 80 mm Hg diastolic. However, unusually low readings should be evaluated for clinical significance.

‡Based on the average of two or more readings taken at each of two or more visits after an initial screening.

hours. This condition requires prompt treatment and follow-up to ensure reduction of blood pressure within 24 to 48 hours.

■ **Managing the Patient** The treatment of hypertension is usually a lifelong process. **Nonpharmacologic interventions** include weight reduction in obese patients, regular exercise, and reduction of dietary sodium. These interventions alone may control mild hypertension.

If nonpharmacologic interventions are ineffective in controlling hypertension within months, many effective **antihypertensive medications** are available. Therapy must be tailored to the individual's needs and complete health profile. Severe or refractory cases may require simultaneous administration of two or more classes of agents. Classes of antihypertensive agents include the following:

- **Diuretics** (e.g., furosemide, bumetanide, thiazide, spironolactone)
- **Beta-blockers** (e.g., atenolol, metoprolol, labetalol, propranolol)
- **Calcium channel blockers** (e.g., nifedipine, verapamil, diltiazem)
- **Angiotensin-converting enzyme (ACE) inhibitors** (e.g., captopril, enalapril, lisinopril)
- **Adrenergic inhibitors** (e.g., clonidine, prazosin)
- **Direct vasodilators** (e.g., hydralazine, minoxidil)

The patient presenting with a hypertensive emergency, showing evidence of acute target organ damage, must have immediate control of blood pressure. Intravenous vasodilators (e.g., nitroprusside or nitroglycerin) and/or intravenous beta-blockers (e.g., labetalol, esmolol, or propranolol) are used. If pheochromocytoma is suspected, beta blockers should not be given alone or else unopposed alpha-adrenergic stimulation may result in paradoxical hypertension. Instead, phentolamine should be used in conjunction with a beta blocker. Intensive care unit treatment and monitoring are appropriate in all patients with hypertensive emergency.

MURMURS

- **Basic Science** Blood flow through the heart is laminar, and therefore the normal cardiac examination reveals only the first and second heart sounds. Additional heart sounds include gallops, opening snaps, ejection and mid-systolic clicks, rubs, and murmurs. **Murmurs are the most common extra sound and represent turbulent flow.**

  Murmurs are designated as functional or structural. Functional murmurs are not associated with any anatomic abnormality, whereas structural murmurs are.

- **History and Physical Examination** When a murmur is found, a directed history should be taken, including duration of the murmur, if known, and the results of any prior workup. If the patient has been advised regarding endocarditis prophylaxis, the presence of a structural murmur is suggested; similarly, a prior history of bacterial endocarditis or rheumatic fever implies a murmur of structural origin.

  The presence or absence of fatigue and dyspnea should be noted, as these imply advanced valvular dysfunction, as does historical evidence of CHF (e.g., edema, orthopnea, paroxysmal nocturnal dyspnea).

  The cardiac examination should include the basics of inspection, palpation, percussion, and auscultation. **By far the most important part of the exam is auscultation. It allows the grading, quality, timing, and pattern of a murmur to be determined.** Additionally, **the radiation and loudest point of a murmur may be used to assess its source.**

- **Laboratory and Diagnostic Studies** **The commonly used diagnostic aids include ECG, chest x-ray, echocardiography, and exercise tolerance test (ETT).** Radionuclide imaging and cardiac catheterization are used less often to assess the effects of and need to intervene in specific murmurs.

  An ECG can reveal evidence of atrial or ventricular enlargement. It is an inexpensive and noninvasive way to assess the effects of a murmur. It is also useful in following the effects of a known structural murmur on a given cardiac chamber. However, ECG has very low sensitivity and specificity when used for these purposes.

  The chest x-ray is a low-risk method of assessing cardiac chamber size. Additionally it provides evidence of pulmonary congestion and is useful in the initial evaluation of murmurs and the follow-up of known structural murmurs.

  Echocardiography can assess chamber size, wall thickness, valve orifice dimension, presence of regurgitant flow, presence of valvular vegetation, and estimation of ventricular ejection fraction. It **is the best noninvasive test for evaluating murmurs.**

The use of exercise testing is best reserved for patients with known structural abnormalities. It is useful to follow exercise tolerance in valvular heart disease to decide when surgical intervention is required.

Formulating the Diagnosis

- **Aortic murmurs Stenotic murmurs are characteristically high pitched, systolic, and crescendo-decrescendo.** They radiate to the neck and are associated with diminution of the pulse and pulse pressure. There may be an associated ejection click prior to the murmur.

  Murmurs of aortic insufficiency are high pitched, occur early in diastole, and are decrescendo. They radiate to the lower sternal border and apex and are associated with a water-hammer pulse and increased pulse pressure.

- **Mitral murmurs Stenotic murmurs are low pitched, crescendo, and rumbling. They occur early in diastole.** There may be a preceding opening snap. **The murmur is best heard at the apex** with the patient in the left lateral decubitus position **and is accentuated by mild exercise.**

  Mitral regurgitation produces a high-pitched, pansystolic murmur that typically blends with and may obscure S_1 and S_2. **It is loudest at the apex and radiates to the ipsilateral axilla.** There is no variation with respiration.

- **Pulmonic murmurs Stenotic murmurs are of medium pitch, systolic, and crescendo-decrescendo. They radiate to the left shoulder** and may vary in intensity with respiration. The second heart sound is often widely split, and there may be a right parasternal thrust.

  Pulmonary insufficiency shares the timing and quality of aortic insufficiency. However, it tends not to radiate and may be associated with a right parasternal thrust. There are no associated pulse or pulse pressure changes.

- **Tricuspid murmurs Tricuspid stenosis shares the timing and quality of mitral stenosis but is usually well localized to the tricuspid area. It tends not to radiate but is strongly accentuated by inspiration.**

  Tricuspid regurgitation has the same timing and quality as mitral regurgitation but is localized to the tricuspid area. It is characteristically accentuated by inspiration and is associated with neck vein engorgement that may be pulsatile. The liver may also demonstrate pulsatile enlargement.

Evaluating the Severity of Problems

New murmurs accompanied by symptoms are those deserving acute medical attention.

Managing the Patient

Management of and prognosis for the patient with a murmur depends on the cause of the murmur and the clinical situation. A patient with mitral regurgitation found during a routine examination will be managed very differently from the patient who develops mitral regurgitation as a consequence of papillary muscle rupture complicating an acute myocardial infarction. It is beyond the scope of this chapter to discuss in depth the management of all murmurs. Certain generalities can be made regarding the stable patient, however.

A patient with a functional murmur should be informed of the innocent nature of the finding. No further intervention is needed unless the murmur is associated with a pathological state such as anemia.

Patients with structural murmurs (e.g., aortic stenosis) should be followed such that a decision to intervene surgically can be made before the onset of cardiac decompensation. It is imperative that the patient is educated regarding the need for endocarditis prophylaxis before undergoing dental or surgical procedures.

Myocardial Infarction

■ Basic Science **Myocardial infarction (MI) is necrosis of heart muscle caused by an inadequate blood supply.** Factors implicated in pathogenesis include progression of coronary artery atherogenesis to the point of total occlusion, subintimal hemorrhage or thrombosis at the site of an existing narrowing or plaque, and coronary artery embolism or spasm.

Infarction, like ischemia, results in alterations in myocardial cell contractility and depolarization. These cellular derangements are the cause of clinical complications. **Impaired contractility can lead to** left ventricular pump failure, resulting in congestive heart failure **(CHF) and cardiogenic shock. Electrical differences between adjacent areas of normal and ischemic myocardium can cause arrhythmias.**

■ History and Physical Examination **Pain is the most common presenting complaint** in patients with acute myocardial infarction (AMI). The classic symptom in AMI is a squeezing or crushing pain **lasting longer than 15 minutes.** The most common location of pain is substernal or epigastric, with 30% of patients complaining of radiating pain down the arms. Less common complaints include pain in the jaw, neck, back, or abdomen. The severity, quality, and location of pain vary significantly among individuals. Nonspecific presentations are more common in the elderly. **Twenty to 25% of known MIs are clinically unrecognized or "silent" (most commonly in diabetic or hypertensive patients).** As with angina pectoris, **pain may be accompanied by other symptoms including dyspnea, diaphoresis, nausea, vomiting, hypertension, lightheadedness, or palpitations.**

Significant **risk factors** for coronary artery disease include **cigarette smoking, hyperlipidemia, diabetes,** and **family history** of cardiac disease. Patients **greater than 40 years old, males,** and **postmenopausal females** are also at increased risk.

The physical examination is most useful in evaluating the complications resulting from myocardial infarction. Although a patient's general appearance may be deceptively normal, most appear anxious and diaphoretic. **Physical findings to note include neck vein dis-**

tension, pulmonary crackles, and new or more pronounced heart murmurs.

■ **Laboratory and Diagnostic Studies** In the emergency setting, there is no specific laboratory test for detecting AMI. Therefore diagnosis is dependent on clinical suspicion in combination with ancillary tests.

The early workup of AMI must include an ECG. **Serial ECGs** must be obtained to support the diagnosis of AMI. A normal ECG does not exclude the diagnosis, since at least 15% of patients ultimately "rule in" for MI with completely normal initial tracings. When the ECG is not normal, abnormalities of the QRS complex, ST segment, or T wave can identify ischemia or infarction.

■ **Formulating the Diagnosis** **With significant clinical suspicion and/or ECG findings, patients should be admitted to a monitored setting for treatment and definitive diagnostic testing.** Serial cardiac enzyme measurement can then be used over a 24-hour period to establish the diagnosis of myocardial infarction. The MB fraction of creatine kinase (i.e., CK-MB) is the most sensitive, specific, and widely used marker for MI because it is found almost exclusively in myocardial cells and peaks within 24 hours after infarction. **When the MB fraction exceeds 4% to 5%, the patient is determined to have "ruled in" for myocardial infarction, even if the total CK is within normal limits.**

■ **Evaluating the Severity of Problems** **The major complications of myocardial infarctions are arrhythmias, congestive heart failure, and shock.**

Continuous cardiac monitoring is necessary in all patients with suspicion for AMI. Ventricular fibrillation is the most dangerous arrhythmia, although bradycardia, tachycardia, and heart block are also common.

Pump failure is the primary cause of in-hospital deaths from AMI and is the pathophysiologic mechanism responsible for CHF and cardiogenic shock.

■ **Managing the Patient** **All patients with suspected MI should be placed on a cardiac monitor and receive supplemental oxygen.** Intravenous access should be established immediately. Pain relief, rest, mild sedation, and a quiet atmosphere can reduce anxiety and thereby lower heart rate, a major determinant of oxygen consumption.

Commonly employed agents in the treatment of AMI include the following:

- ***Morphine*** acts as both a sedating and analgesic agent, reducing oxygen demand.
- ***Nitroglycerin*** relaxes vascular smooth muscle of the venous system (decreasing preload), dilates coronary arteries, antagonizes vasospasm, and increases coronary blood flow to ischemic myocardium. Because nitrates can also induce hypotension, blood pressure must be closely monitored.
- ***Aspirin*** decreases platelet aggregation and limits arterial constriction by blocking the formation of thromboxane A_2. Aspirin significantly reduces mortality and should be given as early as possible.

- ***Heparin*** prevents recurrence of thrombosis after thrombolysis has occurred.
- ***Beta-blockers*** reduce the size of infarction by reducing the oxygen consumption and demands on ischemic areas of myocardium. They are recommended in all patients unless they are contraindicated by CHF, bradycardia, bronchospasm, or hypotension.
- ***Thrombolytics*** revolutionized treatment of AMI; they act to dissolve clot, thus reopening blocked coronary arteries and preventing the further progression of myocardial injury. Greatest benefit is achieved under the following conditions: (1) patients with ST elevation greater than 1 mm in two contiguous leads, (2) cases in which pain was not responsive to nitroglycerin, and (3) cases in which pain began less than 6 hours prior to initiation of therapy. Absolute contraindications for thrombolytics include (1) active bleeding, (2) major surgery or trauma in the past 3 weeks, (3) recent head trauma or stroke in the past 3 months, (4) suspected aortic dissection, (5) known traumatic CPR, or (6) pregnancy.
- ***Percutaneous transluminal coronary angioplasty (PTCA),*** in selected patients, can be performed, as it may provide a more reliable and rapid method of reperfusion.

Pericarditis

■ **Basic Science** The pericardium is made up of two layers, the visceral pericardium and the parietal pericardium. The visceral pericardium is a serous membrane overlying the epicardium. The parietal pericardium consists of dense collagenous tissue that forms a sac surrounding the heart. The space between these two structures normally contains up to 50 ml of fluid.

Pericarditis, the most common pathologic process involving the pericardium, may result from local and systemic diseases. There are three **etiologic categories:**

- ***Infectious:*** viral (coxsackievirus and echovirus), bacterial (staphylococcus, *Streptococcus pneumoniae*, β-hemolytic streptococci), tuberculous, and mycotic (*Histoplasma capsulatum*)
- ***Noninfectious:*** idiopathic, neoplastic (lymphoma, leukemia, metastatic carcinoma of lung and breast), uremia, myxedema, and postirradiation
- ***Hypersensitivity or autoimmune-related:*** collagen vascular disease (lupus, rheumatoid arthritis, scleroderma), drug induced (procainamide, hydralazine), postmyocardial infarction syndrome (Dressler's syndrome)

Finally, pericarditis can present as an acute (<6 weeks), subacute (6 weeks to 6 months) or chronic (>6 months) process. Subacute and chronic pericarditis can result in **constrictive pericarditis** and place the patient at high risk for congestive heart failure.

■ **History and Physical Examination** In acute pericarditis **the most common symptom is burning or sharp substernal chest pain** of sudden or gradual onset that may radiate to the neck, back (especially trapezius), epigastrium, and left arm; it is often exacerbated by deep inspiration, coughing, and positional changes (worse when supine, relieved when leaning forward). It is often difficult to differentiate this symptom complex from that of myocardial infarction. The most common associated symptoms include dyspnea, odynophagia (i.e., pain on swallowing), and low-grade temperature, especially with infectious etiologies.

The **hallmark of pericarditis is the pericardial friction rub.** A pericardial rub is best heard at the left sternal border with the patient leaning forward or in the hands-and-knees position. The quality is similar to that of sandpaper and is often irregular or triphasic. This sign is frequently transient, necessitating serial examinations when the diagnosis is highly suspect. When an effusion is present, heart tones may be muffled and the rub absent. Additionally, with an effusion the patient may develop pulsus paradoxus. Other less specific signs include tachycardia, tachypnea, and fever.

- **Laboratory and Diagnostic Studies** **Primary diagnostic studies include serial ECGs, PA and lateral chest radiographs, and echocardiography. The ECG will classically show diffuse ST elevations across most leads.** Serial studies can pick up an evolving pattern of such elevation. With a large effusion present, electrical alternans may be present. Chest radiography can help in ruling out other diagnoses such as pneumonia, pneumothorax, and pneumomediastinum. Its value in positively diagnosing pericarditis is limited, especially if no previous radiographs exist for comparison. **Echocardiography is the best study for diagnosis of acute pericarditis with an effusion. It is the study of choice for serial follow-up of patients.**

  The remaining laboratory evaluation is aimed at discovering the etiology for pericarditis: **CBC** with differential cell count—infectious etiologies; **BUN/creatinine**—uremia; **viral titers; serological studies** such as ANA and RA latex fixation; **blood cultures**—bacterial etiologies.

  Creatine kinase and CK-MB may be elevated when associated myocarditis exists.

- **Formulating the Diagnosis** The diagnosis of acute pericarditis demands ruling out other, more dangerous entities such as myocardial infarction, pulmonary thromboembolism, pneumothorax, esophageal rupture (i.e., Boerhaave's syndrome), and endocarditis. This must be done by carefully weighing pertinent historical facts, physical findings, and the diagnostic studies just described. **The presence of a pericardial rub virtually confirms the diagnosis.** The remainder of the workup can only point toward the diagnosis while hopefully ruling out the rest of the differential diagnosis.

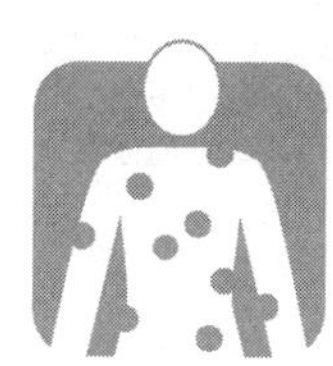

- **Evaluating the Severity of Problems** **The major acute complication of pericarditis is that of pericardial tamponade, although tamponade rarely occurs.** CHF can occur in patients with constrictive pericarditis, which is seen primarily in subacute and chronic cases. Therefore careful evaluation and ongoing monitoring of vital signs, cardiac rhythm, and respiratory status are mandatory in patients diagnosed with pericarditis.

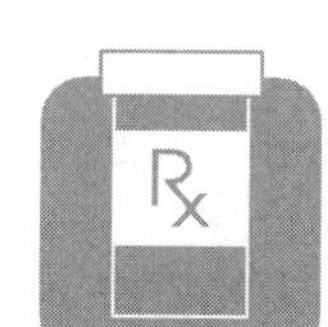

- **Managing the Patient** Early management is expectant with attention to the ABCs. In stable patients, **high-dose NSAIDs** (e.g., aspirin or indomethacin) are recommended for pain. If an underlying etiology is discovered, treatment should then also be aimed at its relief. In patients who develop tamponade, **pericardiocentesis** is indicated. **Patients who develop constrictive pericarditis will require pericardial resection,** with medical management (dietary sodium restriction and diuretics) serving only as an interim measure.

Rheumatic Heart Disease

■ Basic Science **Rheumatic heart disease can be defined as the acute and chronic effects of rheumatic fever on the heart.** Rheumatic fever is characterized by inflammatory lesions involving the joints, heart, skin, and nervous system. **Occurring in 3% of people after an episode of untreated group A streptococcal pharyngitis, rheumatic fever manifests acutely as a migratory polyarthritis, pancarditis, Sydenham's chorea, erythema marginatum, and subcutaneous nodules.** The latent period between antecedent pharyngitis and rheumatic fever ranges from 1 to 5 weeks. **Rheumatic fever does not occur after nonpharyngeal streptococcal infections.**

Rheumatic fever most commonly occurs between ages 5 and 15 but may occur at any age. Because crowding promotes the spread of disease, it is the major environmental factor relating to the occurrence of rheumatic fever.

Forty percent to 50% of individuals with rheumatic fever have a pancarditis. **Rheumatic heart disease most commonly affects the mitral (in females) and aortic valves (in males).** Healing of the valvulitis may result in thickening of the leaflets, fusion at the commissures, and adhesions of the chordae tendineae, leading to various degrees of stenosis and regurgitation, the most common cardiac sequelae. **Chronic rheumatic carditis presents as chronic heart failure, usually secondary to mitral regurgitation.**

Almost all cases of rheumatic fever resolve within 12 weeks. Death can occur as a result of a fulminant acute course, or from permanent residual valvular damage. Chronic persistent carditis may assume a fatal course over months to years or resolve spontaneously with little residual cardiac pathology.

■ History and Physical Examination **Rheumatic fever most commonly presents as an acute febrile illness associated with polyarthralgias affecting the large joints.** Obtaining a history of recent streptococcal pharyngitis is important, although 30% of patients do not recall an antecedent infection. **Only rarely are cardiac symptoms the presenting complaint.** Far more commonly cardiac involvement is an incidental finding. Therefore **a careful cardiac exam focusing on new murmurs (or a change in a preexisting murmur) is essential.** Signs of congestive heart failure may be subtle. A pericardial friction rub may be present. Three characteristic murmurs may occur: an apical systolic, apical mid-diastolic, and basal diastolic murmur.

It is important to examine the patient for erythema marginatum and subcutaneous nodules, two findings common in acute rheumatic fever. Erythema marginatum is a nonpruritic, erythematous, centrally clearing, macular rash involving the trunk and proximal extremities, sparing the face and hands. Subcutaneous nodules range in size from several millimeters to 2 cm and may be found over areas of bony prominence and along extensor tendons.

■ Laboratory and Diagnostic Studies **No laboratory tests are pathognomonic for rheumatic fever or rheumatic heart disease, but an elevated erythrocyte sedimentation rate (ESR), elevated C-reactive protein (CRP), and leukocytosis are helpful in establishing the persistence of rheumatic inflammation.** Anemia secondary to suppression of erythropoiesis is common. Streptococcal antibodies such as antistreptolysin-O (ASO) are increased, except when the diagnosis is delayed by several months. The most common ECG finding is a prolonged PR interval.

Doppler flow echocardiography can be useful in identifying pericardial effusion and valvular lesions. It is also useful in establishing the significance of flow disturbances and following the long-term progression of valvular pathology.

■ Formulating the Diagnosis **Establishing the diagnosis of rheumatic heart disease depends on making the diagnosis of rheumatic fever.** One must rule out other causes of polyarthritis, such as gonococcal, septic, and rheumatoid arthritis. The Modified Jones criteria are useful guidelines. Two of the major criteria (carditis, polyarthritis, chorea, erythema marginatum, and subcutaneous nodules) or one major criterion and two minor criteria (fever, arthralgia, previous rheumatic fever–rheumatic heart disease, elevated ESR or positive CRP, and prolonged PR interval) indicate a high probability of rheumatic fever.

Pericarditis and myocarditis do not occur as isolated findings in rheumatic fever, and the associated murmurs of endocarditis are only rarely absent. Thus a clinical diagnosis of rheumatic heart disease can be made in a patient with rheumatic fever with a new (or changed) murmur, an increase in heart size, pericardial friction rub, effusion demonstrated by echocardiogram, or signs of heart failure.

■ Evaluating the Severity of Problems The severity of the cardiac complications cannot be predicted at the onset of the disease. Prognosis is related to the severity of the carditis, the degree of valvular involvement, and the significance of hemodynamic flow abnormalities by doppler flow echocardiogram. Recurrent rheumatic fever portends a bad prognosis.

■ Managing the Patient **Prevention of rheumatic fever by treatment of streptococcal pharyngitis is successful if instituted within 7 to 9 days of onset. Therapy should be continued for a minimum of 10 days.**

There are no treatments that cure rheumatic fever or change the course of the acute attack. Aggressive supportive therapy may, however, alter the morbidity and mortality of the disease. Upon diagnosis, rheumatic fever should be treated with a course of penicillin to ensure eradication of streptococcus from the pharynx. Parenteral therapy is preferred, and a single dose of 1.2 million units of benzathine penicillin IM is adequate.

Suppressive therapy using salicylates is the main treat-

ment. The dose of aspirin can be increased until resolution of symptoms occurs, or systemic toxicity such as tinnitus, headache, dizziness, or blurred vision develops. Other nonsteroidals are also effective. Prednisone may offer benefit to patients who lack significant response to NSAIDs. Either steroids or NSAIDs must be continued for several weeks following the return of the ESR to normal.

Since rheumatic fever may recur, prophylaxis with penicillin, sulfonamides, or erythromycin is indicated. It must be continued for 5 years following an episode; prophylaxis must also be continued in patients under 25 years, and in those at high risk for recurrent streptococcal infection (e.g., military recruits, parents of young children, and health care workers). Prophylactic antibiotics should also be given before an invasive procedure (e.g., tooth extraction) to prevent endocarditis.

Surgical valve replacement and balloon valvuloplasty remain the only definitive therapeutic options when significant valvular stenosis, regurgitation, and flow abnormalities exist.

Superficial Thrombophlebitis

- Basic Science **Superficial thrombophlebitis is a venous disorder characterized by inflammation and thrombosis of a superficial extremity vessel.** It is generally **caused by minor trauma** or an indwelling catheter.
- History and Physical Examination The vessel involved (most commonly the saphenous vein) is **exquisitely tender to palpation** and appears red and edematous. A cord may be felt. Characteristically, there is no swelling in other parts of the involved extremity.
- Laboratory and Diagnostic Studies Doppler examination of the superficial vessel is not routinely performed but may be useful in revealing extension of thrombophlebitis to deeper vessels.
- Formulating the Diagnosis **The differential diagnosis includes lymphangitis, cellulitis, and gout.** Gout is easily excluded by the history and negative examination of the synovial fluid. Migratory superficial thrombophlebitis is an accompanying manifestation of Buerger's disease. Its presence in a young man who smokes cigarettes and complains of Raynaud's phenomenon and, possibly, pain should make the physician suspicious of the diagnosis.
- Evaluating the Severity of Problems Although one has to be very careful about excluding a deeper pathology, **this condition rarely progresses to deep venous thrombosis** (and embolization to the lungs).
- Managing the Patient **The treatment for this condition is mainly directed at the control of the inflammation, with the application of warm compresses, elevation of the limb, and the use of nonsteroidal antiinflammatory medications. Mobility** is important, as deep venous thrombosis might actually follow forced bed rest. A **follow-up** examination should be scheduled in 2 days, and the patient should be asked to return immediately if the area of inflammation enlarges in spite of therapy.

Varicose Veins

- **Basic Science** **Varicose veins represent a disorder characterized by abnormally prominent, distended, tortuous veins that mainly affect the lower extremities.** Varicose veins are present in approximately one fifth of the population. They develop either on a familial basis, as a consequence of deep venous occlusion or increased intraabdominal pressure (e.g., during pregnancy), or, in the majority of cases, idiopathically. Women are affected in 80% of cases.

- **History and Physical Examination** Varicose veins may remain asymptomatic for a long time; often patients report to their physician for **cosmetic reasons.** Patients may seek help for a sense of **heaviness in the lower extremities,** accompanied by **edema, especially after prolonged standing.** These symptoms often disappear if the patient lies down and elevates the limbs. The symptoms may be more severe when the venous distention is caused by a deep occlusion. **As the condition becomes chronic, persistent edema, induration, and cutaneous pigmentation** (secondary to hemosiderin deposition) **may occur.** Eventually cutaneous atrophy and ulcers may develop.

- **Laboratory and Diagnostic Studies** The **Trendelenburg test** may be used to **assess the competency of the communicating veins.** The patient is placed in the supine position with the legs elevated. A tourniquet is applied around the thigh, close to the groin. The patient is then placed in the upright position and filling of the veins is watched. Filling within 30 seconds suggests incompetence.

  Perthes' test assesses the competency of the deep vessels. A tourniquet is placed high in the calf to temporarily occlude the superficial vessels. Following this, the patient is asked to exercise for a short time. Pressure and discomfort in the calf suggest occluded deep vessels.

- **Formulating the Diagnosis** **The diagnosis of varicose veins is made by the history, physical examination, and objective findings.** Generally, inspection will suffice. The Trendelenburg and Perthes' tests will aid the physician in assessing the etiology.

- **Evaluating the Severity of Problems** If deep venous insufficiency is suspected, continuous-wave doppler plethysmography or descending venography may be used to confirm the diagnosis.

- **Managing the Patient** Young patients often need only reassurance. They should be advised to increase mobility and avoid long periods of standing if possible.

  The therapy of varicose veins should be primarily directed to the control of edema. That can be accomplished by the application of compression stockings and the elevation of the lower extremities whenever possible. The stockings should be tailored to the patient, applied when the swelling is at a minimum, and worn always.

  If dermatitis is present, the application of low-dose steroid creams may improve symptoms. Antibiotics are indicated in the presence of cellulitis. If ulcers are present, then wet dressings should be used.

This often accomplishes healing. If the ulcer is refractory, then surgical stripping of the superficial veins may be considered. However, the superficial veins must not be stripped if the varices are caused by a deep venous thrombosis, since they work as collaterals. In that event, excision of the ulcer followed by skin grafting should be considered.

Finally, in the presence of arteriosclerotic disease, no therapy is often advised. The decreased flow through the arteries actually limits the degree of venous engorgement.

Multiple Choice Review Questions

1. A 50-year-old woman presents with palpitations described as "skipped heartbeats" that cause a momentary lump in her throat. The palpitation occurs several times per day, most commonly when she is lying quietly while trying to go to sleep. An ECG and rhythm strip show a sinus bradycardia with a rate of 58 beats/min and rare premature atrial contractions (PACs). Your next step is to:

a. Take a thorough cardiac history.
b. Start quinidine gluconate 324 mg 3 times a day to suppress the PACs.
c. Arrange admission to a telemetry-monitored bed.
d. Administer potassium chloride 20 mEq intravenously.
e. Schedule formal electrophysiologic testing.

2. A longstanding alcoholic man presents with classic symptoms of heart failure, including exertional dyspnea and easy fatiguability, orthopnea, and PND. His heart is enlarged on examination and chest x-ray. He has rales, elevated jugular venous pressure, and leg edema. Appropriate therapy would include all of the following *except:*

a. Digoxin.
b. Verapamil.
c. Enalapril.
d. Furosemide.
e. Cessation of alcohol use.

3. A previously healthy, young college student complains of extreme fatigue, exertional dyspnea, orthopnea, PND, and weight gain. She recalls having a "bad flu" 2 weeks ago with gastroenteritis. On examination she has classic finding of heart failure. Your working diagnosis is:

a. Acute myocardial infarction.
b. Viral cardiomyopathy.
c. Systemic amyloidosis.
d. Constrictive pericarditis.
e. Papillary muscle dysfunction.

4. A 50-year-old accountant complains of severe substernal pain that radiates to his back. He feels better when sitting up. He has a scratchy systolic and diastolic extra heart sound, and his ECG shows diffuse ST segment elevation. Your working diagnosis is:

a. Acute papillary muscle rupture.
b. Acute pericarditis.
c. Viral cardiomyopathy.
d. Acute myocardial infarction.
e. Acute esophagitis.

5. Over the course of 48 hours you observe the patient in the previous question to develop shortness of breath, tachycardia, pulmonary and venous congestion, and poor urine output. He no longer has a scratchy murmur. You suspect:

a. His MI is extending.
b. He has developed acute bacterial endocarditis.
c. He has developed acute tubular necrosis.
d. He has developed pericardial tamponade.
e. He is recovering from his illness as the murmur is gone.

Chapter 6

Gastroenterology

Acute Diarrhea

■ Basic Science **Acute diarrhea** is most commonly **infectious in etiology.**

Infectious diarrhea is usually **caused by** one of the following mechanisms:

- ***Enterotoxins* stimulate** the **bowel to increase secretions** into the bowel lumen. (***Staphylococcus aureus, Vibrio cholerae, Clostridium,*** and certain strains of ***E. coli*** will cause this type of diarrhea.)
- ***Cytotoxin*** causes **direct injury to bowel.** (It is produced by strains such as *E. coli* O157:H7 and ***Clostridium difficile*—pseudomembranous colitis.)**
- ***Invasive bacteria* invade the mucosa,** causing edema and ulceration. (This includes ***Salmonella* and *Shigella*.**)

Although acute diarrhea is most commonly infectious in etiology, **other disease processes** that should be considered include **inflammatory bowel disease, irritable bowel syndrome, ingestions of laxatives** or **magnesium**-containing **antacids, diverticulitis,** and **stress-related diarrhea.**

■ History and Physical Examination

- **History** Patients with infectious diarrhea may present with a history of **recent travel** or **contacts with persons** who have **had similar** flu-like **illnesses.**

  It is important to determine if there is a **history of inflammatory bowel disease,** or a history of **diverticulitis. Tenesmus** (a persistent, ineffectual spasm of the rectum, accompanied by the desire to empty the bowel) is a **common complaint in patients with irritable bowel syndrome** or **diverticulitis.**

  The patient should be asked about ingestion of magnesium-containing antacids, laxatives, antibiotics, or cholinergic agents.

- **Physical Examination** Patients with diarrhea present with varying degrees of volume loss. Dizziness or **lightheadedness** may be an initial complaint. Vital signs may reveal **orthostatic hypotension,** and patients may be **tachycardic** and have **fever.** The **elderly** can **present with mental status changes.**

  The diarrhea may be bloody, and stool guaiac may be positive.

  Abdominal examination may reveal **diffuse tenderness,** although **peritoneal signs** are **not** usually **found.** One notable exception is in ulcerative colitis with toxic megacolon.

■ **Laboratory and Diagnostic Studies** Laboratory and diagnostic studies should be guided by the severity of the patient's symptoms and potential underlying causes. If the patient is well, observation may be an appropriate initial course of action.

Serum electrolytes are important if the diarrhea is severe and the patient appears dehydrated. A CBC may be helpful if the patient is febrile or has bloody diarrhea.

Examination of stool is vital. **Stool** should be **evaluated for** presence of **blood and leukocytes.**

— ***Invasive bacterial and cytotoxic diarrheas*** commonly present with **gross blood** and **leukocytes.** Stool cultures are indicated when the diarrhea contains leukocytes.

— ***Blood and leukocytes*** are **uncommon in enterotoxin-induced diarrhea.**

Sigmoidoscopy should be performed **if** there is **no improvement with** conservative **treatment** after 1 week.

■ **Formulating the Diagnosis** History, physical examination, and results of diagnostic testing should point to infectious diarrhea.

■ **Evaluating the Severity of Problems** **Physical appearance** and **vital signs** (including **orthostatic blood pressures** and **pulse rate) indicate** the degree of **severity.** Individuals who present with resting tachycardia are significantly volume depleted. The elderly presenting with acute change in mental status may be volume depleted.

■ **Managing the Patient** **Initial treatment** should include **assessment of fluid status** and, if indicated, **intravenous fluid resuscitation.** Intravenous hydration should continue until the patient is able to take sufficient fluids by mouth.

Antibiotics are **given** only **in** cases of **invasive bacterial** or cytotoxic **infectious diarrhea.** Patients with bloody, leukocyte-positive stool should be started on empiric antibiotic therapy.

Antispasmodic agents may be beneficial for symptomatic relief but should be **used with caution.** Antispasmodic agents should **never be used in patients with fever or bloody diarrhea.**

Lomotil and bismuth can be used sparingly under the guidelines just described.

Acute Gastritis

■ **Basic Science** **Gastritis refers to inflammation of the gastric mucosa characterized by erosions and**/or subepithelial **hemorrhages.**

Injury to the gastric mucosa **can occur from a variety of sources, including the following:**

- **Local injury** due to **trauma from nasogastric suction catheters,** therapeutic endoscopic techniques, or movement of a hiatal hernia.
- **Generalized injury** due to the **disruption of mucosal barrier.** This may occur from
    - **Nonsteroidal antiinflammatory agents** (inhibit prostaglandin synthesis)

- **Ethanol**
- **Ischemia** (decreased mucosal blood flow)
- **Radiation**
- **Reflux** of **bile acids,** pancreatic secretions, and duodenal contents, especially from a poorly functioning pyloroplasty or antrectomy

Other predisposing factors include **sepsis, shock, burns, and organ system failure** (heart, kidneys, lungs, liver). Acute gastritis is estimated to occur in 80% to 90% of critically ill hospitalized patients. It is **very common in patients with severe burns, head injury, or severe hemorrhage.**

***H. pylori* has also been implicated as a cause of gastritis.**

Injury to the gastric mucosa allows hydrochloric acid access to unprotected gastric tissue. This **results in tissue injury,** which leads to mucosal friability, erosions, and extravasation of blood into the mucosa and stomach lumen.

- History and Physical Examination **Patients commonly relate a history of alcohol abuse or nonsteroidal use.**

  Gastritis usually **causes mild dyspepsia.** Less common symptoms include **nausea and midepigastric pain.**

  Because hospitalized patients may be too ill to complain about dyspepsia or nausea, the **first sign of gastritis in the hospitalized patient may be blood from the NG aspirate, occult blood in stool, anemia, or abrupt GI hemorrhage** with a falling hematocrit.

  Physical examination may reveal evidence of blood loss such as **pallor, tachycardia, hypotension, and dizziness.**

- Laboratory and Diagnostic Studies **The diagnosis is best made by endoscopy,** which allows inspection of the gastric mucosa. **Barium contrast studies are usually not helpful** because the lesions are superficial.

- Formulating the Diagnosis Acute gastritis should be **suspected based on clinical symptoms.** Evidence of blood loss in a hospitalized individual should prompt evaluation for gastritis.

- Managing the Patient **Treatment of gastritis is via H_2 blockers, sucralfate, and antacids.** Proton pump inhibitors such as **omeprazole** may also be utilized.

  Patients with life-threatening gastritis may require vasopressin infusion. Endoscopic coagulation may be a temporizing measure.

  Surgery is performed for intractable hemorrhage (antrectomy with vagotomy or oversowing a regional lesion). Surgical morbidity and mortality is high owing to the severity of the associated illnesses.

  Documented ***H. pylori* infection** should be **treated with bismuth subsalicylate, ampicillin, and metronidazole.**

  In the high-risk patient **(ICU setting) prophylactic use of intravenous H_2 blockers or sucralfate** is standard. Shock should be treated aggressively with fluid resuscitation, blood replacement, and so on. The mucosa is capable of rapid healing when predisposing factors are corrected.

CIRRHOSIS

- Basic Science **Cirrhosis is defined as irreversible damage to the liver.** Several disease processes cause liver damage. Cirrhosis is the common endpoint of the injury process. The **most common causes of liver injury are** as follows:
  - **Alcoholism** is the **most common cause of cirrhosis. Cirrhosis presents with jaundice, ascites, palmar erythema, spider angiomata, varices, and testicular atrophy.**
  - **Biliary tract disease** can lead to cirrhosis **due to obstruction of bile ducts.**
  - **Primary biliary cirrhosis** is usually **associated with autoimmune disorders** and is caused by inflammation and destruction of bile ducts. **IgM antimitochondrial antibody** is present in 95% of patients. There is a predilection for **middle-aged women.** Early symptoms include pruritus. Jaundice and other signs of chronic liver injury occur years later.
  - **Secondary biliary cirrhosis** is **caused by gallstones, strictures, or cholangitis.** Jaundice and pruritus occur early in presentation. Signs of cholangitis may be present.
  - **Postviral cirrhosis** is due to viral **hepatitis B and C.** Diagnosis is usually made on biopsy.
  - **Passive congestion due to right-sided heart failure** can lead to cirrhosis.
- History and Physical Examination
  - **History** Important historical points include the following:
    - **Alcohol ingestion** (usually greater than a pint of liquor per day for more than 10 years)
    - **Biliary colic or right upper quadrant pain**
    - **History of hepatitis**
    - **Vomiting blood or melena**
    - **Congestive heart failure**

  On presentation, patients may be asymptomatic. Clinical findings such as **palmar erythema, spider angiomata, varices, encephalopathy, jaundice, and ascites** may be present.

  Most symptoms result from organs or tissues that are affected by chronic liver damage.

    - **Gastrointestinal bleeding**—veins that feed into the liver dilate **as a result of portal hypertension. Esophageal varices** may bleed, causing uncontrolled hemorrhage.
    - **Ascites—fluid** builds **in the peritoneum** due to a combination of portal hypertension, low albumin levels, and increased renal sodium reabsorption.
    - **Hepatic encephalopathy—*mental status changes due to multiple toxins*** that normally are removed by the liver. Increased ammonia levels may be detectd.
    - **Abnormal bleeding parameters—increased PT and PTT** cause easy bleeding or bruising due to thrombocyto-

penia and abnormal and diminished production of the vitamin K dependent clotting factors (V, VII, IX, and X).

■ **Laboratory and Diagnostic Studies** Screening laboratory studies include a complete blood count, serum electrolytes, and liver function studies. Liver biopsy may be helpful if there is no clear etiology for the cirrhosis.

- **General**
  - **Thrombocytopenia and anemia** are common findings in cirrhosis.
  - **Liver function enzymes** such as AST **may not be elevated because little viable tissue exists** to cause elevations.
- **Alcoholic cirrhosis** AST is commonly elevated in alcoholic cirrhosis and is usually **greater than ALT** by a 2:1 ratio.
- **Postviral cirrhosis** ALT is significantly elevated compared to AST in postviral cirrhosis.
- **Biliary cirrhosis**
  - **Alkaline phosphatase** and **bilirubin** will be **elevated** in biliary tract disease.
  - **Abdominal ultrasound,** endoscopic retrograde cholangiopancreatography (ERCP), or cholangiography should be obtained.
  - If primary biliary cirrhosis is suspected, testing should include IgM antimitochondrial antibody.
  - Abdominal CT scan or angiography is indicated for Budd-Chiari syndrome (hepatic vein thrombosis).
- **Hepatic encephalopathy** **Elevated ammonia levels** indicate hepatic encephalopathy (sample should be arterial blood placed on ice).
- **Spontaneous bacterial peritonitis** **Ascitic fluid** analysis reveals a **leukocyte count greater than 500** and a polymorphonuclear count greater than 250. Gram's stain may be positive for bacteria.
- **Gastrointestinal bleeding** **Endoscopy** should be obtained in all patients with evidence of cirrhosis and gastrointestinal bleeding.
- **Spontaneous bacterial peritonitis** **Spontaneous bacterial peritonitis** presents with **fever, diffuse abdominal discomfort, and leukocytosis.**

■ **Formulating the Diagnosis** **The diagnosis of cirrhosis** should be **entertained in** any **patient** that **presents with jaundice.** Clinical findings such as **palmar erythema, spider angiomata, varices, encephalopathy, and ascites** may be present to variable degrees and help confirm the clinical suspicion of cirrhosis.

The cause of cirrhosis may be less obvious.

Pruritus suggests biliary disease.

■ **Evaluating the Severity of Problems** Morbidity and mortality are most commonly due to complications of cirrhosis. Treatment should be directed at correcting these complications.

Significant worsening liver function studies may suggest worsening cirrhosis, but improvement of these tests does not correlate with improvement of disease and may, in fact, signal end-stage disease.

Hepatorenal syndrome involves worsening renal insufficiency of unknown etiology. Renal biopsy is unremarkable.

■ Managing the Patient **Management is directed initially at treating the complications of cirrhosis.**

- **Variceal bleeding** must be **treated aggressively with volume and blood resuscitation,** vitamin K, and **endoscopy** and **sclerosis if necessary. Intravenous vasopressin** and similar class agents or **balloon tamponade (Sengstaken-Blakemore tube)** may be used temporarily, especially if esophagoscopy is not available, or the patient is too unstable to undergo endoscopy. Long-term care of varices includes β-**blocking agents, elective sclerosis, and consideration of portal-systemic shunts.**
- Treatment of **hepatic encephalopathy** should be directed at correcting any causative illness and decreasing toxins, including ammonia. **Protein restriction** and **oral lactulose or neomycin** are mainstays of treatment.

  Other causes of mental status changes should be eliminated, including intracranial bleeding, drug ingestion, alcohol ingestion or withdrawal, and meningitis.
- Diagnosis of **spontaneous bacterial peritonitis** is confirmed by peritoneal fluid analysis. Treatment with **ampicillin (or cephalosporin) and an aminoglycoside** should be instituted.
- **Ascites should be managed conservatively. Salt restriction and bed rest** may improve ascites. **Diuretics,** particularly potassium-sparing ones **(spironolactone),** may be given.

  Liver transplantation should be considered for patients with the hepatorenal syndrome.

Colorectal Cancer/Familial Adenomatous Polyposis

■ Basic Science **Colorectal cancer** is the **second leading cause of cancer death** in the United States, with 160,000 Americans newly diagnosed each year. Most colorectal cancer is termed *sporadic,* occurring from a variety of environmental causes. **Familial adenomatous polyposis (FAP) is one of two well-recognized causes of hereditary colorectal cancer.** FAP accounts for about 1% of all colorectal cancer cases. **FAP** is an **autosomal dominant syndrome** caused by a germline mutation.

FAP **patients develop** hundreds to **thousands of colorectal adenomas,** usually **in adolescence. Polyps** occur diffusely **throughout the colorectum.**

■ History and Physical Examination FAP patients are initially asymptomatic. **Symptoms** usually occur in later stages of this syndrome and **include rectal bleeding, change in bowel habits, and abdominal pain.**

Patients can develop colonic obstruction. Colorectal cancer, primarily in the left colon, **occurs in** virtually **all FAP patients if colectomy is not performed.** The average age at diagnosis of colorectal cancer is 39 years.

FAP can also be associated with **various extracolonic lesions,** which can be appreciated on physical examination. These include **skin epidermoid cysts,** osteomas noted particularly on the mandible, pigmented ocular fundus lesions (POFLs) found on ophthalmologic examination, and desmoid tumors found on skin or abdominal examination. FAP **patients are** at **increased risk for intestinal cancers,** especially duodenal and periampullary carcinomas, and for rectal cancer in those with subtotal colectomy. Extraintestinal cancers reported in association with FAP include tumors of the thyroid gland, adrenal gland, biliary tree, brain, and pancreas.

- **Laboratory and Diagnostic Studies** With a family history of FAP, the diagnosis of **FAP is confirmed by finding 100 or more** adenomatous **polyps on endoscopic examination of the colorectum.** In an at-risk patient, the finding of greater than three POFLs on indirect ophthalmoloscopic examination has a 100% positive predictive value for the diagnosis of FAP. Genotyping by direct detection for mutation of APC gene is now available and has virtually 99% sensitivity and specificity for FAP.

  Suspicion of hereditary colorectal cancer should be raised in individuals with multiple family members with colorectal cancer or adenomatous polyps, members with colorectal cancer at young age (less than 50 years of age), or family members with multiple adenomatous polyps.

  The diagnosis of FAP is made in patients that have mutation of the FAP gene and/or have greater than 100 adenomatous polyps on endoscopic examination of the colorectum.

- **Evaluating the Severity of Problems** Major complications in FAP consist of the extracolonic manifestations of FAP just described. After colorectal cancer, the most common cause of death is duodenal cancer.

- **Managing the Patient** **Colectomy** is the **only effective therapy** that eliminates the development of colorectal cancer **in FAP patients. Surgery** should be **done at the time of diagnosis** of FAP to **minimize risk** of **colorectal cancer.**

  Postoperative follow-up is important. Patients with subtotal colectomy **require routine endoscopic surveillance** of the remaining rectum for recurrent adenomas and/or carcinoma. Also, in patients with ileoanal pull-through, endoscopic biopsy surveillance of the pouch should be considered. Most authorities recommend upper endoscopic surveillance (with biopsy and brushing) of the stomach, duodenum, and periampullary region with front and/or side viewing endoscopes.

  Since FAP is a genetic disease, **screening of** at-risk **first-degree relatives by endoscopy** of the colorectum **and genotyping** is imperative.

Constipation

- **Basic Science** Constipation is defined as a **decrease in defecation frequency coupled with** an **increase in stool hardness.** The patient often has extreme difficulty with defecation and often has a feeling of incomplete evacuation.

Constipation is divided into acute and chronic states.

- **Acute constipation** often **indicates an organic disease state (e.g., mechanical bowel obstruction, adynamic ileus from trauma or peritoneal irritation).**
- **Chronic constipation** may indicate **(1) a neurological problem (e.g., irritable bowel disease, Hirshsprung's disease), (2) an electrolyte disorder (hyperglycemia, hypercalcemia), or (3) a psychological disorder.**

■ History and Physical Examination

- **History** The **physician should determine whether**
  - Constipation is **acute or chronic.**
  - **Family history** of constipation or GI problems exist.
  - **Neurological disease** or CNS injury exists.
  - Patient has **changed** his or her **diet or exercise** regimen.
  - Patient is using **new medications—opiates, tranquilizers, anticholinergics,** and other drugs may cause constipation.
  - Patient has **attempted self-treatment.**

  Recent onset of constipation, bloody stools, pain, fever, anemia, and/or weight loss may indicate a tumor.
- **Physical examination** The examination may demonstrate **hemorrhoids, mucosal prolapse,** and **fecal impaction.** These states may **all lead to constipation.**

■ Laboratory and Diagnostic Studies **TSH, T4 and standard serum chemistries** should be obtained to **determine if other abnormalities** exist that may cause constipation **(e.g., hypothyroidism, diabetes mellitus, hypercalcemia, and hypokalemia).**

■ Formulating the Diagnosis History, physical examination, and the results of diagnostic studies will usually reveal the cause of the patient's constipation.

■ Evaluating the Severity of Problems The **cause of the patient's constipation should be sought.** Constipation may be a symptom of a life-threatening disease state (e.g., tumor).

■ Managing the Patient If history and physical examination indicate a **possible organic cause of constipation, then diagnostic testing (e.g., sigmoidoscopy, serum chemistries) should be instituted.**

The patient should be placed on a **high-fiber diet. Increasing dietary fiber (fruits and vegetables)** increases stool bulk and makes stools softer, ensuring easier stool elimination.

Agents that increase stool bulk, such as **psyllium,** may make defecation easier. These are the only medications that should be used on a long-term basis for the relief of constipation.

If the **patient** feels **better** after 1 month **on** the **high-fiber diet** and the diagnostic testing is negative, the **workup should end** and the patient told to continue his or her present regimen.

After **failure of a trial of conservative treatment, colonoscopy, sigmoidoscopy, or barium enema should be ordered** to rule out obstructive anatomical lesions (e.g., tumor).

Laxatives and cathartics should be used carefully. Patients often **become dependent** on these medications and develop an inability to defecate without their use. Patients may also develop problems absorbing nutrients. **Laxatives and cathartics are classified as follows:**

- **Detergent laxatives** These agents **(docusate) act by allowing water to enter into the stool,** increasing the stool size and thereby stimulating peristalsis.
- **Osmotic agents** These agents **(sorbitol, laculose)** are **poorly absorbed large molecules and act by osmotically drawing water into the colon.** The excess water in the colon softens the stool and stimulates peristalsis. These agents may contain large amounts of sodium, magnesium, or phosphate and must be used carefully in patients with heart and renal failure. These agents may cause electrolyte abnormalities if abused.
- **Stimulant cathartics** These agents **(bisacodyl, castor oil) act by irritating the bowel.** Abuse leads to severe electrolyte abnormalities.

Crohn's Disease

■ Basic Science **Crohn's disease** is a **chronic inflammatory bowel disease. The disorder is immune mediated** and **results in nonspecific ulcerations** in the gastrointestinal tract. The ulcers may occur **anywhere in the GI tract** but tend to **cluster around the distal ileum** and **right colon.** The **ulcers are deep** and often **transmural.**

The disease is thought to be immune mediated with environmental triggers.

The disease usually appears in patients **before the age of 30.** There appears to be a **genetic predisposition** to Crohn's disease.

■ History and Physical Examination

- **History** Crohn's disease is characterized by frequent attacks of **diarrhea, severe abdominal pain, weight loss, anorexia, fever, and vomiting.** Additionally, patients may have other immune-related symptoms such as **arthritis, erythema nodosum, uveitis, or kidney stones** (due to changes in the absorptive ability of the gut).
- **Physical Examination** Patients may present with few symptoms or may manifest **fever, right lower quadrant pain, abdominal distention,** or **signs of obstruction.**

■ Laboratory and Diagnostic Studies **Barium x-rays** (enema or swallow) may show a **narrowed lumen and skip lesions.** In severe cases a string sign may be present. If barium x-rays are not diagnostic, colonoscopy may be employed for diagnostic purposes.

CBC may demonstrate a **leukocytosis.** Serum chemistries may demonstrate a low albumin. The **ESR** may be **elevated.**

- Formulating the Diagnosis **History, physical examination,** and **barium x-ray** studies are **diagnostic.**

  Crohn's disease may be **difficult to differentiate from ulcerative colitis.**

  Other diseases that produce an acute abdomen may be confused with Crohn's and include **appendicitis, pelvic inflammatory disease,** and ***Yersinia* enteritis.**

- Evaluating the Severity of Problems **Fistulas and abscess** formation are the **major complications of Crohn's** disease.

  Life-threatening complications include **perforation and electrolyte imbalances.**

- Managing the Patient **Specific treatment of Crohn's includes** the following:
  - **Anticholinergics, loperamide (to relieve cramps and diarrhea).**
  - **Sulfasalazine (to decrease chronic inflammation)** Sulfasalazine is not effective in acute exacerbations of Crohn's.
  - **Corticosteriods (for acute exacerbations)** Prednisone 60 mg per day with a tapering dose after 4 weeks should be prescribed. However, steroids are contraindicated in patients with abscess or infection.
  - **Immunosuppressive agents** Azathioprine, 6-mercaptopurine **may be utilized.** These agents take months to work.
  - **Elemental diet or hyperalimentaion (used to rest the bowel).**
  - **Surgery Removal of the involved bowel segment is the last resort.** Surgery is **not curative,** and the **disease will return** in the vast majority of patients.
  - **Antibiotics (for abscess or infection)** Triple antibiotic coverage includes **ampicillin, gentamicin, and flagyl.**

DIVERTICULAR DISEASE

- Basic Science **Diverticula** are **pouchlike herniations through the muscular layer of the colon.** Diverticula form as a result of elevated intraluminal colonic pressure. They tend to occur in narrow regions of the colon where intraluminal pressure is elevated. Sigmoid and descending colon are the most narrow regions of the colon. Because of this fact **diverticula** tend to **localize in the sigmoid** and the **descending colon.**

  Diverticula are **common** and **usually benign.** However, **two disease states** may occur as a result of diverticula:
  - **Diverticulitis—inflammation of the diverticula.** The penetration of fecal matter through the thin-walled diverticula causes **inflammation and abscess formation in** the **tissues surrounding** the **colon.** With **repeated inflammation,** the **lumen** of the colon **narrows** and **may become obstructed.**

    Inflamed diverticula may form an **abscess,** form **fistulas** to adjoining structures (e.g., bladder), **obstruct** the **colon,** or **perforate** into the peritoneum.

- **Diverticular bleeding The diverticula erode through the vasa recta (vessels)** on the outside of the colon, causing bleeding. **Bleeding may become massive.**

■ History and Physical Examination

- **History**
  - ***General*** Patients should be asked about **previous disease, change in bowel habits, and previous** colonic **diagnostic studies.**
  - ***Diverticulitis*** Patient often describes **cramping** or **dull left lower quadrant pain, fever, nausea, vomiting, constipation,** or **diarrhea.**
  - ***Diverticular bleed*** Patient often describes **painless rectal bleeding;** blood is a **maroon or deep red** color. The patient may feel lightheaded or dizzy, indicating potential hemodynamic instability.
- **Physical** Physical examination should include **vital sign assessment** and orthostatic measurements if bleeding is suspected. **Abdominal and rectal (with stool guaiac) examination** must be performed. Pelvic examination must be performed in females.

  Tender mass on examination is **indicative of diverticulitis.**

■ Laboratory and Diagnostic Studies

- **Diverticulitis CBC with differential** should be drawn. A **leukocytosis with left shift** is often found. Perform **abdominal free air x-ray series** to **rule out perforation.**
- **Diverticular bleed CBC, type, and crossmatch** should be drawn. **Angiography** is the best method for diagnosing bleeding. Colonoscopy is indicated if a mass is palpated or the cause of bleeding is not identified.

  Exploratory laparotomy may be necessary if the cause of bleeding is in question.

■ Formulating the Diagnosis

- **Diverticulitis History and physical examination** usually **point to the diagnosis.** Diseases that may be confused with diverticular diseases include Crohn's disease, irritable bowel syndrome, colon cancer, and appendicitis.

  A **left lower quadrant mass, fever, leukocytosis and left shift suggest abscess or perforation.**
- **Diverticular bleed** Diverticular bleeding must be differentiated from upper GI bleeding.

■ Evaluating the Severity of Problems

- **Diverticulitis** Diverticulitis may lead to sepsis, perforation, abscess, or intestinal obstruction.
- **Diverticular bleed** Patients may exsanguinate. Vitals signs must be closely monitored.

■ Managing the Patient

- **Diverticulitis**
  - **Patients who do not appear ill** may be treated with

oral antibiotics (cephalexin), liquid diet, and **follow-up barium enema.**

- **Patients who appear ill** should be treated with **IV antibiotics (gentamicin and third-generation cephalosporin), NPO,** and **monitored closely. If** the **patient does not improve** or worsens, a **CT scan of** the **abdomen** should be performed to rule out abscess or other abdominal pathology. Any **patient with peritoneal signs** should have an **immediate operation.**

- **Diverticular bleed Most** diverticular **bleeding** will **cease with conservative treatment (IV hydration, monitoring of hematocrit).** In cases where **bleeding does not stop,** infusion of intraarterial **vasopressin** may be necessary. If vasopressin is unable to stop the bleeding, **surgery with diverticular resection** may be necessary.

Esophageal Cancer

- Basic Science **Esophageal cancer** accounts for approximately **1% of U.S. cancers.** Over **85%** of cases are due to **squamous cell** carcinoma and 15% occur from adenocarcinoma. **Elderly black males** in urban areas account for the **majority of cases.**

  Several **factors** are **implicated** in the pathogenesis of this malignancy, including **alcohol; tobacco; dietary carcinogens** (i.e., nitrosamine, fungal toxins); and predisposing conditions such as **Barrett's esophagus** (adenocarcinoma), lye stricture, achalasia, celiac disease, and radiation-induced strictures.

- History and Physical Examination **Patients** with esophageal cancer **present with difficulty swallowing. At first** patients report **difficulty swallowing solids,** but **as the disease progresses patients cannot swallow liquids.** Other symptoms can include **weight loss, chest pain, odynophagia** (a burning sensation while swallowing), **gastrointestinal bleeding,** and **aspiration of food and secretions.**

  Extension of cancer into soft tissues surrounding the esophagus can result in laryngeal paralysis due to recurrent laryngeal nerve involvement, superior vena cava obstruction, or Horner's syndrome. In addition, spread of cancer into the mediastinum may cause tracheal-esophageal or aortic-esophageal fistula.

- Laboratory and Diagnostic Studies **Barium contrast studies** of the esophagus **usually make the diagnosis** of esophageal cancer. Typically, marked **narrowing of the esophageal lumen** with an irregular and ragged pattern to the esophageal mucosa **is noted.**

  Esophagoscopy should follow an abnormal barium contrast. **Biopsy** and **cytologic brushing** of the esophagus **should be collected.** This will enable the physician to make a definitive diagnosis of esophageal cancer.

- Formulating the Diagnosis **Diagnosis** is **made after tissue biopsy examination.**

- Evaluating the Severity of Problems **Additional information** concerning the extent and spread of cancer can be **obtained by CT of**

the thorax and abdomen. Bronchoscopy may be necessary to determine if tracheoesophageal fistulization is present.

- Managing the Patient **Patients with dysphagia should always undergo diagnostic testing.**

Most patients are diagnosed with esophageal cancer only after experiencing symptoms. Since difficulty with swallowing does not occur until the development of significant narrowing of the esophageal lumen, **most symptomatic patients have advanced and incurable disease. The prognosis for esophageal cancer is poor** (5% survival rate 5 years after diagnosis).

Surgical resection is the mainstay of attempts at curative treatment. Surgery is sometimes combined with chemotherapy and radiation. Over **one third of patients** have inoperable disease due to tumor involvement of mediastinal structures or distant metastases.

Most patients will ultimately **receive palliative care,** and attention should focus on general supportive care with restoration of nutritional status. In order **to allow the patient to eat and swallow saliva** the esophageal lumen must be restored to an adequate size. This can be accomplished by peroral dilatation or radiotherapy. After dilation, an esophageal stent prosthesis can be placed endoscopically through the tumor to attempt to maintain lumenal patency. In addition, the stenotic lumen can be enlarged by using a laser or heater probe to ablate the tumor.

Food Poisoning

- Basic Science **Food poisoning is gastroenteritis of infectious etiology. Causative agents include** enterotoxigenic bacteria, invasive bacteria, viruses, and some parasites.

Enterotoxigenic bacteria include *Escherichia coli,* enterotoxigenic *Staphylococcus aureus, Bacillus cereus, Vibrio, Clostridium perfringens,* and *C. botulinum.* **Enterotoxins cause illness by producing a secretory diarrhea.** Traveler's diarrhea is most commonly due to enterotoxins.

Invasive organisms include *Salmonella, Shigella, Campylobacter, Yersinia enterocolitica, Vibrio parhaemolyticus,* certain strains of *Escherichia coli,* and *Helicobacter pylori.*

These **organisms,** with the exception of *Salmonella,* **invade the bowel wall, causing mucosal ulcerations.**

***Salmonella* causes an inflammatory response in the bowel mucosa** without ulcer formation.

Parasitic infections include *Giardia lamblia, Entamoeba histolytica,* and various worm infections.

***Cryptosporidium* and cytomegalovirus** commonly **cause chronic diarrhea in AIDS** patients and immunocompromised patients.

- History and Physical Examination Signs and symptoms are based on the etiology of the gastroenteritis.

  - **Enterotoxigenic bacteria**

    - Patients with enterotoxigenic gastroenteritis complain of **diarrhea,** usually occurring **within a day of ingestion.**

- ***Staphylococcal* poisoning** typically causes **nausea, vomiting, abdominal cramping,** and **diarrhea.** Spoiled **milk products** are most commonly **implicated.**
- **Most enterotoxigenic food poisoning is self-limited** and **treatment is supportive. Stool** is usually **not bloody,** and **fecal leukocytes** are **rarely found.** Viral gastroenteritis presents similarly.

Invasive organisms

- ***Salmonella*** is an invasive organism and typically causes **low-grade fevers, crampy abdominal pain,** and **diarrhea. Fecal leukocytes** are **commonly found. Symptoms** usually **resolve in 3 to 5 days, unless** the **patient develops enteric fever,** in which the fever lasts for weeks and systemic symptoms occur, such as headaches, hepatosplenomegaly, and possibly even abdominal bleeding, perforation, and bacteremia. **Antibiotic treatment** should be **initiated with enteric fever,** but should be withheld in self-limited gastroenteritis, as treatment seems to enhance a carrier state.
- ***Shigella*** is another invasive bacteria and causes **severe abdominal pain, fevers, blood diarrhea,** and **fecal leukocytes.** Treatment with **antibiotics should be initiated.**
- ***Yersinia*** enterocolitis is difficult to culture and may cause a mesenteric adenitis, which can be **confused with acute appenditicitis.**
- ***Campylobacter*** causes symptoms similar to *Shigella,* including **abdominal pain, fevers, and diarrhea.**
- ***Giardia*** is a parasite that has been found in **contaminated water** throughout the United States. Symptoms of **abdominal pain, diarrhea, nausea** with **vomiting,** and **fevers** usually occur **2 weeks after ingestion** of the cysts.
- ***Entameoba*** is transmitted through **contaminated water** or **fecal-oral contact** and may cause **nausea, abdominal pain, and bloody diarrhea. Liver abscesses** may occur and are manifested by **hepatomegaly, fever, and abdominal pain.**

■ **Laboratory and Diagnostic Studies** Initial evaluation should include assessment of hemodynamic status. A complete blood count, serum electrolytes, liver function studies, and pancreatic enzyme analysis may be indicated.

Radiologic studies to diagnose perforation or bowel obstruction may be appropriate.

Stool examination is used to guide therapy. The **absence of blood** and **fecal leukocytes suggest** either an **enterotoxigenic bacteria or virus** as the source of the gastroenteritis. The **presence of fecal leukocytes** strongly **implies that an invasive organism** is present and **requires antibiotic therapy.**

■ **Formulating the Diagnosis** Gastroenteritis is a common illness. The presence of **fecal leukocytes implies** that the **diarrhea** is **due to** an **invasive organism** and should prompt **stool cultures** specific for invasive organisms.

Appropriate testing for stool that is negative for leukocytes includes wet mount examination for ova and parasites and stool cultures. Serologic testing for *Salmonella,* amebiasis, and *Trichinella* may be indicated.

A scotch tape test may be necessary to isolate pinworms in children.

- **Evaluating the Severity of Problems** Most enterotoxigenic gastroenteritides are self-limited, and severity is based on volume loss. Clinical deterioration tends to be more dramatic in patients with an invasive etiology.
- **Managing the Patient**
  - **Supportive therapy** is the mainstay for most gastoenteritidies and includes **rehydration** and **antiemetics.**
  - **Antispasmodics** and **anticholinergics** are relatively **contraindicated in viral and enterotoxigenic diarrhea** and are **absolutely contraindicated in invasive diarrhea.**
  - **Treatment** of **traveler's diarrhea** may also include **bismuth salicylate (Pepto-Bismol).** Recent evidence suggests that **antibiotic therapy with sulfa antibiotics or ciprofloxacin may shorten symptoms.**
  - **Antibiotic therapy is indicated for treatment of *Shigella,* enteric fever due to *Salmonella,* and parasitic infections.** Treatment should be initiated if organisms are seen on smear or cultures are positive. Empiric therapy for enteric fever may be necessary with a quinolone or cephalosporin.
  - ***H. pylori* infections** should be diagnosed by symptoms and cultures of ulcers and **treated with** a three-drug combination, including **bismuth, metronidazole, and ampicillin.**

Irritable Bowel Syndrome

- **Basic Science** Irritable bowel syndrome (IBS) is **caused by altered intestinal motility.** This is thought to occur **due to abnormalities in intestinal electrical rhythm. Intestinal motility may decrease** abnormally, **leading to constipation,** or **increase** abnormally, **leading to diarrhea.**
- **History and Physical Examination** **IBS presents in one of three ways:**
  - **Chronic abdominal pain with constipation:** patients may have a bloating sensation; heartburn and back pain are common. The pain is often relieved with flatus.
  - **Intermittent diarrhea:** this is usually worse in the morning hours.
  - **Alternating constipation and diarrhea:** IBS occurs most commonly in **young to middle-aged females.** Many individuals have a **concomitant psychiatric disorder (depression or anxiety). Stress plays a major role** in many patients.

    Physical examination is usually **normal.**
- **Laboratory and Diagnostic Studies** Laboratory and diagnostic studies should **eliminate other causes for the patient's symptoms.**

If diarrhea is present, stool studies, including examination for blood, fecal leukocytes, and cultures, are indicated.

If bleeding is present, the patient should undergo colonoscopy.

Consider testing for lactose intolerance and gluten sensitivity. The patient's TSH level should be checked to rule out hyperthyroidism.

- **Formulating the Diagnosis** The diagnosis of **irritable bowel syndrome** should be **considered after the exclusion of other diseases.**
- **Evaluating the Severity of Problems** **Irritable bowel syndrome does not** usually **cause significant morbidity or mortality.**
- **Managing the Patient** Irritable bowel syndrome is a **chronic condition.** The **patient** should be **reassured that** the **disease is not life-threatening. Changes in lifestyle** and **decreases in stress** may **alleviate symptoms.**

  Management includes **symptomatic treatment** such as **Lomotil for diarrhea and fiber or anticholinergic medication for constipation.**

LACTOSE INTOLERANCE

- **Basic Science** **Lactase** is an **enzyme** located in the brush border of the small intestine. Lactase **converts** the disaccharide **lactose into** the monosaccharides **glucose and galactose.** Glucose and galactose are actively transported by the gut epithelium into the body.

  A **deficiency of lactase** results in the **inability to digest lactose.** Lactose is an osmotically active sugar and draws water into the gut lumen. Gut bacteria break down the undigested lactose producing gas and acid by-products. This **causes bloating, flatus,** and **abdominal discomfort.**

  Ninety percent of Asians, 45% of blacks, and a **significant proportion of whites develop lactase deficiency** after childhood.
- **History and Physical Examination** A history of **upset stomach, bloating, flatus, and diarrhea after the consumption of milk products** is classic for lactase deficiency. Patients often have a **family history of lactose intolerance.** Physical examination is usually benign.
- **Laboratory and Diagnostic Studies** **High hydrogen gas levels,** measured by a gas breath analyzer, **after a lactose test dose** supports the diagnosis of lactase deficiency. Hydrogen is a by-product of bacterial breakdown of lactose.
- **Formulating the Diagnosis** **History, physical examination,** and results of **breath gas analysis** are **diagnostic** for **lactase deficiency.** Lactase deficiency is often confused with irritable bowel syndrome.
- **Evaluating the Severity of Problems** Lactase deficiency is a **benign disorder.**
- **Managing the Patient** Patients should **avoid milk products** or other lactose-containing compounds. Patients may eat lactose-containing products if they first add **artificial lactase** to those products.

Lower GI Bleeding

- **Basic Science** **Lower GI bleeding** can originate from a **number of sources:**
  - **Upper GI bleeding**
  - **Diverticulosis**—vessel erosion through thinned diverticular wall
  - **Infection**—*Shigella, Salmonella,* amebiasis, or *Campylobacter jejuni*
  - **Cancer**—chronic bleeding
  - **Inflammatory bowel disease**—Crohn's or ulcerative colitis
  - **Anal fissures, hemorrhoids, AV malformation**
- **History and Physical Examination** **History** should determine **amount, duration,** and **frequency of bleeding, color of stools, whether pain is present,** and **past medical history** of **cirrhosis** or **other GI pathology.**

  Signs to look for on physical examination include **caput medusa, ascites, abdominal distension, absent bowel sounds,** or **peritoneal signs.**

  Digital rectal examination detects the **presence and character of blood** (gross, occult, color).

  - **Upper GI—stools** usually **appear melenic.**
  - **Diverticulosis—stools are usually dark red. Pain is mild or absent.** Diverticular bleeding frequently recurs.
  - **Infection—pain** and **bloody diarrhea** occur.
  - **Cancer—** cancer presents with **weight loss, anemia, change in bowel habits, and bowel obstruction.**
  - **Inflammatory bowel disease—pain, tenesmus, and bloody diarrhea occur.**

  If an **upper GI bleed is suspected,** a nasogastric tube should be placed and the **gastric contents sampled** for bleeding.
- **Laboratory and Diagnostic Studies** A CBC, PT, PTT, serum chemistry, and type and cross are done if severe bleeding is present.

  Depending on the degree of bleeding, several different diagnostic modalities may be employed:

  - Anoscopy—generally the first diagnostic modality utilized
  - **Flexible sigmoidoscopy**
  - **Colonoscopy**
  - Angiography—used when colonoscopy fails to identify a source of bleeding
  - Labeled RBCs (^{99m}Tc)—used for patients unable to undergo angiography, or if angiography is nondiagnostic.

  If upper GI bleeding is found, endoscopy should be performed.
- **Formulating the Diagnosis** Diagnosis should be evident after history, physical examination, and diagnostic testing.
- **Evaluating the Severity of Problems** Most lower GI bleeds are not life-

threatening and do not require emergency treatment. However, if the patient appears to be hemodynamically unstable (i.e., tachycardic, hypotensive), volume resuscitation should be instituted.

■ **Managing the Patient** The treatment of lower GI bleeding is dependent on the cause. Diverticulosis is the most common cause of massive lower GI bleeding. Its management is discussed previously.

Arteriovenous malformations are a common cause of lower GI bleeding in the elderly. The lesions are usually located in the right colon. A hemicolectomy may be necesary for patients with bleeding requiring transfusion.

Ischemic colitis is a GI manifestation of atherosclerotic vascular disease. The bleeding is generally mild and self-limited.

The management of other causes of lower GI bleeding (e.g., inflammatory bowel disease, hemorrhoids, cancer) is discussed elsewhere.

Peptic Ulcer Disease

■ **Basic Science** **Food is degraded in the stomach by pepsin** and **hydrochloric acid.** The partially digested food is then passed through the pyloric valve and into the duodenum.

Hydrochloric acid is secreted by parietal cells in the stomach. **Hydrochloric acid production** is **regulated by** both neurohumeral **(vagus nerve)** and chemical factors **(e.g., gastrin, histamine, and acetylcholine).** Hydrochloric acid production lowers stomach pH, which activates the enzyme pepsin.

The **stomach** and **duodenum** are **covered by** a thick coat of **bicarbonate-laden mucus** that **protects** the mucosa **from enzymatic degradation. Prostaglandins,** produced by the mucosa, **enhance** the **stability of** the **mucous** layer.

Peptic ulcers arise when hydrochloric acid and **pepsin overwhelm mucosal defense factors.** This may occur by either of **two mechanisms:**

- **Overproduction of hydrochloric acid** There is a **genetic component** to hypersecretion. Extrinsic triggers such as **caffeine, ethanol,** and **intravenous calcium** also stimulate hydrochloric acid secretion.
- **Ineffective defensive barrier** Extrinsic triggers such as **smoking** and **nonsteroidal antiinflammatory agents inhibit prostaglandins** and **damage the mucous lining of** the **stomach** and duodenum.

  In addition, it has been shown that infection with the bacteria ***Helicobacter pylori*** **causes peptic ulcers.**

  Peptic ulcers occur in regions where the enzyme pepsin is found, namely the gastric and duodenal mucosa.

  - **Duodenal ulcers** are **more common than gastric ulcers.** Duodenal ulcers occur in the first few centimeters of the duodenal bulb.
  - **Gastric ulcers** are mostly located in the lesser curvature of the stomach.

■ **History and Physical Examination** Ulcers are common in patients between **40 and 50 years of age.**

History associated with peptic ulcer includes the following:

- **Cigarette smoking**
- **Aspirin or nonsteroidal antiinflammatory medication use**
- **Stress**
- **Family history of peptic ulcer**
- **Patients commonly present with epigastric pain** that often radiates to the back. The pain is described as **boring, burning, dull,** aching, or nagging. Most patients have **pain relief when** they **eat food.** Pain usually **recurs several hours after eating or after** they retire to **bed.**
- **Physical examination** may reveal some **abdominal tenderness;** however, the examination may also be benign. **Weight loss** occurs in 50% of those with benign ulcers.
- Patients with **active bleeding ulcers** may **present with anemia, melena,** or **hematemesis. Perforation,** a complication of duodenal ulcers, may present as **abrupt, severe pain radiating to** the **back** with **peritoneal signs (e.g., rigid abdomen, patient** lying absolutely **still).**

■ **Laboratory and Diagnostic Studies** **Endoscopy** is the **preferred** initial **study** for the detection of peptic ulcer disease. If a **gastric ulcer** is found, a **biopsy should be performed to rule out malignancy. Duodenal ulcers rarely lead to malignancy.** Gastric biopsies should be taken to detect *H. pylori* infection.

If endoscopy fails to reveal an ulcer, barium contrast x-ray studies may be utilized to locate the ulcer site.

Initial laboratory studies should include a complete blood count (to determine if anemia is present). If there is significant clinical suspicion, pancreatic enzyme levels and free air series x-rays to rule out perforation should be performed. Analysis of gastric acid and gastrin levels are appropriate if there is concern for gastric cancer or Zollinger-Ellison syndrome.

■ **Formulating the Diagnosis** Peptic ulcer disease should be considered in any patient with complaints of abdominal pain. An active GI bleed should be considered in the patient presenting with hematemesis, bloody stools, unexplained anemia, or orthostatic symptoms.

Zollinger-Ellison syndrome can be hereditary and leads to ulcer formation due to increased acid secretion.

■ **Evaluating the Severity of Problems** **Massive GI bleed** is a **medical emergency** calling for **fluid resuscitation, packed red blood cells, and emergent endoscopy.**

GI perforation is a medical emergency calling for immediate surgical consultation and operation.

■ **Managing the Patient** All patients with **active GI bleeding** must be monitored closely for signs of hypovolemia and exsanguination. **Vital signs, orthostatic blood pressures,** and **complete blood count must be monitored closely. Fluid resuscitation** and use of **packed red blood cells** are indicated **in unstable patients.**

Gastric aspiration via NG tube should be performed. **Red blood indicates** an **active bleed. Coffee grounds indicate**

bleeding that has **stopped at present. Active bleeding** may **require emergent endoscopy to sclerose any bleeding vessels.** In the event that medical or sclerotherapy does not stop the bleeding, emergent surgical gastrectomy may be indicated.

Medical treatment strives to **reduce stomach acid** production allowing the mucosa time to heal. **Treatment** should begin with the drugs listed in the following section and continue **for 4 to 6 weeks.** Treatment is **then discontinued and** the **patient observed for recurrence** of **ulcers and symptoms.** Recurrent ulcers should prompt further investigation and consideration for *H. pylori* infection or malignancy.

- **Medications Antacids** directly reduce the amount of acid in the stomach. Antacids should be given 1 **hour after meals and at bedtime.** Poor compliance due to taste, frequency of medication, and diarrhea limit the use of antacids as single therapy.

  H_2-receptor antagonists are the **most common form of therapy.** H_2 antagonists, such as cimetidine, ranitidine, nizatidine, and famotidine, work to block the action of histamine and subsequently acid secretion. Decreased hydrochloric acid production inhibits the activation of pepsin.

  Sucralfate works by **forming a protective coat over an ulcer.** Sucralfate has been shown to be effective in treating duodenal ulcers, but is less effective with gastric ulcers. Dosage is 1 g qid.

  Proton pump inhibitors such as omeprazole and lansoprazole work by preventing acid secretion. Both are **effective in Zollinger-Ellison syndrome.**

  Certain **habits and medications** should be modified or **eliminated:**

  - **Alcohol intake** should be **stopped.**
  - **Smoking should stop.**
  - Use of **nonsteroidal antiinflammatories should stop.**
  - **Caffeine** and **coffee** should be **avoided.**
  - **Stress reduction** is beneficial.

- **Eradication** of *H. pylori* in affected individuals dramatically reduces ulcer recurrence. Currently used regimens involve combination therapy

  - 14 days of Pepto-Bismol, metronidazole, and ampicillin, or
  - 14 days of omeprazole and clarithromycin (less effective, better patient compliance)

Ulcerative Colitis

- Basic Science **Ulcerative colitis** is a **chronic, episodic, inflammatory disease** of the large intestine and rectum. It is **characterized by** profuse watery **diarrhea containing** various amounts of **blood, mucus, and pus.**

  Ulcerative colitis has a bimodal peak of onset **occurring in** the **second and sixth decade.**

- History and Physical Examination

  - **History** Attacks of diarrhea are accompanied by **tenesmus, se-**

vere abdominal pain, fever, chills, anemia, and weight loss. Children with this disease may suffer retarded physical growth. The debilitating symptoms often **prevent the patient from carrying on the normal activities of daily living.**

Additionally patients **may have other non-GI symptoms** such as **arthritis, erythema nodosum, or uveitis.**

- **Physical** Physical exam reveals **bloody, diarrhea stool; significant WBCs; and mucus.**

■ Laboratory and Diagnostic Studies Serum chemistries demonstrate low albumin. Erythrocyte sedimentatioin rate (ESR) is elevated.

Barium x-rays will demonstrate loss of haustra in the colon. **Flexible sigmoidoscopy** or **colonoscopy with biopsy,** is utilized to cement the diagnosis.

Stool cultures must be sent to **rule out infectious diarrhea** as the source of the patient's symptoms.

■ Formulating the Diagnosis Diagnosis of the disease is based on clinical signs, the results of barium x-ray films of the colon, and colonoscopy with biopsy. It is difficult to differentiate ulcerative colitis from Crohn's disease.

Infectious diarrhea must be ruled out before the patient is diagnosed with ulcerative colitis.

■ Evaluating the Severity of Problems **Toxic megacolon** is a **life-threatening complication of ulcerative colitis.** The colon dilates and loses tone, causing an ileus. Toxic megacolon **may lead to perforation, septicemia, and death.** If treatment is not instituted immediately, 40% of patients will die.

■ Managing the Patient Patients need periodic colonoscopy to detect occult GI cancers.

Specific therapy for ulcerative colitis includes the following:

- **Avoid raw fruits and vegetables**
- **Anticolenergics and loperamide (to relieve cramps and diarrhea)**
- **Sulfasalizine (to decrease chronic inflammation)**
- **Corticosteriods (for acute exacerbations)**
- **Immunosuppressive agents** Azathioprine, 6-mercaptopurine **may be utilized.** These agents take months to work.
- **Surgery** Patients with severe disease of **life-threatening complications (toxic megacolon, cancer)** usually require surgery. **Total proctocolectomy with ileostomy is** a **permanent cure for ulcerative colitis.**

  Patients with **toxic megacolon** should be **started on ampicillin** and **gentamicin,** made **NPO,** placed on **IV hydration** and **IV steroids,** and watched closely for signs of peritonitis.

Upper GI Bleeding

■ Basic Science **Upper GI bleeding** may be **due to** any of the following:

- ***Capillary bleeding,*** *(e.g., due to* ***gastritis from ethanol***

or nonsteroidal usage). Bleeding is usually controllable and benign.

- ***Venous bleeding,*** (*e.g., due to* ***esophageal varices***). This indicates probable liver pathology with associated portal hypertension. It is usually associated with coagulopathies and other serious medical problems. High mortality is associated with this type of bleeding.
- **Arterial bleeding** due to **ulcer erosion into one of the gastric arteries.** This may also result from fistula formation between aorta and duodenum. High mortality is associated with arterial bleeding.

■ History and Physical Examination Elements of the **history** that are important are the following:

- **Duration of the bleeding**
- **Hematochezia** (defecating blood)—may indicate upper GI bleed; more likely indicates a **lower GI bleed**
- **Hematemesis** (vomiting blood)—indicative of **upper GI bleed**
- **Ethanol**—chronic heavy drinking associated with **gastritis, alcoholic cirrhosis,** and **esophageal varices**
- **Nonsteroidals**—chronic nonsteroidal usage is associated with **gastritis** and **peptic ulcer disease**
- **Pain**—raises the possibility that the lesion is inflammatory (e.g., **gastritis, esophagitis, peptic ulcer disease)**
- **Painless** GI bleeding—indicative of noninflammatory lesions such as **varices or cancer**
- **History of severe vomiting—may indicate a Mallory-Weiss tear**

Past medical history of cirrhosis, varices, bleeding disorders, epistaxis, peptic ulcer disease, or cancer may point to a particular pathological state.

- **Physical examination**
  - ***Vital signs*** **Tachycardia** (pulse above 100), **decreased blood pressure** (SBP below 100), or **orthostatic blood pressure changes** are **signs of** significant blood loss and **hemodynamic instability.**
  - ***Mass*** A mass may be indicative of a **cancer.**
  - ***Signs of cirrhosis*** Splenomegaly, caput medusae, and ascites may indicate **esophageal varices.**
  - ***Rectal examination and stool guaiac*** Check for rectal masses or active or **occult bleeding.**

■ Laboratory and Diagnostic Studies **Endoscopy** is the **diagnostic study of choice** for **identifying the cause of upper GI bleeding. Angiography may be used if endoscopy does not detect the** bleeding **lesion.**

All patients require a CBC and serum chemistries including liver function tests, PT, and PTT.

An NG tube should be placed into the patient's stomach and gastric fluid aspirated to **check for bright red blood (active bleeding) or coffee grounds (intermittent bleeding).** A negative aspirate does not eliminate upper GI bleeding. Be careful placing rigid NG tubes into patients with suspected esophageal varices.

- **Formulating the Diagnosis** History, physical, and results of endoscopy will establish the cause of the GI bleed. Concurrently with the diagnosis of GI bleeding, the physician needs to assess the degree of GI bleeding.

  Bleeding may be **classified as** follows:

  - **Occult bleed—no hemodynamic instability.** The active bleeding is not detectable. Hematocrit is within normal limits.
  - **Overt bleed—acute bleeding with stable vital signs** and hematocrit
  - **Massive bleed—acute bleeding** with **unstable vital signs**

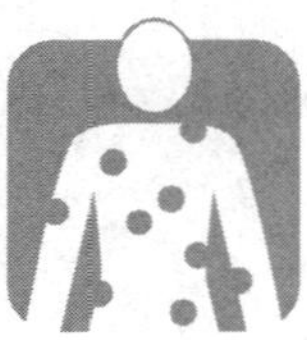

- **Evaluating the Severity of Problems** **Active GI bleeding is a medical emergency. Major complications of upper GI bleeding occur with aspiration of blood. Therefore, patients with no gag reflex or who are unconscious should be intubated.**

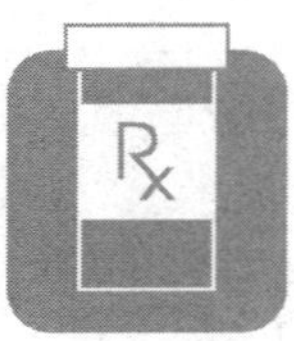

- **Managing the Patient**
  - **Occult bleeding** Endoscopy or barium contrast studies, on a nonemergent basis, should be done to establish the diagnosis. Treatment depends on the pathologic entity.
  - **Overt and massive GI bleeding** The **patient should be stabilized;** this includes the following:
    - **High flow oxygen**
    - **Vital signs frequently measured**
    - **IV fluids (two large bore IVs)**
    - **Packed red blood to keep hematocrit above 30**
    - **Continuous gastric aspiration (NG tube)**
  - **Endoscopy** should be performed **to localize the bleeding site. Treatment** depends on the pathologic entity.
    - **Gastritis** is treated with **H_2 blockers and antacids.** *H. pylori* infection is treated when identified.
    - **Bleeding ulcer** is treated by **heat cautery, injection sclerosis,** or, in cases of recalcitrant bleeding, **surgical partial gastrectomy.**
    - **Esophageal varices** are treated with **vasopressin, endoscopic sclerotherapy,** or a **Sengstaken-Blakemore tube** (esophageal tamponade tubes). In cases of recalcitrant bleeding, **portocaval shunt surgery** should be considered.

Multiple Choice Review Questions

1. A 20-year-old man presents with 3 days of fevers, chills, nausea, vomiting, and diarrhea. Symptoms began with nausea, and he has had protracted vomiting with some flecks of blood. He does feel light-headed upon standing and almost passed out this morning. The diarrhea is green in color, loose, and occurs at least every other hour. He denies any significant past history, and states that he does not remember anyone else eating the same food that he ate. On arrival, his blood pressure lying was 109/40, heart rate was 114, and respiratory rate was 23. Temperature is 101.5° F (38.6° C). Stool is green and guaiac positive. Emesis is bilious and has flecks of blood. The most likely etiology for this patient's acute illness is:

a. Gastric ulcer.
b. Ulcerative colitis.
c. Viral gastroenteritis.
d. *Staphylococcus aureus.*
e. Giardia.

2. The proper treatment plan for the patient in Question 1 includes all of the following *except:*

a. Intravenous hydration.
b. Checking stool for fecal leukocytes.
c. Lomotil to control diarrhea.
d. Complete blood count and SMA-7.
e. Stool cultures.

3. A 20-year-old woman was visiting Cancun last week. She presents to the ED with crampy abdominal pain, fever, and diarrhea. The symptoms also affected her "significant other.'' She denies vomiting. She denies any significant past history. Most likely organisms include:

a. *Trichonella spiralis.*
b. Cytomegalovirus.
c. *Helicobactor pylori.*
d. *Clostridium botulinum.*
e. *Bacillus cereus.*

4. For the patient in Question 3, stool is guaiac negative. Vital signs are within normal limits. Appropriate management includes:

a. Performing upper endoscopy with biopsy.
b. Isolating patient.
c. Discharging patient with instructions to drink lots of fluids.
d. Performing emergent colonoscopy.
e. Performing red cell–tagged scan.

5. A 55-year-old man presents with a 1-month history of difficulty swallowing. He denies any significant weight loss, fevers, or pain. He states that the problem swallowing has gotten gradually worse. He notes that he has been under lots of stress lately with work, his impending divorce, and the fact that his daughter is about to marry a Hell's Angel. His only significant history is anxiety disorder. Next step in management is:

a. Urgent psychiatric counseling.
b. CT scan of chest.
c. Anxiolytic medication, close monitoring, and office counseling.
d. Barium swallow.
e. Esophageal dysmotility study.

Chapter 7

Psychiatry

Adjustment Disorders

- **Basic Science** An **adjustment disorder** is defined as the **development of significant emotional or behavioral symptoms in response to an identifiable stressor.** Onset is **within 3 months** of the stressor, and **duration does not exceed 6 months,** unless the stressor becomes chronic or has enduring consequences. The adjustment **reaction is in excess of what would be expected given the circumstances** or leads to marked dysfunction in occupational or interpersonal areas.

  Disturbed emotions occurring in the course of an adjustment disorder include **depressed mood, anxious mood,** or a **combination of both. Disturbances of conduct also can develop** (violence, truancy, reckless driving, and so on). Other possible manifestations include physical complaints, social withdrawal, or a decline in work or academic performance. These patients do not meet the criteria for major depression or a full-blown anxiety disorder. If the disturbance lasts for more than 6 months in the absence of enduring stressors, other diagnoses should be considered.

- **History and Physical Examination** A careful psychiatric history and mental status examination must be performed. A physical examination is also helpful to rule out potential medical etiologies for the presenting symptoms.

- **Laboratory and Diagnostic Studies** As with all acute presentations, an organic etiology must be excluded; however, there are no laboratory evaluations routinely used to make the diagnosis of adjustment disorder.

- **Formulating the Diagnosis** Adjustment disorders should be **distinguished from normal bereavement** and from the expectable **emotional and behavioral disturbances that patients with personality disorders** exhibit when exposed to stress. Also, they differ from posttraumatic stress disorders. Complications of adjustment disorders include the development of a major depressive episode, substantial psychosocial impairment, and suicidality.

- **Evaluating the Severity of Problems** Suicidal patients require aggressive therapy, generally including admission.

- **Managing The Patient** The **treatment of adjustment disorders varies depending on the circumstances.** Supportive individual psychotherapy, group therapy, and marital and family therapies can all be of substantial help. Social interventions (e.g., changing work or school environments or temporary disability) may aid in speeding resolution of the disorder. **Pharmacologic interventions,** mostly **ben-**

zodiazepines, may assist in the **management of severe anxiety or agitation.** The use of **antidepressants** for **"adjustment disorder with depressed mood"** is frequently **practiced** but **remains controversial.** Psychiatric hospitalization is reserved for suicidal patients who require close supervision.

Anxiety Disorders

- **Basic Science** Anxiety disorders are a group of illnesses that afflict approximately 8% of the population. They are among the most common and most successfully treated of the mental illnesses. The major anxiety disorders are panic disorder, social phobia, obsessive-compulsive disorder, posttraumatic stress disorder, and generalized anxiety disorder. A blend of physical and environmental triggers combine to create anxiety disorders. Theories include genetic, biologic, psychoanalytic, and learning. **First-degree relatives of individuals with panic disorder are four to seven times more likely to develop panic disorder.**

- **History and Physical Examination** **Panic disorder is characterized by recurrent panic attacks.** These are discrete periods in which there is sudden onset of intense apprehension, fearfulness, or terror, often associated with a sense of doom. Physical symptoms, such as **shortness of breath, heart palpitations, chest pain, choking, or smothering sensations** are present. Often the patient **fears losing control or going crazy.** This may be **associated with agoraphobia,** which is anxiety about or avoidance of places and situations from which escape might be difficult. Agoraphobia can also occur without panic attacks. **Social phobia** is the **resultant anxiety provoked by exposure to certain types of social or performance situations** and often leads to avoidant behavior. Individuals with **obsessive-compulsive disorder** attempt to **cope with their anxiety by associating it with obsessions,** defined as persistent ideas, thoughts, impulses, or images that are intrusive, **or compulsive behaviors,** defined as ritualistic or repetitive behaviors or mental acts that serve to neutralize anxiety. The most common obsessions are repeated thoughts about contamination, repeated doubts, a need to have things in order, horrifying images, and sexual imagery. Common compulsions include hand washing, ordering, and checking. Mental acts typically are praying or counting or repeating words silently. Individuals usually recognize that the obsessions and compulsions are excessive and unreasonable. **Posttraumatic stress disorder** is characterized by the **reexperiencing of an extremely traumatic event, accompanied by symptoms of increased arousal, such as hypervigilence, irritability, and an exaggerated startle response,** and by avoidance of any stimuli associated with the trauma. **Generalized anxiety disorder** is characterized by **long-term, persistent, and excessive anxiety** and uncontrollable worry. Somatic symptoms and symptoms of increased arousal are present.

- **Laboratory and Diagnostic Studies** There are no specific lab tests for anxiety disorders. Testing for diagnoses of exclusion typically includes **thyroid function tests, CBC, glucose, electrolytes, and calcium.** An **ECG** should be considered. Panic attacks can be precipitated by carbon dioxide inhalation or sodium lactate infusion. Some studies suggest a higher incidence of mitral valve prolapse in those with panic disorder.

- **Formulating the Diagnosis** Medical illness and iatrogenic and toxicologic etiologies may be the cause of the patient's anxiety and need to be ruled out. **Differential diagnoses** are numerous. **Drug use and intoxication** with central nervous stimulants such as cocaine, amphetamines, and caffeine should be considered. **Medication side effects** may mimic symptoms of an anxiety disorder. **Drug withdrawal** states from central nervous system depressants such as barbiturates and alcohol may precipitate anxiety symptoms. General medical conditions can manifest clinically significant anxiety. Endocrine conditions include **hyperthyroidism, hyperparathyroidism, and pheochromocytoma.** Respiratory conditions that may manifest as anxiety are **chronic obstructive pulmonary disease, pneumonia, hyperventilation, and pulmonary embolus.** Cardiac disturbances, such as **arrhythmias, supraventricular tachycardia, and congestive heart failure, are among the differential.**
- **Evaluating the Severity of Problems** Coexistence of several anxiety disorders is common. There is strong comorbidity of major depression and substance abuse disorders in individuals with anxiety disorders. Suicide evaluation is always performed in these patients.
- **Managing the Patient** Psychologic and somatic treatments are known to be effective. Psychotherapy, behavioral therapy, cognitive therapy, and group therapy may be utilized alone or with medications. **Benzodiazepines** such as diazepam, lorazepam, alprazolam, and clonazepam effectively treat anxiety-related symptoms of hyperarousal. However, there exists potential of abuse and development of tolerance with this type of medication. **Antidepressants** are the **treatment of choice by psychiatrists for the majority of the anxiety disorders. Obsessive-compulsive disorder** can be successfully treated with **clomipramine, fluoxetine, or fluvoxamine.** Tricyclic antidepressants such as **imipramine, desipramine, or nortriptyline are frequently used for panic disorder.**

Attention Deficit Hyperactivity Disorder

- **Basic Science** Attention deficit hyperactivity disorder (ADHD) affects **3% to 5% of all school-aged children.** The prevalence of this disorder is **four to five times higher in boys than in girls.** ADHD is classified as a **disruptive disorder.** ADHD is thought to be a childhood disorder, however, long-term prospective studies indicate that a significant subset of children with ADHD continues to manifest symptoms into adulthood that are severe enough to warrant medication.

  ADHD is felt to be a disorder of the neurotransmitter systems in the brain that control sustained attention, impulsivity, and motor activity. The neurotransmitters dopamine and norepinephrine have been implicated as the primary modulators of attention and arousal in the central nervous system (CNS). The dopamine hypothesis of ADHD suggests that the **depletion of dopamine,** or the unavailability of dopamine to receptors, results in the alteration of sustained attention and arousal. There appears to be a genetic component to this disorder.

- **History and Physical Examination** Most children presenting for evaluation of possible ADHD symptoms do so between the ages of **6 and 12 years.** Typically symptoms of motor hyperactivity and impulsivity result in problematic behaviors in the classroom setting and at home. Core symptoms of ADHD are the following: **excessive or inappropriate motor activity, difficulty with tasks requiring sustained attention, and difficulties with impulse control.**

  Evaluation may consist primarily of obtaining clinical history, observing the patient, and obtaining collaborative data from other sources such as school.

- **Laboratory and Diagnostic Studies** There are no medical or psychological tests specifically designed for the evaluation of ADHD. The **diagnosis** must always be **made by combining the clinical presentation and history of the patient with supportive data obtained from other resources such as the classroom teacher or psychometric tests.**

- **Formulating the Diagnosis** The diagnosis of ADHD is based on the presence of core clinical symptoms. The existence of an underlying learning disability needs to be considered in every child presenting with ADHD symptoms. This may be assessed by careful consultation with the classroom teacher or may require more extensive psychological testing such as an IQ and/or achievement test. Children also will need assessment for underlying medical conditions, which may include lead toxicity, fetal alcohol syndrome, thyroid dysfunction, other endocrine abnormalities, and brain injury. Children with mental retardation, pervasive developmental disorder, and other developmental disorders are also frequently distracted and inattentive.

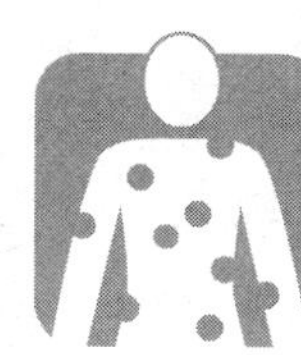

- **Evaluating the Severity of Problems** Consultation with a pediatric neurologist is indicated for those children who may have complex symptoms such as seizures or tic disorders. Child psychiatrists are frequently consulted when there are complex underlying psychosocial issues or other psychiatric diagnoses present.

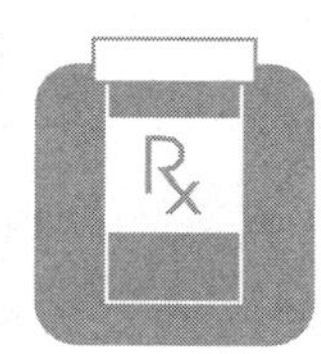

- **Managing the Patient** The treatment of ADHD requires an approach that coordinates school-based services, parents, physicians, and therapists. Most experts agree that **treatment should always be initiated with nonpharmacological interventions. Medications should be considered when these initial therapies are insufficient to allow for improved functioning of the patient.**

  Nonpharmacological treatments include classroom changes such as smaller teacher-to-student ratios, social skills training, and remedial services for any learning disabilities. Therapies that teach the child to think of alternative behaviors before acting on impulse have proven successful. Parent education and training focused on parenting skills can be extremely useful and are always a necessary part of any treatment program.

  For those children requiring pharmacologic interventions, medications can be highly effective in alleviating symptoms. **Stimulant medications are the mainstay of therapy and include methylphenidate, dextroamphetamine, and pemoline.** The most **common side effects of stimulant treatment may include in-**

somnia, decreased appetite, tics, irritability, and decreased growth rate.

Delirium and Dementia

- Basic Science **Delirium is characterized by an alteration in consciousness that is a direct consequence of a general medical condition.** Causes of delirium include substance intoxication or withdrawal, medications, metabolic abnormalities, infections, or frequently a combination of these factors. Patients with CNS pathology (e.g., dementia or CVAs) are more susceptible to a delirium.

  Dementia is an acquired, persistent syndrome of intellectual impairment with memory deficits and at least one of the following: aphasia, apraxia, agnosia, or a disturbance in executive function. The deficits represent a decline from a previous level of functioning and cause significant impairment in social or occupational functioning. Alzheimer's disease is the most common cause of dementia, representing 45% of those diagnosed. Vascular dementia represents approximately 15% of cases. Psychiatric disorders account for many potentially reversible dementias (10% of patients presenting with cognitive deficits). Other etiologies of dementia include alcoholic dementia, metabolic abnormalities (hypothyroidism, B_{12} deficiency), HIV, hydrocephalus, neoplasms, Huntington's disease, trauma, infections, Parkinson's disease, postanoxia, and combinations of these. Dementias can also be categorized as cortical (Alzheimer's, frontal lobe, degeneration), subcortical (Parkinson's, Huntington's, HIV, multiple sclerosis, Wilson's disease) or mixed (vascular).

- History and Physical Examination A complete history and thorough physical exam is imperative in evaluating any change in mental status. Onset, clinical course, and baseline functioning must be established. The history often differentiates delirium from dementia. **Delirium usually has an acute onset with a fluctuating course, sleep-wake alteration, impairment of attention, and disorientation.** Hallucinations, delusions, and psychomotor agitation or retardation may also be present. **Alcohol withdrawal delirium (i.e., delirium tremens) occurs 2 to 7 days after cessation of alcohol use and is marked by confusion, visual or tactile hallucinations, tremor, agitation, and autonomic hyperactivity.**

  Dementia commonly has an insidious onset and is chronically progressive with predominant deficits in memory. Hallucinations, psychomotor agitation, disorientation, and attentional difficulties occur less frequently. Delirium superimposed on dementia is not uncommon.

  A careful physical examination must be performed, including complete neurological and mental status exams, with emphasis on cognitive functioning. Mini-mental status exams may be helpful as a screening tool and gauge of severity of impairment. Formal neuropsychological testing may also be of value.

- Laboratory and Diagnostic Studies Laboratory data are useful in determining potential etiologies of mental status changes. The underlying cause of a **delirium** must be established to render appropriate treatment. The **workup may include assessment of vital signs,**

CBC with differential, sedimentation rate, complete blood chemistries, liver and renal function tests, arterial blood gas, urinalysis, urine drug screen, drug levels, appropriate cultures (e.g., urine, blood), an ECG, a chest x-ray, a CT scan of the head, and, if indicated, a lumbar puncture.

In evaluating **dementia,** potentially reversible causes must be identified. Laboratory evaluation includes the **workup just described with the addition of B_{12} and folate levels, thyroid function tests, HIV tests, heavy metals, RPR, and EEG as clinically appropriate.**

- **Formulating the Diagnosis** The differential diagnosis of an acute change in mental status includes delirium, dementia, and primary psychiatric disorders such as schizophrenia, mania, and dissociative disorders. Previous psychiatric history, course of illness, and impairment in level of consciousness and cognition will differentiate delirium from other psychiatric conditions.

  Dementia must be differentiated from benign senescent forgetfulness, which lacks the impairment in social or occupational functions of dementia; delirium that has an acute onset, and fluctuating level of consciousness; and depression, or "pseudodementia." Depressed patients may be distinguished from those with dementia due to history of previous depression, complaints about cognitive deficits, lack of effort in responses, and inconsistent results in cognitive testing. Patients with dementia may also develop depressive disorders. For these patients, treatment includes the use of antidepressants and/or electroconvulsive therapy (ECT).

- **Evaluating the Severity of Problems** Medical illness often complicates delirium and dementia; the magnitude of organic disease should be aggressively sought as symptoms may be masked by the patient's mental state.

- **Managing the Patient** The primary **treatment of delirium is the identification and correction of the underlying cause.** The patient should be placed in a safe environment with close monitoring. Restraints may be necessary to ensure safety. Vital signs, fluid input and output, and appropriate labs should be monitored. **Discontinue all nonessential medications. High-potency neuroleptics are used to treat target symptoms,** including hallucinations, delusions, paranoia, anxiety, and agitation. Doses of **haloperidol ranging from 0.5 to 5 mg** (depending on patient's age and severity of symptoms) **po, IM, or IV are given every 30 minutes for acute agitation.** The total dose in a 24-hour period necessary to calm the patient is then given in divided doses on a routine basis. Once the delirium has cleared, the neuroleptic should be tapered over the next 3 to 5 days. Occasionally lorazepam is added to the regimen in severely agitated patients. **Alcohol withdrawal delirium is treated using adequate doses of benzodiazepines** (higher doses than those used for uncomplicated alcohol withdrawal). Multivitamins, folate, and thiamine should also be given. Neuroleptics are sometimes needed to control severe agitation, delusions, or hallucinations.

  Reversible causes of dementia must be treated. Other etiologies are managed through supportive interventions. Treat any concurrent medical conditions. Behavioral difficulties may be successfully treated through nonpharmacological interventions such as frequent reorientation and the provision of a consistent, safe, familiar environment. When medications must be used, low-dose, high-potency neuroleptics are effective. Avoid low-potency neuroleptics and benzodiazepines. Educational and supportive interventions are helpful to the family and caregivers.

Eating Disorders: Anorexia, Bulimia, and Obesity

- Basic Science **Eating disorders** are characterized by **a pattern of disturbed or unconventional feeding habits.** They frequently begin in childhood and adolescent years, potentially lasting into adulthood. These disorders include anorexia nervosa (starvation) and bulimia nervosa (gorging and vomiting). Obesity is classified in the DSM-IV as a physical condition affected by psychological factors, which may result from overeating to reduce anxiety or a failure to perceive appetite, hunger, and satiety. A combination of genetic, biological, and environmental factors may predispose one to eating disorders.
- History and Physical Examination **Anorexia nervosa** is characterized by a **body weight 15% below expected normal for age and height.** Serious cognitive distortions exist, including descriptions of **intense fears of gaining weight** or becoming fat, even though underweight. **A disturbance of perception** of **one's body weight,** shape, or size leads to claims of "feeling fat" even when emaciated. In females, **amenorrhea,** or evidence of three consecutive missed menstrual cycles, is included diagnostically. This illness is common to middle and upper socioeconomic groups, models, and ballerinas. The onset of anorexia is in late adolescence; 90% of patients are female.

Anorexia nervosa is **potentially fatal in nearly one fifth of cases.** It is more common in industrial societies with an abundance of food and where attractiveness is linked to being thin. Resulting amenorrhea is usually secondary to weight loss, and in prepubescent females menarche may be delayed by this illness. It is rare for the victim of anorexia to complain of weight loss; most individuals are brought to professionals by friends or family after a substantial weight loss has occurred. Patients with anorexia nervosa **maintain considerable denial and** are prone to a **lack of insight.** Therefore it is necessary to obtain further history from the family or other outside sources to assess the degree of weight loss.

Bulimia nervosa combines the **morbid fear of obesity with an equally compelling passion for eating.** It is manifested by **recurrent episodes of binge eating, lack of control, overeating, and routine self-induced vomiting** (purging). To prevent weight gain, bulimic patients follow strict diets, **abuse laxatives,** and exercise excessively. Along with marked over-concern with body shape and weight, these individuals ingest sweet, high-caloric food such as ice cream and cake.

Individuals with bulimia nervosa typically are female, are within normal weight range, and have a higher rate of mood and anxiety disorders than average. **Substance abuse involving alcohol and stimulants** occurs in about one third of cases.

Self-induced vomiting can lead to **finger lacerations** and **calluses, dental caries,** and characteristic "moth-eaten" teeth. Further common findings include **salivary gland swelling** and potentially lethal **metabolic alkalosis** as a result of repeated vomiting. Potentially **fatal consequences** include **esophageal tears and cardiac arrhythmias.** Cardiac myopathies may occur with the abuse of syrup of ipecac.

Obesity has been defined as a **body weight 20% over that found in standard height-weight tables.** Obesity may contribute to various medical ailments, including orthopedic abnormalities, hypertension, glucose intolerance, and reduced respiratory capacity, which may progress to Pickwickian syndrome.

■ Laboratory and Diagnostic Studies Laboratory abnormalities found in **anorexia nervosa** include subtle hematological (anemia) and chemical profile changes. **Elevated kidney function studies** due to dehydration, **metabolic abnormalities,** and **ECG changes** (sinus bradycardia) may also be present.

Purging behavior in bulimia nervosa can lead to **electrolyte abnormalities** including **lowered serum sodium, potassium, and chloride.** Loss of stomach acid via emesis produces a **metabolic alkalosis;** and **elevated serum amylase** may be found with elevated salivary enzymes from excess vomiting.

Diagnostic studies (including skeletal imaging, pulmonary function tests, and a cardiovascular workup) may be necessary to evaluate the potential consequences of **obesity.** Thyroid function studies may help to define the etiology of obesity.

■ Formulating the Diagnosis **Anorexia nervosa** must be **differentiated from bulimia** as well as **other causes of weight loss.** Organic etiologies include **hypothalamic-pituitary axis tumors, Addison's disease, hyperthyroidism,** and various **GI disorders.** In addition, psychiatric conditions such as depression, drug dependence, somatization disorders, and psychotic disorders, with delusions regarding food, may result in significant weight loss. **Bulimia** must be **distinguished from anorexia and psychiatric disorders** including depression and dysthymia. **Organic causes of obesity include hypothalamic lesions and hypothyroidism.**

■ Evaluating the Severity of Problems Patients with anorexia nervosa can become so malnourished that admission is necessary for parenteral hyperalimentation. Serious electrolyte derangements may occur in bulimics secondary to laxative abuse and vomiting; these should be sought and corrected.

■ Managing the Patient Depending on the degree of weight loss in anorexia, admission to an inpatient psychiatric or medical unit may be required. The mainstay of treatment focuses on **assured weight gain** and monitoring and correction of metabolic complications. Treatment includes setting goals of desired weight with supervision of meals. In severe cases **nutritional supplements and nasogastric feedings may be in order.** Pharmacotherapy has limited efficacy; periodic benefits can be made with antidepressants if depression coexists. **Psychotherapy is aimed at changing attitudes and distorted body images.**

In **bulimia nervosa,** hospitalization due to life-threatening electrolyte and metabolic abnormalities may be required. This treatment mode often provides greater benefits than with anorexia. **Psychotherapy is aimed at normalizing eating habits and attitudes about food** and the individual's pursuit of ideal body weight.

In the treatment of **obesity, low-calorie diets** offer immediate, accessible weight loss methods. Unfortunately, these results are often temporary. Such diets lower the body's resting metabolic rate to an energy-conserving level that thwarts further desirable weight loss. One way to avoid this problem is **regular exercise** to return the resting metabolic rate to normal levels. A combination of **social supports, group therapy, self-help programs (such as Weight Watchers), and behavioral therapies** have shown to be more effective than pharmacologic treatment. Unfortunately, after 5 years most people, despite treatment, have regained all the weight they had lost.

Studies reveal that it is easy to take off weight but very difficult to keep it off. The short- and long-term effects of therapies for obesity vary greatly. **Appetite suppressants are moderately successful, though only while being taken.** In the long term, behavior therapies and cognitive treatments tend to be more efficacious than medication, but all three treatments generate similar and modest outcomes.

Mood Disorders

- **Basic Science** Mood disorders are characterized by a disturbance in mood along with changes in activity level, cognitive abilities, and vegetative functions. There are primarily two types of mood disorders: **major depressive disorder (unipolar depression) and bipolar disorder (manic-depressive illness).** Whereas patients suffering from major depressive disorder experience single or recurrent episodes of depression, those suffering from bipolar disorder experience episodes of mania with or without episodes of depression. The lifetime prevalence for **major depressive disorder** ranges from **10% to 25% for females and 5% to 12% for males. The onset is frequently between the ages of 20 and 50.** The lifetime prevalence for **bipolar disorder** is equal for men and women and ranges from **0.5% to 1.6%. The onset usually occurs in the second or third decade of life.**

  There are three additional categories of mood disorders: **hypomania, cyclothymia, and dysthymia.** Hypomania consists of manic symptoms that do not meet the full criteria for a manic episode. Cyclothymia and dysthymia are disorders that represent less severe forms of bipolar disorder and major depressive disorder, respectively.

  The precise causes of mood disorders are unknown. Both biologic and psychosocial factors have been implicated. Promising but unproven biological theories focus on brain **norepinephrine and serotonin abnormalities.** Mood disorders tend to **run in families.** The evidence of heritability of bipolar disorder is stronger than that for major depressive disorder. However, the mode of genetic transmission remains uncertain.

- **History and Physical Examination** The central feature of a **major depressive disorder** is either **depressed mood or loss of inter-**

est or pleasure in nearly all activities for at least **2 weeks.** Other symptoms associated with depression include **anxiety or agitation, changes (increase or decrease) in appetite, changes in weight, insomnia or hypersomnia, and psychomotor agitation or retardation. Decreased energy, feelings of worthlessness or guilt, difficulty thinking, concentrating, or making decisions, and recurrent thoughts of death or suicide may be present.** Occasionally depression may be associated with **psychotic symptoms in the form of delusions and hallucinations.** The most serious consequence of a major depressive episode is attempted or completed **suicide.** Suicide risk is especially high for individuals with **psychotic features, a history of previous suicide attempts, a family history of suicide, or a history of substance abuse.** Between 10% and 15% of patients with mood disorders commit suicide. Additional signs and symptoms associated with depression include decreased libido, inability to experience pleasure (anhedonia), and poor concentration. These symptoms frequently result in neglect of work, family, and friends and lead to social withdrawal and isolation.

Patients with **bipolar disorder** may suffer from depression as well as manic episodes. The hallmark of a manic episode is **elevated, expansive, or irritable mood.** During the period of elevated mood, patients may also experience **inflated self-esteem or grandiosity, decreased need for sleep, excessive and pressured speech, flight of ideas, distractibility, and psychomotor agitation.** Expansiveness, unwarranted optimism, grandiosity, and poor judgment often lead to an imprudent involvement in pleasurable activities such as buying sprees, reckless driving, poor business investments, and indiscriminate sexual behavior. Impairment is usually severe enough to cause marked impairment in functioning or to require hospitalization to protect the individual from the negative consequences of actions that result from poor judgment. Patients may also develop **psychotic symptoms in the form of delusions and hallucinations.** Occasionally patients may develop **catatonic symptoms in the form of stupor, mutism, negativism, and posturing.**

Acquiring a history of past psychiatric illnesses, including response to medications or treatment, is essential. A family history regarding psychiatric disorders is important. Information about alcohol and drug abuse must also be included. The presence of medical problems or recently prescribed (or nonprescribed) medications may suggest a nonpsychiatric etiology for the mood disorder. Childhood events including physical or sexual abuse, as well as current life stressors, might play an important role in the current mood disorder.

Careful attention must be paid to the vital signs and neurological examination. A detailed mental status examination must search for any signs of altered sensorium or prominent cognitive deficits. The clinician must address the issues of suicide, homicide, insight, and judgment.

■ **Laboratory and Diagnostic Studies** The clinician's first task is to separate primary mood disorders from those due to physical diseases and drug ingestion. Routine lab studies including **CBC, electrolytes, liver function tests, and urinalysis** must be obtained. **Urine**

drug screen, T_4, and TSH levels must be obtained. An **ECG** is useful especially prior to starting medications. Radiological studies such as **CT or MRI** may be obtained in certain cases to rule out intracranial causes of depression and in cases where electroconvulsive treatment (ECT) will be given.

Psychological testing can be useful in the diagnosis of mood disorders. These can help clinicians detect the presence of suicidal ideation and psychotic thought processes and give a profile of the patient's personality style. They can also be helpful in differentiating depression from dementia.

■ **Formulating the Diagnosis** The clinician must always consider possible organic etiologies for mood disorders. A variety of medical and neurological conditions and pharmacological agents cause symptoms of both depression and mania. The following is a list of causes of mood disorder symptoms:

- **Neurologic disorders:** Parkinson's disease, Huntington's chorea, Wilson's disease, stroke, infections, multiple sclerosis, and epilepsy
- **Endocrine diseases:** thyroid disorders (hypothyroidism and hyperthyroidism), adrenal gland disorders (Cushing's and Addison's disease), and other metabolic abnormalities
- **Infectious and inflammatory diseases:** mononucleosis, systemic lupus erythematosus, acquired immune deficiency syndrome (AIDS), and pneumonia
- **Miscellaneous:** cancer (especially pancreatic), liver disease, renal disease, and vitamin deficiencies (B_{12}, thiamine, and folate)
- **Pharmacologic agents:** antihypertensive medications and other cardiac drugs, sedatives and hypnotics, steroids and other hormones, antibacterials, and antifungals

Depression as a symptom can be present in a variety of psychiatric disorders. The common psychiatric conditions associated with mood symptoms are **alcohol and drug dependence, anxiety disorders, dementia, delirium, adjustment disorder with depressed mood, and bereavement.**

■ **Evaluating the Severity of Problems** The foremost decision a physician must make is whether to treat the patient on an outpatient basis or through hospitalization. The **indications for hospitalization include uncertainty of diagnosis, suicide or homicide risk, inability to care for oneself, and lack of a social support system.** Not infrequently patients with mood disorders refuse hospitalization on a voluntary basis. This may necessitate involuntary psychiatric hospitalization.

■ **Managing the Patient** The mainstay of treatment of depression is antidepressant medication. Antidepressants include **tricyclic and tetracyclic antidepressants, monoamine oxidase inhibitors (MAOIs), and selective serotonin reuptake inhibitors (SSRIs).** Until recently, the tricyclic antidepressants were used most frequently. These drugs include amitriptyline, nortriptyline, imipramine, desipramine, and doxepin. These drugs differentially **prevent reuptake of norepinephrine and serotonin.** The common side effects of tricyclic antidepressants include dry mouth, inability to urinate, constipation, hypotension, cardiac arrhythmias/conduction delays, and sexual dysfunction. Most of the tricyclic antidepressants have a **2- to 3-week onset of action.** Patients are usually started on a low dose that is gradually increased over a few weeks, according to clinical response and tolerance of side effects. The **monoamine oxidase inhibitors (MAOIs)** are usually not first-line treatment for depression because of the possibility of serious side effects. The drugs most commonly used are phenelzine, tranylcypromine, and isocarboxazid. MAOIs may be particularly useful for depressed patients who **present with atypical features of marked anxiety, excessive sleep, and increased appetite.** Patients taking MAOIs have to be on a special **tyramine-free diet.** These patients must avoid consumption of **aged cheese, red wine, pickled herring, chocolate, and other tyramine-containing food.** If MAOIs are given and tyramine is not avoided, patients can **develop severe hypertension,** which can lead to complications **including death.** Although a hypertensive crisis is the most serious side effect, hypotension is the most common side effect of the MAOIs. In recent years, **SSRIs (fluoxetine, sertraline, and paroxetine)** have become available for treatment of depression. SSRIs are associated with fewer side effects and are particularly safe for patients with cardiac disease. Common side effects associated with the SSRIs include **gastrointestinal side effects and sexual dysfunction.**

The choice of an antidepressant for a particular patient is determined by a history of response to a particular agent and the side effect profile of the medication. The most common clinical mistake leading to **an unsuccessful trial of an antidepressant drug is the use of too low of a dosage for too short of a time.** Unless adverse effects prevent it, the dosage of an antidepressant should be raised to the maximum recommended dosage and maintained at that level for **at least 4 weeks before a drug trial is considered unsuccessful.** However, if a patient is improving clinically on a low dose, then the dosage should not be raised unless clinical improvement stops. In those patients who fail to respond to a particular antidepressant, a different class of antidepressant medication should be tried. In treatment-resistent patients, a trial of a combination of antidepressants as well as lithium, thyroid hormone, or other adjunctive therapies should be considered. **ECT is recommended for medication-resistant or acutely suicidal patients.**

Lithium carbonate and citrate are the standard drugs used for the treatment of manic episodes. Most manic patients, however, may require **short-term use of antipsychotics or benzodiazepines** at the initiation of treatment to control psychotic symptoms and agitation. Noncompliance is a particularly difficult problem in the treatment of manic patients, and the clinician should always be aware of this potential cause for lack of response. The **second line of drugs in the treatment of mania** is **carbamazepine and valproic acid.** Patients with bipolar disorder benefit from prophylactic treatment with any of the medications just mentioned.

Extensive research has shown that a **combination of pharmacotherapy and psychotherapy is the most efficacious treatment for mood disorders. Interpersonal and cognitive therapists** have developed approaches specifically for the treatment of depression. **Insight-oriented psychoanalytically based psychotherapy, family therapy,** and occasionally **behavior therapy** can also be useful.

Personality Disorders

- **Basic Science** At present little can be definitively stated about the exact etiology of personality disorders. The cause is likely multifactorial, including developmental and genetic factors. With certain disorders these relationships are better understood. For example, persons with a diagnosis of borderline or antisocial personality disorder are more likely than the general population to have a history of childhood abuse, neglect, and poor parenting. On the other hand, twin and adoption studies with antisocial personality disorder indicate the presence of heritable genetic factors.
- **History and Physical Examination** Although the general medical history and physical examination should be performed in all individuals, there are no pathognomonic findings in patients with isolated personality disorders. The psychiatric history is discussed in the following sections.
- **Formulating the Diagnosis** All **personality disorders** demonstrate **enduring** and **inflexible patterns of experiences and behavior** that are **inconsistent with cultural expectations.** These must occur in at least two of the following areas: cognition, emotional response, impulsivity, and interpersonal relationships. They also must cause significant dysfunction and distress for the individual in respect to academic, occupational, and social functioning. A pervasive pattern of maladaptive functioning must be present. The existence of a few isolated personality traits is not sufficient to make a diagnosis of a personality disorder. The symptoms often **begin in adolescence or early adulthood.** They cannot be accounted for by another mental disorder, substance abuse, or a general medical or neurologic condition that affects CNS function.

  Personality disorders are divided into three primary clusters that contain disorders with similar symptoms.

  - ***Cluster A*** disorders include **paranoid, schizoid, and schizotypal personality disorders.** As a group, individuals with a **cluster A** diagnosis are **suspicious, eccentric,** and **socially isolative.**
  - ***Cluster B*** diagnosis individuals are distinguished by **impulsive, dramatic** behaviors and a **decreased capacity for empathy.** This cluster includes **antisocial, narcissistic, borderline, and histrionic personality disorders.** Substance abuse and mood disorders are common comorbid conditions, especially with borderline and antisocial personality disorder.
  - ***Cluster C*** diagnosis individuals are typically excessively **anxious, fearful, or perfectionistic. Avoidant, dependent,** and **obsessive-compulsive personality disorders** are included in the C cluster. Individuals with a

cluster C diagnosis may suffer from a comorbid anxiety or mood disorder.

In this review, only key features of each disorder will be emphasized.

- **Paranoid personality disorder** ***Paranoid personality disorder*** is characterized by **pervasive suspiciousness, distrust, and perception of being attacked.** These individuals are typically cold and aloof and are prone to becoming excessively angry. They may bear grudges against others or become litigious. The diagnosis differs from delusional disorder and schizophrenia by the **absence of frank delusions.**
- **Schizoid personality disorder Schizoid personality disorder** includes persons who are **socially withdrawn** exhibiting **decreased emotional responsiveness and ability to relate to others.** They frequently do not experience their lack of intimate relationships as disturbing and prefer solitary activities.
- **Schizotypal personality disorder Schizotypal personality disorder** is similar to schizoid personality disorder in that these persons are **socially isolated.** They experience discomfort with interpersonal relationships and demonstrate **odd and eccentric behavior, speech,** and **thoughts,** and they **may experience perceptual disturbances** such as telepathy or ideas of reference. They may present with frank psychotic symptoms, but these are brief and lack severity in comparison with schizophrenia. Family history studies support a relationship between this disorder and schizophrenia.
- **Antisocial personality disorder Antisocial personality disorder** is characterized by a pervasive **disregard** and **violation** of the **rights of others** with little or **no remorse.** Individuals with this diagnosis often behave impulsively and irresponsibly. They exhibit outbursts of **anger** and frequently have **legal difficulties.** This pattern of behavior must have started prior to age 15 (history of conduct disorder); however, the diagnosis requires that the patient be at least 18 years of age.
- **Borderline personality disorder Borderline personality disorder** is similar to antisocial personality disorder in that there is a history of a **pattern of instability and lack of control over impulses.** In addition, individuals with this disorder suffer from **disturbed interpersonal relationships, self-image, and affect** (especially anger). They may **frequently threaten or attempt suicide** and make excessive, often manipulative demands in close interpersonal relationships or of care providers. They may complain of chronic feelings of emptiness and boredom and may suffer from dissociative or transient psychotic episodes.
- **Histrionic personality disorder Histrionic personality disorder** is characterized by a pattern of **excessive attention seeking and exaggerated emotional responses,** which are frequently superficial and fleeting. Often they exhibit seductive, uninhibited, and lively behaviors.
- **Narcissistic personality disorders Narcissistic personal-**

ity disorder** includes individuals with a **heightened need for admiration.** The disorder is further characterized by a **sense of uniqueness, grandiosity, entitlement,** and decreased empathy and exploitation of others. These individuals usually lack the impulsivity seen in antisocial personality disorder.

- **Avoidant personality disorder Avoidant personality disorder** includes patients who exhibit a pattern of **feelings of inadequacy, excessive sensitivity to criticism** or evaluation, and **social inhibition.** Unlike the person with schizoid personality disorder, these individuals **desire close relationships** and are distressed by their interpersonal difficulties.
- **Dependent personality disorder Dependent personality disorder** is often coexistent in individuals with a diagnosis of avoidant personality disorder. It is given when there is a pervasive pattern of **submissive behavior** and an **exaggerated need to be cared for,** or **fears of separation.** They frequently rely on others to make decisions and require excessive reassurance to do so on their own.
- **Obsessive-compulsive personality disorders Obsessive-compulsive personality disorder** is characterized by a **preoccupation with perfection, orderliness, rules, and interpersonal control,** which, because of rigidity, **ultimately interferes with efficiency,** task completion, flexibility, and personal relationships.

■ **Evaluating the Severity of Problems** Nearly all patients with personality disorders are treated as outpatients.

■ **Managing the Patient** Individuals with a cluster A personality diagnosis frequently do not seek psychiatric treatment. In the general medical setting they are managed by maintaining a straightforward, supportive stance. **Antipsychotic medications** may be helpful with **paranoid and schizotypal personality disorders.** There is **little evidence** that individuals with **antisocial personality disorder respond to conventional psychiatric treatment.**

Cluster B disorders may respond to **limit setting,** a **clear explanation of required expectations** for treatment, and **strict adherence to professional boundaries** by all health care providers. Individuals who suffer from **narcissistic, borderline, and histrionic personality disorders** can improve with **psychotherapy.** A variety of medications such as tricyclic antidepressants, serotonin reuptake inhibitors, monoamine oxidase inhibitors, neuroleptics, anticonvulsants, and benzodiazepines have been useful in persons suffering from borderline personality disorder. These medications help with mood instability, suicidality, impulsive behaviors, and dissociative and psychotic symptoms.

Individuals with a cluster C diagnosis generally respond well to a variety of standard psychotherapies.

In individuals suffering from a personality disorder, one should be careful to **avoid attributing all of their complaints to the personality disorder.** An attempt must be made to diagnose other comorbid psychiatric conditions, such as anxiety and mood disorders, which respond well to somatic treatment.

SCHIZOPHRENIA

- **Basic Science** Schizophrenia is a major mental disorder of uncertain etiology that afflicts approximately **1% of the population.** Men and women are equally vulnerable. **Increased activity of the neurotransmitter dopamine in the brain** has been implicated etiologically, leading to the "dopamine hypothesis" of schizophrenia. Other neurotransmitters are likely also involved. Imaging of the brain with various techniques suggests involvement of the frontal, prefrontal, and temporal areas of the brain in the development of the disorder. **Evidence of genetic risk is clear** but incompletely understood. Concordance among monozygotic twins is about 50%, implicating both genetic and environmental factors. The illness most often **presents as the result of converging genetic, neurobiologic, and psychological vulnerabilities and psychosocial stressors.**

- **History and Physical Examination** Schizophrenia is a clinical diagnosis made by a combination of history and physical examination. It is a pervasive disorder that involves multiple areas of psychological function, may permanently change personality, and often leads to a gradual **deterioration of motivations and abilities to work, play, and relate to others.** Psychological processes affected include disturbances in the content of thought (e.g., **delusions**); thought disorder (**abnormal flow of ideas**); perceptual disturbances (e.g., **hallucinations**); loss of ability to generate or regulate affect (e.g., **flattening, blunting, gross inappropriateness**); abnormal psychomotor behavior (e.g., **odd mannerisms, agitation, catatonia**); **amotivation;** and losses of a sense of self, volition, and relatedness.

  Disturbances in the form and content of thought are central dimensions of the illness. Classically, schizophrenic thought is characterized by **loose associations,** meaning that the normal connections between ideas are lost. Other types of disordered thought found in schizophrenia include **clang associations, neologisms,** and **blocking.**

  Delusions are false beliefs that are unshaken by reason. They may have a single theme or multiple themes, including **religious, grandiose, somatic, and sexual preoccupations.** Schizophrenic **delusions tend to be bizarre;** they are usually unconnected to mood and are termed *mood incongruent.* **Ideas of reference** (perceiving random events as intended to affect only oneself) and beliefs that people or supernatural forces can directly access or **control thoughts** are common.

  Hallucinations in schizophrenia may involve any of the five senses, although **most often are auditory.**

  Many patients are plagued by a collection of symptoms called *negative* or *deficit symptoms* of schizophrenia. Anergy, amotivation, apathy, and ambivalence are typical to lesser or greater degrees. **Negative symptoms** are **less likely to improve with antipsychotic medications** than are positive symptoms such as hallucinations and delusions.

- **Laboratory and Diagnostic Studies** There are no pathognomonic features or specific tests to define the illness. Evaluation should include careful assessment to exclude organic causes. Complete general physical and neurologic exams and routine laboratory tests (including CBC, chemistries, thyroid and syphilis testing, sedimentation rate, urinalysis, and drug screen) are essential. Most psychiatrists today would agree that MRI scanning to exclude masses, vascular lesions, and evidence of white matter or degenerative processes is necessary in first presentations or if there are atypical features. Other mimics of schizophrenia include amphetamine- or cocaine-induced psychosis, subcortical dementias (e.g., Wilson's and Huntington's disease), and mood disorders (e.g., major depression and bipolar disorder with psychotic features). Other major psychiatric illnesses share features of schizophrenia, including schizoid, schizotypal, and paranoid personality disorders, the delusional disorders, and schizoaffective and schizophreniform disorders.

■ **Formulating the Diagnosis** Schizophrenia typically presents in late adolescence or early adulthood. There are three distinct phases of the illness: prodromal, active, and residual. To **diagnose schizophrenia** a patient must meet **two criteria: (1) an active phase of the illness with prominent psychotic symptoms lasting more than 1 week, unless symptoms are interrupted by effective treatment; and (2) a total duration of symptoms from all phases of at least 6 months.** The active phase of the illness is defined by the active psychosis: the typical delusions, hallucinations, grossly disorganized thought, affective disturbance, and abnormal psychomotor behavior previously described. Prodromal and residual phases share many signs and symptoms. The prodromal phase may last many weeks or months, often so insidious as to be recognized only in retrospect. Individuals gradually become more aloof, perplexed, preoccupied, peculiar in behaviors and beliefs, and less willing or able to participate in school or work. **Attention to grooming, dress, and other social norms may be lost.** Families may be more puzzled than worried. The residual phase is present when active phase symptoms are essentially absent, but oddities in speech and behavior and serious disturbances in social and role functioning persist. Negative symptoms are often prominent in this phase and represent much of the enduring disability of schizophrenia.

The course of schizophrenia is variable. Common complications include major depression, chemical dependency, suicide, chronically poor social function, and homelessness.

There are **five subtypes of the disorder:** paranoid, catatonic, disorganized, undifferentiated, and residual.

- **Paranoid** The **paranoid type** is characterized by preoccupation with systematized delusions or auditory hallucinations with a single theme. Classic themes include persecution by agencies or supernatural forces and religious, scientific, and sexual ideas. This type carries the **best prognosis.**
- **Catatonic** The **catatonic type** shows prominent abnormalities of psychomotor behavior, such as posturing, mutism, and stupor.
- **Disorganized** The **disorganized type** is typified by grossly disorganized delusions and hallucinations of various themes, and grossly disordered thought and speech that can lead to incoherence. This subtype carries the **poorest prognosis.**

- **Undifferentiated** The **undifferentiated subtype** is used to describe cases that do not clearly fit the other patterns.
- **Residual** The **residual subtype** describes individuals that no longer have active features of the illness but are disabled with negative symptoms and may continue to have considerable oddities of speech and behavior.

■ **Evaluating the Severity of Problems** Patients with homicidal or suicidal ideation require immediate evaluation.

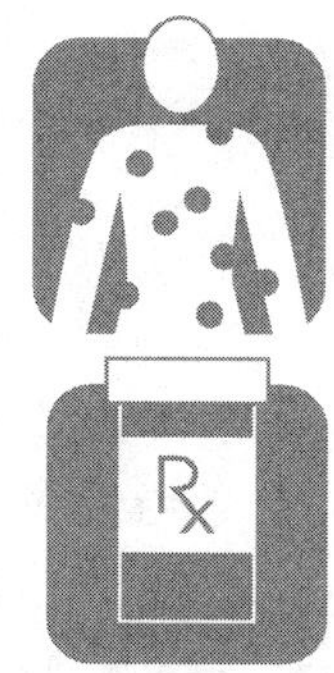

■ **Managing the Patient** Pharmacological interventions in the active phase are critically important. Acute treatment usually includes antipsychotic medications. **Most antipsychotics block postsynaptic D_2 dopamine receptors.** Newer atypical antipsychotics, such as clozapine and risperidone, do not have this mechanism of action, creating uncertainty about necessary pharmacologic effects. To establish acute antipsychotic effects, a dose the equivalent of 300-400 mg of **chlorpromazine** is usually needed. Benzodiazepines are useful adjuncts for sedation and may allow lower doses of antipsychotics to be effective in acute psychosis.

Many antipsychotics are available. They are arranged on a continuum from **high potency (e.g., haloperidol) to low potency (e.g., chlorpromazine). High-potency agents** have a low incidence of sedation, blood pressure, and anticholinergic effects but are more likely to **cause extrapyramidal side effects and akathisia** (a syndrome of motor restlessness). **Low-potency agents conversely may cause considerable sedation, orthostasis, and uncomfortable anticholinergic side effects** but less commonly muscular problems. **Extrapyramidal side effects** are often easily **treated with anticholinergic agents** (e.g., benztropine and trihexyphenidyl) by opting for a lower-potency agent **or lowering the dose of the antipsychotic.**

All **antipsychotic side effects** are important, but three are especially so. **Acute dystonia** is usually an early side effect that may be quite dramatic. Muscular males are particularly at risk. Signs may develop rapidly, including general muscular rigidity, torticollis, and oculogyric crisis. Laryngeal spasm can lead to airway obstruction. IM or **IV diphenhydramine (25-50 mg) or IM benztropine (2 mg) will generally rapidly reverse dystonia.** Following an episode of acute dystonia, oral anticholinergic agents should be continued for at least several days to prevent recurrence. **Tardive dyskinesia** is a syndrome of involuntary movements typically involving muscles of the lips, face, and tongue but potentially affecting any skeletal muscle. Risk factors include high-potency antipsychotics, total dose and duration of exposure, older age, and female sex. If identified early, the dyskinesias most often remit with reduction or elimination of drug treatment. The movements may be permanent; 15% to 20% of patients exposed for more than 1 year will eventually show signs of tardive movements. **Neuroleptic malignant syndrome is a rare but potentially fatal side effect that may present any time during antipsychotic treatment.** It should be considered in any patient exposed to dopamine blockade or withdrawal of a dopamine agonist. Presentation includes fever, muscle rigidity, unexplained elevation of white cell counts and serum creatine kinase, and delirium. There is no definitive or reliable treatment beyond hydration and symptomatic management, however, bromocriptine and dantroline can be helpful. Complications can lead to renal and hepatic failure, seizures, coma, and death.

Psychotherapies focus on helping the patient and family understand the nature of the illness, supporting them in learning to live with and around disabilities, coping with a chronic disease, and connecting with and monitoring rehabilitation efforts.

Relapse prevention is a critical part of treatment. Antipsychotics clearly prevent return of symptoms and may need to be continued indefinitely. **Family interventions aimed at decreasing** undue amounts of **expressed emotion,** which lead to tension in the environment, have been associated with reduced rates of relapse.

Somatoform Disorders

■ Basic Science ***Somatoform disorders* consist of a group of disorders characterized by physical symptoms resembling medical disease but without sufficient organic pathology** or a known pathophysiological mechanism. These include somatization disorder, conversion disorder, pain disorder, hypochondriasis, and body dysmorphic disorder. Syndromes that appear somatoform in nature but do not conform to the disorders just listed are usually termed *undifferentiated.*

- In ***somatization disorder,* the patient has a long history of physical complaints affecting multiple organ systems** (e.g., pain, GI, GU, pseudoneurological) that remain unexplained and lead to distress, disability, and/or medical care. The illness is chronic in nature and has a female preponderance. Onset is early in life. A fluctuating course is common, and complete remissions are rare. Morbidity is significant in occupational, interpersonal, and medical spheres.
- ***Pain disorder* is characterized by a preoccupation with pain for more than 6 months, without a clear organic explanation.** Disability or distress is disproportionate to the pain. Low back pain is most common. The age of onset is usually in the 40s or 50s. The disorder often leads to depression or dysthymia.
- Patients with ***conversion disorder* exhibit a loss of or alteration in physical functioning, suggesting a physical disorder** (usually pseudoneurological: pseudoseizures, pseudosyncopes, pseudoparaplegia, mutism, and the like). The disorder is more common in females and often correlates with a known acute psychosocial stressor. The symptoms usually resolve spontaneously, although they may become chronic and result in substantial sequelae (e.g., chronic pseudoepilepsy, prolonged limb weakness leading to atrophy).
- Patients with ***hypochondriasis*** exhibit a long-standing **preoccupation with, fear of, or belief that they have a serious disease.** They exhibit a ruminative, anxious, obsessive quality to health concerns, with only temporary response to reassurance. It is slightly more common in men and peaks in the 40s and 50s. These patients may also be prone to depression.
- ***Body dysmorphic disorder*** patients have an **excessive preoccupation with an imagined defect in appearance,** causing

marked distress and psychosocial impairment. They often pursue dental or surgical interventions to correct the "defect," leading to more dissatisfaction. In severe cases, morbidity is significant.

- **History and Physical Examination** The evaluation of somatoform disorder must include a **careful search for a possible medical explanation for the patients' symptoms.** A lack of objective signs of disease matched with a disproportionate magnitude of distress or disability should alert the physician to the possibility of a somatoform disorder. Other "somatizing" psychiatric disorders (e.g., depression, panic disorder) should be identified and treated. However, it is essential to remember that somatoform disorders **often coexist with medical illness and other psychiatric disorders.** Many of these patients are resistant to psychiatric care. They should be informed about the possibility of a psychological "component" to their symptoms early in the workup, thus minimizing the chances of resistance and denial. The exploration of family dynamics and a realization of the patient's unconscious motives will greatly assist in the diagnosis and management.

- **Laboratory and Diagnostic Studies** Patients with somatoform disorders often undergo extensive, unnecessary testing. Laboratory investigation should be performed as indicated by the history and physical examination.

- **Formulating the Diagnosis** These disorders must be differentiated from other "somatizing" illnesses, such as adjustment disorder (with physical complaints), major depression with somatic complaints ("masked" depression), anxiety disorders with physical symptoms (particularly panic disorder and generalized anxiety disorder), and psychotic disorders presenting with somatic delusions (e.g., schizophrenia or delusional disorder, somatic type). In the latter, somatic concerns are often bizarre and are impervious to logic or reason.

  Also important is the understanding that **somatoform symptoms are not willfully or deliberately generated** but are the **product of unconscious conflicts or psychological needs.** This clearly separates them from **malingering (deliberate feigning for specific gain) and from factitious disorders (intentional deception of health care providers for unclear motives other than to obtain medical care).**

- **Evaluating the Severity of Problems** Although these patients often feign illness when it is not present, they can present with legitimate serious medical disease, and therefore this possibility should not be dismissed.

- **Managing the Patient** Guidelines for treatment involve the development of a stable and trusting relationship with a **primary physician who can then "centralize" the patient's care.** A vital part of the physician's role is the **reduction of iatrogenic morbidity through *reasonable* workups and treatments.** A negotiated contract involving an acknowledgment of the physician's inability to "cure" the patient, an invitation to more realistically redefine goals, and an ongoing reassurance that medical care will not be withdrawn despite negative objective findings is helpful. Occasionally, regular-interval, symptom-independent visits can help reduce the patient's need for renewed somatization to obtain care.

Substance Abuse

- **Basic Science** Adverse consequences of alcohol and drug use include not only their well-known medical sequelae, but also psychiatric problems. Most psychiatric symptoms can be caused or exacerbated by alcohol or drugs.

  About 10 million people in the United States have alcohol dependence. About 9 million report using marijuana regularly, 1.6 million report using cocaine, and less than 1 million use heroin. Slightly more than 1 million people report using hallucinogens, inhalants, or other drugs.

  In response to equivalent doses of ethanol, women have higher blood ethanol concentrations than men, in part because of decreased gastric mucosal alcohol dehydrogenase activity. Alcoholic women develop cirrhosis more rapidly than do alcoholic men.

  Intravenous drug users are at risk for many infectious diseases, including human immunodeficiency virus (HIV), hepatitis, syphilis, gonorrhea, cellulitis, skin abscesses, endocarditis, pneumonia, and tuberculosis. Alcohol and drug users are also at risk for accidental and traumatic injuries.

  Familial patterns are often noticed for both alcohol and drug dependence. Genetic factors are involved in the transmission of alcohol dependence and are suspected in the transmission of other substance dependence.

- **History and Physical Examination** **Most alcohol and drug users use more than one drug,** and it is extremely common for **drug users to deny their use.** The clinician should ask about types of drugs used, amounts used, route of administration, age at onset of use for each drug used, means of obtaining drugs (or the money for them), settings in which drugs are used, sequelae of use, family history, past treatments, and periods of abstinence. The clinician should determine whether there has been progression of use, including increasing severity of symptoms and increasing frequency and amounts used over time.

  It is important to determine how long the patient has remained abstinent from drugs and alcohol and how he or she accomplished such abstinence. This will be helpful in deciding which treatment level the patient needs when referral is made. If during abstinence psychiatric symptoms gradually improve without medication, these symptoms might be secondary to alcohol or drug use.

  Physical examination should include special attention to signs of active drug use (needle tracks) and signs of treatable but easily overlooked illnesses related to alcohol and drug use, such as Wernicke's encephalopathy or hyperthermia. The examiner should also look for rhinorrhea, changes in pupil size, nystagmus, cutaneous abscesses, piloerection (gooseflesh), tremor, and an altered level of consciousness.

- **Laboratory and Diagnostic Studies** When evaluating a potential substance abuser, a urine drug screen should be considered. Presence of drugs (or their metabolites) in blood or urine is diagnostic. Elevation of gamma glutamyl transferase (GGT) is a useful indicator of excessive drinking. Mean corpuscular volume (MCV) may be elevated in alcoholics as well. Creatine kinase (CK) levels should be done when cocaine-induced rhabdomyolysis is suspected. Electrocardiograms and serial CK measurements can be helpful in evaluating the cocaine user who complains of chest pain.

■ **Formulating the Diagnosis** Substance use disorders are divided into substance abuse and substance dependence.

- **Substance abuse** includes impairment or distress from one or more of the following within 12 months: failure to fulfill social or familial role because of use, use in physically hazardous situations, legal problems related to use, and social or interpersonal problems related to use.
- **Substance dependence** includes three of the following symptoms within 12 months: tolerance, withdrawal, loss of control of use, efforts to cut down or control use, increased time spent to get or use drugs, activities given up because of drug use, and use despite medical or psychological consequences.

■ **Evaluating the Severity of Problems** **Patients who have needed medical management for alcohol withdrawal in the past are likely to need it again** in subsequent episodes of withdrawal. **Patients with unstable vital signs or medically complicated withdrawal should be admitted for detoxification.** Uncomplicated alcohol withdrawal can be accomplished on an outpatient basis. Consider psychiatric admission for patients whose substance abuse or overdose has caused psychotic symptoms, severe depression, suicidal risk, or has exacerbated a coexisting psychiatric illness. In emergency situations involving alcohol and drugs, the staff must rapidly assess the patient for vital sign instability and evidence of bleeding, head trauma, or coma. CNS depression is common with overdoses of sedative-hypnotics, opiates, and large amounts of alcohol.

■ **Managing the Patient** Withdrawal symptoms can develop within a few hours of last use when substances with short half-lives are used. Withdrawal symptoms from substances with long half-lives may develop after a much longer period, sometimes weeks, and might be misdiagnosed.

Alcohol and sedative-hypnotic withdrawal is serious and potentially life-threatening. Symptoms should be **treated early with benzodiazepines,** thiamine, folate, and multivitamins. One must continue to monitor vital signs and response to doses of administered benzodiazepines. Minor withdrawal syndrome can progress to major withdrawal syndrome and delirium tremens. Such progression can be prevented by administering appropriate doses of medication and providing supportive care. Waiting until symptoms become more severe or giving suboptimal doses of benzodiazepines can make major withdrawal more likely. **Phenytoin is ineffective for prevention of recurrent alcohol-related seizures.**

Outpatient treatment for benzodiazepine withdrawal is appropriate for those patients who have been using low to moderate doses of benzodiazepines on a long-term basis. A slow benzodiazepine taper over several weeks or months is often recommended. For barbiturate or mixed sedative-hypnotic withdrawal, phenobarbital is often used.

Opiate withdrawal is not life-threatening but may cause significant discomfort. For opiate withdrawal supportive care is usually all that is necessary. Administration of the α_2-adrenergic agonists clonidine or guanabenz can be effective. Nonsteroidal antiinflammatory drugs can be given for abdominal cramping. It is seldom necessary to give opiate analgesics.

Cocaine or other stimulant intoxication symptoms often resolve without any pharmacologic intervention. Severe agitation accompanying stimulant intoxication or withdrawal is often treated with benzodiazepines or with low doses of haloperidol. **Cocaine withdrawal can include severe dysphoria and suicidal ideation, requiring psychiatric admission.**

Treatment of substance use involves abstinence while participating in a program of education about addiction, coping skills training, family support and education, and sometimes adjunctive medications. Patients are particularly vulnerable to relapse in the first 12 months of abstinence from alcohol or drugs. Adjunctive medications that can be helpful in maintaining abstinence include disulfiram and naltrexone for alcoholism, and naltrexone for opiate dependence.

Multiple Choice Review Questions

1. A 19-year-old college freshman is brought to the hospital by his advisor. The student is convinced that a military surgeon has implanted a "memory incapacitator" in the brains of his dorm mates. The "waves" given off by these devices disrupt his concentration and sap his energy, leading to failing grades. He also hears the voices of two faculty members commenting on his thoughts. He becomes agitated in the emergency room and requires restraints. The differential diagnosis at this point includes all of the following *except:*

a. Amphetamine-induced psychosis.
b. Schizophrenia.
c. Malingering.
d. Brief reactive psychosis.
e. Schizoid personality disorder.

2. Which of the following steps is least likely to help clarify the diagnosis?

a. Dexamethasone suppression test
b. Further history from the family
c. Urine toxicology screen
d. Neurologic examination
e. HIV testing

3. All of the following are reasonable next steps in the management *except:*

a. Haloperidol 2 mg intramuscular.
b. Lorazepam 2 mg intramuscular.
c. Chlorpromazine 50 mg orally.
d. Nortriptyline 25 mg orally.
e. Trifluoperazine 5 mg orally.

Chapter 8

Rheumatology

Bursitis

■ **Basic Science** Bursitis represents **inflammation of a bursa,** the thin-walled pouches lined with synovial tissue that surround many articular joints. Bursae act to reduce friction on tendons or muscle as they move over bony prominences. The cause of bursitis is unknown but is **thought to be related to trauma, overuse, infection, or chronic joint disease such as rheumatoid arthritis.**

Acutely, the inflamed bursal lining secretes a serous effusion. Repeated trauma or many attacks of acute bursitis can lead to chronic bursitis. In **chronic bursitis,** the synovial lining proliferates and thickens, and **calcifications or adhesions may develop.** Pain from bursitis may limit the use of a joint; over time, this can result in restricted range of motion and disuse muscle atrophy.

Bursitis **most commonly affects the subacromial (subdeltoid) bursa** and the **trochanteric bursa** but may also strike the olecranon, prepatellar, calcaneal, ischial, and iliopsoas bursae. The bursa at the first metatarsal head and the bursae of the pes anserine may also be affected. Bursitis is **more common in males.**

■ **History and Physical Examination** **Pain, swelling, and limited range of motion are the signs of acute bursitis. Chronic bursitis** may present acutely or with **chronic pain and limitation of motion.**

■ **Laboratory and Diagnostic Studies** The evaluation of patients with acute olecranon or prepatellar bursitis should include aspiration to rule out septic bursitis. A synovial fluid white blood cell count exceeding 1000/μl indicates inflammation secondary to gout, rheumatoid arthritis, or infection. A white blood cell count in excess of 50,000/μl is diagnostic for septic bursitis. *Staphylococcus aureus* is the most commonly identified etiologic organism.

■ **Formulating the Diagnosis** **Bursitis must be distinguished from arthritis. Unlike arthritis, bursitis is more likely to begin abruptly, with swelling and focal tenderness.**

Subacromial bursitis presents as tenderness in the shoulder with pain aggravated by movement, especially abduction. Subacromial bursitis may be confused with supraspinatus tendinitis, which has a similar clinical presentation.

Trochanteric bursitis presents as gradually worsening pain in the area over the greater trochanter, which patients may describe as "hip pain." Pain is worsened in external rotation and by abduction against resistance. Tenderness over the entire lateral thigh may be seen. Patients prone to trochanteric bursitis include those with osteoarthritis of the hip or spine, scoliosis, or leg-length discrepancies.

Olecranon bursitis presents with swelling and tenderness (i.e., a "goose egg") over the elbow with normal range of motion.

Pes anserine bursitis is seen in middle-aged or older, overweight women who may also have osteoarthritis of the knee. Pain is often described as "knee pain" and is worsened by climbing stairs and rising from a chair.

- **Evaluating the Severity of Problems** Septic bursitis is more worrisome in patients compromised by immunosuppression. These patients should be admitted for aspiration and parenteral antibiotics.
- **Managing the Patient** **Immobilization, joint rest, and NSAIDs** may be all that are needed to treat acute bursitis. Weight loss is helpful for bursitis affecting weight-bearing joints. **Injection of the bursa with corticosteroid and anesthetic** followed by joint rest for 1 or 2 days is another method of treatment. **Chronic bursitis may also be treated with steroid injections,** but these become less helpful after repeated use; they **should not be repeated more than once every 3 months. Physical therapy is beneficial** in acute and chronic bursitis by helping to prevent joint contracture and by conditioning the joint to prevent reinjury. **Bursitis due to infection** is treated with antibiotics and aspiration. Septic bursitis **must always be ruled out before an area is injected with corticosteroid.**

FIBROMYALGIA

- **Basic Science** Fibromyalgia (fibrositis) is a disease marked by **fatigue and muscle aches.** The etiology is uncertain, but it is thought to be caused by a disorder of stage 4 sleep. Fibromyalgia is **worsened by stress or lack of sleep,** and stress often elicits or worsens pain. Fibromyalgia is sometimes seen with dysmenorrhea, headaches, or irritable bowel syndrome. Some patients have psychological difficulties, apparent before or after the onset of symptoms.

  Most patients with fibromyalgia are **young women.** Teenage girls are occasionally affected, and older adults may develop fibromyalgia. Men also may develop fibromyalgia, but it is often localized and related to overactivity. No histologic or biochemical changes have been found in patients with fibromyalgia.
- **History and Physical Examination** The typical patient is a **young, healthy female** who complains of the **gradual onset of achiness, stiffness, and fatigue** present upon awakening and worsening as the day goes on. Awakening at night and having trouble falling back to sleep are also common. Some patients may give a history consistent with sleep apnea, such as loud snoring or spells of apnea ended by loud gasps or snores (as noticed by the patient's partner). Patients with fibromyalgia may also complain of anxiety and nonspecific gastrointestinal problems.

  Eighteen possible **"tender points"** are identifiable in patients with fibromyalgia. **Pain** at these spots is **elicited by pressing with one finger using moderate force.** Nontender control points, such as the forehead, are also tested and generally do not elicit pain. Tender points should not be confused with trigger points seen in myofascial pain; pressure on these areas elicits referred pain some distance from the point of pressure.

- **Laboratory and Diagnostic Studies** The erythrocyte sedimentation rate, complete blood count, and screening blood work can eliminate other causes of a patient's symptoms. Lyme titer, monospot, thyroid function tests, and creatine kinase may also be done when indicated by history. Sleep studies are done in patients with a history consistent with sleep apnea.

- **Formulating the Diagnosis** The **diagnosis** of fibromyalgia is one **of exclusion** since there are **no known laboratory or x-ray abnormalities.** Psychogenic disease should also be ruled out.

  The American College of Rheumatology has proposed **criteria for diagnosing fibromyalgia.** By these criteria, a patient with fibromyalgia will have at least **11 of the 18 tender points besides diffuse, widespread pain.**

- **Evaluating the Severity of Problems** Some patients with fibromyalgia become debilitated as a result of their pain and fatigue; however, these patients should never be given narcotics.

- **Managing the Patient** Mild cases of fibromyalgia may spontaneously resolve, but **chronic, recurrent disease is common.** The **sleep disturbance may be improved by low doses of a tricyclic antidepressant** (e.g., **amitriptyline**) given at bedtime. If morning drowsiness occurs, the dosage may be decreased, given a few hours before bedtime, or another tricyclic may be tried. Cyclobenzaprine (Flexeril), a tricyclic antidepressant usually used as a muscle relaxant, may be more acceptable to patients who feel stigmatized by taking antidepressant medication.

  Heat, stretching, exercise, and massage may help improve symptoms. A small number of patients experience pain relief with NSAIDs. It is important to acknowledge that the pain is not "all in one's head" and to reassure the patient of the benign nature of fibromyalgia. Some patients benefit from psychological counseling or stress-reduction therapy, but the primary concern is the restoration of restful sleep.

Gout

- **Basic Science** **Gout is a metabolic disease characterized by a recurrent acute arthritis secondary to the deposition within joints of monosodium urate crystals.**

  Untreated gout will go through four stages:

  - **The first stage is asymptomatic hyperuricemia,** in which serum urate levels are elevated but crystal deposits have not formed.
  - **The next stage is acute gouty arthritis, usually monoarticular, commonly affecting the metatarsophalangeal joint of the great toe.**
  - **Stage 3,** the intercritical period, **is the time between acute attacks;** during this time, the patient is completely asymptomatic.

- In **stage 4 (chronic tophaceous gout),** urate crystals are deposited in soft tissue in masses called tophi. Common locations are the pinna, olecranon bursa, hands, feet, and within tendons. Patients with chronic gout may also develop uric acid kidney stones and subsequent kidney damage.

  There are **three causes of hyperuricemia: overproduction** (metabolic) or **undersecretion** (renal) of uric acid **or a combination of the two.** Most gout patients are undersecreters.

■ **History and Physical Examination** The **onset** of an acute gouty attack **is abrupt;** the patient typically goes to bed feeling fine and is awakened by **excruciating joint pain,** often **in the big toe** ("podagra"). The patient cannot bear weight on the joint and may not even tolerate the weight of bedsheets over his or her feet. Fever and malaise may be present. The affected **joint is erythematous, swollen, and exquisitely sensitive.** Classically, the patient is an overweight male who reports overeating or drinking heavily before an attack.

If gout goes untreated, joint erosion and deformity may occur. Therefore patients with long-standing gout may appear to have rheumatoid arthritis; tophi may be seen. Patients may have limitation of joint motion, usually in the hands or feet, but occasionally in the shoulders or sacroiliac joints.

■ **Laboratory and Diagnostic Studies** Microscopic examination of the synovial fluid from an affected joint will reveal **negatively birefringent needle-shaped crystals.** A Gram's stain should be done to evaluate for infectious causes.

■ **Formulating the Diagnosis** The history and physical examination usually suggest the correct diagnosis, but **definitive diagnosis is made by synovial fluid examination.** Serum urate elevations of >7 mg/dl are usually seen in patients with gout, but most people with hyperuricemia never develop gout. Additionally, some patients will have normal serum urate levels during an acute attack.

■ **Evaluating the Severity of Problems** Although most forms of gout can be controlled medically, certain patients experience a progressively worsening destructive arthropathy.

■ **Managing the Patient** **The treatment of an acute attack may include** the following:

- ***NSAIDs,*** typically indomethacin, are given until symptoms resolve (usually 5 to 10 days).
- ***Colchicine*** interferes with the inflammatory response to urate crystals. It may cause **nausea, vomiting, diarrhea,** and bone marrow suppression. It is administered orally or intravenously. Use of colchicine has largely been supplanted by NSAIDs, but it may be beneficial if given within several hours after the onset of an acute exacerbation.
- ***Bed rest*** is useful during an acute attack since ambulation may precipitate a recurrence.
- ***Analgesics***—narcotics may be necessary initially for pain relief. **Aspirin is contraindicated.**
- ***Corticosteroids***—intraarticular injections are especially useful when only a single joint is affected.

Treatment between acute exacerbations may include colchicine, uricosuric drugs, allopurinol, and a diet free of substances known to cause an attack (e.g., certain foods and alcohol). Uricosuric drugs (e.g., probenecid, sulfinpyrazone) block tubular reabsorption of filtered urate. They cannot be used in patients with renal insufficiency. Allopurinol, a xanthine oxidase inhibitor, blocks production of uric acid and is therefore useful in overproducers. The dosage of allopurinol is guided by the serum urate and creatinine concentrations.

Infectious Arthritis

■ Basic Science **Infectious (septic) arthritis is an inflammation of the synovium caused by infection with bacteria, viruses, or fungi.**

- **In young children, infectious arthritis is most likely due to *Staphylococcus aureus.*** Other causes include group B streptococcus, *Haemophilus influenzae,* and gram-negative bacteria.
- **The most common cause of infectious arthritis in adolescents and adults is *Neisseria gonorrhoea.*** Other common etiologies in older children and young adults are *Staphylococcus aureus,* group A streptococcus, and the pneumococcus.

  Hematogenous spread is the most common cause of infectious arthritis, but direct inoculation of the joint by injury, during surgery, or from overlying infected tissue may also be responsible. Joint inflammation from other causes (e.g., rheumatoid arthritis) increases susceptibility to bacterial arthritis.

■ History and Physical Examination Bacterial arthritis presents as **acute monoarticular joint pain.** The infected joint is swollen, warm, and tender. Signs of systemic illness, such as fever and malaise, may be present. Severely ill patients may present in septic shock.

The **knee is the most common site of bacterial arthritis,** although the hip is also commonly affected. **Joint effusions are commonly seen in nongonococcal arthritis, but are usually minimal in gonococcal arthritis.**

Gonococcal arthritis often affects the interphalangeal joints. **Patients with gonococcal arthritis often show signs of systemic infection** (e.g., genitourinary symptoms, gonococcal skin rash, or splinter hemorrhages in the distal extremities).

Viral arthritis usually presents as a **symmetric polyarthritis;** the presence of fever is variable. The onset of arthritis may herald systemic illness (e.g., hepatitis B), may occur immediately after systemic illness (e.g., mumps or rubella), or may manifest weeks to months after systemic illness (e.g., certain coxsackieviruses).

■ Laboratory and Diagnostic Studies Since untreated bacterial arthritis can lead to joint destruction, any joint that may be infected should be aspirated for Gram's stain and culture. Synovial fluid from an infected joint has a WBC count >100,000/µl, (>90% neutrophils) and a synovial fluid glucose/blood glucose ratio <0.5. Blood cultures are indicated in any ill patient in whom a suspicion of joint infection exists.

- **Formulating the Diagnosis** A septic joint should be considered in any patient presenting with an acute or subacute arthritis and in patients with a chronic monarthritis. Synovial fluid examination and culture are not always diagnostic, but often narrow the differential diagnosis considerably.

  Viral arthritis rarely leads to joint destruction, but systemic viral infection can lead to chronic or recurrent arthritis. Diagnosis may be difficult since patients often present after the acute viral illness has resolved and can no longer be diagnosed via acute phase serology or viral isolation. A history of recent viral illness (e.g., rubella) can aid diagnosis, but systemic symptoms are absent in some viruses. For example, arthritis associated with hepatitis B usually occurs in the prodromal phase of the illness.

- **Evaluating the Severity of Problems** A delay in the diagnosis of a septic joint carries a morbidity and mortality approaching 25%.

- **Managing the Patient** **Bacterial arthritis should be treated with intravenous antibiotics and daily therapeutic taps of the affected joint.** The choice of antibiotic should be made on the basis of Gram's stain and, later, culture and sensitivity of the organism. If no organism is recovered, antibiotics should be chosen to cover the most likely (i.e., *S. aureus*) and most dangerous (i.e., *N. gonorrhoea*) pathogens. Intravenous antibiotics should be administered for 1 to 2 weeks, followed by oral antibiotics (at two to three times the normal dose to ensure adequate joint penetration). **The duration of treatment is guided by the clinical response.** Arthrotomy and **open drainage may be necessary for loculated fluid collections or joints with difficult access (e.g., the hip). Intraarticular antibiotics** have not been shown to be helpful and **should therefore be avoided.**

  Additionally, the **joint should be immobilized in a position that reduces stretch on the joint capsule.** While immobilized, the patient should receive prophylactic anticoagulation to prevent the development of a deep venous thrombosis. Passive range-of-motion exercises should begin as the inflammation resolves. Weight bearing should be avoided until all pain and inflammation are gone.

  Viral polyarthritis is treated symptomatically and is usually self-limited.

JOINT EFFUSION, SYNOVIAL CYSTS, AND GANGLIA

- **Basic Science** **Joint effusion is a nonspecific finding of many joint diseases.** Excess fluid in the synovial joint can result from noninflammatory or inflammatory conditions, hemorrhage into the joint space, and infection.
  - **Noninflammatory** causes include **trauma, osteoarthritis,** neurogenic arthropathy, Ehlers-Danlos syndrome, and certain metabolic disorders.
  - **Inflammatory causes include rheumatoid arthritis, systemic lupus erythematosus,** systemic sclerosis (scleroderma), **crystal arthritis,** ankylosing spondylitis, psoriatic arthritis, Reiter's syndrome, and inflammatory bowel disease.

- **Hemorrhagic causes include trauma, anticoagulation therapy, and hemophilia.**
- **Infectious causes include *Staphylococcus aureus* and gonococcal and streptococcal joint infections.**

Synovial cysts are the result of herniation of the joint space. This is most common when a large synovial effusion is present but may also occur with minimal effusion. Joint alteration and destruction are other predisposing factors. The **most frequent site for a synovial cyst is the knee joint and is termed a *Baker's cyst.*** Knee motion creates a functional one-way valve, allowing fluid into the cyst but preventing its return to the joint space. Baker's cysts are seen in patients with large synovial fluid effusions (e.g., rheumatoid arthritis) and can occasionally occur in osteoarthritis. Uncommonly, synovial cysts rupture, and fluid tracks through muscle and soft tissue.

A ganglion is a small cyst (usually <1-2 cm in diameter) that occurs in a joint capsule or tendon sheath. The cause is thought to be due to myxoid degeneration of joint connective tissue with surrounding soft tissue enclosing the collagenous degenerative tissue. Trauma may precipitate ganglia formation. Ganglion cysts are found **most commonly in the wrists.**

■ **History and Physical Examination** Joint effusion presents with joint swelling, pain, and erythema in varying degrees, **depending on the underlying cause.** Accompanying **symptoms will** similarly **vary.**

A **Baker's cyst presents as a palpable, discrete swelling in the popliteal space,** occasionally visible when the patient is standing. Initially, this swelling causes little pain, but pain worsens as more fluid seeps into the cyst. If the cyst ruptures, diffuse painful swelling of the calf results. These symptoms may mimic thrombophlebitis.

A **ganglion cyst is a palpable, firm, movable subcutaneous nodule.** A ganglion presents most commonly as a **painless swelling,** but larger ganglia may cause pain with joint motion. They are **most often found at the scaphoid-lunate articulation.**

■ **Laboratory and Diagnostic Studies**

- Joint aspiration and examination of the fluid is important in determining the cause of unknown joint effusion. Normal joint fluid is a viscous, colorless, transparent fluid with white blood cells (WBC) <200/µl; polymorphonuclear leukocytes (PMN) represent <25% of the total.
- Fluid from a joint with bacterial arthritis will be cloudy, with variable color and viscosity, and a WBC >100,000/µl, PMN >75%.
- Noninflammatory arthritis results in joint effusions that are viscous, yellow, and transparent with WBC 200-2,000/µl, PMN <25%.
- Fluid from an inflammatory joint has low viscosity, is yellow and transparent, and has a WBC 2,000-100,000, PMN >50%.
- Fluid from a hemorrhagic effusion usually contains frank blood. Joint fluid should be Gram's stained, cultured, and examined for crystals.

■ **Formulating the Diagnosis** The diagnosis of an effusion rests heavily on examination of the synovial fluid as just outlined. If an underlying disease process is suspected of causing the effusion, further workup for that disease is indicated.

An unruptured synovial cyst can be diagnosed by ultrasound, but a ruptured cyst may not show up on ultrasound. Arthrography can be used to diagnose a ruptured cyst.

■ **Evaluating the Severity of Problems** A ruptured Baker's cyst may be confused with thrombophlebitis, the latter having significant potential for morbidity and mortality if missed.

■ **Managing the Patient**

- **The treatment of joint effusions is specific to the underlying cause.**
- **Synovial cysts** may be **treated by aspiration and injection with corticosteroids.**
- Treatment of ganglia may be unnecessary in the absence of symptoms. Aspiration of the central collagenous fluid alone may be sufficient. Sometimes injection with corticosteroid after aspiration is helpful. If this is unsuccessful, the ganglion may be removed surgically. Splinting the joint may help prevent recurrence.

OSTEOARTHRITIS

■ **Basic Science** **Osteoarthritis (OA, degenerative joint disease) is the most common form of arthritis.** It is characterized by **gradual onset (usually after age 40),** progressive worsening of symptoms, and lack of systemic involvement.

Traditionally, **OA is classified as** follows:

- **Primary**—affects the distal interphalangeal joints (Heberden's nodes), the metacarpophalangeal and carpometacarpal joints of the thumb, the knee, the hip, the metatarsophalangeal joint of the great toe, and the cervical and lumbar spine
- **Secondary**—as a sequela to acute or chronic joint injury (e.g., occupational overuse, a fracture, rheumatoid arthritis)

In OA there is **degeneration of the articular cartilage** with **hypertrophy of bone at the articular margins. Inflammation, if present, is minimal.** The synovial membrane becomes thickened, but does not form adhesions.

■ **History and Physical Examination** The **onset of symptoms in osteoarthritis is usually gradual.** The first symptoms noted may be **decreased mobility and mild, fleeting (<30 minutes) morning stiffness.** Initially, **pain occurs with movement and is relieved by rest;** as the disease progresses, pain may occur at rest and is common at night. Pain on passive joint movement, limitation of joint motion, small effusions, and/or crepitus are noted on physical examination.

Involvement of the hand is commonly seen in the DIP, PIP, and trapezioscaphoid joints. **Heberden's nodes (osteophytes of the**

DIP) and, less commonly, Bouchard's nodes (osteophytes of the PIP) may develop.

Hip involvement is characterized by a gradual onset of pain in the groin or inner thigh. Pain may occasionally be referred to the knee. A limp may be present, and internal and external rotation are limited.

OA of the knee causes pain and crepitus with motion. Osteophytes can sometimes be palpated. Valgus or varus deformities may be apparent when lateral and medial cartilage loss is disproportionate.

Osteoarthritis of the spine involves the apophyseal joints and is most common at the level of L3-L4. Local pain and stiffness may be accompanied by radicular pain caused by compression of the nerve roots by osteophytes. Spinal stenosis, most common in the lumbar spine, results from a combination of degenerative osteophytes, soft tissue inflammation, and spondylolisthesis; it causes constant or recurring leg pain.

■ **Laboratory and Diagnostic Studies** There are no characteristic lab abnormalities in the patient with isolated, primary OA. **X-ray findings include narrowed joint space, osteophytes, increased density of subchondral bone, and bone cysts.** The vast majority of all people will have evidence of OA in weight-bearing joints by age 40.

■ **Formulating the Diagnosis** **The diagnosis is based on the clinical exam combined with radiographic findings.** Radiographic findings can show evidence of OA but do not always correlate with clinical severity. The absence of inflammation excludes many other forms of arthritis from the differential.

■ **Evaluating the Severity of Problems** More serious diseases such as osteoporosis, metastatic neoplasia, and multiple myeloma may all be confused with OA.

■ **Managing the Patient** Treatment for osteoarthritis is individualized; patients with minimal symptoms may need no treatment at all. **Analgesics, exercise and physical therapy, steroid injections, weight loss (if obese), surgery, or combinations of these therapies can provide relief.**

Regular use of **nonnarcotic analgesics** can alleviate pain. **Acetaminophen** often provides sufficient pain relief. NSAIDs can also be used successfully for analgesia, but their antiinflammatory action is often not necessary, and side effects from chronic use can be troublesome. Enteric coated aspirin is better tolerated by some patients, and the nonacetylated salicylates also cause fewer gastrointestinal side effects.

Physical therapy and exercise can improve functional status without aggravating symptoms. Moist heat may be of benefit to some patients.

Surgery (e.g., arthroscopy or joint replacement) is an option for treating severe joint damage or deformity that cannot be adequately treated conservatively.

Osteoporosis

■ **Basic Science** **Osteoporosis is the most common metabolic bone disease. Although bone mineralization is normal in osteoporosis, all elements (i.e., bone matrix and mineral**

content) are present in decreased amounts (osteopenia). Osteoporosis results in the loss of cortical bone and a decrease in the number of trabeculae. Bone is constantly undergoing remodeling (i.e., formation and resorption). When the rate of resorption exceeds the rate of formation, bone mass is decreased. This decrease leads to a loss of structural strength; therefore fractures are common.

There are two major, primary forms of osteoporosis:

- **Type I** occurs in **postmenopausal women from age 50 to 75.** This form is thought to be **hormone related.**
- **Type II** occurs in **men and women over age 70 and is related to aging.**

Risk factors for osteoporosis are as follows:

- **Age** (>50 for women, >60 for men)
- **Heredity**
- **Gender** (F > M)
- **Hormonal factors** (decreased estrogen in women, decreased androgen in men)
- **Associated disease**
- **Drugs** (e.g., steroids, furosemide)
- **Poor nutrition**
- **Weight** (thin patients are at increased risk)
- **Activity levels** (weight-bearing exercise is protective)
- **Cigarette smoking**
- **Alcohol use**

■ History and Physical Examination **Patients with osteoporosis** frequently **present with fractures** and their sequelae. **Commonly fractured bones include the proximal femur and the vertebrae.** Patients with vertebral fractures have spinal deformity and back pain. The pain usually begins suddenly, may radiate to the flank and abdomen, and is worsened by movement. Trauma may precede the pain, as may bending or lifting. Sharp pain gradually subsides within a week, but dull pain may last a month or more. Some patients with vertebral fractures never experience acute pain but will complain of backache worsened by sudden movement. Fractures usually occur in the anterior portion of the vertebra, leaving **the vertebra wedge-shaped. The result is a loss of height, increased thoracic kyphosis, and increased cervical lordosis ("widow's hump").**

■ Laboratory and Diagnostic Studies **Blood tests are not helpful in diagnosing osteoporosis.** Serum calcium, phosphorus, and alkaline phosphatase are usually within normal limits, although alkaline phosphatase may be acutely elevated after a fracture.

■ Formulating the Diagnosis **Bone density measurements are used to diagnose osteopenia. X-rays are not a reliable way of identifying decreases in bone density since they may be normal despite a bone loss of 30%.** X-rays can, however, reveal asymptomatic fractures. Radiographically, vertebral bodies in osteoporosis appear biconcave due to subchondral plate weakening and enlargement of the discs.

- **Evaluating the Severity of Problems** Hip fractures in the elderly carry a significant risk of mortality. Every effort should be made to prevent falls in the elderly.

- **Managing the Patient** The **best way to treat osteoporosis is to prevent its occurrence. Young people can increase bone mass with weight-bearing exercise and sufficient calcium** intake. **Type I osteoporosis is preventable with estrogen replacement therapy beginning at menopause.** Progestins should also be given; although they do not affect bone density, they ameliorate the increased risk of uterine cancer seen in patients receiving estrogen alone.

  Once bone loss has begun, treatment is more difficult. Calcium supplements are often the first step. Vitamin D is also important and may be required in elderly patients with impaired vitamin D formation. Large doses of vitamin D should be avoided since they can cause toxicity and worsen bone loss.

  Pharmacologic antiresorptives are important in treating osteoporosis. **Estrogen replacement therapy begun after bone loss has developed can slow further bone loss and may initially cause a small increase in bone mass.** If estrogen cannot be given, calcitonin may be used to block osteoclasts and slow bone resorption, but it is expensive and must be administered by subcutaneous injection three times per week. Bisphosphonates decrease bone turnover and resorption.

  Further treatment of osteoporosis involves weight-bearing exercise, back extension exercises, and fall prevention.

Polymyalgia Rheumatica and Temporal Arteritis

- **Basic Science** **The vasculitides are a group of disorders that share a common feature: inflammation and necrosis of blood vessels. Examples of primary vasculitides include** Behçet's syndrome, polyarteritis nodosa, Wegener's granulomatosis, Henoch-Schönlein purpura, **polymyalgia rheumatica (PMR), and temporal arteritis (TA).** Vasculitides are rare, can involve any vessel in the body, and may overlap one another in pathogenesis and symptomatology.

  PMR is a disorder that primarily affects elderly females, who develop shoulder and pelvic girdle pain and stiffness, associated with fever, malaise, and weight loss. It is frequently seen in association with TA.

  TA is a form of giant cell arteritis affecting the elderly, which, if untreated may result in blindness. Although it classically occurs in the temporal artery, any branch of the aortic arch may be affected.

- **History and Physical Examination** **Patients with TA** typically complain of severe **temporal headaches, jaw claudication, and changes in vision.** The **temporal region may be tender to palpation,** with areas of nodular swelling. Bruits are often heard over the carotids. As with PMR, **fever, malaise, and weight loss are often present.**

Patients with PMR have **difficulty lifting their arms** to brush their hair, and they complain of **leg weakness** resulting in difficulty rising from a chair or climbing stairs. The gelling phenomenon (**stiffness** after a period of inactivity) is a common complaint. Patients have tenderness and limited range of motion in their shoulders. **No true muscle weakness** is present.

- Laboratory and Diagnostic Studies **An elevated erythrocyte sedimentation rate (ESR), anemia, and increased alkaline phosphatase are common** nonspecific findings **in both TA and PMR. Temporal artery biopsy** is indicated in all patients in whom TA is suspected. A finding of granulomatous arteritis confirms the diagnosis, whereas its absence should prompt other diagnostic considerations. In PMR creatine kinase and muscle biopsy are both normal.
- Formulating the Diagnosis Proximal weakness, fever, malaise, and weight loss in persons over age 50 should prompt consideration of PMR. Similar symptoms, with (or without) temporal headache, jaw pain, or visual disturbance, suggest TA. A suspicion of TA mandates temporal artery biopsy. The diagnosis of TA should be sought in all patients with PMR, since the two frequently occur together.
- Evaluating the Severity of Problems The **risk of untreated TA is permanent visual impairment** secondary to occlusion of the ciliary and ophthalmic arteries.
- Managing the Patient **TA must be promptly treated (while awaiting biopsy results) with high-dose corticosteroids to prevent the possible sequelae of blindness, stroke, myocardial infarction, and death.** Treatment consists of prednisone 60 mg daily continued for at least 1 or 2 months. If symptoms resolve, the steroid may be slowly tapered. Recurrent symptoms should immediately be treated with increased steroid dosage. **Symptoms, rather than ESR, guide treatment.** Some patients are weaned from steroids within 1 year, but many require longer treatment. Patients who do not respond to prednisone are treated with azathioprine or methotrexate.

  Patients with PMR often respond dramatically to small doses of prednisone, usually 10 to 20 mg per day, tapered slowly after clinical response. Patients may require a small maintenance dose, and those who cannot be successfully tapered, may receive weekly methotrexate.

PSEUDOGOUT

- Basic Science **Pseudogout is a disease characterized by the deposition of calcium pyrophosphate precipitates in articular cartilage, synovium, and periarticular tendons and ligaments.** Pseudogout occurs in older people who often already have degenerative joint disease, which may form a nidus for crystal formation. Pseudogout may occur in association with other diseases such as gout, hyperparathyroidism, amyloidosis, hypothyroidism, and hemochromatosis. Attacks of pseudogout are usually self-limited. Repeated attacks of pseudogout may lead to destructive arthritis.
- History and Physical Examination Pseudogout may manifest acutely, subacutely, or chronically. An acute attack of pseudogout **may mimic**

gout in its presentation and severity, but is usually less severe. Joint pain, tenderness, and warmth are seen. An attack is sometimes precipitated by trauma and **typically involves the knee,** but may also be seen in the hand, wrist, elbow, shoulder, or ankle. Patients may have a low-grade fever.

- Laboratory and Diagnostic Studies **Joint space aspiration reveals weakly positive birefringent and rod-shaped or rhomboid-shaped crystals** when examined under polarized microscopy. X-ray examination may reveal linear calcification in the articular cartilage.
- Formulating the Diagnosis **Pseudogout may mimic many arthritides; however, a definitive diagnosis is made by microscopic examination of the joint aspirate.**
- Evaluating the Severity of Problems Although a destructive arthritis may follow repeated attacks of pseudogout, it does so much less often than in gout.
- Managing the Patient Acute pseudogout may be treated by **joint aspiration and intraarticular injection of** microcrystalline **corticosteroid. NSAIDs, typically indomethacin, work well to relieve pain.** Oral or intravenous colchicine is sometimes used to treat acute attacks, and low-dose **colchicine may be given prophylactically to prevent pain recurrence.** However, there is no good way of removing or dissolving the crystals deposited in the joint tissue, and therefore no satisfactory method of preventing joint destruction in patients with repeated attacks.

Rheumatoid Arthritis

- Basic Science **Rheumatoid arthritis (RA) is a chronic systemic inflammatory disease of unknown cause. The synovial membranes and periarticular structures of peripheral joints are primarily affected, although systemic manifestations also exist.** The disease has a female predominance with an onset usually between the second and fifth decade.

  Joint involvement is characterized by a proliferative synovitis. The hyperplastic synovium (pannus) becomes edematous, and folds of surplus synovium protrude into the joint space, forming villi. The joint cartilage, subchondral bone, joint capsule, and ligaments are eroded by the autoimmune process. Continued damage leads to deformity and loss of function. Rheumatoid nodules typically occur on extensor surfaces and in periarticular structures, although they may affect nearly any organ system.
- History and Physical Examination **Patients usually present with the gradual onset of fatigue, weakness, anorexia, increasing pain, and stiffness. Morning stiffness lasting at least 1 hour is characteristic;** patients also experience stiffness after periods of inactivity (gelling). Initially the pain may be poorly localized, but later, pain is felt in the affected joints and is worsened with movement. **Arthritis is usually symmetric, and the joints most often affected are the wrist, metacarpophalangeal (MCP), proximal interphalangeal (PIP), elbows, knees, ankles, and feet.** The cervical spine may also be affected (e.g., C1-C2 subluxation is some-

times seen). The distal interphalangeal joints and lumbar spine are only rarely involved.

Affected joints are swollen and warm and have limited range of motion. With prolonged active disease, flexion deformities, subluxation, fibrosis, and ankylosis may occur. Entrapment of the median nerve at the wrist leads to carpal tunnel syndrome.

Rheumatoid nodules are seen in one fifth to one third of patients. They present as movable lumps of varying size and consistency located under the skin. Nodules are usually asymptomatic but can occasionally break down.

Systemic involvement may consist of vasculitis, pericarditis, polyneuropathy, pulmonary fibrosis, pneumonitis, or splenomegaly.

■ **Laboratory and Diagnostic Studies** **Laboratory abnormalities include an elevated erythrocyte sedimentation rate and the presence of rheumatoid factor (RF),** an antibody to gamma globulin. RF is seen in 80% of patients with RA but may also be present in patients with other rheumatic conditions. Synovial fluid from affected joints contains 2,000-100,000 WBCs/µl, reflecting variable degrees of inflammation.

Early changes on x-ray include juxtaarticular demineralization and soft tissue swelling. Later, joint space narrowing, erosions, and characteristic deformities are seen.

■ **Formulating the Diagnosis** A clinical presentation accompanied by the typical joint symptoms suggests the diagnosis of RA. The American College of Rheumatology has proposed the following criteria for the diagnosis of RA **(four of the seven must be present):**

- **Morning stiffness lasting >1 hour, present >6 weeks**
- **Arthritis of three or more joints, present >6 weeks**
- **Arthritis of the wrist, MCP, or PIP joints, present >6 weeks**
- **Symmetric arthritis, present >6 weeks**
- **Rheumatoid nodules**
- **Serum rheumatoid factor**
- **Radiographic changes—bony erosion or bony decalcification**

During the early phase of the disease the diagnosis is not excluded in patients who fail to meet the strict criteria.

■ **Evaluating the Severity of Problems** Severely affected joints, especially the hip and knee, may be amenable to joint replacement.

■ **Managing the Patient** The treatment of RA is based on controlling the disease process, preventing deformity, relieving pain, and maintaining joint function. Patient education and conservative measures (e.g., physical therapy, joint rest) form the basis of therapy.

The **first line of pharmacologic treatment aims at reducing pain and swelling. NSAIDs (e.g., aspirin, indomethacin, ibuprofen) and corticosteroids provide good pain relief.** Corticosteroids, although antiinflammatory and immunosuppressive, do not usually modify the disease process. They are used to control pain during flares and in treating RA vasculitis but are not indicated for long-term disease treatment.

Disease-modifying drugs are used in patients with aggressive destructive disease and in patients who do not respond to antiinflammatory drugs. There is growing support for much earlier use of these agents. Most agents in this class are slow acting and have significant side effects; therefore the agent is chosen based on its toxicity and the severity of the patient's illness. Following is a list of modifying drugs and their side effects:

- **Gold**—dermatitis and stomatitis often result. Renal and hematologic side effects require urinalysis and complete blood count (CBC) before each treatment.
- **Methotrexate**—gastric irritation and stomatitis are common. CBC and liver function tests must be checked at least monthly due to serious side effects, including marrow suppression and hepatotoxicity.
- **Hydroxychloroquine**—onset of action may take months. Potential toxicity to retina mandates a biannual eye exam.
- **Penicillamine**—dermatitis is a common side effect. Monthly CBC and urinalysis are necessary to evaluate for hematologic and renal toxicity.
- **Sulfasalazine**—CBC is necessary to monitor for neutropenia and thrombocytopenia.
- **Azathioprine**—chronic use results in immunosuppression. CBC, liver function tests, and urinalysis are required at routine intervals.
- **Cyclophosphamide**—this is reserved for the most recalcitrant cases. It is a potent immunosuppressive with side effects involving several organ systems. Close follow-up is necessary to monitor leukopenia.

Systemic Lupus Erythematosus

■ Basic Science **Systemic lupus erythematosus (SLE) is an autoimmune disorder that occurs mainly in young women. Nearly all patients have joint symptoms, and multiple system involvement is common.** Some symptoms result secondary to antigen-antibody complex deposition. The course of the disease may be benign or fulminant and depends on the organ system(s) involved. **Although the cause of SLE is unknown, recipients of certain drugs (e.g., hydralazine, procainamide, isoniazid) may manifest a lupus-like illness.** A genetic predisposition exists for SLE. Transfer of maternal antibodies to the fetus may result in transient lupus-like manifestations with permanent congenital heart block.

■ History and Physical Examination Patients with SLE may present with an acute exacerbation, or symptoms may develop gradually. Typically, acute exacerbations are followed by full or partial remission. Systemic symptoms include **fever, malaise, fatigue, anorexia, and nausea.**

Most patients have **musculoskeletal complaints. Arthralgias, myalgias, and arthritis are often the first signs of SLE.** Pain is often out of proportion to the physical examination.

Cutaneous manifestations include the typical malar ("butterfly") rash; it may be **painful and pruritic** and is **often pre-**

cipitated by exposure to sunlight. Photosensitivity may also result in a diffuse maculopapular rash. Patchy alopecia is common at the beginning of an acute exacerbation. Discoid lesions begin as erythematous plaques that become raised and scaly and resolve with central scarring and depigmentation.

Renal manifestations occur in almost all patients but are asymptomatic in half. Patients with lupus nephritis have persistent proteinuria, though renal disease may range from benign proteinuria to end-stage renal disease.

The central and peripheral nervous system, including the meninges, may be involved. **Depression and anxiety are frequent. Seizures are common,** and organic brain syndromes, psychosis, headaches, cranial nerve palsies, transverse myelitis, cerebellar dysfunction, mononeuritis or polyneuritis, optic neuritis, and aseptic meningitis are other possible neuronal manifestations. Organic brain syndrome, with **delirium, emotional lability, and memory deficits, occurs in most patients with central nervous system disease.**

Cardiopulmonary involvement is seen in over half of patients. Cardiac manifestations include pericarditis, myocarditis, and endocarditis. Patients may present with arrhythmias, conduction defects, cardiomegaly, murmurs, subacute bacterial endocarditis, or CHF. Pleurisy, pleural effusion, pneumonia, and pneumonitis are frequent pulmonary manifestations.

Peritonitis often results in nonspecific gastrointestinal symptoms such as diffuse **abdominal pain, nausea, vomiting, and anorexia.**

Vascular and hematologic manifestations include **vasculitis and thrombosis.** Autoantibodies, the so-called **lupus anticoagulants, render some patients hypercoagulable.**

Other manifestations of SLE include Raynaud's syndrome, splenomegaly, and lymphadenopathy. Affected patients become pregnant at the same rate as unaffected individuals but have a higher **incidence of miscarriage, intrauterine death, and prematurity.**

■ **Laboratory and Diagnostic Studies** Several autoantibodies are present in the sera of SLE patients. The autoantibody most reliably present is the antinuclear antibody (ANA). However, ANA is nonspecific since it is seen in several other autoimmune disorders. **Anti-dsDNA and anti-Sm antibodies are more specific for SLE.** Patients usually show **false positivity on VDRL secondary to antiphospholipid antibodies.** LE cells, in which nuclei have been phagocytized by neutrophils, are sometimes found on microscopic examination of affected tissue. Complement (C_3, C_4, and CH_{50}) levels may be low, especially during disease exacerbations.

■ **Formulating the Diagnosis** The diagnosis of SLE may be apparent in patients with typical manifestations. Criteria proposed for the classification (not diagnosis) of SLE by the American College of Rheumatology require 4 of the following 11 criteria:

- **Malar rash**
- **Discoid rash**
- **Photosensitivity**
- **Oral ulcers**

- **Arthritis**
- **Serositis**
- **Renal disorder**
- **Leukopenia <4000/µl, lymphopenia <1500/µl, hemolytic anemia, or thrombocytopenia <100,000/µl**
- **Neurologic disorder**
- **Positive LE cell, anti-dsDNA, anti-Sm antibody, or false-positive VDRL**
- **High ANA titer**

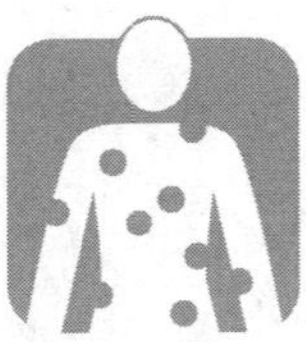

■ Evaluating the Severity of Problems The primary cause of death in lupus patients is opportunistic infection, followed by renal or CNS involvement secondary to active SLE.

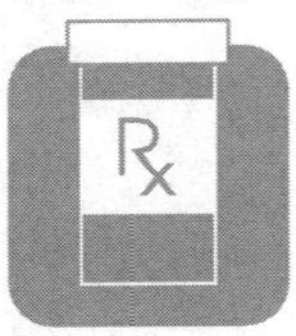

■ Managing the Patient **Since SLE is incurable, the main goals of treatment are to reduce the number and severity of acute flares and to minimize and treat end-organ disease.** Drugs responsible for lupus-like disease should be stopped.

Salicylates and other NSAIDs **are the first line of treatment.** NSAIDs are often effective against musculoskeletal manifestations, serositis, and fever, but response varies among patients. Adverse reactions, particularly renal dysfunction, can limit NSAID use.

Corticosteroids are required for severe systemic symptoms; topically applied steroid is often effective for the dermatological manifestations of SLE. In the treatment of CNS involvement, steroids may precipitate psychosis.

A **patient who does not respond to corticosteroids** or who cannot be tapered **may need more aggressive therapy.** Common **alternatives include antimalarials** (e.g., **hydroxychloroquine or quinacrine) and more potent immunosuppressives** (e.g., **methotrexate, azathioprine, and cyclophosphamide**), all of which have significant side effects.

Multiple Choice Review Questions

1. A 67-year-old woman with a history of osteoarthritis presents to the emergency room complaining of moderate right elbow pain and swelling. She denies injury to or overuse of the elbow. Otherwise she feels well. An x-ray of the joint is normal except for possible minimal calcification of the joint cartilage. You suspect pseudogout and aspirate the joint. What would you expect to find on examination of the synovial fluid?

a. Positively birefringent crystals, low viscosity, 75,000 WBCs and negative gram stain
b. Negatively birefringent crystals, low viscosity, 75,000 WBCs and negative gram stain
c. Positively birefringent crystals, transparent fluid, low viscosity, and <200 WBCs
d. Negatively birefringent crystals, transparent fluid, high viscosity, and <200 WBCs

2. A 24-year-old woman comes to your office for her annual Pap smear. While she is there, she comments that her 76-year-old grandmother recently broke a hip and was diagnosed with osteoporosis. Your patient wants to know how she can avoid developing osteoporosis as she gets older. You suggest:

a. Estrogen replacement at menopause.
b. Bone densometry measurements.
c. Calcitonin therapy.
d. Calcium supplements and weight-bearing exercise.

3. A 31-year-old man with a history of IV drug use presents to the emergency room with a 1-day history of fever and an acutely swollen, painful right knee. Purulent fluid is obtained with joint aspiration. Gram stain of this fluid reveals gram-positive cocci in clusters. After admitting him to the hospital, you should treat him with:

a. Daily joint aspiration, followed by intraarticular injection of antibiotics and IV antibiotics.
b. Daily joint aspiration coupled with IV antibiotics.
c. High-dose IV antibiotics and splinting the joint to prevent hematogenous spread of infection.
d. High-dose (double to triple normal dose) PO antibiotics and joint aspiration on a PRN basis for pain.

4. A 16-year-old girl is admitted from the ED with a suspected onset of SLE. She had been seen in her pediatrician's office the week before with possible atypical virus; laboratory tests at that time showed a mild anemia and a creatinine of 1.0. On presentation to the ED, the patient has a malar rash, arthralgias, confusion, a pericardial rub, and pedal edema. Laboratory tests showed an ANA of 1:1600, positive titer anti-dsDNA, low complements, normal PT/PTT, and a creatinine of 4.8. What is the most appropriate treatment at this time?

a. Corticosteroids and high-dose NSAIDS
b. Hemodialysis
c. Corticosteroids and low-dose heparin
d. Corticosteroids and pulsed cytoxan

5. A 36-year-old man with no prior history of joint symptoms developed a sudden onset of left ankle pain, redness, and swelling. You suspect that he may have septic arthritis and aspirate fluid from his ankle. What finding most likely supports the diagnosis of septic arthritis?

a. PMNs >75%
b. Transluscent fluid
c. WBC 80,000
d. Fragmented crystalline structure

Chapter 9

Orthopedics

Carpal Tunnel Syndrome

■ **Basic Science** Carpal tunnel syndrome is the **most common compression neuropathy** of the **upper extremity.** Affliction with this disorder is responsible for substantial decreased worker productivity and lost days from work.

The carpal tunnel is a fibro-osseous canal on the volar aspect of the wrist through which pass the long flexor tendons and the median nerve (Fig. 9.1). There are **nine flexor tendons,** four from the flexor digitorum superficialis, four from the flexor digitorum profundus, and one from the flexor pollicis longus, that pass through the canal. The **median nerve** carries **motor** function to the three muscles of the **thenar eminence** (flexor pollicis brevis, abductor pollicis brevis, opponens pollicis) as well as the **radial two lumbricales. Sensation** to the **radial three and one-half fingers** is also provided by the median nerve. The palmar cutaneous branch of the median nerve provides sensation to the palm of the hand but is superficial to the carpal tunnel and is usually not affected by this syndrome. Extrinsic or intrinsic pressure to the median nerve is responsible for the symptoms of this syndrome.

■ **History and Physical Examination**

- **History** Carpal tunnel syndrome most often affects **middle-aged women.** The classic complaint is that of pain, tingling, or **numbness** in the thumb, index, long, and radial aspect of the ring fingers. It is often **first noticed at night** and will awaken the patient from sleep, especially in those who sleep with their wrists flexed. Repetitive wrist flexion or extension activities also precipitate symptoms. The symptoms are usually intermittent, but severe cases may have persistent numbness or motor weakness.

- **Physical examination Decreased sensation to light touch and two-point discrimination** are usually seen within the median nerve distribution. **Atrophy and weakness of the muscles of the thenar group** are present in severe cases. A positive **Tinel's sign** (reproduction of the symptoms with tapping on the nerve at the wrist) and **Phalen's test** (reproduction of the symptoms with forced wrist flexion; Fig. 9.2) are usually found.

■ **Laboratory and Diagnostic Studies**

- **Laboratory** Labs are usually not indicated unless an endocrine abnormality is suspected. Hypothyroidism may produce swelling within the canal resulting in compression neuropathy. Collagen-

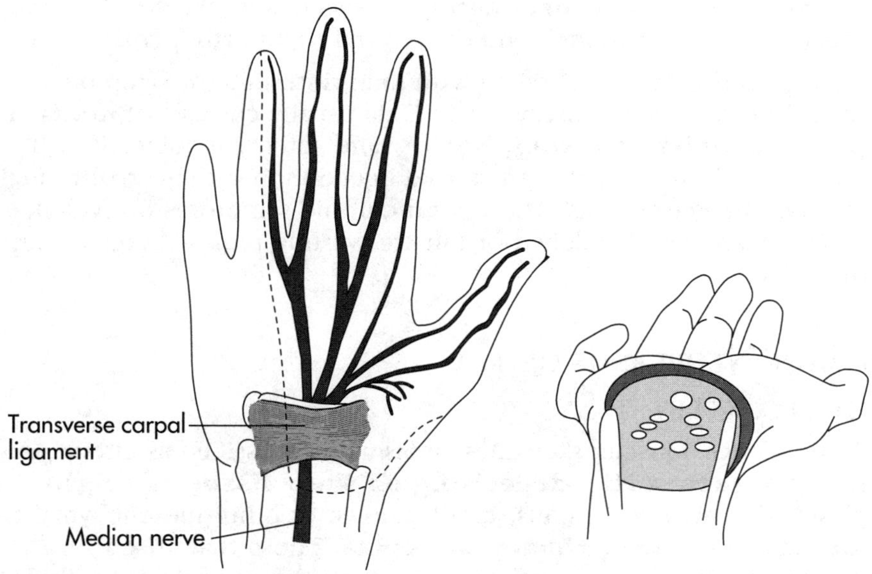

Fig. 9.1 Carpal tunnel with median nerve under the transverse carpal ligament. *(Modified from Simon R:* Emergency orthopedics—the extremities, *Norwalk, Conn, 1995, Appleton C. Lange.)*

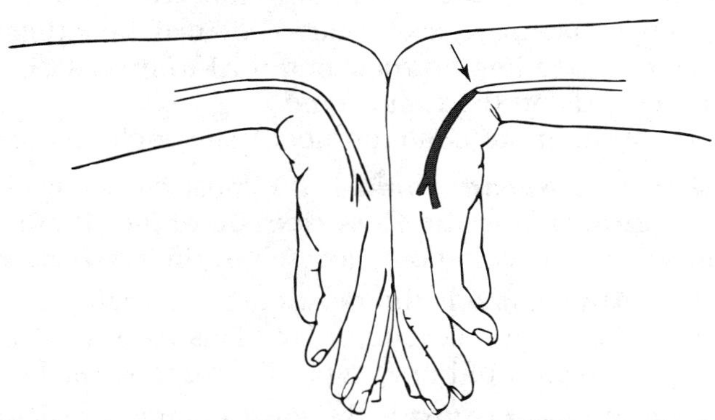

Fig. 9.2 The Phalen's test is performed by compressing the opposing dorsal surfaces of the hand with the wrist flexed together as shown. This causes tingling over the median nerve distribution as shown. *(Modified from Simon R:* Emergency orthopedics—the extremities, *Norwalk, Conn, 1995, Appleton C. Lange.)*

vascular disorders may also produce this syndrome from synovitis within the tunnel.

- **Diagnostic studies Nerve conduction velocities (NCVs)** are useful in confirming the diagnosis. **Sensory** fiber conduction **slowing** is most predictive of this syndrome.

  X-rays are not indicated unless an extrinsic compression is suspected.

■ **Formulating the Diagnosis** An accurate history and physical usually reveal the diagnosis. NCVs are helpful to document abnormal conduction. The differential diagnosis of carpal tunnel syndrome includes thoracic outlet syndrome, mononeuropathy or polyneuropathy, cervical radiculopathy, intraspinal canal lesions, and multiple sclerosis.

- **Evaluating the Severity of Problems** Persistent sensory deficit or muscle wasting is an indication for urgent referral to an orthopedist.
- **Managing the Patient** Patients with mild, intermittent symptoms should be managed conservatively. This usually consists of **neutral position wrist splinting,** NSAIDs, and activity modification. If symptoms do not improve, a steroid injection into the carpal tunnel may be of some benefit. Those with chronic symptoms or evidence of significant sensory deficit or muscle wasting require surgical carpal tunnel release.

Common Neoplasms of the Musculoskeletal System

- **Basic Science** Musculoskeletal neoplasms are classified as either **primary or metastatic, depending on their tissue of origin.** Neoplasms that arise in osseous, cartilaginous, or other mesenchymal tissues are considered **primary sarcomas.** Those that arise from distant sites and spread to these tissues are considered **metastatic** and are usually **carcinomas.** Unregulated cellular proliferation is common to all neoplasms, regardless of their origin.
  - **Primary musculoskeletal neoplasms** **Primary** musculoskeletal neoplasms are **rare,** with approximately 5000 to 10,000 cases being newly diagnosed each year. They may be either benign or malignant, depending on their potential to metastasize. Age distribution depends on the tumor type.

    Some of the more common neoplasms include the following:

    - ***Rhabdomyosarcoma*** is the most common malignant **soft tissue neoplasm** in the **first decade** of life. It arises from skeletal muscle and is commonly located in the **truncal regions.**
    - ***Osteosarcoma*** is the most common malignant **bone neoplasm** in the second decade of life. This usually arises within the metaphyseal region of bone, typically in the **knee** area.
    - ***Chondrosarcoma*** is the most common malignant **cartilage neoplasm.** These usually occur within the **shoulder and hip regions** within the later decades of life.
    - ***Malignant fibrous histiocytoma*** is the most common malignant **soft tissue neoplasm** in adults, typically occurring in the **fifth to sixth decade.**
    - ***Multiple myeloma*** is the most common malignant **bone neoplasm** in adults. It arises from **bone marrow precursor cells.**
    - ***Giant cell tumor*** is a common benign tumor occurring in the **epiphyseal region of long bones,** especially in the second and third decades.
    - ***Osteochondroma*** is a common benign tumor arising from a **remnant of the physis** (growth plate). This is usually present in the metaphyseal or diaphyseal areas of long bones.
    - ***Enchondroma*** is a common benign cartilaginous tumor arising in **small tubular bones,** typically the hands and feet.
  - **Metastatic musculoskeletal neoplasms** **Metastatic** neoplasms are the **most common tumors of the musculoskel-**

etal system. They typically **arise from carcinomas** in other areas of the body. The **axial skeleton** as well as the proximal extremities are the most common areas involved. These neoplasms generally are lytic, destroying the bone with their growth. Some are able to form bone, although this bone is not normal.

The origin of the most common metastatic carcinomas, in **decreasing frequency,** are **(1) breast, (2) prostate, (3) lung, (4) kidney,** and **(5) thyroid.**

- History and Physical Examination
  - **History** A **painless, rapidly enlarging mass** is often the presenting complaint. In those cases where the neoplasm has significantly weakened the bone, pain may be associated with a pathologic fracture after relatively minor trauma. A **prior history of cancer** is one of the most important questions to be asked.
  - **Physical examination** A palpable, nontender mass is usually present. A careful neurovascular examination is important.
- Laboratory and Diagnostic Studies
  - **Laboratory** General screening labs are rarely diagnostic. Usually elevations are seen in the sedimentation rate and lactate dehydrogenase. Studies must be tailored with a particular tumor type in mind.
  - **Diagnostic studies** Plain radiographs, in two planes, of the affected area must be obtained. Further investigation of the specific area may require CT, MRI, and bone scan. **CT** is best for evaluating **intraosseous lesions,** whereas **MRI** is best for **soft tissue masses or assessing the intramedullary extent of the lesion.** A **bone scan** is useful for assessing the activity of the lesion as well as the possibility for **other lesions.** A **chest x-ray or CT** is **mandatory to rule out metastatic disease** when it is suspected. A **biopsy** of the neoplasm is usually obtained for **tissue diagnosis** and treatment options. This should be performed by the surgeon who will perform the definitive treatment.
- Formulating the Diagnosis History, physical, radiographic testing, and biopsy results will usually yield the diagnosis.
- Evaluating the Severity of Problems If the physician is suspicious of a neoplasm, the patient should be referred to an orthopedic surgeon and oncologist immediately. The remainder of the workup is best accomplished by the treating physician.
- Managing the Patient Surgical or nonsurgical treatment depends on the specific tumor type, location, and severity. Adjunctive modalities, such as radiation therapy and chemotherapy, are commonly utilized.

Fractures, Sprains, and Strains

- Basic Science The term ***fracture*** refers to a **break** in the bone. In adults, this is always complete, involving both cortices of the bone. However, in children, the pliability of their immature trabeculae allows for the potential of plastic deformation (bending) without true fracture. Fractures result from significant transmission of energy to bone (trauma). The higher the energy, the more comminuted (fragmented) the fracture.

The term ***sprain*** **refers to a tear of ligamentous tissue.** Ligaments provide the bone-to-bone connection across joints. Sprains also result from trauma. The severity is based on the amount of tearing and the resultant instability of the involved joint.

The term ***strain*** **refers to a tear of muscular tissue.** The severity is also based on the amount of tearing, which usually occurs at the musculotendinous junction.

History and Physical Examination

- **History** Some form of trauma is necessary to produce injury to the bones, ligaments, or muscles. A careful questioning of the exact mechanism of injury is necessary in order to formulate a differential diagnosis. However, the same mechanism of injury may result in different structures being compromised in different individuals, depending on the patient characteristics.

  The usual complaints are **pain, swelling, discoloration, and disuse** from the injury. Deformity may be a complaint if a fracture has occurred.

- **Physical examination** A complete examination of the specific area involved must be performed. Skin changes, such as swelling, ecchymosis, abrasions, and contusions, should be noted. Palpation of these areas usually reveals tenderness of the specific anatomic structures involved.

  Range of motion of the injured area is usually decreased secondary to pain and swelling. Stability of the joints may be compromised if sprains or fractures are present. Range of motion and stability must be recorded for the **joint above and below the injury.**

  Neurologic testing of both motor and sensory components should be performed. The vascular status must also be noted.

Laboratory and Diagnostic Studies

- **Laboratory** Labs are generally not useful unless a suspected pathologic process is involved. A severe injury with only a minor mechanism would raise suspicion for this consideration.

- **Diagnostic studies** Plain x-rays, in two planes, are needed in cases where fracture is suspected. The joints above and below the fracture site should be included on the radiographs. CT or MRI may be a consideration if further evaluation is needed.

Formulating the Diagnosis A careful history and physical examination, with an excellent knowledge of anatomy, will usually reveal the diagnosis. Radiographs are necessary in cases of fracture.

Evaluating the Severity of Problems Fractures are associated with significant trauma, and severe bleeding may be present, depending on the involved bone. The possibility of a **compartment syndrome (tissue ischemia secondary to elevated compartment pressures)** always exists. These demand urgent evaluation and surgical treatment. Similarly, severe strains and sprains demand immediate evaluation. Joint dislocation, following severe ligamentous compromise, must also be acutely addressed.

■ **Managing the Patient** Minor sprains and strains, and even some fractures, may be treated by the emergency or primary care physician. This usually includes elevation, compression, ice, rest, and NSAIDs. Significant sprains should be splinted.

All severe sprains, strains, and fractures demand urgent orthopedic evaluation. These are managed differently depending on the involved area. Usually some sort of splinting is utilized. Surgery is performed for many fractures and some strains and sprains.

Low Back Pain

■ **Basic Science** Low back pain is the **most frequent cause** of worker **disability** in America, resulting in substantial decreased productivity and days lost from work. Approximately 40% of all people will have complaints of significant back pain at some time in their life. About 95% of these cases resolve spontaneously within 6 weeks.

Most acute low back pain is attributable to injuries to the back musculature or ligamentous structures. These are classified as strains or sprains, according to which structure is involved. **Strains** refer to **muscular** injuries, which usually occur at the musculotendinous junction. **Sprains** refer to **ligamentous** injuries and may occur at any point along their bone-to-bone course.

A small proportion of acute low back pain may have significant, serious pathology as its underlying cause. Examples include fracture, neoplasm, infection, or other conditions with potential neurologic compromise. The challenge to physicians is differentiating these potentially serious etiologies from the remainder.

■ **History and Physical Examination**

- **History** Generally patients relate an episode of **lifting using improper body mechanics.** Their pain may be of sudden onset or delayed, the latter being the more common scenario. Low back pain and an inability to move without a substantial increase in this pain are typical complaints. **Radicular symptoms** may be present if there is nerve root irritation. These are usually manifest as **electric-type sensations** in a certain nerve root distribution. Numbness and weakness may be related by the patient as well. **Sciatica** refers to radicular symptoms within the **sciatic nerve distribution.** Ninety percent of cases involve the L5 or S1 nerve root (Fig. 9.3).

  The history should attempt to rule out potentially serious spinal conditions. Significant trauma should raise suspicion for fracture. Constitutional symptoms such as weight loss, fever, and chills may indicate infection or neoplasm. Significant pain not associated with activity may also be indicative of these latter two conditions. **Saddle distribution numbness with bowel and bladder dysfunction (cauda equina syndrome),** with or without lower extremity symptoms, is a neurologic emergency and demands prompt surgical intervention.

- **Physical examination** The examination begins with a careful observation of the patient. Those strapped to a long spine board, verses those pacing the room waiting for the examiner to arrive, demand different approaches and levels of suspicion.

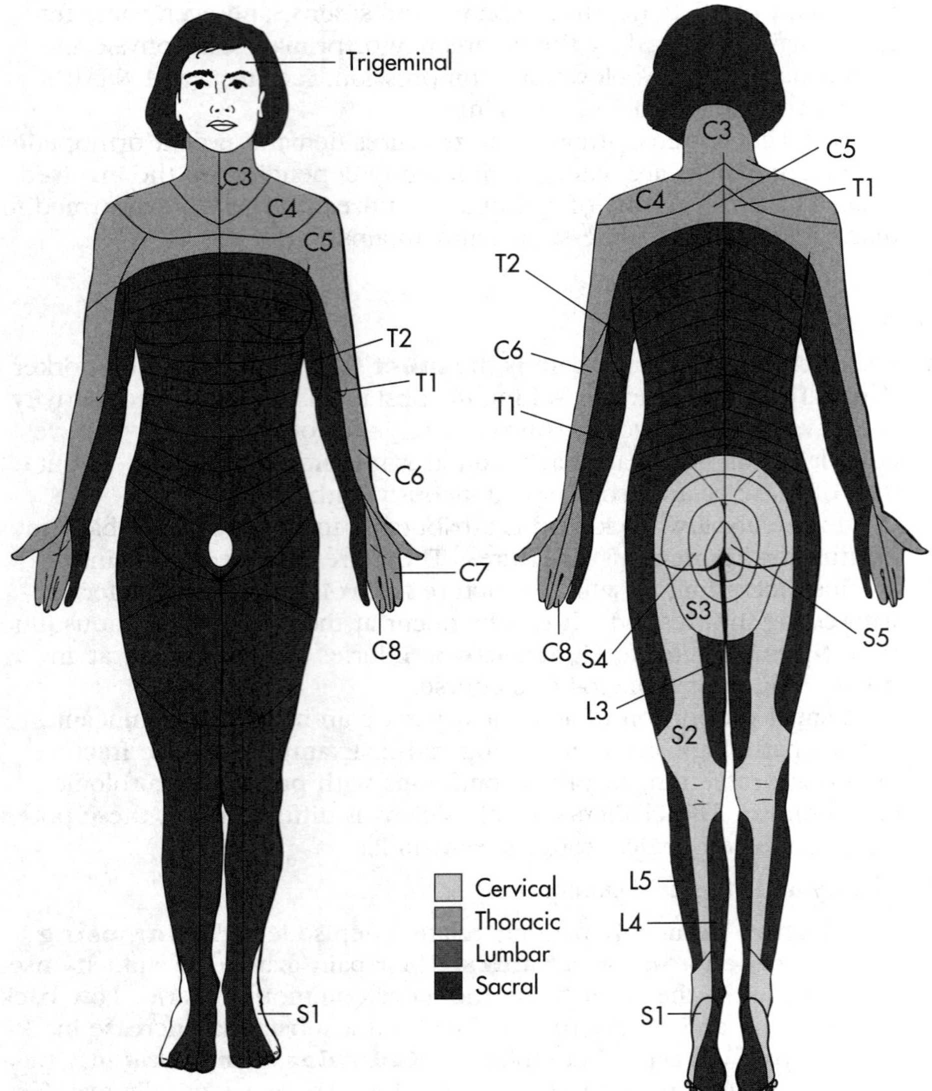

Fig. 9.3 Segmental innervation is illustrated by dermatomes, regions of skin from which afferent fibers converge at a single spinal root. The dermatomes in which sensory disturbances are localized correlate relatively well with the level of anatomic vertebral injury and spinal cord injury. The dermatome maps may vary in different individuals.

A survey of the skin condition, with note of any abrasions, ecchymoses, lacerations, or swellings, should be noted. Careful palpation of the spine should be performed, with attention to severe tenderness, crepitus, and deformity. Often several areas may be tender, and multilevel involvement may be present.

A complete neurologic evaluation is a necessity. This should involve not only the bilateral lower extremities but the perineal area as well. Careful motor and sensory testing of each radicular and dermatomal level, with comparison to the uninvolved side, is mandatory. **Motor** testing should be based on the five-point strength scale, with **0** = no active twitch, **1** = visual or palpable twitch, **2** = contraction with gravity eliminated, **3** = contraction against gravity, **4** = submaximal strength, and **5** = maximal strength. Motor groups of the lower extremity include **L2/3 iliopsoas** (hip flexors), **L4 quadriceps** (knee extension), **L5 extensor hallicus longus** (great toe extension), and **S1 gastrocnemius/soleus** (ankle plantar flexion). Dermatomal sensory charts are available in many texts. Specific isolated areas include **L2 proximal anterior thigh, L3 middle anterior thigh, L4 anterior shin, L5 medial dorsal foot, S1 lateral dorsal foot, S2 plantar foot and posterior leg, and S3/4 perianal area.** Reflexes, including **patellar (L4), achilles (S1), and bulbocavernosus (lower sacral),** are also needed.

Placing the sciatic or femoral nerve under stretch is a useful test for irritability. A positive **straight leg raise,** indicated by reproduction of the radicular symptoms, is useful in testing the sciatic nerve. Hip extension similarly tests the femoral nerve.

- **Laboratory and Diagnostic Studies**
  - **Laboratory** Generally laboratory tests are not needed or useful in the diagnosis of low back pain. However, in those cases where infection or neoplasm is a concern, CBC, ESR, and other studies may be needed.
  - **Diagnostic studies** Radiographs are not indicated in the majority of acute low back pain patients, especially within the first 6 weeks. If symptoms persist, or there is concern for fracture, neoplasm, infection, or other serious condition, then plain x-rays in two planes are indicated. Bone scan, CT, or MRI may be needed, especially in those cases where surgical intervention is anticipated.
- **Formulating the Diagnosis** History and physical examination are usually sufficient for formulating the diagnosis. Radiographs will usually exclude any serious bony pathology.
- **Evaluating the Severity of Problems** Low back pain without neurologic deficit or evidence of radiographic abnormalities can be treated in the routine fashion. If neurologic signs are present, prompt referral to an orthopedic surgeon or neurosurgeon is needed. Cauda equina syndrome demands emergent surgical intervention.
- **Managing the Patient** Conditions without neurologic involvement can be managed by the primary care or emergency physician. Usually **short-term bed rest (2-3 days),** NSAIDS, and early mobilization are recommended. Patient education is also an important aspect of treatment. Most patients improve within 6 weeks.

  For those patients with neurologic impairment or persistent pain, prompt referral is indicated. Surgical intervention depends on the specific underlying condition.

Osteomyelitis and Septic Arthritis

- **Basic Science** The term ***osteomyelitis*** refers to an **infection of the bone. *Septic arthritis*** refers to an **infection of the joint.** These may be classified into acute or chronic based on the duration of the infection. Generally those cases of **shorter duration than 6 weeks are considered acute.** Chronic osteomyelitis or septic arthritis usually results from undiagnosed or inadequately treated acute infections. The etiology of the infection may be bacterial, including mycobacterial or fungal. Some of the more common causative organisms include ***Staphylococcus,*** *Streptococcus, Pseudomonas, Salmonella, Mycobacterium,* and *Aspergillus.*

The method of **inoculation** into the bone or joint may be **hematogenous** (via the vascular supply), by **direct extension** (via an overlying soft tissue infection), or by **penetration** (as in an open fracture).

History and Physical Examination

- **History** The most common complaints include **pain, swelling, warmth,** and **erythema** of the affected site. It is important to inquire about any coexisting infections, including those of the respiratory-pulmonary, genitourinary, gastrointestinal, integument, or other systems. Most patients with hematogenous inoculation have one of these systems as an identifiable source, either concurrently or in the recent past.

  Patient factors such as age, medical health, significant past medical history, and social factors are extremely important. Diabetics commonly acquire osteomyelitis via direct extension from an overlying foot ulcer. Intravenous drug abusers provide direct access for the microorganisms to the circulatory system. Immunodeficient patients may be susceptible to any organisms, especially those uncommon to normal individuals.

- **Physical examination** Patients usually appear **incapacitated** by the infection. They may **limp** or not use the affected extremity.

  Systemic fever is often present, in addition to **local warmth** at the affected site. Tenderness, swelling, and erythema are usually present and guide the examiner to the specific bone involved. In cases of joint involvement, **range of motion will be decreased and painful.** An **effusion** is always present. Drainage may or may not be present in the acute setting, but is typical in chronic cases.

Laboratory and Diagnostic Studies

- **Laboratory** The **erythrocyte sedimentation rate,** C-reactive protein, platelets, and other acute phase reactant tests are **elevated.** The white blood cell count is usually elevated and may show a left shift. Blood cultures may be positive if the infection has spread systemically. Other studies are tailored according to the suspected source of the infection.

- **Diagnostic studies Plain radiographs** in two planes of the affected area is the initial study. Changes observed include **destruction of the normal bony architecture,** evidence of soft tissue swelling, and possible periosteal reaction (depending on the duration of the infection). **Mixed radiolucent and radiodense areas** are common in **chronic cases.** Radiographic evidence of joint effusion is present in septic arthritis. Areas not accessible to plain radiographs or those requiring further investigation may utilize bone scanning, CT, or MRI. These greatly facilitate surgical planning.

  A **biopsy of the infected bone or aspirate of the infected joint** is recommended for organism identification. This is usually easily accomplished percutaneously under indirect or direct visualization. The biopsied material should be sent for Gram's stain and culture analysis. If possible, the biopsy should be performed **prior to antibiotic initiation.**

- **Formulating the Diagnosis** A careful history and physical are key to incorporating this diagnosis into your differential. A tissue biopsy or aspirate will provide the definitive organism.
- **Evaluating the Severity of Problems** Acute osteomyelitis and septic arthritis are severe conditions and must be recognized promptly. Emergent referral to an orthopedic surgeon is necessary as many cases require surgical intervention.

  Systemic sepsis is possible and demands emergent treatment.
- **Managing the Patient** **Acute osteomyelitis** recognized early can usually be managed with **intravenous antibiotics.** Those cases associated with significant bony destruction or **abscess** require **surgical irrigation and debridement** to reduce the organism load. Antibiotics are then initiated and continued for a varied period of time, generally 6 weeks. Laboratory and radiographic follow-up are utilized to monitor progress.

  Chronic osteomyelitis is difficult to manage. Extensive surgical debridement and long-term antibiotics are required. Generally the duration approaches 6 months to years. Amputation is an acceptable treatment in refractory cases.

  Septic arthritis demands emergent **surgical irrigation and debridement** of the joint. Irreversible destruction of the cartilaginous surface occurs within a relatively short time. Postoperative antibiotics are usually continued intravenously for several days and then orally for several weeks.

Pediatric Orthopedics

- **Congenital Torticollis**
  - **Basic science** Torticollis refers to a **unilateral contracture** of the **sternocleidomastoid muscle,** resulting in the head being **tilted toward the side of the shortened muscle** and the **chin rotated toward the opposite side.** The exact etiology of the fibrosis producing this contracture is not completely understood.
  - **History and physical examination**
    - ***History*** Presentation is usually at birth or during the first few weeks of life. The mother notes an inability to rotate the chin to the affected side or tilt the head toward the contralateral side.
    - ***Physical examination*** Observations consistent with the complaints just described are usually found. A **palpable mass** is found within the affected muscle.
  - **Laboratory and diagnostic studies** Laboratory and diagnostic studies are not useful in making this diagnosis.
  - **Formulating the diagnosis** A history and physical examination are all that is required to make the diagnosis. Neurologic, spinal, traumatic, and inflammatory conditions must be excluded.

- **Evaluating the severity of problems** Recognition of this condition must be made promptly with referral to an orthopedic specialist. If the contracture is not treated, secondary deformities of the face and head develop.
- **Managing the patient** Treatment consists of **passive stretching** of the contracted sternocleidomastoid begun immediately after diagnosis is made. The majority of cases respond to this modality. Surgical intervention is reserved for those cases discovered late or those not responding to stretching.

Developmental Dislocation of the Hip

- **Basic science** DDH refers to a wide spectrum of abnormalities of the hip, ranging from simple instability to frank dislocation. It is relatively common, occurring in 2.5 to 10 infants per 1000 live births, with females being affected six to eight times more often than males. The etiology is a combination of genetic and environmental influences.
- **History and physical examination**
  - ***History*** Usually no history is elicited unless the condition is detected late. A limp may be the chief complaint if this is the case.
  - ***Physical examination*** A thorough evaluation of the hips during the newborn period is essential for making this diagnosis. The findings depend on the age at which time the condition is discovered. The **Ortolani test** is a **sign of reentry of the femoral head into the acetabulum from the dislocated position.** The **Barlow test** is a **sign of dislocation of the femoral head from the located position.** Asymmetric gluteal creases may be another sign of an already-dislocated hip.
- **Laboratory and diagnostic studies** Laboratories are not useful in making the diagnosis. **Ultrasound** is utilized to assess the amount of femoral head coverage by the acetabulum in cases of dislocatable hips.
- **Formulating the diagnosis** The physical examination is the key to the diagnosis.
- **Evaluating the severity of problems** Any question of abnormalities with the hips demands urgent orthopedic referral. Delay in treatment can increase the severity of the condition.
- **Managing the patient** In hips that can be located, early treatment consists of a **Pavlic harness,** which keeps the hips in an **abducted and flexed position.** Surgical procedures to locate the hips may be required in cases of late diagnosis or those of severe nature.

Nursemaid's Elbow

- **Basic science Radial head subluxation, or nursemaid's elbow,** is one of the most common orthopedic injuries in young children. Almost all cases occur between 1 and 4 years of age. It is caused by a sudden longitudinal traction on a wrist of a child whose elbow is extended and pronated (Fig. 9.4). The **annular**

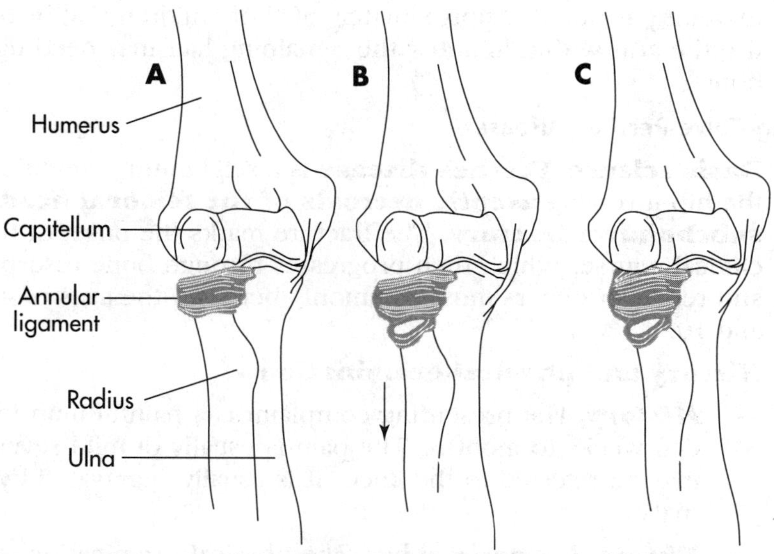

Fig. 9.4 Pathology of pulled elbow: **(A)** normal annular ligament that attaches the radius to the ulna; **(B)** traction on the radius, causing a tear in the attachment of the annular ligament; **(C)** when traction is released, fibers of the ligament are caught between the radial head and the capitellum. *(From Hardwood-Nuss A:* The clinical practice of emergency medicine, *Philadelphia, 1991, J.B. Lippincott.)*

ligament becomes trapped between the radial head and the capitellum, preventing reduction of the radial head.

- **History and physical examination**
  - ***History*** The child is usually brought to the ER with the complaint of disuse and pain of the affected extremity. **A forceful pull** on the extremity by the caregiver (nursemaid) is the presentation.
  - ***Physical examination*** The child **holds the painful extremity in a pronated and slightly flexed position.** There is local tenderness over the radiocapitellar joint. Supination of the forearm is limited and resisted.
- **Laboratory and diagnostic studies** Laboratory studies are not useful. Radiographs may show a volarly subluxated radial head. However, often the subluxation is **reduced unknowingly by the x-ray technician** while trying to obtain a true AP view of the elbow with the forearm in supination.
- **Formulating the diagnosis** History and physical examination are all that is needed to establish the diagnosis.
- **Evaluating the severity of problems** Urgent referral to an orthopedist is necessary if the radial head does not spontaneously relocate. Delay increases the difficulty of relocation secondary to swelling.
- **Managing the patient Reduction** is accomplished by **gentle flexion of the elbow to 90 degrees with subsequent rapid, full supination.** The examiner's thumb may aid by pushing the radial head back into position. A clunk may be palpated during the reduction. Immediate relief of pain and use of the

extremity is noted. Approximately 50% of children will begin to use the arm within minutes; the remaining half may need up to 24 hours.

Legg-Calvé-Perthes Disease

- **Basic science Perthes disease** is a self-limiting condition of the hip involving **aseptic necrosis of the femoral head with subchondral fracture.** The fracture marks the onset of the clinical course, which then progresses through bone resorption and repair. It occurs most commonly between the ages of 4 and 10.
- **History and physical examination**
  - ***History*** The presenting complaint is a painful limp for several weeks to months. The pain is usually of mild severity and may be referred to the knee. It is usually aggravated by activity.
  - ***Physical examination*** The physical examination is remarkable for an antalgic gait on the affected side and **decreased internal rotation and abduction.** A flexion-adduction contracture may be noted.
- **Laboratory and diagnostic studies** Plain radiographs should be obtained of the affected hip. Imaging of the contralateral hip is useful for comparison. The radiographic findings depend on the stage in the natural course of the disease and the extent of involvement. **Areas of necrosis, resorption, or repair** may be identified.
- **Formulating the diagnosis** Although history and physical raise the possibility of this disease, radiographs make the diagnosis.
- **Evaluating the severity of problems** Although not emergent, this disease does require close orthopedic follow-up over a period of several years. Urgent referral should be made.
- **Managing the patient Maintenance of range of motion and femoral head coverage** by the acetabulum are the primary goals of treatment. Protected weight-bearing is usually instituted during the repair phase.

Slipped Capital Femoral Epiphysis (SCFE)

- **Basic science** SCFE describes a condition where the **proximal femoral epiphysis slips off of the metaphysis.** This occurs through the physis and usually results in the epiphysis being posterior and inferior to the normal location. It most commonly occurs in **adolescent, overweight, hypogonadal boys.**
- **History and physical examination**
  - ***History*** The presentation may be acute or chronic. Pain within the hip region and an inability to bear weight in the acute cases is the common scenario. Some precipitating trauma can usually be elicited.
  - ***Physical examination*** On physical examination, range of motion is painful and limited. The patient holds the **hip in external rotation, and shortening is noted.** If the pain

is vague, it may be referred to the knee or thigh instead of being localized to the hip.

- **Laboratory and diagnostic studies** Plain radiographs of the hip demonstrate the slip. The severity is based on the degree of slippage.
- **Formulating the diagnosis** The diagnosis is confirmed by radiographic analysis.
- **Evaluating the severity of problems** SFCE is an emergent condition and requires immediate referral to an orthopedist. The patient should be placed immediately in a wheelchair on a non-weight-bearing status.
- **Managing the patient** Treatment consists of **pin fixation of the slip in situ.** Rarely is reduction of the slip prior to pin fixation performed due to the risk of aseptic necrosis. Since bilaterality is present in about 30% of cases, close observation of the unaffectedhip is indicated.

Juvenile Rheumatoid Arthritis

- **Basic science** Rheumatoid arthritis in children is a **generalized systemic disease of which arthritis is but one component.** The average age of onset is 6 years but may range from 1 to 14. The many variants of this disease may be grouped into three categories: **pauciarticular, polyarticular, and systemic.** The systemic variety is the most severe type, and involvement of multiple organ-systems is common.
- **History and physical examination**
  - ***History*** **Painful, swollen joints** are the usual presenting complaints in the pauci and polyarticular varieties. The systemic variety adds fever, rash, < ocular and other systemic complaints.
  - ***Physical examination*** Tender, swollen joints are found in all varieties. Fever, hepatosplenomegaly, generalized lymphadenopathy, pericarditis, rash, subcutaneous nodules, ocular inflammation, and amyloidosis may occur in the systemic variety.
- **Laboratory and diagnostic studies**
  - ***Laboratory*** The CBC may be normal or have an elevation in the WBC. The ESR is usually elevated, and antinuclear antibodies may be present. A joint aspirate should be sent for cell count.
  - ***Diagnostic Studies*** Plain radiographs should be obtained. **Soft tissue swelling, capsular distention,** and **periarticular osteopenia** are the initial findings. With progression of the disease, articular space narrowing and juxtaarticular erosions occur.
- **Formulating the diagnosis** When one joint is involved on presentation, the differential diagnosis includes osteomyelitis, septic arthritis, and trauma. When more than one joint is involved, other infections and inflammatory processes such as gonococcal arthritis, rheumatic fever, and SLE must be considered.

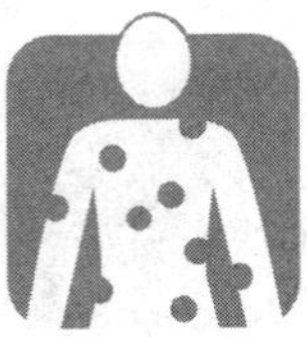

- **Evaluating the severity of problems** A multidisciplinary approach is imperative, with pediatricians, orthopedic surgeons, hematologists, physical and occupational therapists, psychologists, and social workers all involved.

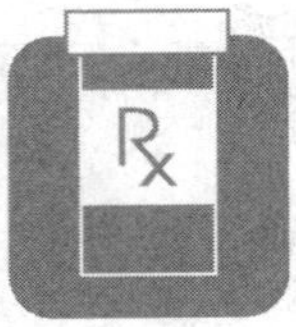

- **Managing the patient** The objectives of orthopedic intervention are to prevent and correct joint contractures and growth deformities of the bones. Conservative treatment involves drug and physical therapy. Surgical intervention may be needed in advanced cases and is tailored to the individual problem.

Multiple Choice Review Questions

1. An obese, 12-year-old, black male presents to the emergency room with a complaint of severe hip pain after stepping off a curb earlier today. He denies any previous hip pain or significant past medical history. On examination, he is afebrile and holds his lower extremity in an externally rotated position. He is unwilling to let the examiner move his hip secondary to pain. Radiographs show the femoral head to be in a slightly posterior and inferior position in comparison to the contralateral side. No joint space narrowing or density changes are noted of the femoral head. The most likely diagnosis would be:

a. Legg-Calvé-Perthes disease.
b. Juvenile rheumatoid arthritis.
c. Septic arthritis.
d. Transient synovitis of the hip.
e. Slipped capital femoral epiphysis.

2. A 50-year-old woman slipped and fell on the ice. After rising to her feet, she noted persistent pain along the lateral aspect of her right foot. On evaluation, she has decreased sensation along the lateral aspect of her foot and weakness in her ankle plantar flexion. Reflex examination shows the ankle jerk to be absent. The nerve root most likely to be involved is:

a. L3.
b. L4.
c. L5.
d. S1.
e. S2.

3. A 2-year-old female is brought to the ER with a complaint of fever to 103.5°, restlessness, and shaking chills. On examination, in addition to the fever and chills, a swelling of the proximal arm is noted. This is erythematous and tender, but range of motion of the shoulder is not decreased. A WBC is elevated to 40 K. Radiographs fail to identify a lytic lesion or evidence of an effusion. Orthopedic consultation is not available for several hours. The next step in the treatment of this patient would be:

a. Broad-spectrum antibiotic administration.
b. Blood cultures.
c. Aspiration of humerus.
d. Bone scan.
e. Observation until an orthopedist arrives.

4. An 18-year-old woman falls off her bicycle onto an outstretched arm. She complains of considerable pain in the wrist, moderate swelling, and decreased motion. On examination, there is a moderate amount of swelling radially. There is no tenderness of the distal radius or ulna, but palpation in the anatomic snuff box is exquisitely tender. Radiographs reveal no fracture. The next step in the treatment of this patient would be:

a. Arm sling with early restoration of motion.
b. Short arm casting.
c. Long arm splinting with repeat radiographs in 10 days.
d. CT scan of wrist.
e. MRI of the wrist.

5. An otherwise healthy 47-year-old woman comes to the office with a complaint of pain in her right hand and wrist, which awakens her at night. She reports that she first noticed the pain about 2 months ago. On physical examination you find an area of hypesthesia involving the thumb and adjacent two fingers. Her strength is normal, and there is no muscle wasting in the hand. The most appropriate management for this patient is:

a. Referral for immediate carpal tunnel release.
b. Observation only, with follow-up in 6 weeks.
c. Splint the wrist in neutral position at night and prescribe NSAIDs.
d. A short course of oral prednisone.
e. MRI of the cervical spine.

Chapter 10

Dermatology

Acne

■ **Basic Science** **Acne vulgaris** is a disease of the hair follicle and associated sebaceous gland. It occurs in at least 85% of adolescents and is considered by some to be physiologic. It is usually self-limited.

The exact etiology of acne is unknown. Key factors include the following:

- Abnormal keratin development within hair follicles resulting in follicular plugging (**retention hyperkeratosis**)
- **Excess sebaceous gland secretion**
- Bacteria, especially ***Propionibacterium acnes***
- **Inflammation**

The first event is follicular plugging; sebaceous gland secretions and *P. acne* accumulate behind the plug. Distention and rupture of the hair structure can then occur with the release of proinflammatory agents. The inflammatory reaction can lead to eventual **scarring.**

Androgens play a role in sebum production and acne formation. **Excess systemic androgens** (as in disease states) or increased localized concentrations **can worsen acne.**

■ **History and Physical Examination** The patient population is primarily **adolescent;** onset usually occurs after age 8 and can persist into late adulthood. **Severe disease is more common in males.** Infants, usually less than 3 months old, can present with acne (i.e., acne neonatorum).

Two general types of lesions exist: noninflammatory and inflammatory. The **comedone** is the prototypic noninflammatory lesion. There are two types of comedones:

- **Open comedones (blackheads)**
- **Closed comedones (whiteheads)**

Closed comedones can progress to inflammatory lesions.

Inflammatory lesions include **papules, pustules, nodules,** and **"cysts."**

Multiple types of lesions in different states of development can be seen in any given patient. The typical locations include the **face, upper chest** and **back,** and **upper extremities.**

Inflammatory lesions may become scarred. Darkened skin at sites of healed acne lesions can occur; lightening requires weeks to months (longer in darkly pigmented individuals).

Acne-like eruptions can occur due to the following:

- Medications (systemic and topical steroids, lithium, and so on)

- Mechanical irritation
- Occupational exposure (tars, cutting oils, chlorinated hydrocarbons)
- Cosmetics and hair care products
- Sun exposure

■ **Laboratory and Diagnostic Studies** There are no routinely used diagnostic studies in the evaluation and treatment of acne.

■ **Formulating the Diagnosis** The diagnosis is clinical. The **existence of polymorphous acne lesions in an adolescent or young adult is virtually diagnostic.** The presence of **monomorphic lesions should raise suspicion of drug-induced, mechanical, or occupationally induced acne.** The **sudden worsening of preexisting acne** should raise concern about possible **gram-negative folliculitis.** The differential diagnosis also includes acne rosacea, folliculitis, perioral dermatitis, sarcoid, and others.

■ **Evaluating the Severity of Problems** Clinical evaluation of the severity of inflammatory lesions and potential for and existence of scarring is crucial. Patients with suspected gram-negative folliculitis should be cultured.

Onset of acne at an older age or severe, persistent acne in females should raise concern about androgen excess, either of adrenal or ovarian origin. Severe, persistent acne in prepubescent children should raise similar concerns. Endocrine consultation may be indicated.

■ **Managing the Patient** Many treatments for acne vulgaris exist; **three general categories are topical agents, systemic agents, and surgical approaches.**

- **Topical** First-line topical agents in current use include tretinoin, benzoyl peroxides, and antibiotics. **Tretinoin** cream (and gel) works by reversing abnormal keratinization; it is the treatment of choice for comedonal acne. **Benzoyl peroxides** exert a potent bacteriostatic effect on *P. acne.* The topical antibiotics in greatest use are **clindamycin** and **erythromycin;** they exert antibacterial and antiinflammatory effects.

- **Systemic** Systemic antibiotics are indicated for most inflammatory lesions. **Tetracycline** and **erythromycin** are first-line agents. **Tetracycline** must be taken on an empty stomach for maximal absorption; it may cause gastrointestinal upset, can increase the risk of vaginal yeast infections, and is **contraindicated in pregnancy and children younger than 12 years old** because of its effects on teeth and bones. Erythromycin can produce gastrointestinal upset; this may be reduced if taken with food.

  Any systemic antibiotic may decrease the effectiveness of oral contraceptives. This risk is small but must be recognized.

  Severe scarring nodular cystic acne is best managed by a dermatologist as systemic retinoids are indicated. **Accutane (13-cis-retinoic acid)** inhibits sebaceous gland activity and can markedly improve severe acne. Potential side effects include **teratogenic birth defects,** liver chemistry abnormalities, hypertriglyceridemia, dry skin and mucous membranes, musculoskeletal complaints, the risk of future scarring at sites of trauma or surgery, and others. Its use requires compulsive laboratory and clinical monitoring. Accutane must not be used by women who are pregnant or those who might become pregnant during treatment.

- **Surgical** Surgical approaches are used during all stages of acne, from comedone extraction to dermabrasion and excision of scars.

Actinic Keratosis

- Basic Science **Actinic keratosis** is a **precancerous lesion** consisting of atypical keratinocytes that have undergone clonal proliferation. Actinic keratosis **may transform into squamous cell carcinoma.**
- History and Physical Examination
  - An **actinic keratosis** presents as a **rough, sandpapery, red or tan patch ≤1 cm in size on sun-exposed skin.**
  - **Actinic keratoses** are most commonly seen on **sun-damaged skin of white adults.**
  - **Immunosuppressed individuals** (such as **renal transplant patients**) are at **high risk.**
- Laboratory and Diagnostic Studies **Lesions over 1 cm in size, nodularity, or ulceration raise the suspicion of squamous cell carcinoma and require biopsy.**
- Formulating the Diagnosis **Clinical inspection** is usually sufficient to establish the diagnosis.
- Evaluating the Severity of Problems Less than 0.1% of actinic keratoses develop into squamous cell carcinoma. However, most patients present with multiple lesions; therefore over 12% of patients with actinic keratosis will develop squamous cell carcinoma.
- Managing the Patient **Cryosurgery, topical 5-fluorouracil, and electrocautery are used to treat actinic keratoses.** Sun protection with wide-brimmed hats, **sun-protective clothing, and** topical **sun block** reduce the development of new lesions.

Atopic Dermatitis

- Basic Science *Atopic dermatitis* and *eczema* are frequently used synonymously. Atopic dermatitis is a **genetically determined skin disorder** in which factors such as antigens, mechanical friction, and irritating materials **produce an excessive reaction characterized by itch.** It is frequently **associated with hay fever, asthma,** and **IgE elevation** in the peripheral circulation. It occurs in up to 20% of the population.
- History and Physical Examination Atopic dermatitis occurs at any age but **usually starts before 5 years of age.** Acutely it presents as edematous areas of red, weepy **blisters.** Upon rubbing and scratching, the blisters break, producing **oozing and crusting. White dermatographism (a white streak left in the skin after pressing on the skin) is a characteristic finding.**

• **Infantile** atopic dermatitis frequently involves the **scalp, face, trunk, and extensor areas** (the forearms and over the knees and legs), which are rubbed and subject to friction during crawling.

• **In later childhood, the dermatitis involves the flexor areas** (e.g., the axillae, and antecubital, and popliteal fossae). This presentation usually lasts from 12 to 20 years of age.

• Most **adults** will have "sensitive skin," and a small minority will continue to have full-blown atopic dermatitis throughout life. In **adult-hood,** "coin-shaped" **skin patches** are **common,** especially during the winter season. Other adult manifestations include pruritic ill-defined skin patches and **hand dermatitis.**

A family or personal **history of atopy** (i.e., dermatitis, hay fever, and asthma) is common. The history needs to focus on **aggravating factors** such as irritating clothing, skin care products, wet work, frequent hot-water baths or showers (in adults), and overheating from prolonged wearing of occlusive gloves. **Chronic atopic dermatitis is characterized by skin thickening and scaling.**

■ **Laboratory and Diagnostic Studies** **Patch testing** for contact allergens is appropriate in recalcitrant cases of atopic dermatitis.

■ **Formulating the Diagnosis**

• **Diagnosis is established clinically. It is based on the physical appearance and duration of the lesions, as well as a family history of atopy.**

• Contact dermatitis and atopic dermatitis have similar appearances. Therefore the physician needs to see the patient several times before the diagnosis of atopic dermatitis is made.

• Secondary infection with yeast, dermatophyte, or bacteria needs to be excluded.

■ **Evaluating the Severity of Problems** Severity is correlated to the extent of dermatitis, presence of infection, and resultant impairment (e.g., inability to use hands or to walk due to severe inflammation). Hospitalization is rarely necessary, and only after failed outpatient management. Cataracts may develop in patents with protracted atopic dermatitis. **Secondary** staph or strep **infection is common.**

■ **Managing the Patient** **Alleviation of inflammation and itching, treatment of any infection, and a careful search for irritants and allergens are necessary.**

• **Topical steroids** are the drugs of choice for the treatment of atopic dermatitis.

• Tepid compresses may be used if crusting is present.

• **Oral antihistamines** (diphenhydramine 25 to 50 mg) are indicated for control of itching. Topical antihistamines should be avoided since contact dermatitis may result.

• Oral corticosteroids or other immunosuppressants have been used as a last resort for severe, unresponsive, generalized atopic dermatitis.

• A short course of hospitalization to remove the patient from environmental etiologic factors is sometimes indicated.

• Finally, the abrupt onset of a **recalcitrant eczema** in an adult can be a sign of **internal malignancy** and should stimulate an appropriate workup.

Basal Cell Carcinoma

- Basic Science **Basal cell carcinoma** is a malignant tumor arising from the basal keratinocytes of the epidermis. It **is the most common skin malignancy,** accounting for 500,000 cases annually in the United States. It is common among fair-skinned whites but is unusual in blacks, Hispanics, and Asians. **It is locally destructive and invasive; however, metastases are extremely rare** (less than 0.1% of cases).
- History and Physical Examination **Chronic sun exposure is the most common predisposing cause** for basal cell carcinoma. Other causes include exposure to arsenic, pesticides, certain medications (e.g., potassium arsenite), and ionizing radiation. Ask about a history of skin cancers, since individuals with a history of basal cell carcinoma are at increased risk for additional lesions. Certain hereditary diseases carry an increased risk for basal cell carcinomas: albinism and xeroderma pigmentosum.

  Basal cell carcinoma is **slow growing;** it occurs most commonly on the **face, especially the nose.** The most frequent type of basal cell carcinoma is **nodulocystic** or **noduloulcerative.** It presents as a raised, pale, white or yellow-tinged lesion with overlying telangiectasia. As it enlarges, the center of the lesion ulcerates (hence the nickname ***rodent ulcer***). Occasionally, basal cell carcinomas are **pigmented** with blue-black coloration. These lesions must be differentiated from melanoma.
- Laboratory and Diagnostic Studies **Biopsy and histologic examination are mandatory** when basal cell carcinoma is suspected.
- Formulating the Diagnosis Consider the diagnosis of basal cell carcinoma when seeing persistent nodulocystic or noduloulcerative lesions. **Skin biopsy is required for diagnostic confirmation.**
- Evaluating the Severity of Problems Basal cell carcinoma is locally invasive and disfiguring but metastasizes extremely rarely. When metastasis does occur, regional nodes, lungs, and bones are most commonly involved.
- Managing the Patient Basal cell carcinoma can be treated with **sharp dissection, electrosurgery and curettage, cryosurgery, ionizing radiation, and Mohs' microscopically controlled surgery.** Mohs' surgery is indicated for recurrent or poorly defined basal cell carcinoma and lesions in anatomic areas such as the eyelid, canthi, and nasolabial or postauricular folds.

  Preventive efforts include sun protection with comfortable clothing, hats, and sun block that provides both UVA and UVB photoprotection. Sun exposure should be limited between 10:00 AM and 4:00 PM, when solar irradiation is most intense.

Cellulitis and Erysipelas

- Basic Science **Cellulitis is an acute cutaneous bacterial infection accompanied by systemic symptoms. Erysipelas is a**

very characteristic type of superficial cellulitis, involving only superficial dermal tissue. Erysipelas involves lymphatic vessels more extensively than cellulitis. Recurrent episodes of erysipelas result in scarring of lymphatic channels.

The causative agents of cellulitis and erysipelas are almost always **group A β-hemolytic *Streptococci* and *Staphylococcus aureus.* Diabetics tend to have a polymicrobial cellulitis** with a greater potential for a septic presentation.

Minor trauma or an existing skin dermatosis provide a portal of entry for bacteria that result in cellulitis and erysipelas. Conditions predisposing to cellulitis and erysipelas include diabetes, malnutrition, immunosuppression, lymphatic or venous obstruction, and prior history of cellulitis.

■ **History and Physical Examination** Patients present with the abrupt onset of fever, chills, malaise, and nausea. Within 1 to 2 days of these symptoms an area of **red, warm, and tender skin appears and begins to grow in size.** A history of trauma to the affected area may be elicited from the patient.

Physical examination reveals a febrile patient.

- **Erysipelas** is distinguished by a hot, bright red skin lesion with a **sharply demarcated border.**
- **Cellulitis** is a hot, red, skin lesion with an **indistinct border.**

Involved areas may develop purpura, blisters, and even small areas of necrosis. Lymphangitis and painful, enlarged lymph nodes may occur.

- Cellulitis occurs most commonly on **the lower extremities,** although it may occur in **any site after trauma** (e.g., puncture wounds, animal bites or scratches) and **in chronically edematous skin** (e.g., an upper extremity after axillary lymph node dissection for breast cancer).
- Erysipelas occurs most commonly on the lower extremities and on the face.

■ **Laboratory and Diagnostic Studies** **Culture or biopsy from the lesion are usually negative. Blood cultures also are usually negative but should be obtained if sepsis is suspected.** In cases of operative wound streptococcal cellulitis, Gram's stain of serious discharge may show chains of gram-positive cocci.

Complete blood counts will reveal an elevated white blood cell count of 15,000-20,000 or higher.

Urinalysis may reveal changes of **poststreptococcal glomerulonephritis** (usually evident **10 to 14 days after the onset of infection**).

■ **Formulating the Diagnosis** The diagnosis of cellulitis and erysipelas is made based on the clinical picture as just outlined. The differential diagnosis includes contact dermatitis, venous thrombosis, and lymphatic obstruction by tumor (e.g., recurrent breast cancer).

■ **Evaluating the Severity of Problems**

- **Involved skin should be marked daily with a pen to ensure that patients are responding to antibiotics.**

- **Systemic infection needs to be recognized when present;** sepsis, pneumonia, and even death may occur in untreated patients.
- There is a **risk of extension of the infection** to muscle, bone, or other tissues.
- **Poststreptococcal glomerulonephritis is a potential complication of cellulitis.**

■ Managing the Patient

- **Antibiotic therapy** with a **penicillinase-resistant penicillin** (e.g., oxacillin) or cephalosporin is the mainstay of treatment.
- Erythromycin is an alternative agent.
- **Bed rest with elevation of the involved site is essential. Cool, clean compresses provide symptomatic relief.**
- With recurrent infection, **coexisting conditions need to be treated.**

Common Warts (Verrucae Vulgaris)

■ Basic Science Warts are **benign epithelial tumors** caused most frequently by **human papilloma virus.**

■ History and Physical Examination Common warts may appear at any age but are most **common in older children.** Warts may occur on the skin individually or in groups. **After several months the wart will commonly disappear; however, warts may persist for many years.** Warts may also reappear after periods of dormancy.

Common warts are raised, **brown, or tan nodules.** They are **firm, and round or irregular.** Warts most commonly appear in areas subject to trauma such as **fingers, elbows, and knees.**

■ Laboratory and Diagnostic Studies If there is concern about potential malignancy, the wart should be removed and sent for histologic evaluation.

■ Formulating the Diagnosis Common warts are recognized by their appearance.

■ Evaluating the Severity of Problems Some papilloma viruses may cause malignant warts (i.e., cutaneous warts in immunosuppressed and transplant patients). Suspicious lesions should be referred to a dermatologist.

■ Managing the Patient Most common warts spontaneously disappear within 2 years of appearance. The **physician may remove the wart by any of the following:**

- Having the patient **apply a collodion solution daily (salicylic and lactic acid solution)**
- Applying **liquid nitrogen** to the wart for 30 seconds
- Electrodesiccation (causes scarring)
- Laser surgery (causes scarring)

About one third of warts will reappear after removal by any means.

Contact Dermatitis

■ **Basic Science** **Contact dermatitis** occurs because of both immediate **chemical irritation** and **type IV delayed hypersensitivity reaction.**

■ **History and Physical Examination** Patients will usually **present with acute or chronic skin irritation and inflammation.** The source of irritation may include **poison ivy, clothing, rubber** and latex products, **metals** (e.g., chains and bracelets), **dyes, cosmetics, and drugs.** Often the patient has used a skin medication for years before contracting an allergy.

Symptoms vary according to the severity of reaction. **Redness, irritation, itching, and sometimes blister formation** are observed. The dermatitis is usually well demarcated and occurs where the allergen contacts the skin. **Once the cause is removed,** the **dermatitis** generally **disappears** within several weeks.

■ **Laboratory and Diagnostic Studies** **If the source of irritation is unclear, patch testing** with a standard group of allergens should be considered.

■ **Formulating the Diagnosis** A history as just outlined, in combination with characteristic skin changes, suggests contact dermatitis. Confirmation is made with patch testing.

■ **Evaluating the Severity of Problems** **Severe reactions** can cause **facial and eyelid swelling.** In rare cases the physician must pay **close attention to** the patient's **respiratory status.**

■ **Managing the Patient** The **allergen should be removed from the skin.** Unless the allergen is removed from the skin, treatment will be ineffective. Mild reactions can be managed by **gently washing the skin** and applying **topical corticosteroids.** If the **reaction is severe (or facial swelling occurs), oral corticosteroids should be given.** Antihistamines are ineffective in contact dermatitis.

In known poison ivy exposure the skin should be washed immediately with soap and water.

Dermatophyte Infections

■ **Basic Science** **Dermatophytes are fungi that exclusively invade the dead tissues of the skin, nails, or hair.** Several fungi exist in this ecological niche. Differentiating between fungi is often difficult, so infections are classified according to the site involved.

■ **History and Physical Examination** Infection is classified according to the area infected:

- Nail fungal infection (tinea unguum)
- **Ringworm of the body (tinea corporis)—annular lesion with raised borders:** pink to red with central clearing
- **Ringworm of the feet (athlete's foot, tinea pedis)—itching, pain, and inflammation occurring between the toes:** areas of maceration, with raised blister-like lesions; occurs more often in warm weather
- **Ringworm of the crotch (jock itch, tinea cruris)**—areas

of **maceration, pain, and erythema in the crotch:** more common in males; recurrence is common; associated with tight clothing and the summer season

- **Ringworm of the scalp (tinea capitis)—patches of dryness, scaliness and hair loss, or boggy inflammation of the scalp:** occurs most commonly in **children**

■ Laboratory and Diagnostic Studies Hairs from affected areas examined with potassium hydroxide may demonstrate the fungus when viewed **microscopically.**

■ Formulating the Diagnosis The diagnosis is made clinically in combination with the result of microscopic analysis and fungal culture.

■ Evaluating the Severity of Problems Most dermatophyte infections are superficial and may be **treated with topical antifungals;** however, some infections may become widespread or resistant to therapy. Ringworm of the nail bed is especially difficult to eradicate.

■ Managing the Patient Most dermatophyte infections are superficial and may be **treated with topical antifungals such as miconazole, clotrimazole, terbinafine, and ketoconazole.** For **infections that become widespread** or are resistant to topical antifungals, systemic therapy with **oral agents (griseofulvin, terbinafine, ketocomazole)** is used. Griseofulvin (and the azoles) is contraindicated during pregnancy.

Genital Warts

■ Basic Science **Genital warts (condyloma acuminata)** are lesions of **epithelial proliferation** due to infection by **human papilloma virus (HPV).**

HPV is a double-stranded, **highly contagious** DNA virus of the papillomavirus family. Viral DNA replication occurs only in a differentiated epithelium, limiting infection to the most superficial layer of skin.

There are **more than 60 HPV types. Genital warts are most often associated with HPV 6 and 11.** An association of HPV with **epithelial dysplasia and carcinoma** has been shown. There is a frequent association of **HPV 16 and 18 with cervical dysplasia** and carcinoma of the cervix, vulva, and penis.

Genital warts are **acquired by contact,** either during sexual intercourse or passage through an infected birth canal. The incubation period after exposure lasts weeks to months.

■ History and Physical Examination A **history of other sexually transmitted diseases may be elicited.** The existence of genital warts in the mother of an affected infant may be demonstrated.

Condyloma acuminata **occurs on both skin and mucus membranes. Genital lesions in men** are often initially **at the frenulum or coronal sulcus, and at the posterior introitus in females** (i.e., areas of greatest contact). Perianal and rectal lesions are seen predominantly in those who practice anal intercourse. Intraoral lesions may occur via sexual transmission.

Genital warts can vary widely in appearance from multiple, coalesced, flesh-colored to pink or grey papules with a ridged, moist surface, to large cauliflower-like growths. Genital warts are usually asymptomatic. Infected patients do not have lymphadenopathy.

- **Laboratory and Diagnostic Studies** Subclinical lesions may be demonstrated by surface whitening with dilute acetic acid **(acetoacid whitening).** A biopsy will aid diagnosis and is essential when dysplasia or carcinoma is suspected.
- **Formulating the Diagnosis** The **diagnosis of genital warts is made clinically** as just outlined. The differential diagnosis includes condylomata lata of secondary syphilis, squamous cell carcinoma, and benign growths.
- **Evaluating the Severity of Problems** The **demonstration of externally located genital warts mandates internal examination.** Females should undergo a thorough pelvic exam with a Pap smear to look for dysplasia. Anoscopy should be performed when perianal involvement exists. Urethral visualization should be done if urethral lesions are suspected. **Sexual partners should be evaluated. Young children with condyloma should be evaluated for sexual abuse,** particularly if new lesions appear after the age of 2.
- **Managing the Patient** Local destructive approaches include the following:
    - **Cryotherapy with liquid nitrogen**
    - Application of 20% to 50% **trichloroacetic acid**
    - **Electrocautery**
    - Application of **podophyllin resin** followed by washing off within 4 to 8 hours

Podophyllin should not be given to pregnant women. Repeat office visits are necessary until lesions are completely gone.

Carbon dioxide laser ablation can be used for larger, recalcitrant lesions. The vapor produced by laser ablation should be evacuated since it contains potentially infectious viral DNA. **Surgical excision** is occasionally required and may be the treatment of choice either for large, pedunculated lesions, or when dysplasia is suspected.

Herpes Zoster

- **Basic Science** The varicella zoster virus causes **varicella (chicken pox)** and **zoster (shingles).** The virus is a double-stranded **DNA virus** that belongs to the herpes virus family. It is very contagious. Primary **varicella is** usually **transferred by respiratory secretions,** though the fluid of chicken pox and **zoster lesions may also transmit the disease.** The incubation period is approximately 14 days. After the incubation phase, the skin lesions of chicken pox appear; three to five successive crops of lesions may appear. The skin then clears. **The virus may remain alive, dormant in sensory nerve ganglia for many decades.**

Reactivation of the virus is felt to be due to a decrease in cellular immunity. The reactivated virus replicates in sensory nerve ganglia, causing local inflammation and destruction. After replication, the virus travels down sensory neurons into the skin. This results in the cutaneous eruption of zoster. The virus may reactivate and produce pain without causing skin lesions.

■ History and Physical Examination **Patients with zoster usually have a history of chicken pox.** Coexisting **immunosuppressive conditions may be noted** (e.g., malignancy, AIDS, and the use of immunosuppressive drugs). Events that may hasten reactivation of the virus include trauma, surgery, or sunburn.

With zoster, a **prodromal syndrome** of sharp or lancing pain, burning, fever, headache, malaise, and tenderness **at the site of future skin lesions is often noted. After 1 to 4 days, grouped, red papules erupt.** The eruption is unilateral and **confined to one or a few dermatomes.** The most commonly affected dermatomes are thoracic, cervical, trigeminal, and lumbosacral. Clearing occurs within 2 to 4 weeks. Patients may have tender regional adenopathy.

Involvement of the first branch of the trigeminal nerve is of special concern, as 50% of these patients will develop ophthalmic involvement. Lesions of the tip of the nose suggest potential eye involvement.

■ Laboratory and Diagnostic Studies A **Tzanck smear** performed by scraping the base of a vesicle or pustule and staining **should reveal multinucleated giant cells.** Cell culture and tests for viral antigens are specific and serve to distinguish varicella zoster from other herpes viruses.

■ Formulating the Diagnosis The **diagnosis is based on the history (prodromal pain) and physical findings (vesicles in a dermatomal distribution).** The prodromal pain syndrome can be confused with multiple conditions including myocardial infarction, cholecystitis, pleuritis, or otitis.

■ Evaluating the Severity of Problems Potential complications of zoster depend on the site of reactivation and the host's immunologic status. Patients with cranial nerve V_1 involvement should be evaluated early by ophthalmology. Patients with ear involvement should be evaluated by an ENT.

The existence of distant lesions removed from the site of primary disease suggests disseminated disease. Immunocompromised patients are at increased risk for disseminated disease that may lead to pneumonitis, hepatitis, pericarditis, and neurologic involvement. Secondary bacterial infections and skin necrosis can occur.

The most common complication of zoster is persistent pain after the skin clears (postherpetic neuralgia). Over one third of the patients with zoster will experience persistent pain for some time (50% of cases resolve in 3 months and 75% of cases within 1 year).

■ Managing the Patient **Acyclovir has been shown to halt the progression of skin lesions, increase viral clearing from vesicle fluid, and reduce disseminated disease. Its effect on postherpetic neuralgia suggests a lower incidence and shorter duration of disease.**

Recommendations for treatment with Acyclovir include the following:

- Immunocompetent patients greater than 50 years old

- Immunocompetent patients with evidence of cutaneous dissemination
- Immunocompromised patients

In immunocompetent patients less than 50 years old, treatment is symptomatic (i.e., analgesics, cool compresses, and calamine lotion). Ointments and topical steroids are contraindicated.

Postherpetic neuralgia is treated with analgesics. Narcotics should be avoided. Systemic corticosteroids may lessen the incidence of postherpetic neuralgia. Topical capsaicin, tricyclic antidepressants (e.g., amitriptyline), and antiseizure medications (e.g., carbamazepine, valproate) may lessen symptoms. Referral to pain care clinics may be helpful.

IMPETIGO

Basic Science

- **Impetigo is a superficial cutaneous infection usually due to group A β-hemolytic streptococci.**
- ***Staphylococcus aureus*** is the sole cause in approximatley 10% of cases and **often is the cause of bullous impetigo.**
- **Newborns can develop group B streptococcal impetigo.**

Children and young adults are most commonly affected; the prevalence is highest in late summer and early fall. Impetigo is **highly contagious;** it is acquired either from persons with impetigo or from bacterial carriers. **Minor trauma is needed for infection to occur.** It is found most commonly on the **face** (especially the nasolabial and perioral areas), extremities, and trunk.

Poststreptococcal glomerulonephritis can complicate impetigo. Chronic carrier states can occur with **colonization of the nasopharynx.**

History and Physical Examination A history of exposure to affected individuals and crowded living situations may be elicited. There are no preceding systemic symptoms of note. Examination reveals **papules, vesicles, and pustules that very rapidly become** crusted and a honey-colored yellow. **Bullous impetigo develops blisters** (0.5-1.0 cm) lasting up to 5 days. Lesions persist for several weeks without treatment and heal without scarring. Enlarged regional lymph nodes may occur.

Laboratory and Diagnostic Studies Culture and Gram's stain of early vesicles (or the base of older lesions) **will demonstrate gram-positive organisms.** Antistreptolysin O (ASO) titers are usually negative.

Formulating the Diagnosis **The clinical picture of honey-colored crusted lesions is very suggestive.** The differential diagnosis includes the following:

- Chicken pox (presents with more extensive distribution)
- Herpes zoster (has a prodromal burning or tingling sensation, more hemorrhagic crusting, and a dermatomal distribution)
- Contact dermatitis

■ Evaluating the Severity of Problems

- **Impetigo usually responds to treatment without complications,** although untreated patients may develop a carrier state.
- The **risk of glomerulonephritis** needs to be **emphasized.**

■ Managing the Patient **Systemic antibiotics** (e.g., penicillin, erythromycin, or an oral first-generation cephalosporin) is the treatment of choice. **Topical treatment with mupirocin ointment** (Bactroban) is often effective for localized infections.

Keloids and Hypertrophic Scars

■ Basic Science **Keloids** and hypertrophic scars **result from hyperproliferation of fibrous tissue following skin injury.** In normal scars, fibroblasts are arranged parallel to the skin surface. In keloids and hypertrophic scars, there is a whorled arrangement of fibroblasts. Keloids are more common in blacks.

■ History and Physical Examination Keloids present with a history of skin injury or inflammation (e.g., cystic acne lesions) followed soon, or up to several months later, by an overgrowth of a raised, well-circumscribed lobular scar. Hypertrophic scars are raised scars that follow the lines of original injury. A keloid is larger than the original injury site; for instance, a large lobulated nodule may occur at the site of ear piercing. Keloids can itch and be painful, and range in color from flesh colored to hyperpigmented, or from pink to purple.

■ Laboratory and Diagnostic Studies Biopsy may be, but is rarely, used to identify keloids.

■ Formulating the Diagnosis The diagnosis is made by appropriate history and physical examination.

■ Evaluating the Severity of Problems Discomfort, disfigurement, and restriction of movement are the main concerns. Ulceration and hyperesthesia can also occur.

■ Managing the Patient

- **Smaller keloids can be injected with steroids.**
- **Larger lesions can be excised with steroid injection at the edges.** Larger lesions may be grafted if necessary with pressure dressings applied to both the donor and graft site.

Intralesional steroids should be used at the first sign of new hypertrophic scarring.

Malignant Melanoma

■ Basic Science **Malignant melanoma** is a melanocyte neoplasm. Melanomas generally present as **pigmented lesions of various size and shape.** These lesions can invade and **metastasize to distant organs** so rapidly that the patient may die within months of diagnosis. Patient **prognosis is correlated with the depth of the melanoma.** Early detection and removal of the tumor are critical to the treatment of malignant melanoma. Melanoma is responsible for 2% of all cancer deaths.

In the United States over 30,000 new cases are diagnosed annually. The incidence has been increasing and is attributed to ultraviolet radiation exposure of unpigmented skin. **Body sites that receive intermittent recreational sun exposure, (i.e., the trunk of men and lower extremities of women) are at increased risk for development of melanoma.** Sites of primary lesions besides the skin include the eye, meninges, oral mucosa, urethra, and anus. Sixty percent of malignant melanomas arise in normal skin; 40% develop from pigmented moles.

Melanoma is much less common in **blacks and Asians.** The most frequent sites are the soles, mucus membranes, palms, and **nail beds.**

There are four major types of melanoma:

- *Superficial spreading*—(over two thirds of melanomas). Patients generally notice mole enlargement or irregular coloration. This type usually presents as a raised plaque with red, blue-black coloration and is found most commonly on the trunk of men and the legs of women.
- *Lentigo maligna—arises on photodamaged skin of older individuals.* It is usually 1-3 cm in size and is a flat, irregularly pigmented tan-brown spot.
- *Nodular-presents as dark, raised growths.* The lesion is usually asymptomatic. Patients seek medical attention because of rapid lesion growth.
- *Acral lentiginous—the most common melanoma in Africans and Asians.* These lesions have an irregularly bordered and pigmented tan-brown-black macular component and may be found under the nail bed, digits, palms, and soles.

■ **History and Physical Examination** Inability to tan and frequent, severe sunburns are associated with a higher risk of melanoma. Other factors associated with increased risk include family history of melanoma, increased numbers of moles, and clinically atypical moles.

Patients with xeroderma pigmentosum, a disease of defective DNA repair, have a several thousandfold increased incidence of malignant melanoma. The cardinal indication of melanoma is a **persistent, changing, pigmented lesion. Common presenting features of malignant melanoma are as follows:**

- **Increase in the size** of a pigmented lesion
- **Change in the color** of mole or lesion (i.e., lightening, darkening, or irregular pigmentation)
- **Change in the lesion's border**
- **Irritation** (i.e., itching, bleeding, pain, or tenderness)
- **Lesion ulceration**
- **Local signs of lesion inflammation**

■ Laboratory and Diagnostic Studies **Excisional biopsy should be done on all highly suspicious lesions.** There is no evidence that incisional biopsy causes "seeding" of tumor.

Two methods are used to histologically measure melanomas: the **Clark level** (measures **depth of invasion** from surface of skin) and the **Breslow thickness** (measures **thickness of lesion**).

■ Formulating the Diagnosis **Diagnosis is established by careful histologic examination.**

■ Evaluating the Severity of Problems Palpable nodes mandate further investigation for metastases, such as node dissection or biopsy. Liver chemistries, liver-spleen scans, and scanning of the chest and abdomen are indicated in these cases.

Prognosis is directly related to the depth of invasion. For tumors <0.85 mm in thickness, 5-year survival is better than 99%; for tumors ≥3.65 mm the 5-year survival is only 42%.

■ Managing the Patient **Melanomas are treated by surgical excision.** Staging is essential for managing the patient. A three-stage system is used:

- **Stage I** refers to **localized disease** with no evidence of regional lymph node involvement. Therapy for stage I melanoma generally consists of **surgical excision.** Patients should be checked at 6-month intervals for several years and annually thereafter (physical examination should include complete skin examination, node palpation, chest x-ray, CBC, and liver function tests).
- **Stage II** indicates **regional node involvement.** Therapy for stage II includes **wide surgical excision and regional lymph node dissection.**
- **Stage III** indicates that **distant metastases** are present. Therapy for distant metastasis is surgical excision with **radiation, chemotherapy, and immunotherapy.** Treatment includes relief of pain and discomfort.

Psoriasis

■ Basic Science The essential pathogenic event in psoriasis is **increased epidermal cell proliferation.** The disease affects 2% to 4% of the population.

■ History and Physical Examination Psoriasis can occur at any age; however, onset is common during the young adult years. A family history of psoriasis is common.

The onset of psoriasis is usually gradual. Precipitating events for the development of psoriatic lesions **include trauma, infection** (especially with β-hemolytic streptococci), emotional stress, and endocrine factors. Precipitation or worsening can also be **drug induced;** known culprits include lithium carbonate and β-blockers.

The classic lesion of psoriasis is a **sharply demarcated, silver-colored scaly patch with a red base.** A variety of raised lesions and **plaques of various sizes** may be present. These **nonpruritic** plaques are usually found on extensor surfaces (i.e., **knees and elbows**).

The nails are frequently involved. Scalp involvement with thick, large, flaky scales is common. **Arthritis is common;** psoriatic arthritis (a rheumatoid factor-negative form of arthritis sometimes indistinguishable from rheumatoid arthritis) is a rare complication of psoriasis.

■ Laboratory and Diagnostic Studies **Skin biopsy is important in uncertain cases.** Biopsy utilizing direct immunofluorescence is indicated when lupus is in the differential diagnosis.

- Formulating the Diagnosis **The diagnosis is made both by history and by clinical inspection of the lesion.** Presence of scalp and nail involvement helps confirm the clinical impression. Drug eruptions and several other skin disease states may present with similar physical findings, making biopsy necessary.
- Evaluating the Severity of Problems Some forms of **psoriasis can be life-threatening.** A **dermatology consultation** and inpatient management are essential in certain cases (e.g., in patients presenting with systemic signs of infection). **Major complications of severe psoriasis include sepsis and electrolyte imbalance.**
- Managing the Patient There are only a few effective therapies for the treatment of psoriasis. These include topical corticosteroids, lubricants, and keratolytics. Since most cases of psoriasis cannot be cured, treatment is aimed at controlling the lesions. Topical treatment should be tried first. Topical agents include the following:

  - **Corticosteroids**
  - **Calcipotriene** (a vitamin D analog)
  - **Coal tar** and its derivatives
  - **Anthralin**

  Use of these agents in combination or sequentially provides the best control.

  Severe psoriasis is treated with the following:

  - **Phototherapy**
  - **Psoralens and high-intensity ultraviolet light**
  - **Retinoids**
  - **Methotrexate** (if recalcitrant and disabling)

Pyogenic Granuloma

- Basic Science **Pyogenic granuloma** is a localized growth of **granulation tissue** surrounded by a collar of skin (epithelium). The lesion generally develops at a site of injury and represents a fibrovascular response to trauma.
- History and Physical Examination Patients generally report a **history of trauma** to the involved area. Pyogenic granuloma presents as a **smooth red nodule. Central ulceration** and bleeding are common. Pyogenic granuloma is most frequently seen in **children and young adults** on the face, hands, forearms, palms, soles, and oral mucosa. **Recurrence is common.**
- Laboratory and Diagnostic Studies **Biopsy is required for definitive diagnosis.**
- Formulating the Diagnosis Pyogenic granuloma **must be differentiated from malignant melanoma.** Therefore excisional biopsy provides the definitive diagnosis.
- Evaluating the Severity of Problems **This is a benign lesion.**
- Managing the Patient Surgical excision serves both diagnostic and therapeutic purposes. **Recurrence can be treated by conservative surgery or electrocautery.**

SCABIES

- **Basic Science** **Scabies** is an intensely **pruritic skin eruption** caused by the mite *Sarcoptes scabiei.* The fertilized female mite tunnels under the stratum corneum into the viable epidermis and **lays eggs within the burrowed skin.** Eggs mature and larvae are released in 3 to 5 days. The larvae become adults in 10 to 14 days and group around the hair follicle. **Lesions result from hypersensitivity to the mite.** Scabies is spread by skin-to-skin contact with infected carriers. Since the mite can survive 1 to 5 days off the human body, clothing and bed linens may act as fomites.

  In immunosuppressed patients, thousands of organisms may be present, creating a crusted form of scabies known as Norwegian scabies.

- **History and Physical Examination** **The cardinal symptom of scabies infection is intense itching.** Itching is **intensified at night.** Since the mite may live away from its host for up to 5 days, a history of contact with known scabies carriers is not necessarily elicited.

  Scabies is commonly present on the hands, feet, axillae, breasts, periumbilical region, hips, buttocks, and **genitalia.** The **primary lesions** consist of raised threadlike lines and **raised erythematous lesions up to 4 or 5 mm in diameter.**

- **Laboratory and Diagnostic Studies** Diagnosis is confirmed by **demonstration of mites on microscopic examination of skin scrapings** obtained from a lesion. Several lesions should be scraped to collect material because it can be very difficult to get a positive scraping.

- **Formulating the Diagnosis** Itching and the presence of dermatitis in close contacts may help establish the diagnosis. Once microscopic analysis of skin scrapings demonstrates the scabies mite, the diagnosis is confirmed.

- **Evaluating the Severity of Problems** Scabies is highly infectious, especially where skin-to-skin contact is present (e.g., within a household, parent-child interactions, and body contact sports such as wrestling). Itching can be very distressing. In areas of excoriation, secondary streptococcal infection may occur.

- **Managing the Patient** Despite a negative scraping, treatment may be required, especially if persistent, generalized itching with a scabies-like eruption is present. The **treatment** is aimed at eradicating the mite and **consists of** the following:

  - **Permethrin 5% cream or lindane (Kwell)** is used. Crotamiton and 10% sulfur in petrolatum have also been used.
  - **Close contacts** need to be **treated** simultaneously.
  - **Clothing and linens should be washed** in hot water, or dry cleaned, or isolated in closed plastic bags for 5 days to prevent reinfection from fomites.
  - **Pruritus should be treated with systemic antipruritics,** or topical corticosteroids, or both.
  - **Secondary infections** need to be treated with appropriate **antibiotics.**

Squamous Cell Carcinoma

- **Basic Science** **Squamous cell carcinoma is a malignancy that arises from keratinocytes.** It can arise suddenly or from preexisting lesions such as **actinic keratoses.** Squamous cell carcinoma is the **second most common skin cancer** in the United States, accounting for 80,000 to 100,000 cases per year.

  Cutaneous squamous cell carcinoma causes **tissue damage by direct tumor invasion.** Cutaneous squamous cell carcinomas **rarely metastasize.**

  Approximately **30% of oral (lingual and mucosal) squamous cell carcinomas metastasize** before diagnosis is made.

- **History and Physical Examination** Squamous cell carcinoma occurs most frequently on **sun-exposed skin areas of fair-skinned people,** particularly those of Northern European ancestry living in sunny areas. Men, particularly farmers and sailors, are more frequently affected.

  Squamous cell carcinoma often arises from a preexisting lesion such as an **actinic keratosis.** Immunosuppressed individuals and those with severe photodamage are at particular risk for developing squamous cell carcinoma. The appearance of squamous cell carcinoma is variable. In the early stage, squamous cell carcinoma presents as a smooth, **red-pink raised lesion with a crusted surface.** As the lesion develops it may become **irregular, wartlike, and ultimately ulcerative.** Other **potential squamous cell carcinoma sites** include the following:

  - *The* ***oral cavity*** (i.e., the floor of the mouth, the side of the tongue, and the soft palate). Oral squamous cell carcinoma is associated with leukoplakia; about 4% of leukoplakia lesions develop into squamous cell carcinoma.
  - *The* ***vulva.*** Squamous cell carcinoma can start here as a small, wartlike lesion or as an erosion with itching and bleeding.
  - *The* ***scrotum.*** Squamous cell carcinoma also presents here as a wartlike lesion or as a friable nodule.
  - *The* ***penis.*** The initial lesion can be a well-defined red patch with an irregular border.

- **Laboratory and Diagnostic Studies** **Biopsy is necessary for diagnosis.**

- **Formulating the Diagnosis** Asymptomatic but persistent oral lesions should undergo biopsy 2 weeks after any irritating factors have been eliminated.

  Clinical differential diagnoses include basal cell carcinoma, fungal infections, and other skin malignancies. **Skin biopsy will provide definitive diagnosis.**

- **Evaluating the Severity of Problems** Squamous cell carcinomas arising on sun-exposed areas are less aggressive than those arising on mucocutaneous surfaces, or on areas affected by radiation dermatitis. The rate of metastasis from lesions arising in sun-damaged skin is between 0.5% and 3%; as the reticular dermis becomes invaded by larger lesions, the rate of metastasis increases.

■ **Managing the Patient** Therapy depends on the age of the patient and on the size and location of the lesion. **Treatment methods** commonly used include the following:

- **Surgical excision**
- **Curettage and electrodesiccation** (of smaller lesions on sun-exposed areas)
- **Irradiation** (in older and debilitated patients)
- **Mohs'** microscopically controlled **surgery** (for difficult, recurrent, or ill-defined lesions)

Syphilis

■ **Basic Science** Syphilis is a contagious disease caused by the spirochete ***Treponema pallidum.*** Acquisition occurs via **sexual contact,** from mother to infant, or from blood transfusions. The organism penetrates through nonintact skin or mucous membranes; it then enters lymphatic vessels and is disseminated hematogenously. Four stages are recognized: primary, secondary, latent, and tertiary.

- The incubation period usually lasts **3 weeks** before a **chancre** occurs at the site of inoculation. This is a painless eruption that erodes to form a firm, clean, **painless ulcer.** Enlarged lymph nodes are present with infection. The **chancre heals in 2 to 8 weeks without treatment.**
- **Secondary** or disseminated syphilis occurs 2 to 12 weeks after contact, during the period of highest treponemal load. **Flulike symptoms, widespread adenopathy, and rash are the three principal findings.**
- Latent syphilis follows secondary syphilis and represents a period of infection without clinical manifestations. Latency may persist indefinitely.
- Tertiary syphilis occurs in approximately one third of untreated patients and dermatologically is characterized by a cutaneous **gumma.** Cardiac **(aortic aneurysms)** and neurologic disease occurs and accounts for significant morbidity and mortality.

The United States is currently experiencing an epidemic of syphilis.

■ **History and Physical Examination** The patient may present with a history of a known syphilis contact or present with another sexually transmitted disease (STD). Often the patient presents with a **chancre;** however, the chancre may go unnoticed or misdiagnosed.

In secondary syphilis the patient often complains of **flulike symptoms:** low-grade fever, headache, malaise, arthralgia, myalgia, and weight loss. Physical examination reveals **widespread adenopathy. Specifically, one finds nontender,** discrete, firm, **rubbery 0.5- to 2-cm nodes** and a **rash.**

The **initial rash of secondary syphilis is** macular **(flat), copper colored, symmetric,** and primarily on the patient's trunk. A copper-colored papular **(raised) eruption occurs later.** The **palms, soles,** face, extremities, genitals, scalp, and mucous membranes **are affected.**

Papules in the groin and other moist skin folds enlarge, forming

condyloma lata (i.e., broad, moist, gray-white plaques teeming with spirochetes). Painless superficial erosions occur in the mouth, genital, and anal area.

- Laboratory and Diagnostic Studies **A rapid plasma reagin** (RPR) or **venereal disease research laboratory** test (VDRL) and titers should be performed; a **positive result indicates** that the patient has or has had **syphilis.** These tests detect anticardiolipin antibodies and false positive results may occur in many disease states. **Specific treponemal tests** (e.g., the **FTA-ABS,** fluorescent treponemal antibody-absorption test) are rarely checked in primary or secondary syphilis, but may be confirmatory.
- Formulating the Diagnosis

• **Primary syphilis is suggested by the occurrence of a chancre and lymphadenopathy.**

• **Secondary syphilis is "the great imitator"** and should be **considered in all patients with a suggestive rash,** especially if there is **palm and sole involvement, unexplained adenopathy, or flulike illness.**

The differential diagnosis for the "rash" of secondary syphilis is extensive.

- Evaluating the Severity of Problems It is important to remember that secondary syphilis is a systemic illness. Central nervous system, ocular, renal, splenic, or hepatic involvement may occur. Examination for other STDs is necessary, especially HIV. Local laws may require notification of public health officials and all sexual partners.
- Managing the Patient **The treatment for all stages of syphilis is penicillin.**

• The recommended regime for "early syphilis" is penicillin (i.e., a single dose of **benzathine penicillin G, 2.4 million units IM).**

• Alternatives include doxycycline 100 mg po bid for 14 days, and erythromycin 500 mg po qid for 14 days, though higher relapse rates occur with erythromycin.

• **Pregnant females with syphilis (all stages) must receive penicillin; penicillin-allergic pregnant females should be desensitized and then treated.**

• An acute self-limited febrile reaction frequently occurs in patients receiving treatment for syphilis. It is marked by the abrupt onset of headache, fever, chills, diaphoresis, gastrointestinal irritation, and exacerbation of cutaneous features. It is treated with rest and aspirin or acetaminophen; it usually resolves within 6 hours. Approximately 60% to 75% of patients with secondary syphilis develop the reaction.

• Patients should have serial evaluations of RPRs or VDRLs to assess response to treatment.

Yeast Infections (Candidiasis)

- Basic Science ***Candida albicans*** exists as part of normal human flora. The organism is usually found in warm, moist areas of the body (between skin folds and on mucous membranes). The organism may become **pathogenic if host defenses weaken** or if the environment becomes favorable to the organism (such as following use of oral antibiotics).

- History and Physical Examination Candidiasis of the skin usually begins in intertriginous zones (between opposing skin surfaces where friction and moisture occur). The lesions appear as **red, well-demarcated patches of various sizes.** The lesions usually are **surrounded by small, red satellites** that contain pus.

  Common sites of infection include the inner aspect of the **thighs,** the **axillae, underneath** pendulous **breasts,** and the **gluteal and inguinal folds of infants (diaper rash).**

  Oral candidiasis (thrush) appears as **white plaques on** the **tongue and buccal mucosa.** The plaques can be scraped from the tongue. In **infants and young children, oral candidiasis is both common and benign. In adults,** however, **oral candidiasis is** often a **presenting sign of AIDS.**

- Laboratory and Diagnostic Studies **Yeast and pseudohyphae** can be demonstrated **microscopically** from scrapings treated with potassium hydroxide, or Gram's stain.

- Formulating the Diagnosis **The diagnosis is made clinically.** Confirmation involves microscopy. In unusual cases, histological examination of the skin specimen is necessary.

- Evaluating the Severity of Problems Superficial candidiasis is common in both infants and AIDS patients. The infection is benign and easily treated.

- Managing the Patient **Topically applied nystatin and imidazoles (e.g., ketoconazole) are the drugs of choice** for yeast infections of the skin and mucous membranes.

  - For **recurrent infections, oral nystatin** tablets are prescribed.
  - In **diaper rash, nystatin powder** is used. Frequent diaper changes to **keep the skin dry** are also important.
  - **Oral candidiasis** is treated with **nystatin liquid.** The patient is instructed to **"swish and swallow"** the liquid.

Multiple Choice Review Questions

1. A 27-year-old housewife presents with a 3-day history of a nonpruritic rash on her trunk and malaise. On physical examination, she has multiple scaly, light red plaques on the trunk; flat, scaly papules on the palms and soles; and enlarged cervical, axillary, and epitrochlear lymph nodes. You suspect secondary syphilis. Initial work-up should include all of the following *except:*

a. An RPR or VDRL.
b. A potassium hydroxide preparation from a plaque.
c. An FTA-ABS.
d. Questioning about other sexually transmitted disease.
e. A genital examination.

2. Treatment of choice for secondary syphilis is:

a. Erythromycin 500 mg every 6 hours for 14 days.
b. Benzathine penicillin G, 2 million units intramuscular ×1.
c. Benzathine penicillin G, 2 million units intramuscular ×3.
d. Observation, as the rash will spontaneously regress.
e. Penicillin VK, orally 500 mg every 6 hours for 14 days.

3. You have been asked to evaluate a 10-year-old boy with "hemorrhoids." On examination, he has several flat papules with a corrugated surface localized to perirectal skin. You suspect genital warts. Further evaluation must include all of the following *except:*

a. Thorough questioning for possible sexual abuse and reporting to social services.
b. A biopsy.
c. Anoscopy.
d. Full skin and mucous membrane examination.
e. An RPR.

4. Treatment may be best managed by:

a. Application of a topical anesthetic 1 hour before liquid nitrogen cryotherapy, repeated every 3 to 4 weeks until clear.
b. Home application of podophyllin on a daily basis.
c. Surgical excision.
d. Systemic penicillin.
e. One-time liquid nitrogen application followed by home salicylic acid application.

Chapter 11

Endocrinology

ADRENOCORTICAL HYPOFUNCTION

■ **Basic Science** The adrenal cortex secretes three classes of hormones: androgens, mineralocorticoids, and glucocorticoids, the last two associated with clinically significant deficiency syndromes. Adrenocortical hypofunction describes the condition in which steroid output from the adrenal cortex is insufficient to meet the body's needs. Adrenal insufficiency may be due to primary adrenal disorders, such as autoimmune destruction, infection, hemorrhage or metastatic invasion; other causes are classified as secondary adrenal insufficiency and include pituitary failure and suppression of the hypothalamic-pituitary-adrenal (HPA) axis from exogenous steroid.

The most common cause of primary adrenal insufficiency (also known as Addison's disease) is idiopathic atrophy; an autoimmune mechanism is thought to be responsible. There is an increased incidence of other endocrine and autoimmune processes in these patients.

Secondary adrenal insufficiency is due to insufficient adrenocorticotropic hormone (ACTH) secretion, either from exogenous steroid (most commonly) or from primary pituitary disorders. Unlike patients with primary adrenocortical insufficiency, these patients often have normal or near-normal levels of aldosterone.

■ **History and Physical Examination** The **most common symptoms** in patients with adrenocortical hypofunction are **weakness, hyperpigmentation, weight loss, anorexia, nausea, and vomiting.** Less often, patients complain of abdominal pain, salt craving, diarrhea, constipation, and syncope. Patients with secondary adrenal insufficiency do not have hyperpigmentation, nor do they complain of symptoms attributable to hypoaldosteronism. The medical history should include questions regarding the duration of symptoms, since hyperpigmentation and weight loss are seen only in subacute and chronic cases. Patients with AIDS are at risk for a necrotizing infection caused by cytomegalovirus and other opportunistic pathogens.

The physical examination is often remarkable for **hypotension and orthostasis. Hyperpigmentation,** when present, is seen diffusely, including areas shielded from sun exposure; mucous membranes may be pigmented as well.

■ **Laboratory and Diagnostic Studies** Early in the course of adrenal insufficiency, basal steroid output may be normal, but secretion during periods of stress may be inadequate. This abnormality is best elicited through the **ACTH stimulation test,** where plasma cortisol is measured 30-60 minutes after the administration of intravenous cosyntropin, an ACTH analog. In advanced disease states, serum sodium, chloride, and bicarbonate are reduced, and serum potassium is elevated.

- **Formulating the Diagnosis** The **diagnosis of adrenal insufficiency is made by demonstrating an inadequate response to ACTH stimulation testing. In primary adrenal insufficiency,** the serum **ACTH is elevated** and the serum aldosterone is below normal. **In secondary adrenal insufficiency,** the serum **ACTH is low** and the serum aldosterone is normal.

- **Evaluating the Severity of Problems** **Adrenal crisis** is the acute exacerbation of chronic adrenal insufficiency. It occurs in patients with chronic adrenal insufficiency during periods of severe stress (e.g., surgery, sepsis) or septicemia (e.g., the Waterhouse-Friderichsen syndrome). However, the **most frequent cause is abrupt withdrawal of exogenous steroids in steroid-dependent patients.** Affected individuals complain of severe abdominal pain, nausea, and vomiting; they become somnolent and hypotensive. **Treatment** consists of **rehydration** with normal saline **and treatment with stress doses of steroid.** Steroid treatment is best accomplished with a 100 mg bolus of hydrocortisone, followed by 100 mg intravenously every 8 hours. Separate mineralocorticoid is unnecessary in patients receiving this dose of cortisol.

- **Managing the Patient** **Treatment consists of** steroid replacement. A typical **hydrocortisone** regimen consists of 25 mg in the morning and 12.5 mg at night; this mimics the normal diurnal variation of steroid secretion. Patients with hypoaldosteronism (i.e., with primary adrenocortical insufficiency) must receive a **source of mineralocorticoid; usually, fludrocortisone** is given once daily in dosages from 0.05 to 0.1 mg. Patients should be given explicit instructions to **supplement steroid during periods of stress or illness.**

  Complications of therapy include electrolyte disturbances; therefore, routine blood work should be checked periodically. Hydrocortisone, which may cause gastritis, should be taken with meals.

Anterior Pituitary Disorders

- **Basic Science** The pituitary gland is located in the sella turcica, below the optic chiasm. Via its portal system, the anterior pituitary receives a mixture of peptide hormones and biogenic amines. These include the following:

  - Thyrotropin-releasing factor (which stimulates TSH [thyroid-stimulating hormone] and, to a much lesser extent, prolactin)
  - Gonadotropin-releasing factor (which, when released in a pulsatile fashion, stimulates LH [leutenizing hormone] and FSH [follicle-stimulating hormone] release)
  - Corticotropin-releasing factor (stimulates ACTH and β-endorphin)
  - Growth-hormone–releasing factor (stimulates GH [growth hormone] release)
  - Somatostatin (inhibits GH release)
  - Prolactin-release–inhibiting factor (dopamine)

Products released by the anterior pituitary are TSH, LH, FSH, ACTH, β-endorphin, GH, and prolactin (PRL). **A pituitary disorder occurs when output is either excessive or insufficient, or when a tumor is present.**

Large pituitary **tumors** may compress either the chiasm (causing **bitemporal hemianopsia**) or the optic tracts (producing visual field cuts). Tumors may also compress the contents of the cavernous sinus (i.e., cranial nerves III, IV, V_1, V_2, and VI, and the carotid artery), which sit lateral to the pituitary.

Nearly all pituitary tumors are benign adenomas (usually prolactin-secreting microadenomas). They may produce products in excessive or deficient amounts; however, most adenomas produce no symptoms. These tumors may occur sporadically or as part of the multiple endocrine neoplasia syndrome, type I (MEN I), which consists of tumors involving the pituitary, pancreas, and parathyroid hyperplasia.

In **children, insufficient growth hormone** results in **growth retardation and hypoglycemia,** whereas **excess causes gigantism.** In **adults, excessive growth hormone** results in **acromegaly,** ketogenesis, and impaired glucose tolerance; the body composition of adults may be affected adversely in adults with insufficient growth hormone.

In men, prolactin serves no function; therefore prolactin-secreting tumors come to attention secondary to chronic headache, compression of adjacent structures, or clinical hypogonadism (e.g., impotence). **Macroadenomas are generally found.**

In women, complaints of **galactorrhea, amenorrhea, and infertility** are often elicited; **microadenomas** are the norm.

TSH-releasing and gonadotropin-releasing hormone tumors also occur but are much less common. ACTH-secreting tumors cause Cushing's disease and are discussed later in this chapter.

Pituitary insufficiency, an insidious disorder, is characterized by the relative absence of one, several, or all the anterior pituitary hormones. Pituitary apoplexy represents spontaneous infarction or hemorrhage of a pituitary tumor that can lead to hypopituitarism.

■ **History and Physical Examination** Complaints of chronic headache concurrent with visual field cuts must prompt one to rule out a pituitary tumor. With regard to clinically significant prolactin excess, an inquiry must be made regarding the use of drugs that inhibit dopamine (e.g., phenothiazines and other antipsychotics).

Patients with pituitary insufficiency are usually lethargic and sluggish. There may be an absence of pubic and axillary hair associated with sexual organ atrophy.

Apoplexy is associated with a severe headache, stiff neck, nausea, vomiting, a depressed sensorium, and visual changes.

■ **Laboratory and Diagnostic Studies** **Studies in patients with suspected tumors must include either computed tomography (CT) or (preferably) magnetic resonance imaging scans (MRI). In patients with GH-secreting tumors,** both GH and insulin-like growth factor I (its functional mediator) should be measured; however, **a glucose challenge test is diagnostic.** In patients with acromegaly or gigantism, GH levels will fail to suppress following a glucose load.

Patients with prolactin excess will manifest this with an increased serum prolactin. Routine chemistries should also be obtained in all patients with suspected pituitary dysfunction.

- **Formulating the Diagnosis** The diagnosis of hormone excess or insufficiency is made by direct measurement. Tumors are diagnosed by imaging studies (e.g., CT or MRI).

  Patients at risk for hypopituitarism include those with a history of trauma, aneurysm, granulomatous disease, hemochromatosis, diabetes mellitus, and sickle cell anemia.

- **Evaluating the Severity of Problems** Patients with disorders of ACTH may develop serious electrolyte disorders. Those with large tumors or apoplexy (both rare) may develop intracranial processes requiring decompression.

- **Managing the Patient** **Children with GH-deficiency** are candidates for **replacement therapy.** For **patients with GH-excess,** options include **inhibitors of GH release (e.g., somatostatin analogs**), transsphenoidal microsurgery, and, less often, radiation.

  Prolactin-secreting tumors may be treated with **bromocriptine;** patients who fail bromocriptine therapy are candidates for surgery, although the relapse rate is high.

  Management of hypopituitarism involves correction of the underlying cause. Hyponatremia (present for a variety of reasons) responds to fluids, as opposed to the syndrome of inappropriate secretion of antidiuretic hormone (SIADH), which responds to fluid restriction. Thyroxine, hydrocortisone, gonadal steroids, and GH (in children) should be replaced.

Cushing's Syndrome

- **Basic Science** Cortisol is a glucocorticoid produced by the adrenal cortex. Normally its manufacture is a result of adrenocorticotropin (ACTH) release by the pituitary, which in turn is released under the influence of corticotropin-releasing hormone (CRH) from the hypothalamus. Whether **excessive cortisol is released** autonomously by the adrenal or is stimulated to do so by the production of ACTH or CRH, **Cushing's syndrome results.** Cushing's disease refers specifically to hypersecretion of pituitary ACTH. **The most common cause of Cushing's syndrome remains** iatrogenic, that is, **exogenously administered steroid.** The spontaneous syndrome is more common in women, particularly in the third and fourth decades.

- **History and Physical Examination** Patients relate a **history of fatigability, weakness, mood changes, and easy bruising.** Women may note menstrual irregularities.

  Hypertension is often present. Clinical manifestations of hypercortisolemia include **truncal obesity, impaired glucose tolerance, hirsutism, acne, abdominal striae, and osteoporosis.**

- **Laboratory and Diagnostic Studies** Useful **screening tests** in patients with clinical evidence of cortisol excess include the measurement of **24-hour urinary free cortisol and the overnight 1 mg dexamethasone suppression test.** A diurnal variation in plasma cortisol (which peaks around 8:00 AM) is normally present; its absence is a strong indicator of Cushing's syndrome.

If these screening tests suggest hypersecretion, there are many tests available to refine the diagnosis. The response to high-dose dexamethasone in patients with an ACTH-secreting pituitary tumor is usually a >50% reduction in plasma cortisol. A decrease of this magnitude is not seen in patients with ectopic ACTH syndrome or primary adrenal disease. In this case an adrenal etiology should be sought; an abdominal CT is most useful for this purpose. If the adrenals are normal, a thorough search for an ectopic ACTH-secreting tumor should be made; over 90% of these tumors are found in the chest. Other tests include the CRH stimulation test, inferior petrosal sinus sampling, and a brain imaging study.

- **Formulating the Diagnosis** It is useful to group patients into ACTH-dependent (Cushing's disease and ectopic ACTH syndrome) and ACTH-independent (adrenal disease) categories. **Measurable plasma ACTH virtually excludes primary adrenal disease as the etiology.**
- **Evaluating the Severity of Problems** Adrenal carcinoma forecasts a poor prognosis. Similarly, the outlook for patients with **oat-cell lung cancer, the most common source of ectopically produced ACTH,** is poor.
- **Managing the Patient** The mainstay of **treatment for Cushing's disease is transsphenoidal microadenectomy.** Pituitary irradiation is an option for nonsurgical candidates. Postoperative patients should be treated for adrenal insufficiency. Additional glucocorticoid should be given for periods of stress and illness.

  Surgical resection is the treatment for ectopic sources of ACTH. Adrenal adenoma and carcinoma are treated surgically as well. Medical adjuncts are available and include mitotane, aminoglutethimide, ketoconazole, and metyrapone.

Diabetes Mellitus

- **Basic Science** Diabetes mellitus is a chronic disorder that is characterized by a relative or absolute deficiency of insulin secretion and activity. There are several classification systems of diabetes; most systems recognize type I (insulin-dependent diabetes mellitus, or IDDM), type II (non-insulin-dependent diabetes mellitus, or NIDDM), gestational diabetes, and diabetes secondary to other causes (e.g., pancreatic disease or drugs).

  The **onset of type I diabetes typically occurs before age 30.** Pancreatic β-cells are gradually destroyed by an autoimmune mechanism; eventually there is insufficient insulin to maintain glycemic control. A viral etiology has been proposed but remains unproven. The **development of symptoms** can be **abrupt** and **includes polydipsia, polyphagia, and polyuria.** A **genetic predisposition** exists **for type I** diabetes, **although it is weaker than that for type II** diabetes; there is a less than 50% concordance rate for monozygotic twins. Dependence on exogenous insulin and susceptibility to ketosis are characteristics of type I diabetes.

  Most patients with type II diabetes are diagnosed after

age 40. Although the pathogenesis remains unclear, hallmarks of this disease include a primary β-cell defect and **resistance to secreted insulin.** The **development of symptoms** is usually **gradual** and may not come to attention until end-organ damage (e.g., retinopathy, nephropathy) has occurred. The concordance rate for monozygotic twins exceeds 90%. Dependence on diet and oral hypoglycemics and resistance to ketosis are characteristics of type II diabetes.

Gestational diabetes is present in up to 2% of pregnancies. Glucose tolerance becomes most impaired in the third trimester. Maintenance of euglycemia is particularly important in the prevention of fetal morbidity and mortality. Most women with gestational diabetes do not require treatment for diabetes after their pregnancy.

■ **History and Physical Examination** A recent **history of weight loss, polydipsia, polyphagia, and polyuria in an adolescent or young adult suggests** the diagnosis of **type I diabetes.** These patients are often at or below normal weight. **Most type II diabetics are obese.** They may present with complaints similar to those of the type I diabetic or may complain only of fatigue and irritability.

In patients with long-standing diabetes (of either type), **end-organ complications** are evident on physical examination (e.g., **retinopathy, neuropathy, peripheral vascular disease**). Hypertension is common in type II diabetics.

■ **Laboratory and Diagnostic Studies** Because plasma glucose levels are subject to fluctuations during periods of disease, severe stress, or drug ingestion, hyperglycemia must be documented on more than one occasion. Specifically, **fasting plasma glucose in excess of 140 mg/dl on two separate occasions is diagnostic. Alternatively, the diagnosis may be made by using the 2-hour oral glucose tolerance test;** two values, one of which must be the 2-hour level, must exceed 200 mg/dl. The classic symptoms of polydipsia, polyphagia, and polyuria are suggestive but not diagnostic of diabetes.

Hemoglobin A_{1c} is an assay that reflects glycemic control over the preceding 3 months; the result is expressed as a function of total hemoglobin. After the diagnosis of diabetes has been made, most practitioners recommend home glucose monitoring kits. Additionally, type I diabetics should routinely test their urine for ketones.

■ **Formulating the Diagnosis** The diagnosis of diabetes mellitus is established by the aforementioned abnormal glucose levels. **Individuals who have glucose levels that exceed normal values but do not meet the criteria for diabetes have impaired glucose tolerance.**

Other illnesses (Cushing's disease, cystic fibrosis, pheochromocytoma, etc.), drugs (thiazide diuretics, steroids, etc.), and severe stress can precipitate hyperglycemia.

■ **Evaluating the Severity of Problems** The **acute metabolic consequences** of diabetes mellitus **include diabetic ketoacidosis and hyperosmolar nonketotic coma.** In diabetic ketoacidosis (DKA) there is insufficient insulin resulting in hyperglycemia, hyperglucagonemia, stress hormone release, and ketone body formation. The increased glucose causes an osmotic diuresis producing losses of sodium, potassium, calcium, and phosphate, in addition to profound dehydration. **Criteria for the diagnosis of DKA are hyperglycemia, hyperketonemia, and metabolic acidosis.** Although infection and failure to take insulin are usual inciting events, often no precipitating cause is found. **Initial treatment should include regular insulin and normal saline.** There is usually a **large total body potassium deficit,** though patients may present with hyperkalemia. **Acidosis** usually corrects with rehydration, and bicarbonate is not routinely indicated.

Hyperosmolar nonketotic coma occurs over a period of days, during which time there is progressively **worsening hyperglycemia, dehydration, and hyperosmolality** (≥325 mOsm/L). This form of diabetic coma is seen more often in the elderly, and unlike DKA, ketonemia and acidemia are not prominent features. The inciting event should be sought and corrected; **initial resuscitation is similar to that for DKA.** Normal saline should be used initially, although the majority of replacement fluid should be 0.5 normal saline, since the total body water deficit exceeds that of sodium.

The **chronic complications of diabetes include macrovascular disease, microvascular disease, and neuropathies.** In diabetics, macrovascular disease is a result of dyslipidemia and accelerated atherosclerosis. Many diabetics suffer prematurely from heart and peripheral vascular disease; silent ischemia is not uncommon, and atypical chest pain often represents advanced coronary artery disease.

The microvasculature of diabetics reveals diffuse capillary basement membrane thickening; nephropathy and retinopathy are two common manifestations of this vascular pathology. Beyond tight glycemic control, angiotensin-converting enzyme (ACE) inhibitors help prevent progression of renal failure. The laser has a major role in the treatment of retinopathy; diabetics warrant at least annual ophthalmologic follow-up.

Diabetic neuropathy worsens in proportion to the duration and magnitude of hyperglycemia. Autonomic neuropathy may involve the gastrointestinal system (e.g., gastroparesis) and cardiovascular system (e.g., orthostatic hypotension). The most common form of neuropathy is a symmetric distal polyneuropathy, although radiculopathies and mononeuropathies also occur. The most common complaint involves painful lower-extremity paresthesias. Neuropathic complications often include Charcot's joints and foot ulcerations; therefore meticulous foot care is advised. Treatment with B-complex vitamins, phenytoin, carbamazepine, and amitriptyline have all been tried with modest success, although control of blood sugar is the primary therapy.

■ **Managing the Patient** The importance of the **diet** in diabetes cannot be overemphasized. The American Diabetes Association has made recommendations regarding the distribution of total energy intake; the absolute number of calories should be appropriate to reach and maintain ideal body weight. **Protein** should represent about **15%** of daily calories, **carbohydrate** should approximate **55% to 60%,** and calories from fat (preferably monounsaturated or polyunsaturated) should represent less than **30%.**

Most insulin-requiring patients may be started on a two-injection-per-day regimen of human recombinant insulin. For example, a lean, active individual may be controlled on 10 units of NPH (neutral protamine

Table 11.1 *Selected Insulin Characteristics*

Preparation	Onset of Action	Peak Effect	Duration of Action
Regular	½-1 hour	2-4 hours	6-8 hours
NPH	1½ hours	4-12 hours	24 hours
Lente	1-3 hours	6-12 hours	18-24 hours
Ultralente	4-6 hours	8-20 hours	24-28 hours

NPH, Neutral protamine Hagedorn insulin.

Hagedorn) before breakfast and 5 units before dinner; dosages must be adjusted upward for obese, sedentary individuals. For tighter control, three or more injections per day may be necessary (Table 11.1).

Often regular insulin is added before meals or to adjust for daily dietary variations. An attempt is made to **maintain fasting and preprandial glucose between 70 and 120** mg/dl. One-hour **postprandial glucose** should ideally be **less than 160** mg/dl. Sites of insulin injection should be varied, since use of the same site leads to tissue fibrosis and impaired absorption.

Hypoglycemia is an important side effect of insulin use (discussed later in this chapter). Since **beta-blockers** mask the autonomic sequelae of hypoglycemia, they should be **avoided.**

Oral hypoglycemic agents are used in type II diabetics who are not controlled adequately by diet and exercise alone. The sulfonylureas result in increased pancreatic β-cell secretion and potentiate glucose transport in the periphery.

Two new oral agents for the treatment of mild Type II diabetes have recently become available in the United States. Metformin is a biguanide which lowers blood sugars both by slowing glucose production from the liver and by lessening of insulin resistance at the level of target tissues. Because their mechanisms of action are complementary, metformin can be prescribed in addition to sulfonylureas for additive glucose lowering effects. Metformin can also be used as monotherapy, particularly in obese Type II diabetics because metformin is less likely to promote weight gain than are the sulfonylureas or insulin. Metformin should not be used in patients with an abnormal serum creatinine or in patients with severe liver disease because of the risk of lactic acidosis. It should be held for patients receiving IV contrast injections and should be avoided in patients who drink alcohol regularly.

Acarbose is an inhibitor of alpha glucoside. Acarbose is an inhibitor of intestinal enzymes which break down complex sugars into simple sugars suitable for absorption. The drug therefore delays carbohydrate absorption, lowering postprandial glucose peaks. It can be used as monotherapy in Type II diabetics or in combination with sulfonylureas. Most frequent side effects include flatulence, abdominal pain, and diarrhea.

Finally, two phenomena deserve mention: the dawn phenomenon and Somogyi effect. The **dawn phenomenon** is an **increase in fasting blood glucose levels and insulin requirements before breakfast.** The cause appears to be an inability to respond to normal nocturnal surges of growth hormone secretion with increased insulin secretion. The **Somogyi effect** represents **rebound hypergly-**

cemia in response to counterregulatory hormones (e.g., epinephrine, cortisol) during unrecognized hypoglycemic episodes.

Disorders of Calcium Balance

■ **Basic Science** Calcium is an essential mineral for neuromuscular function and bone metabolism. Total serum calcium represents the bound (primarily to albumin) and free (clinically significant) fractions. Although **the total serum calcium is often altered by changes in albumin concentration, the fraction of free calcium usually remains constant.**

Hypercalcemia and hypocalcemia are disorders of calcium balance. **Hypercalcemia,** the more common of the two, is **caused predominantly by hyperparathyroidism and malignancy.** Other causes include sarcoidosis, vitamins A and D toxicity, thiazide diuretics, lithium ingestion, the milk-alkali syndrome, immobilization, and acute renal failure. Parathyroid hormone and vitamin D serve to raise serum calcium; calcitonin, a hormone secreted by the parafollicular cells of the thyroid, has weak actions that contribute to lowering serum calcium.

Hypercalcemia and hypophosphatemia are constant **features of hyperparathyroidism.** With routine blood tests, asymptomatic hypercalcemia secondary to hyperparathyroidism is **often discovered incidentally.** Other clinical manifestations of hyperparathyroidism include nephrolithiasis, peptic ulcers, mental status changes, and bone resorption. **Solitary parathyroid adenomas** represent the **most common etiology for hyperparathyroidism** and present primarily between the third and fifth decades.

Malignancy (most often a squamous cell tumor) is the second most common cause of hypercalcemia. It is caused both by humoral mediators (in the absence of bone metastasis) and by local destruction of bone by tumor.

Hypocalcemia, specifically a low free serum calcium, is caused by chronic renal failure, hypoparathyroidism (postsurgical or idiopathic), hypomagnesemia or hypermagnesemia, acute pancreatitis, vitamin D deficiency, and, rarely, multiple blood transfusions.

■ **History and Physical Examination** Clinical manifestations of **hypercalcemia** include the following:

- Renal (polyuria and **nephrolithiasis**)
- Gastrointestinal (anorexia, nausea, vomiting, and constipation)
- Neurologic (weakness, fatigue, stupor, and coma)
- Electrocardiogram (ECG) manifestations (shortened QT interval)

Clinical manifestations of **hypocalcemia** include the following:

- Paresthesias and tetany (e.g., carpopedal spasm)
- **Chvostek's sign** (twitching of the facial muscles with tapping over the facial nerve)
- **Trousseau's sign** (carpal spasm with inflation of a blood pressure cuff over systolic blood pressure for 3 minutes)
- ECG manifestations (prolonged QT interval)

■ **Laboratory and Diagnostic Studies** Hypercalcemia usually becomes symptomatic when the level exceeds 12 mg/dl. **Elevated calcium levels** in asymptomatic patients **should be repeated** to eliminate the possibility of lab error. **Ionized calcium should be measured** to determine the amount of free (or active) calcium. **Parathyroid hormone should be measured;** it will be **elevated in** patients with **hyperparathyroidism** and will be **suppressed in malignancy-related hypercalcemia** (assuming normal renal function). Routine chemistries and a complete blood count should be included in the workup; they are especially useful in identifying patients with hematologic malignancies. Bone scans and x-ray examinations are indicated if evidence of bone involvement exists.

Laboratory studies in patients with hypocalcemia should include serum ionized (free) calcium, magnesium, phosphorus, creatinine, and parathyroid hormone. **Hypoalbuminemia is a frequent cause of low total calcium;** however, the ionized fraction is typically unaffected.

■ **Formulating the Diagnosis** Hyperparathyroidism and malignancy account for more than 90% of individuals with hypercalcemia. Over half of patients with hyperparathyroidism are asymptomatic; in contrast, **patients with hypercalcemia secondary to a malignancy typically manifest evidence of malignancy before the discovery of hypercalcemia.** Additionally, most patients with malignancy-related hypercalcemia will live less than 6 months. Therefore the duration of hypercalcemia and the presence of symptoms are the two historical factors that are most helpful in the diagnosis.

The differential for **hypocalcemia** includes disease states characterized by relative or absolute deficiency of PTH, magnesium, and vitamin D.

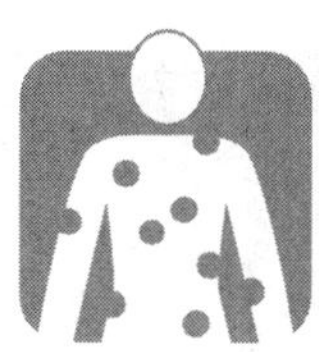

■ **Evaluating the Severity of Problems** Severe hypercalcemia (>15 mg/dl) requires aggressive treatment including invasive hemodynamic monitoring, and frequent assessment of electrolytes. Dialysis may be necessary if hypercalcemia fails to respond to conventional therapy.

Severe hypocalcemia may cause laryngeal spasm, convulsions, and respiratory arrest. Repletion of calcium is best achieved with an intravenous solution of calcium gluconate; hypomagnesemia, if present, must also be corrected.

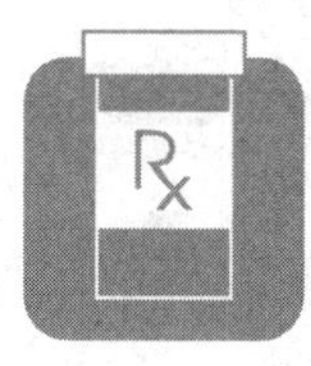

■ **Managing the Patient** The treatment of hypercalcemia should be directed at its cause. Nearly all patients with **hyperparathyroidism should be managed surgically;** exceptions to this tenet are older (>50 years old) asymptomatic patients with normal bone mass and renal function, who may be closely monitored.

Acutely, the **lowering of serum calcium is** best achieved **through saline diuresis;** this serves to expand the intravascular volume. Once this has been achieved, a loop diuretic (e.g., **furosemide**) **may offer additional benefit.** Other effective therapies include pamidronate (inhibits osteoclasts, slow onset, lasts for days), calcitonin (rapid onset, useful in renal failure, rapid tolerance to effect a problem), oral phosphate (used in a chronic setting), and glucocorticoids (slow onset, useful in certain malignancies and granulomatous diseases).

Correction of chronic hypocalcemia due to hypoparathyroidism or renal failure is through replacement of calcium (e.g., $CaCO_3$) and vitamin D (e.g., calcitriol).

Hyperthyroidism

- **Basic Science** **Hyperthyroidism** is a pathophysiologic state resulting from an **excess amount of free thyroid hormone** in blood and tissues produced by a hyperfunctioning thyroid gland. *Thyrotoxicosis* is a broader term describing a condition of thyroid hormone excess irrespective of thyroid gland function. Although there are **many causes** of this condition, **Graves' disease is the most common.** Other common causes include iatrogenic (overmedication), toxic nodular goiter, and thyroiditis.

  Graves' disease is an autoimmune disorder characterized by the presence of **gamma immunoglobulin (IgG) directed at the TSH receptor.** The IgG serves to stimulate thyroid function. Graves' disease most often affects middle-aged women.

  Toxic nodular goiter refers to single or multiple autonomously functioning nodules, seen more often in the elderly.

  Thyroiditis occurs in several variants and results in inflammation-induced release of thyroxine (T_4) and triiodothyronine (T_3).

- **History and Physical Examination** Patient **complaints often include nervousness, increased sweating, hypersensitivity to heat, palpitations, fatigue, increased appetite, weight loss, insomnia, weakness, and frequent bowel movements.**

  Family members of patients with Graves' disease have an increased incidence of other autoimmune disorders.

  Dramatic physical examination findings are more common in younger patients; signs of thyroid disease in the elderly may be masked. The **general signs include goiter, tachycardia, widened pulse pressure, warm moist skin, tremor, and eye signs (stare, lid lag, and lid retraction).** There are certain signs specific to Graves' disease: infiltrative ophthalmopathy (often resulting in exophthalmos and ocular muscle weakness) and pretibial myxedema (a nonpitting edema in the pretibial region).

- **Laboratory and Diagnostic Studies** When hyperthyroidism is suspected, **serum TSH and T_4 measurements** should be made. The presence of a normal TSH level almost always excludes the diagnosis of hyperthyroidism. The presence of a **low TSH and a high T_4 level confirms the diagnosis.** Although most patients with hyperthyroidism manifest a high T_4, certain patients have only an elevated T_3.

  Other tests, such as the TSH receptor antibody, antithyroid antibodies, T_3 resin uptake, and radioiodine (^{123}I) scanning, are available but have more limited indications.

- **Formulating the Diagnosis** The diagnosis is usually evident from the history and physical examination and is confirmed by the laboratory.

- **Evaluating the Severity of Problems** The risk of atrial fibrillation is increased in patients with hyperthyroidism (especially the elderly) and deserves treatment.

  Thyrotoxic storm is a special complication of Graves' disease characterized by **fever, tachycardia, marked weakness,**

mental status changes, and shock. It is often precipitated by an infection or stressor.

- **Managing the Patient** The **antithyroid drugs propylthiouracil and methimazole** inhibit thyroid hormone synthesis and **are first-line therapy** in patients with Graves' hyperthyroidism. Beta-blockers, such as propranolol, are useful adjuvants in patients with Graves' disease in that they ameliorate some symptoms of thyroid hormone excess. **Radioiodine (^{131}I) is another first-line therapy for patients with Graves' disease,** although it **often results in permanent hypothyroidism.**

  Thyrotoxic storm is a medical emergency treated with supportive measures that include intravenous fluids, steroids, an antithyroid drug (e.g., propylthiouracil), potassium iodide or ipodate sodium (administered after antithyroid drugs to block the release of thyroid hormones), and a beta-blocker.

HYPOGLYCEMIA

- **Basic Science** **Hypoglycemia is** a clinical syndrome manifest by a **symptomatically low plasma glucose.** It is useful to **classify** hypoglycemia **as** either **postprandial or fasting.**

  Postprandial hypoglycemia most commonly occurs in postgastrectomy patients where hypoglycemia results from insulin-glucose imbalance after rapid glucose absorption; less often, it is seen in patients with rare enzymatic defects such as galactosemia and hereditary fructose intolerance. Idiopathic hypoglycemia, a rare condition, occurs in the otherwise normal person in whom symptoms develop following a carbohydrate load.

  Fasting hypoglycemia occurs when glucose use exceeds glucose production. **Most commonly,** this **occurs during therapy of diabetes mellitus with either insulin or sulfonylureas.** Other drugs, such as salicylates, beta-blockers, and alcohol (during periods of glycogen depletion), are also known offenders. Less frequently, severe liver disease, enzymatic defects, and tumors (e.g., insulinomas) are to blame.

- **History and Physical Examination** Patients typically complain of either **autonomic** (sweating, anxiety, palpitations) or **neuroglycopenic** (lethargy, incoordination, headache) **symptoms.** The history should include questions about recent food, alcohol, and drug intake.

  Physical examination findings associated with hypoglycemia are often remarkable but are generally too nonspecific to be diagnostic.

- **Laboratory and Diagnostic Studies** Although controversial, a 5-hour oral glucose tolerance test is sometimes used, during which time a symptom log is recorded by patients in whom postprandial hypoglycemia is suspected.

  The **first diagnostic step** in patients with suspected fasting hypoglycemia is the **measurement of plasma glucose after an overnight fast;** the fast is continued either until symptomatic hypoglycemia (<45 mg/dl) occurs or 72 hours have passed. Should symptomatic hypoglycemia occur, the following additional tests should be performed: plasma insulin and C-peptide, and a urine screen for sulfonylureas.

In the fasting hypoglycemic patient, an insulinoma must be ruled out.

- **Formulating the Diagnosis** Symptoms that occur in relation to meals (<4 hours afterward) suggest postprandial hypoglycemia. Those occurring after exercise or with no relation to meals suggest fasting hypoglycemia. The diagnosis is established when the criteria for **Whipple's triad** are met: **symptoms of hypoglycemia, a concomitant low plasma glucose, and a relief of symptoms with normalization of plasma glucose.**

  The diagnosis of postprandial hypoglycemia can be established only after fasting hypoglycemia has been carefully excluded and symptomatic hypoglycemia has been documented after a "normal" mixed meal.

- **Evaluating the Severity of Problems** Lethargy, seizures, or coma suggests severe or prolonged hypoglycemia and must be treated immediately. Symptoms in diabetics suffering from autonomic neuropathy or those on beta-blockers may be masked.

- **Managing the Patient** The **initial management of serious hypoglycemia is** either intramuscular **glucagon** (for outpatients) **or** an **IV** bolus of 25 or 50 g of glucose as a 50% solution (**D_{50}**), followed by a constant infusion (for inpatients) until the patient is able to eat. Frequent measurement of plasma glucose may be necessary to ensure adequate replacement. After initial management, attention may be turned to determining the cause of hypoglycemia.

  Hypoglycemia following drugs or certain nutrients is managed by controlling or removing the causative agent. Diabetics should be encouraged to carry with them a source of glucose, such as a piece of candy.

  Postprandial hypoglycemia is less easily managed, but eating frequent small meals high in protein and restricted in carbohydrates may be useful.

Hypothyroidism

- **Basic Science** Hypothyroidism is the hypometabolic state that results from a deficiency of thyroid hormone at the tissue level. **Hypothyroidism** is a condition **seen** predominantly **in two groups,** patients with **Hashimoto's thyroiditis and those who have received treatment** (i.e., ^{131}I or surgery) **for thyrotoxicosis or cancer.**

  Hashimoto's thyroiditis is a **chronic inflammatory disorder** of the thyroid with a significant **autoimmune component.** It commonly afflicts middle-aged women, is often familial, and tends to be chronic and progressive. This disorder frequently coexists with other autoimmune diseases such as systemic lupus erythematosus, IDDM, rheumatoid arthritis, Sjögren's syndrome, and pernicious anemia.

- **History and Physical Examination** The signs and symptoms of hypothyroidism are often insidious in onset and may be quite subtle.

  Patient **complaints** may include **fatigue, weakness, decreased appetite, weight gain, cold intolerance, constipation, and paresthesias in the hands and feet. Menorrhagia and infertility may be present.**

  These patients may have **bradycardia and hypothermia.** They frequently have **slow dysarthric speech and a hoarse voice.** The skin is dry, thick, and doughy; the face appears puffy and expressionless. The heart may be enlarged, with a serous pericardial effusion. The **hair** tends to be **dry and sparse.** The **reflexes reveal** a **prolonged relaxation time.** In patients with Hashimoto's disease, a goiter is present that has a rubbery consistency, which, over years becomes atrophic and fibrotic.

- **Laboratory and Diagnostic Studies** The **single most useful test** in making the diagnosis of hypothyroidism is the measurement of **TSH.** It is elevated without exception in patients with thyroid gland failure and allows the diagnosis to be made often before the thyroid hormones (T_4 and T_3) become low. Detection of antimicrosomal and antithyroglobulin antibodies often aids in the diagnosis of Hashimoto's thyroiditis.

- **Formulating the Diagnosis** Hypothyroid patients may present with a clinical picture consistent with chronic renal failure, nephrotic syndrome, or severe anemia, although a classic presentation is difficult to mistake.

- **Evaluating the Severity of Problems** **Myxedema coma is a rare but serious complication** of hypothyroidism seen mostly in colder climates. Precipitants include cold exposure, infection, and trauma. Patients have severe hypothermia, hypoxia, hyporeflexia or areflexia, seizures, and bradycardia. Treatment must include rapid hormone replacement with either T_3 or T_4, in addition to gentle rewarming, respiratory support, steroids, and antibiotics when necessary.

- **Managing the Patient** The preferred treatment for hypothyroidism is **replacement therapy with L-thyroxine.** The average dose is 125 µg/day, orally, less in the elderly and more during pregnancy; the appropriate dose is that which restores the serum TSH level to normal. The effect of replacement becomes maximal after about 1 month, although clinical response may be seen after several days.

  The starting dose must be individualized. In the elderly or those with coronary artery disease it is prudent to begin replacement with 25 µg/day and increase the dose in similar increments every 4 weeks until the TSH has normalized.

Posterior Pituitary Disorders

- **Basic Science** Oxytocin and vasopressin (antidiuretic hormone) are small peptides produced in the hypothalamus and released by the posterior pituitary. In the female, oxytocin is involved in breast milk release and uterine contraction; it has no known function in the male. **Vasopressin,** a hormone primarily responsible for maintaining serum osmolality, **causes SIADH when released in excess** and diabetes insipidus **(DI) when** released in an **insufficient** amount. Of the two syndromes, **SIADH is more common.** The **primary stimulus for vasopressin release is osmolality;** to a lesser degree, hypovolemia results in the release of vasopressin. Lung cancer (i.e., small cell carcinoma) is a frequent source of ectopic release of vasopressin.

  The two **main forms of DI** are **central** (where an insufficient amount of vasopressin is released) and **nephrogenic** (where the distal nephron fails to respond to released vasopressin). Psychogenic polydipsia is a psychiatric disease characterized by excessive freewater ingestion. It can appear clinically like DI and should be ruled out by dehydration testing.

- **History and Physical Examination** **Several drugs,** including diuretics, chlorpropamide, carbamazepine, clofibrate, and vincristine, have all been **implicated as causes of SIADH.** Trauma, pain, nausea, pulmonary tuberculosis, and **small cell lung cancer** are also potential causative factors. The physical examination is usually remarkable for the **absence of edema.**

  Diabetes insipidus may occur following **head trauma,** brain surgery, brain tumors, severe anoxia, pregnancy, and other **tumors** (e.g., breast cancer) that are metastatic to the pituitary; therefore questions should be asked regarding a history of these events. Ingestion of **ethanol, lithium, or phenytoin may cause transient DI.** Patients with DI complain of **intense thirst and increased urine output;** frequently a history of nocturia is present. The physical examination may reveal evidence of hypovolemia and dehydration.

- **Laboratory and Diagnostic Studies** Common laboratory findings in patients with **SIADH** include hypo-osmolality in the presence of **concentrated urine** (>100 mOsm/L), hyponatremia, a low blood urea nitrogen, and hypouricemia.

  In **DI** a **dilute urine** (specific gravity <1.005, osmolality <200 mOsm/L) is found.

- **Formulating the Diagnosis** The **diagnosis of SIADH** is made in the presence of appropriate laboratory findings **only after eliminating** the more **common causes of physiologic vasopressin release,** which include the use of diuretics; dehydration; and renal, anterior pituitary, thyroid, and adrenal disease.

  Before making the **diagnosis of DI, other causes of polyuria must be excluded;** the diagnosis is then made following the demonstration of a **dilute urine in the presence of increased serum osmolality.** Water deprivation may be used to demonstrate an inability to concentrate the urine. If administration of vasopressin significantly increases urine osmolality, central DI is confirmed. Failure to respond to vasopressin administration supports the diagnosis of nephrogenic DI.

- **Evaluating the Severity of Problems** In the setting of SIADH, **severe SIADH may produce** lethargy, headaches, **seizures, and coma.** Delirium or dementia may be manifestations of chronic hyponatremia.

  Water deprivation in patients with **DI** may lead to **profound dehydration;** replacement fluids and electrolytes must be administered to prevent symptomatic hypovolemia.

- **Managing the Patient** The **treatment of SIADH is free water restriction;** although this is not curative, it masks the manifestations of the disease until the underlying etiology is found and corrected. For patients with chronic SIADH who may be unable to tolerate fluid restriction for extended periods, **demeclocycline, an inhibitor of ADH action, may offer benefit.** Severely hyponatremic patients who manifest coma or seizure should receive a cautious infusion of hypertonic saline until the serum sodium has reached 120 mEq/L or the neurologic symptom has resolved; fluid restriction should then be carried out. If serum sodium is corrected too quickly, central pontine myelinolysis may occur.

The **treatment of** patients with symptomatic central DI consists of intranasal **DDAVP (desmopressin),** a vasopressin analog. The **treatment of nephrogenic DI is more difficult** but includes adequate hydration; sodium restriction, thiazide diuretics, and indomethacin may also be helpful.

Thyroid Neoplasia

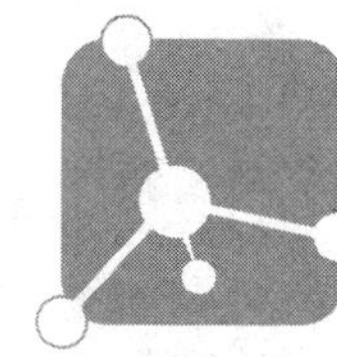

- **Basic Science** Benign adenomas represent the majority of solitary thyroid tumors. Of the malignant thyroid tumors, **four major types** exist: **papillary, follicular, medullary, and anaplastic.**

  Papillary carcinoma is the **most common of the primary thyroid cancers,** and like the others, is more common in women. It is a slow-growing cancer that **spreads by** the **lymphatics.** Overall, mortality is quite low.

  Follicular carcinoma is the next most common variety; this tumor, when compared to papillary, is smaller and **metastasizes earlier,** usually **by** a **vascular route.** Again, mortality is low.

  Medullary carcinoma arises from the parafollicular C-cells that produce calcitonin. It may occur sporadically or as part of another syndrome, such as the multiple endocrine neoplasia syndrome, type II (MEN-II). It **metastasizes both** locally **via** the **lymphatics and to distant sites** such as the lung, bone, and liver. Mortality is relatively low for most varieties of medullary carcinoma.

  Anaplastic carcinoma represents a small fraction of thyroid cancers. It is different from the others in that its clinical course is **distinctly aggressive,** usually resulting in death within months of diagnosis.

- **History and Physical Examination** The history should contain questions regarding hyperfunction and hypofunction of the thyroid; a **hyperfunctioning gland** is **less likely to be cancerous.** The **presence of pain, rapid rate of growth, and a change in voice are all suspicious for carcinoma.** A **history of** head, neck, or upper chest **irradiation is very important,** as these individuals are at increased risk for carcinoma later in life. Patients with a family history of MEN are also at great risk (for medullary carcinoma). Finally, **nodules** appearing in the young or the elderly, or in males, all **raise suspicion of malignancy.**

  The physical examination may reveal features suggesting cancer: associated **lymphadenopathy;** a **fixed, nontender lesion;** or one **of stony-hard consistency.**

- **Laboratory and Diagnostic Studies** The diagnostic approach to a solitary nodule should begin with **fine needle aspiration** (FNA). It is a sensitive and specific technique often useful in directing management. When FNA is inconclusive one may choose to follow the lesion, perform a trial of suppressive therapy with thyroxine, or employ scintiscanning with technetium (^{99m}Tc) pertechnate, an anion concentrated by the thyroid. Most nodules (both benign and malignant) are hypofunctioning, or "cold." Those that are "hot" are almost always benign. Ultrasound may provide information on size and the presence of cystic elements, though it is rarely useful in differentiating benign from cancerous masses. Similarly, **traditional laboratory studies offer little information to discriminate benign from malignant masses. The serum thyroglobulin is used to follow patients as a marker of recurrence of differentiated thyroid cancers** after complete surgical and radioactive iodine ablation of the thyroid.

- **Formulating the Diagnosis** The cornerstone of diagnosis remains FNA.
- **Evaluating the Severity of Problems** Lesions with distant metastases represent advanced disease and portend a poor prognosis.
- **Managing the Patient** **Benign lesions** of the thyroid may be **treated with suppression and observation** at 6-month intervals. Growth of the lesion or a positive FNA at follow-up are indications for surgery.

In general, **malignant lesions are treated surgically.** ^{131}I may be useful postoperatively in destroying residual cancer. **Thyroxine is often used postoperatively to treat iatrogenic hypothyroidism;** additionally, all papillary and follicular cancers are TSH dependent and benefit from suppression. Selected patients may require scintiscanning. **Chemotherapy and radiotherapy are ineffective in the treatment of most thyroid cancers.**

Multiple Choice Review Questions

1. A 33-year-old female reports to you for a routine office visit with the complaint of fatigue and generalized weakness, occasionally accompanied by nausea. Her vital signs are significant, with a supine blood pressure of 102/76, and a standing blood pressure of 84/52. You notice that she has a bronze darkening of her skin. An electrolyte panel, including calcium and magnesium, is normal. Of the following, the most appropriate test to order next is:

a. ACTH Stimulation Test.
b. Serum aldosterone.
c. 24-hour urine cortisol.
d. Abdominal CT scan.
e. Toxicology screen.

2. A 52-year-old woman is seen in your office because of depression. She takes no medicines and has no significant past medical history. Her vital signs are significant, with a blood pressure of 164/94. She is an obese woman with noticeable facial hair. She has thin skin, which is bruised in several areas. The patient undergoes a 1 mg dexamethasone suppression test, which reveals an elevated serum cortisol that fails to suppress. The next most appropriate step in the management of this patient is:

a. Pituitary MRI.
b. Perform a high-dose dexamethasone suppression test.
c. Adrenal CT scan.
d. Measurement of urinary free cortisol.
e. Empiric treatment with metyrapone.

3. A 13-year-old male with a 3-year history of insulin-dependent diabetes is seen in your office because he is becoming lightheaded in the afternoon. He is noncompliant with fingerstick glucose measurements. He reports no change in his diet. His insulin schedule is as follows:

Before breakfast: 10 units NPH, 5 units regular

Before dinner: 5 units regular
At bedtime: 5 units NPH

He recently joined a swim team where he practices in the morning. He appears normal on physical examination. Based on these findings, the most appropriate initial therapy is:

a. Decrease NPH insulin dosage at bedtime.
b. Decrease regular insulin dosage before breakfast.
c. Increase NPH insulin dosage at bedtime.
d. Decrease NPH dosage before breakfast.
e. Decrease regular insulin dosage before dinner.

4. You are called by the nursing home about a new resident, a 76-year-old man who suffers from dementia. The nurse reports that his urine output has been in excess of 10 L/day for his first two days at the home. Which of the following is *least likely* to be responsible?

a. Neurogenic diabetes insipidus
b. Nephrogenic diabetes insipidus
c. Psychogenic polydipsia
d. Severe diabetic nephropathy
e. Diabetes mellitus

5. A 39-year-old woman with Graves' disease is being admitted to the hospital for a thyroidectomy. Which of the following is *least likely* to be present during postoperative day 1?

a. Thyroid storm
b. Wound hemorrhage
c. Recurrent laryngeal nerve injury
d. Hypothyroidism
e. Hypoparathyroidism

Chapter 12

Urology

Acute Renal Failure

■ Basic Science Acute renal failure is defined as the **rapid onset of decreased renal function.**

The cause of **acute renal failure** may be **categorized as follows:**

- **Prerenal insults** Prerenal damage can result from **decreased perfusion to the kidney.** It may be **secondary to** the following conditions:
  - **Hypovolemia**
  - **Heart failure**
  - **Liver failure**
  - **Sepsis**
  - **Renal vascular obstruction** (renal artery stenosis, renal artery, or vein thrombosis)
- **Intrinsic renal insults** Damage to the kidney parenchyma may occur from the following:
  - **Ischemia** (major trauma, vascular disease)
  - **Nephrotoxins** (heavy metals, x-ray contrasts, antibiotics)
  - **Disease affecting the glomerulus** (e.g., glomerulonephritis, hypertension)

  The areas of the kidney that may be damaged include the following:

  - **Glomerulus**—acute glomerulonephritis
  - **Tubules**—acute tubular necrosis, ischemia, antibiotics, contrast agents
  - **Interstitium**—acute interstitial nephritis, drugs
- **Postrenal insults** Postrenal insults include disease states that include:
  - **Urethral obstruction**
  - **Bladder obstruction**
  - **Enlarged prostate**
  - **Kidney stones**

  Once the initial insult has occurred, **kidney function can remain diminished.** This may be **due to:**

  - **Tubular debris** causing obstruction

- **Continued decreased renal blood flow**
- **Decreased glomerular permeability**
- **Back leak of glomerular filtrate** through damaged tubular cells

■ History and Physical Examination The **first sign of renal failure in the hospitalized patient may be oliguria** (less than 400 ml of urine output per day; this amount of urine is the minimum daily output necessary to excrete daily waste products of metabolism).

Patients may present with lethargy, fatigue, nausea, and itching. Renal failure may also be discovered incidentally on a routine urinalysis.

Acute renal failure is classified into three broad categories: prerenal failure, postrenal failure, and intrinsic renal failure. **History and physical examination** should be **directed at establishing the category of renal failure, since both prerenal and postrenal failure may be corrected immediately.**

- **Prerenal failure**
  - ***Impaired cardiac function***
    - **History** CHF, myocardial infarction, etc.
    - **Physical examination** Elevated JVD, edema, lung rales
  - ***Liver failure***
    - **History** hepatitis, cirrhosis, etc.
    - **Physical examination** jaundice, liver tenderness, etc.
  - ***Renal vascular obstruction***
    - **History** usually asymptomatic but may cause back pain
    - **Physical examination** bruit may be heard
  - ***Hypovolemia***
    - **History** trauma, dehydration, presyncope
    - **Physical examination** tachycardia, hypotension, decreased capillary refill, orthostatic blood pressure changes
  - ***Sepsis***
    - **History** high temperature, shaking chills, abdominal pain, vomiting, diarrhea, altered mental status
    - **Physical examination** hypotension, tachycardia, tachypnea, warm skin, patient appears toxic
- **Postrenal failure**
  - **History inability to urinate, straining to urinate, postvoid dribbling**
  - **Physical examination palpable bladder, enlarged prostate**

  Once prerenal and postrenal obstruction have been ruled out, intrinsic renal causes of ARF are investigated.
- **Intrinsic renal failure**
  - ***Glomerular disease***
    - **History sore throat, skin rash, weakness,**

fatigue, fever, edema, dark urine (blood), joint pain, foaming urine

- **Physical examination hypertension, edema**

— ***Tubular diseases***

Direct toxicity

- **History** iodine containing radiographic **contrast agents, aminogylcosides** (e.g., gentamicin), anticancer drugs, amphotericin, heavy metals
- **Physical examination fever**

Hypersensitivity

- **History NSAID**-acetaminophen, salicylates, **penicillin, sulfa drugs**
- **Physical examination rash**

Pyelonephritis

- **History nausea, vomiting, high temperature, dysuria, burning, previous UTIs**
- **Physical examination flank pain, fever**

■ Laboratory and Diagnostic Studies Serum chemistries, **BUN, creatinine, CBC,** and **urinalysis** should be obtained. **Urinary sediment** should be examined, checking for RBC or WBC casts.

If postrenal failure is **suspected,** a **GU ultrasound** is obtained in order to image the GU tract and detect dilatation of the bladder, ureters, or kidneys.

Heme positive dipstick without red blood cells in the urine is indicative of rhabdomyolysis.

■ Formulating the Diagnosis **The physician must classify renal failure into prerenal, postrenal, or intrinsic renal categories.** The **workup** should be performed **in the following order:**

1. **If obstruction** is **possible, a renal ultrasound should be ordered after thorough physical examination.**
2. **Once obstruction** has been **ruled out,** the physician **must differentiate between intrinsic** renal disease **and prerenal causes** of acute renal failure. Differentiation of these diseases may be aided by BUN and creatine.
   - **Prerenal failure—BUN/Cr ratio greater than 20:1**
   - **Intrinsic renal failure—BUN/Cr ratio less than 20:1**

- **Urine sodium**
  - **Prerenal failure—urine sodium less than 20 mEq/L** (the kidney tubules are able to resorb sodium)
  - **Intrinsic renal failure—urine sodium is greater than 20 mEq/L** (kidney is unable or does not need to resorb sodium)

Urine osmolality and urine specific gravity

- **Prerenal failure—urine osmolality greater than 500; urine specific gravity greater than 1.020** (kidney resorbs all the water it can from the tubles)

- **Intrinsic renal failure—urine osmolality less than 400; urine specific gravity less than 1.010** (kidney cannot or has no reason to resorb water)

Fractional excretion of sodium
The **most useful parameter in differentiating between prerenal and intrinsic renal disease** is as follows:

- **Prerenal—less than 1%**
- **Intrinsic renal disease—greater than 1%**

Urinary sediment

- **Prerenal disease—normal sediment pattern** (hyaline casts may be seen)
- **Intrinsic renal disease**
  Acute glomerulonephritis-hematuria, RBC casts
  Acute tubular necrosis-granular cell casts

These findings may not be accurate if patients recently received diuretics, glucose, water or if there is metabolic alkalosis.

■ **Evaluating the Severity of Problems** Severity of acute renal failure depends on the urine output, electrolyte disorders, acid-base disturbances, volume status, and the etiology of the renal failure.

Major causes of death are:

- **Gastrointestinal hemorrhage**
- **Sepsis**
- **Heart failure**

Fluid and electrolyte complications during this phase include the following:

- **Hyponatremia**—secondary to the kidneys' inability to excrete water
- **Hyperkalemia**—note that a **serum potassium > 6.5** is **life-threatening** and **requires immediate therapy** with **Kayexalate, insulin, and glucose.**
- **Metabolic acidosis—**secondary to the inability to excrete acidic metabolic end products
- **Hyperphosphatemia—**secondary to the inability to excrete phosphate
- **Hypocalcemia**—by product of high blood phosphate levels

Hematologic problems are as follows:

- **Anemia**
- **Platelet malfunction**
- **Coagulopathy**

■ **Managing the Patient** **Preventing acute renal failure is key.**

- Patients who are **hypovolemic should be hydrated** to promote normal kidney perfusion. This is **especially true of surgical patients before and after procedures.** When urine output falls the patient must be evaluated for volume status.
- Patients on **nephrotoxic drugs such as gentamicin should**

have drug serum levels drawn (after the fourth dose) to ensure that acute tubular necrosis will not occur.

- **Medications** potentially **causing acute renal failure should be stopped, if possible.**
- If **urinary obstruction** is discovered it should be **corrected immediately.**

If ARF does occur, treatment is geared toward reducing complications.

- The patient's **daily volume input/output and serum electrolytes** should be **monitored.**
- **Water overload manifests** itself **as a decrease in serum sodium. If the patient is water overloaded,** he or she **should be water restricted.**
- Because of reduced ability to excrete potassium, all **potassium intake should be limited.**
- **Serum potassium > 6.5** is **life-threatening** and **requires immediate therapy** with **Kayexalate, insulin, and glucose.**
- Because of reduced ability to excrete phosphate, **phosphate binders** (e.g., aluminum hydroxide, and calcium carbonate) are **employed.**
- Hypercalcemia can be treated with potassium exchange resins.
- **All medications that are renally excreted must be redosed** so that the patient does not attain toxic levels **(e.g., digoxin).**
- **Patients** with renal failure are **prone to infections. Any sign of infection** (e.g., increased temperature) should be **treated aggressively** with **blood, urine, and line culturing; CXR; as warrented, searching for source of infection.**
- Patients with **severe hyperkalemia, pericarditis, mental status changes,** or other complications **recalcitrant to medical therapy require dialysis.**

Benign Prostatic Hypertrophy

■ Basic Science **Benign prostatic hypertrophy (BPH)** is a **benign nodular neoplasia of the prostate** that **arises in** the **tissue surrounding the urethra. Enlargement of this tissue results in compression** and **narrowing of** the prostatic **urethra.**

Compression of the prostatic urethra **obstructs urine flow, causing the patient to retain urine.** Retained urine **predisposes the patient to urinary tract infections and kidney stones.** Increased urinary tract pressures lead to hydronephrosis, with **potential for renal failure.**

The predominant active hormone in prostate tissue is dihydrotestosterone, not testosterone.

■ History and Physical Examination

- **History** Prostatic hyperplasia becomes a **common condition in the fourth or fifth decade.**
  - Symptoms of obstruction are common (e.g., **slow stream,**

hesitancy, a sensation of incomplete emptying, and postvoid dribbling).

- Symptoms of bladder irritability are also common (e.g., **frequency, urgency, and nocturia**). Hematuria may be associated with BPH, although more serious underlying causes must be excluded. The **onset of symptoms is progressive** but may be **worsened by medications with anticholinergic or α-adrenergic effects.**

- **Physical examination** Physical examination of the prostate usually demonstrates an **enlarged gland without hard nodules or localized tenderness.** In patients with relatively advanced chronic obstruction, the bladder may be palpable or percussible. This indicates significant postvoid residual volumes.

■ Laboratory and Diagnostic Studies

- Assessment of **BUN and creatinine** is important. **Increased levels may indicate renal failure due to prostatic obstruction.**
- A urinalysis should be sent to determinine if hematuria, pyuria, or bacteria are present. These findings indicate other potential causes for the patient's symptoms.
- Estimation of postvoid residual volume is important.
- **Imaging of the urethra with contrast to evaluate the anatomy** of the prostate and exclude urethral strictures, stones, or tumor, which may cause irritative voiding symptoms, should be done.
- **Prostatic-specific antigen levels may be increased** in patients with significant amounts of prostatic hypertrophy.

■ Formulating the Diagnosis **Benign prostatic hypertrophy is the most common cause of decreased stream** associated with irritative voiding **in older men.**

■ Evaluating the Severity of the Problem **Patients with obstruction that has caused deterioration of renal function,** or hydronephrosis **should undergo medical or surgical treatment.**

Patients presenting with significant azotemia (renal failure) **should be admitted.** A significant **postobstructive diuresis may occur following catheter drainage** requiring fluid monitoring and intravenous replacement of electrolytes.

Urinary retention or recurrent **UTIs are indications for medical or surgical treatment.**

Postvoid residuals exceeding 100-200 cc indicate the need for monitoring and treatment.

■ Managing the Patient Patients should **avoid medications with α-adrenergic or anticholinergic effects.** These medications are commonly found in **over-the-counter cold formulations. Avoiding caffeine and alcoholic beverages** may reduce symptoms of frequency and urgency. The bladder should be emptied regularly to prevent overdistension of the bladder.

Monitoring the patient's symptom is appropriate if the symptoms are not severe.

Medical therapies include treatment with **alpha-blocking agents.** Alpha-blocking agents **relax the smooth muscle of the prostatic urethra.**

Hormonal therapy with **5-α-reductase** inhibition **prevents conversion of testosterone to dihydrotestosterone within the prostate.** This **causes the prostate to shrink.**

Medical castration, or antiandrogens, are a **second-line therapy.** Antiandrogens have greater toxicity and are primarily used in treatment of prostate cancer.

If medical treatment is **unsuccessful** at controlling the patient's symptoms, **surgical intervention is indicated.**

Chronic Renal Failure

■ **Basic Science** The kidney acts as:

- A filter removing toxins from the bloodstream
- A regulator of the body's water, electrolyte balance
- A regulator of acid-base homeostasis
- An activator of vitamin D, which promotes gut absorption of calcium
- A producer of erythropoietin, a hormone that stimulates RBC production

Most diseases that effect renal parenchyma gradually destroy the kidney. This leads to chronic renal failure, leading to loss of the usual kidney functions just listed.

Major **causes of chronic renal failure include** the following:

- **Diabetes mellitus—the most common cause;** chronic renal failure within 7 to 15 years of proteinuria
- **Hypertension**
- **Glomerulonephritis**

Chronic **renal failure is a gradual process.** The kidney attempts to compensate for loss of nephron function by increasing the workload of the remaining nephrons. This strategy works until less than 25% of nephrons are functioning. After this point is reached the kidneys lose the ability to filter toxins effectively. This leads to **buildup of toxins, leading to a condition called the uremic syndrome.** The uremic syndrome was so named because of the increase in measured serum BUN and creatinine levels, although it is accompanied by many other findings.

■ **History and Physical Examination**

- **History** Patients may be **asymptomatic or** may present complaining of **fatigue, itching, nausea, vomiting, anorexia, paresthesias, sensory deficits,** and **mental status changes.**
- **Physical examination Physical examination** may demonstrate **wasting, skin lesions, edema,** and **hypertension.** Urine output can be low, normal, or high.

■ **Laboratory and Diagnostic Studies** **CBC** may demonstrate **anemia** (normocytic, normochromic, hypochromic microcytic)

Serum chemistries such as **BUN and creatinine** are **elevated.** Chemistries may demonstrate **high phosphate, low calcium, low bicarbonate,** and **high potassium levels.**

Urinalysis shows **proteinuria** and **abnormal tubular casts. Specific gravity** is usually approximately **1.010** (the serum specific gravity) because the kidney has lost its ability to concentrate urine.

Postrenal causes of kidney failure should be ruled out by KUB and abdominal ultrasound. A 24-hour analysis of urine should be performed, and the **creatine clearance** in that time period measured **to determine the glomerular filtration rate.**

■ Formulating the Diagnosis **History, physical examination, serum chemistries,** and **analysis of urine sediment** secure the diagnosis.

Renal biopsy may be helpful in defining specific lesions such as glomerulonephritis. However, once end-stage renal failure has developed, the results of biopsy are often nonspecific and not helpful.

The cause of chronic renal failure should be ascertained to determine if the condition causing the renal failure may be stopped to slow progression of the disease. Some of the reversible conditions include infection, obstruction, fluid loss, nephrotoxic agents, congestive heart failure, and hypertension.

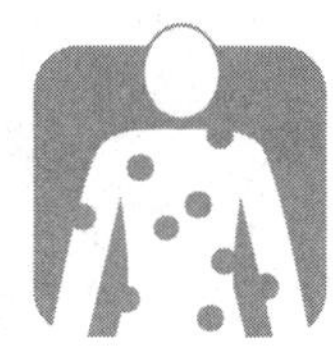

■ Evaluating the Severity of Problems **Uremic syndrome is manifested by the following symptoms:**

- **Cardiac effects**
  - **Pericarditis**
  - **CHF**
  - **Hypertension**
- **Miscellaneous**
  - **Metastatic calcification**
- **Hematologic effects**
  - **Anemia**
  - **Prolonged bleeding times—platelet dysfunction**
  - **Increased susceptibility to infection—impaired lymphocyte function**
- **GI effects**
  - **Nausea**
  - **Vomiting**
  - **Anorexia**
  - **Bleeding**
- **Neurological effects**
  - **Polyneuropathy**
  - **Uremic encephalopathy**
  - **Seizures**

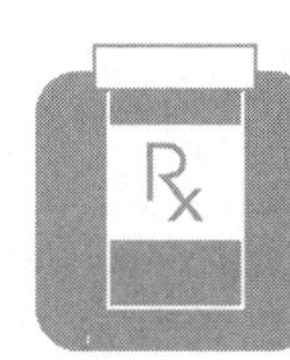

■ Managing the Patient

- **Emergency management** Patients with **potassium levels demonstrating ECG changes of hyperkalemia** must be **treated** emergently **with glucose, insulin, and calcium chloride to prevent arrhythmias.**

  Immediate dialysis is indicated for:

- **Life-threatening hyperkalemia**
- **Life-threatening acidosis**
- **Uremic pericarditis**
- **Unmanageable volume overload (hypertension, pulmonary edema)**

- **Chronic management** Therapy in chronic renal failure is designed to prevent further renal deterioration and to prevent the uremic syndrome.

  Treatment should include the following:

  - **Controlling hypertension** to prevent further deterioration
  - **Reduced phosphate intake** (dairy foods are high in phosphate)
  - **GI phosphate binders** (e.g., calcium carbonate) to reduce phosphate levels
  - Anemia treated by recombinant **erythropoietin therapy and iron**
  - Treatment of chronic electrolyte disturbances (i.e., hyperkalemia with low potassium diet)

  When end-stage renal disease is not responding to medical therapy, dialysis and possibly renal transplantation should be considered.

CYSTITIS

■ Basic Science Cystitis is an **inflammatory condition of the bladder.** The most **common cause of cystitis is** infection of the bladder with **gram-negative bacteria (e.g., *E. coli*).** Cystitis may also be caused by tumor or kidney stones.

Cystitis is much **more common in women than in men.** This is **because of** the **short** female **urethra. Most cases of female cystitis occur after** sexual **intercourse.** In the **male, cystitis** is generally **caused secondary to prostatitis,** urethritis, or urethral instrumentation.

■ History and Physical Examination Most **patients report** the following:

- **History**
  - Pain on urination **(dysuria)**
  - Increased frequency of urination **(frequency)**
  - Urge to urinate after voiding **(urgency)**
  - **Urinating small volumes**
  - **Low back pain**
  - **Previous UTIs**
  - **Hematuria** in up to one third of cases
- **Physical examination**
  - **Suprapubic discomfort**
  - **Fever** may be present

- **Laboratory and Diagnostic Studies** Urine should be sent for **urinalysis (UA). If infection** is **present UA will demonstrate** the following:
  - Detectable levels of **leukocyte esterase** (WBC by-product)—**most sensitive indicator**
  - Increased levels of **nitrites** (bacterial by-product)
  - **Increased pH—**if urea-splitting organisms are present in the urine

  Urine sediment may also demonstrate WBCs and RBCs. Large numbers of **bacteria may also be present.**

  If the **UA is positive a urine culture** and sensitivity **should be obtained.** Cultures that grow greater than 100,000 colonies of a single organism indicate probable urinary tract infection. In symptomatic women 100 to 10,000 organisms indicate a probable UTI.
- **Formulating the Diagnosis** History, physical examination, and laboratory results will secure the diagnosis.
- **Evaluating the Severity of Problems** **Diabetes mellitus, immunosuppression, urologic abnormalities, and pregnancy may complicate cystitis.**
- **Managing the Patient** **For uncomplicated infections a course of oral antibiotics** (3 to 7 days) will suffice. Antibiotics shown to be effective in simple cystitis include the following:
  - **Trimethoprim-sulfamethoxazole (first choice)**
  - **Amoxicillin**
  - **Nitrofurantoin**
  - **Ciprofloxacin**

  In men the **possibility of prostatitis makes 3 weeks of treatment necessary.** The patient should be referred to a urologist for further workup.

  Phenazopyridine (Pyridium 200 mg three times per day) will help with dysuria.

  Patients with frequent cystitis may be placed on a prophylatic course of antibiotics.

Kidney Stones (Nephrolithiasis)

- **Basic Science** **Kidney stones are composed of** any of the following compounds:
  - **Calcium** The majority of kidney stones contain **calcium oxalate.** These **stones may be seen on x-ray.**

    Conditions that predispose to the development of **calcium stones include** the following:

    - ***High levels of blood and urine calcium*** *resulting* ***from*** intestinal **overabsorption of calcium** (vitamin D intoxication, sarcoidosis), increased bone metabolism (**primary hyperparathyroidism**), and decreased renal reabsorption of calcium.
    - ***High levels of urine oxalate*** *resulting from* increased endogenous production and increased intestinal absorption (**Crohn's disease** or other ileal disease).

- ***High levels of urine urate***—urate crystals provide a nidus for calcium precipitation

- **Uric acid** Uric acid stones form as a result of high levels of urine uric acid. In order for these stones to form there must be a low urine pH and low urine volume. These stones **cannot be seen on x-ray.**
- **Cystine** Cystine stones are formed as the result of a genetic disorder that leads to excessive amino acid excretion through the renal tubules. These stones **may be seen on x-ray.**
- **Struvite stones** These stones are called **infectious stones.** These stones occur as the **result of urea-splitting bacteria** (*Proteus,* ***Pseudomonas,*** and *Klebsiella*) that produce a high urinary pH. Infectious stones are often large, filling the entire renal pelvis (i.e., staghorn calculi).

  Foreign bodies in the urinary tract can also lead to struvite stone formation.

■ History and Physical Examination

- **History** Patients present with **severe, intermittant, unilateral flank pain (renal colic) that radiates to the groin or testicle.** Patients will often experience **nausea, vomiting, and low-grade fever.**

  Patients may have a **previous history of kidney stones.** Many have a **family history of kidney stones.**

  Patients should be questioned about previous urinary tract infections, medications, diet, bowel disease (Crohn's), gout, and renal disease.
- **Physical examination** Patients have various degrees of costovertebral tenderness. **Patients are often writhing in pain.**

■ Laboratory and Diagnostic Studies

- **Urinalysis**—will usually demonstrate **hematuria and crystals.**
- **Urine chemistry—increased BUN and creatine indicate** renal failure from urethral **obstruction, or dehydration.**
- **X-ray (KUB)**—a significant number of stones can be seen on x-ray.
- **Intravenous pyelogram—will establish the diagnosis** in patients with normal renal function.
- **Abdominal ultrasound**—may be used to visualize the urinary tract **in cases of suspected obstruction.**
- **If infection is possible a CBC should be drawn.**

■ Formulating the Diagnosis History, physical examination, and results of KUB, IVP, or ultrasound will yield the diagnosis.

■ Evaluating the Severity of Problems **Indications for hospitalization include obstruction, high fever, uncontrollable pain, solitary kidney,** and **dehydration (not responsive to hydration).**

■ Managing the Patient

- Most **stones less than 5 mm in diameter** will **often pass spontaneously. Treatment includes oral hydration** and **narcotic analgesics** (e.g., Demerol, Dilaudid). The nonsteroidal medication, Toradol, is also an effective analgesic.

• **Patients with infection** should be admitted to the hospital and **started on IV antibiotics.**

• Patients with **high-grade obstruction,** persistent pain, and complicated infection may **require drainage of the urinary tract.**

• Patients with **unpassably large stones,** located in the kidney, **may require surgical treatment.** Surgical treatment is often performed by **extracorporeal shock wave lithotripsy (ESWL).** With this method, shock waves are directed at the stone, resulting in stone fragmentation. Stones are then passed spontaneously.

• **Surgical treatment of lower ureteral stones** is accomplished through **urethroscopy and direct extraction.** Bladder stones are almost uniformly removed cystoscopically.

• Patients with struvite stones can be treated with antibiotics to prevent progression of infection or stone formation.

• For **cystine stones and uric acid stones the mainstay of therapy involves hydration and oral alkalization.** Patients with cystine stones can also be treated by D-penicillamine to decrease the cystine excretion in the urine.

• Since **kidney stones are often recurrent** further studies are necessary to uncover metabolic problems.

• **Passed stones should be collected and analyzed for composition.** Screening laboratory data should include a CBC, SMA 7, serum calcium, phosphorus, and uric acid. A 24-hour urine collection for calcium, phosphorus, oxalate, uric acid, and creatine should be obtained.

Nephritic Syndrome

■ Basic Science The **nephritic syndrome** is **defined by** the following:

- **Active urine sediment with RBC casts**
- **Proteinuria (usually <3 g per day)**
- **Azotemia**
- **Edema**
- **Hypertension**

The nephritic syndrome is a disease state that results in glomerular damage from many different processes, including primary renal diseases, or may be secondary to systemic processes.

Common causes of the nephritic syndrome include the following:

• **Acute poststreptococcal glomerulonephritis**—appears most often in childhood; develops between 1 and 4 weeks after strep throat or impetigo. Pathological examination of the kidney demonstrates immune complexes deposited in the glomerular basement membrane. This disease state is usually self-limited.

- Post infectious, including *Streptococcus*
- Vasculitis syndromes, including systemic lupus erythematosus, Wegener's granulomatosus, and Goodpasture's syndrome
- MPGN (membranoproliferative glomerulonephritis)
- **Systemic lupus erythematosus**

- History and Physical Examination
  - **History** Patients may complain of **recent illness (e.g., strep throat or skin rash—impetigo), edema, dark or bloody urine, oliguria, or hemoptysis.** Patients may also have a medical history of **lupus or other autoimmune diseases.**
  - **Physical examination** Physical examination may demonstrate **hypertension, edema,** and other disease-specific symptoms.
- Laboratory and Diagnostic Studies **Urinalysis** will demonstrate the following:
  - **Hematuria**
  - **Proteinuria**—total protein lost per day will be less than 3 g.
  - **High BUN and creatinine**—initial BUN and creatinine may be obtained to compare to previous labs drawn and to follow during the course of the disease.
  - **Urinary sediment** Demonstration of **RBC casts are pathognomonic for glomerulonephritis; RBC, WBC, and granular casts are common.**
  - **Disease-specific laboratory studies may be ordered.**
    - Poststreptococcal GN—ASO titers
    - Complement levels—for vasculitis
    - Goodpasture's syndrome—antiglomerular membrane antibodies

  Definitive diagnosis may be obtained through renal biopsy. Biopsy specimens are analyzed by conventional, electron, and immunofluorescent microscopy.
- Formulating the Diagnosis **History, physical examination,** and results of **urinalysis help confirm** the diagnosis of the **nephritic syndrome.** Demonstration of **RBC casts** are **pathognomonic for glomerulonephritis.**

  About half of all patients will have diagnosis of nephritis made incidentally after a routine urinalysis.
- Evaluating the Severity of Problems **Rapidly progressive glomerular nephritis (RPGN)** is a **severe form of** the **nephritic syndrome associated with renal failure.** The overall prognosis of these patients is poor. RPGN does not usually respond to medical management. Irreversible anuria commonly occurs. Hemodialysis may be required for the patient to survive.
- Managing the Patient **Complications of the nephritic syndrome must be managed** while the cause of nephritis is explored. Complications include **hypertension,** which is treated with **antihypertensive medication,** and **edema,** which is treated with **diuretics.**

  Disease-specific treatment includes **acute post-infectious GN (possibly antibiotics), and SLE (cytotoxic agents and steroids). Goodpasture's syndrome may require admission.**

Nephrotic Syndrome

- Basic Science **Nephrotic syndrome occurs because the glomeruli lose the ability to retain protein.** Pathologically it appears that either of two processes occur:
  - The glomerular epithelial cells lose surface charge. This leads to the inability of the kidney to retain small charged molecules.
  - The glomerulus loses the ability to act as a size selective barrier. Large, less densely charged molecules are unable to be retained in the vascular system.
  - The **classic features of nephrotic syndrome include:**
    - **Edema**
    - **Proteinuria (more than 3 grams)**
    - **Hypoalbuminemia**
    - **Azotemia**
    - **Hyperlipidemia**
  - **Renal disease states that may lead to nephrotic syndrome include:**
    - **Minimal Change Disease**
    - **Membranous glomerulonephritis**
    - **Focal Sclerosing glomerulosclerosis (FSGS)**
    - **Mesangial Proliferative glomerulonephritis**
  - **Systemic diseases that may lead to nephrotic syndrome include:**
    - **Diabetes mellitus**
    - **Amyloidosis**
    - **Neoplasms**
    - **Vasculitis**
    - **Infectious causes (e.g., hepatitis)**
- History and Physical Examination
  - **History** The patient may complain of **weakness, pruritus, abdominal pain, nausea, vomiting, or edema.** Patients may also have **difficulty breathing, ascites,** chest pain (pericardial effusion). Patients often describe **frothy urine.** History of previous infectious diseases (e.g., syphilis, hepatitis, HIV) may indicate the cause of nephrotic syndrome.
  - **Physical examination** Patients may be **hypertensive** and often have **edema.** Edema may present in a specific area (e.g., scrotal swelling).
- Laboratory and Diagnostic Studies
  - **Urinalysis**
    - Significant **proteinuria (>3.0 g/day).** Twenty-four hour collection of urine should be sent for creatinine, creatinine clearance, and total protein measurements.
    - **Urine sediment** may demonstrate **broad, waxy casts.**

- **Serum chemistry**
  - **Hypoalbuminemia (<2.5 mg/dl in adults)**
  - **Lipidemia**
  - **CBC**
  - **Anemia—caused by loss of erythropoietin production**

  A **renal biopsy may define the cause of nephrotic syndrome.**

■ **Formulating the Diagnosis** History, physical examination, and results of UA will be diagnostic for nephrotic syndrome.

■ **Evaluating the Severity of Problems** Inability to control volume and electrolytes, bleeding diathesis (e.g., platelet dysfunction from uremia, hypercoaguable state) suggests severe disease. **Patients with severe disease may require hemodialysis.**

■ **Managing the Patient** **Complications of the nephrotic syndrome must be managed** while the cause is explored. Complications include:

- **Hypertension** and **edema**—treated with **diuretics, fluid and salt restriction, appropriate antihypertensives**
- Multiple **electrolyte abnormalilties—reduced intake of potassium, sodium, and phosphorus-containing products**
- Proteinuria—ACE inhibitors slow the progression of urine protein loss in **diabetes**
- **Acidoses—**treated with **sodium bicarbonate supplements**
- Thrombosis **(DVT and pulmonary emboli)—heparin and coumadin** when indicated
- **Specific treatment** of the variety of causes of nephrotic syndrome often **includes corticosteroids and cytotoxic agents.**
  - Minimal Change Disease usually responds to steroid therapy alone and has an excellent prognosis in 80% to 90% of patients, often within six weeks of treatment.
  - Idiopathic Membranous GN—the mainstay of treatment includes steroids, cytotoxic agents, including cyclophosphamide, chlorambucil, and cyclosporine.
  - The majority of patients with FSGS progresses to renal failure regardless of medical treatment.
  - The treatment of nephrotic syndrome caused by systemic disease (e.g., diabetes, amyloidosis, SLE, HIV) is directed at the underlying disease.

Prostate Cancer

■ **Basic Science** Prostate cancer is a **common disease of elderly men.** Prostate cancer is an **adenocarcinoma.** Tumors originate from the peripheral zone of the prostate, which is located near the rectal wall. Prostate cancer most **commonly metastasizes to bone.**

PSA is a normal protein produced by the prostate that is involved in the liquefaction of semen. Prostate cancer is associated with an increased secretion of PSA into the circulation. This provides the basis of its use as **a tumor marker.**

- History and Physical Examination **Risk factors** for prostate cancer include **family history** and **black race.**
  - **History Prostate cancer** is **silent until advanced. Local symptoms** may **occur at that time.** Symptoms attributed to prostate cancer are **decreased stream, frequency, feeling of incomplete emptying, and back pain.**
  - **Physical examination** The prostate may have areas of **localized nodularity, induration, or hardness.** Note that the prostate **exam may be normal despite** a significant volume of **cancer.**

    Metastatic prostate cancer can **present with weight loss, lymphadenopathy, hematuria,** or **bone pain.**

- Laboratory and Diagnostic Studies **Prostatic-specific antigen (PSA)** has become the predominant laboratory marker associated with prostate cancer. A PSA **within** the **normal range does not exclude prostate cancer. Palpable abnormalities** of the gland **should be investigated.**

  The normal process of aging includes benign enlargement of the prostate, which is also associated with increasing PSA levels. A **modest elevation of PSA** above the reported standard limit **cannot be reliably used to indicate** the presence of **cancer. If** the serum **PSA** is **3-4 times** the upper limit of **normal, however, prostate cancer** becomes much more **likely.** Other factors that may cause a substantial transient rise in the PSA include prostate biopsy, urethral instrumentation, or prostatic infarct.

  Diagnosis of prostate cancer is **made by needle biopsy** of the prostate **with histopathological evidence of cancer.**

  In patients with an elevated PSA, systematic biopsies are performed over various regions of the prostate to attempt to detect nonpalpable diffuse disease. Pathologic grading of the biopsies provides some correlation to the aggressiveness of the tumor.

- Formulating the Diagnosis History, physical examination, and **histopathologic evidence from needle biopsy secure the diagnosis** of prostate cancer.

- Evaluating the Severity of Problems **In patients with** proven **prostate cancer, bone scans** are **performed to rule out metastatic disease. CT or MRI** scans **of the pelvis** are **performed to determine if** the **lymph nodes** are **enlarged, suggesting metastatic disease.**

  Evidence of widespread **metastatic disease** and a poorly differentiated pathology are associated with a **poor prognosis.**

- Managing the Patient **Cancer that is well differentiated and of microscopic volume** (less than 1-2 mm) is usually **observed.** The patient should have repeat **physical examinations and PSA** levels **several times per year.**

  Prostate cancer that is localized within the gland and of sufficient volume or histologic grade to be life-threatening

should be treated. Treatment includes either **radical surgical excision or radiation therapy.**

Prostate cancer is present in many elderly patients with significant other illness. **The mere presence of prostate cancer** alone **is not** a sufficient **indication to initiate treatment. Many elderly patients can be observed.**

Hormonal therapy by **surgical castration or hormonal agents** (e.g., diethylstilbestrol, LH-RH agonist, or antiandrogens) **may be utilized. Hormonal therapy is palliative,** not curative. There is **no effective chemotherapy** at present for disease that is refractory to hormone therapy.

In patients with **bone metastasis, x-ray therapy** may make the patient more comfortable.

Prostatitis

- Basic Science **Bacterial prostatitis** is usually **caused by** the same types of **gram-negative organisms** commonly **found in UTIs (e.g., *E. coli*).** Less common organisms include tuberculosis, fungus, and gonorrhea. Bacterial prostatitis **occurs in older men.** Bacterial prostatitis is **categorized into acute disease** and **chronic disease.**

  Nonbacterial prostatitis is the **most common inflammatory disorder of the prostate.** The exact etiology of this condition is unknown at present time. Nonbacterial prostatitis primarily **affects younger men.**
- History and Physical Examination
  - **Bacterial prostatitis**
    - ***Acute bacterial prostatitis***
      - ***History*** Rapid onset of **high fever, chills, back pain, dysuria, frequency, urgency, hesitancy, low back pain, and perineal discomfort**
      - ***Physical examination*** Patient is **febrile** and may appear septic. Rectal exam shows evidence of **prostate inflammation, swelling, warmth, tenderness, and induration.** Urethral discharge is absent.
    - ***Chronic bacterial prostatitis***
      - ***History*** **Recurrent urinary tract infections** with the same organism, **low back pain, testicular pain (epididymitis), urgency, frequency, and dysuria. Fever** is usually **not present.**
      - ***Physical examination*** **Prostate is usually not tender or inflamed.**
  - **Nonbacterial prostatitis**
    - ***History*** **Dysuria, frequency, urgency, perineal discomfort, low back pain**
    - ***Physical examination*** **Prostate** is **usually not tender or inflamed.**
- Laboratory and Diagnostic Studies
  - **Acute bacterial prostatitis** **Urinalysis** should be performed and a **urine culture** collected. **Prostatic massage** is **con-**

traindicated in patients who present with symptoms and fever suggestive of acute bacterial prostatitis **due to the risk of causing bacteremia and sepsis.** Since cystitis often accompanies prostatitis, a urine culture and sensitivity will usually reveal the offending organism.

- **Chronic bacterial prostatitis Urinalysis** should be performed and a **segmental urine culture** collected (as described in the following text).

  Three sets of urine specimens are obtained for culture:

  - A first void sample (the first few milliliters of the urinary stream)
  - A midstream sample
  - A sample of expressed prostatic fluid (expressed by prostatic massage)

  For the patient to be diagnosed with chronic bacterial prostatitis the **expressed prostatic secretion** should **have** more than **10 times the number of bacteria than** either the **initial or midstream samples.** The organism isolated should be the same one as has been causing recurrent infections.

  Microscopic examination of the expressed prostatic secretions will usually show an **increased number of WBCs.**

- **Nonbacterial prostatitis Urinalysis** should be performed. **Cultures** are **collected as in chronic bacterial prostatitis. No organisms are identified by culture,** although **microscopy of expressed prostatic secretions demonstrate large numbers of WBCs.**

■ Formulating the Diagnosis Age, history of previous urinary tract infections, prostate examination, microscopic analysis of prostatic secretions, urinalysis, and urine culture guide diagnosis.

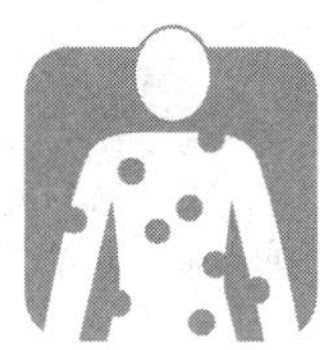

■ Evaluating the Severity of Problems **Acute bacterial prostatitis can lead to sepsis.** The patient must be hospitalized and monitored for signs of sepsis.

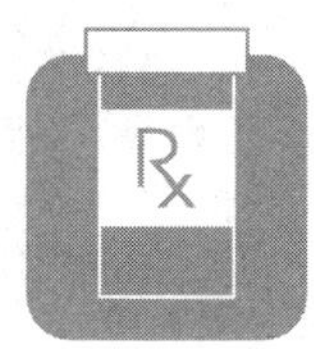

■ Managing the Patient

- **Acute bacterial prostatitis** Patients with acute bacterial prostatitis require **hospital admission.** A substantial degree of **bladder outlet obstruction may occur with the need for drainage.** These patients require **monitoring for sepsis.**

  Oral sulfamethoxazole (Bactrim) should be started empirically. If the **patient** is **thought to be septic, IV ampicillin** and **gentamicin** should be **started.** Antibiotic **therapy** is **modified when culture results** and organism antibiotic sensitivities are **determined.** Treatment continues for about 1 month.

  The newer **fluoroquinolone (e.g., ciprofloxacin) antibiotics** are often utilized because of **good prostate tissue penetration.**

  If urine retention occurs, suprapubic drainage is **preferable to** placement of a **Foley** catheter.

- **Chronic bacterial prostatitis Treatment is based** on urine **culture** and sensitivity **results.**

  Ciprofloxacine (a fluoroquinalone) is the **drug of choice.**

Treatment lasts for 6 weeks. Trimethoprim/sulfamethoxazole (Bactrim) may also be used.

Chronic bacterial prostatitis is **treated on** an **outpatient basis.**

- **Nonbacterial prostatitis** Nonbacterial prostatitis is difficult to treat due to the lack of a definite etiology.

  Treatment involves the following:

  - A **trial of tetracycline** for 2 weeks
  - **Nonsteroidal antiinflammatory agents**
  - **Sitz baths**
  - **Anticholinergic medication**

  Nonbacterial prostatitis is **treated on** an **outpatient basis.**

Pyelonephritis

- Basic Science Pyelonephritis is a **diffuse pyogenic infection of the** pelvis and parenchyma of the **kidney.** It is usually the **result of an infection that ascends from** the **lower urinary tract** to the kidney. ***E. coli*** contamination of the urethral meatus is a **common cause in females.** Infection may spread to the kidney from other locations in the body.
- History and Physical Examination
  - **History** The **onset** of pyelonephritis is **rapid.** Patients complain of **fever, chills, nausea, vomiting, and flank pain.** Additionally, up to one third of patients complain of **dysuria and frequency.**
  - **Physical examination** Physical examination demonstrates **costovertebral tenderness on the infected side.** Occasionally abdominal rigidity may be demonstrated.
- Laboratory and Diagnostic Studies Urine should be sent for **urinalysis.** If infection is present, UA will demonstrate the following:

  - Detectable levels of **leukocyte esterase** (WBC by-product)
  - Increased levels of **nitrites** (bacterial by-product)
  - **Increased pH, if urea splitting organisms** are present in the urine

  Urine sediment may also demonstrate **WBCs** and **RBCs** and **white cell casts. Large numbers of bacteria** may also be present.

  If the **UA is positive,** a **urine culture and sensitivity should be obtained.** Cultures usually show greater than 10,000 colony-forming units.
- Formulating the Diagnosis **History, physical examination, and laboratory studies suggest the diagnosis of pyelonephritis.**
- Evaluating the Severity of Problems **Chronic pyelonephritis** develops slowly after repeated bacterial infection of the kidney and may progress to renal failure in **3% of patients.** Many cases are associated with some form of obstruction, such as stone or stricture in the ureter.
- Managing the Patient Antibiotic treatment should be instituted **immediately after blood and urine have been collected for cul-**

ture. Most patients are started on a regimen of **intravenous antibiotics.** One to two days after the patient is afebrile, he or she may be switched to oral antibiotics. Treatment should continue for 2 weeks. (For patients who do not appear very ill, treatment with oral antibiotics may be initiated as the primary treatment.)

Typical antibiotics include **trimethoprim-sulfamethoxazole, second- or third-generation cephalosporin (e.g., ceftriaxone).**

In patients with **diabetes mellitus, immunosuppression, urological abnormalities,** and **pregnancy, two antibiotics should be used.** Antibiotic choice should be guided by the results of culture data.

If an obstruction is **identified, surgical correction** should be considered.

Patients with frequent pyelonephritis may be placed on a prophylactic course of antibiotics.

URETHRITIS

■ Basic Science Urethritis is defined as **infection of the urethra.** The **disease is classified into** one of the following categories:

- **Gonococcal urethritis** ***Neisseria gonorrhoea*** is a **gram-negative intracellular diplococci. Infection** occurs **through sexual contact,** usually with an asymptomatic carrier. **Untreated infections lead to fibrosis** and **urethra stricture,** or **tubal abnormalities** that **may** in turn **lead to tubal pregnancy** and **infertility.**
- **Nongonococcal urethritis** Several different organisms are responsible for this disease state. ***Chlamydia trachomatous*** causes **half of** the **infections.** ***Ureaplasma urealyticum*** causes 30%. **In female *chlamydia*** infections can **lead to salpingitis,** which may lead to **infertility.**

■ History and Physical Examination

- **Gonococcal urethritis**
  - ***History*** Patient may describe **new sexual partner,** sudden onset of **dysuria,** and **purulent discharge.** If the infection is left untreated, symptoms will subside over months.
  - ***Physical examination*** Physical examination will demonstrate a **purulent discharge from** the **urethra.**
- **Nongonococcal urethritis**
  - ***History*** Patient may describe a **new sexual partner.** The patient will report **dysuria *or* discharge** rather than both. Symptoms may increase over 2 to 3 weeks. **Symptoms** may also **fluctuate, sometimes disappearing for days** and then returning.
  - ***Physical examination*** **Urethral discharge** of a more **mucous or mucopurulent quality** will be present.

■ Laboratory and Diagnostic Studies **In men** a **Gram's stain and culture of urethral discharge** should be performed.

In females swabs of the cervical mucus should be obtained for **Gram's stain and culture. Vaginal swabs are not satisfactory** since *N. gonorrhea* does not grow in the vagina.

- **Gram's stain results**
  - **Gonococcal urethritis: gram-negative intracellular diplococci.**
  - **Nongonococcal urethritis: white blood cells** and **no gram-negative intracellular diplococci**

In addition, **rectal and pharyngeal swabs for Gram's stain and culture** should be obtained since ***N. gonorrhea*** can **grow in these areas.**

Tests for chlamydia trachomatous and **RPR for syphilis** should also be sent.

- Formulating the Diagnosis In men **history, physical examination, and Gram's stain usually form the basis for treatment.**

In females identification of *N. gonorrhea* may be difficult from Gram's stain, necessitating the **identification of *N. gonorrhea* by culture** to confirm clinical suspicions.

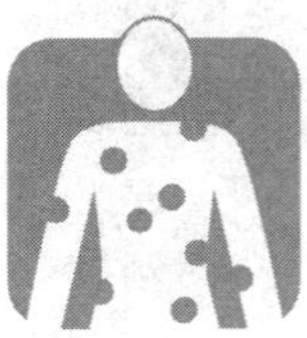

- Evaluating the Severity of Problems **Pelvic inflammatory disease** may develop in up to **20% of women with gonococcal infection.** Gonococci ascend from the cervix to the fallopian tubes. Infection may further spread into the abdomen, causing peritonitis. This is especially common during menstruation.

Infection may cause the fallopian tube to enlarge, making it prone to obstruction and scarring. This raises the patient's chance of **infertility, tubal pregnancy, or tubal abscess. PID is also commonly caused by *Chlamydia* infection.**

Disseminated gonococcal infection may occur by hematogenous spread of the organism. **Fever, polyarthralgias, and skin lesions** may occur. The gonococcus may seed joints, causing **septic arthritis,** which may eventually destroy the joint. **In young adults gonococcal infection is the most common cause of septic arthritis.**

- Managing the Patient The **treatment for gonococcal urethritis is intramuscular ceftriaxone.** Penicillin, which was used in the 1950s, is no longer preferred because of organism resistance.

Nongonoccal urethritis is **treated with** tetracycline, **doxycycline,** erythromycin, or **zithromax. If** the **patient** is **poorly compliant, a single dose of zithromax** is used.

All patients with gonococcal infection should also be treated for nongonoccal urethritis.

Multiple Choice Review Questions

1. A 65-year-old woman with a history of end-stage renal disease is admitted to the emergency department with a potassium level of 6.8. Her ECG shows peaked T waves. Administration of insulin, glucose, calcium, and kayexalate fails to decrease serum potassium. Repeat serum potassium is 6.9. The most appropriate therapy would be:
   - **a.** Administration of 30 g kexalate.
   - **b.** 10u insulin, one ampule of glucose.
   - **c.** Repeat ECG.
   - **d.** Bicarbonate administration.
   - **e.** Hemodialysis.

2. Your patient with ESRD may have any of the following complications as a result of her chronic renal failure, *except:*
   - **a.** Coagulopathy.
   - **b.** Macrocytic anemia.
   - **c.** Hyperphosphatemia.
   - **d.** Metabolic acidosis.
   - **e.** Hyperkalemia.

3. A 35-year-old man noted a purulent discharge at the tip of the penis, as well as dysuria. He notes that this occurred after an encounter with a new sexual partner. *Usual* organisms cultured from this patient could include all but:
   - **a.** Gonococcus.
   - **b.** *Chlamydia.*
   - **c.** *Ureaplasma urealyticum.*
   - **d.** Syphilis.
   - **e.** *Staph aureus.*

4. Complications of this disease may include all of the following *except:*
   - **a.** Infertility.
   - **b.** Urethral structure.
   - **c.** Tubal pregnancy.
   - **d.** Vaginitis.
   - **e.** PID.

5. A 22-year-old woman is seen in the ED with chief complaints of abdominal pain, dysuria, hematuria, and frequency. She has a low-grade fever. No flank pain is found on physical examination. Risk factors for this condition may include all of the following *except:*
   - **a.** Pregnancy.
   - **b.** Sexual intercourse.
   - **c.** Use of diaphragm.
   - **d.** Delayed micturition after intercourse.
   - **e.** Use of tampons.

6. Treatment regimens for cystitis may include all of the following *except:*
   - **a.** Trimethoprim-sulfamethoxazole.
   - **b.** Amoxicillin.
   - **c.** Nitrofurantoin.
   - **d.** Pyridium.
   - **e.** Vancomycin.

Chapter 13

Obstetrics and Gynecology

Abnormal Pap Smear

■ Basic Science **The Pap smear is a screening test used to detect premalignant lesions of the cervix.** The introduction of the Pap smear has reduced the incidence of invasive cervical cancer by 50%. **Pap smears are used to collect cells at the squamocolumnar junction of the ectocervix. This is the site of neoplastic growth in the cervix.**

Abnormal Pap smears can result from either dysplasia (i.e., precancerous change) or infection. Neoplastic changes can range in severity from atypia to carcinoma in situ. Cervical lesions found on Pap smear may resolve (either on their own or with treatment), **remain dysplastic,** or **progress to invasive cancer.**

■ History and Physical Examination Screening is very important since **cervical dysplasia rarely causes symptoms.** The history may reveal **risk factors** significant for the development of cervical dysplasia. These include the following:

- **Early intercourse**
- **Multiple sex partners**
- **A history of sexually transmitted diseases (STDs)** (especially human papilloma virus)
- **Tobacco use**

The physical examination may reveal abnormalities (e.g., genital warts) associated with cervical dysplasia and carcinoma. A Pap smear is obtained by rotating a spatula on the cervix. This obtains cells from the squamocolumnar junction. The sample is then smeared on a microscope slide and fixed.

■ Laboratory and Diagnostic Studies The Pap smear is only a screening test. Both false negative and false positive results can occur. Further evaluation of an abnormal result is required. This is usually done with colposcopy (a magnified examination of the cervix). Colposcopy provides a 10x to 16x magnification of the transformation zone. **Colposcopy is considered adequate only if the entire transformation zone is seen.** During the examination a dilute solution of acetic acid is applied to the cervix. Abnormal areas often appear white and may display prominent blood vessels in a mosaic pattern. **A biopsy of abnormal areas should be obtained.**

■ Formulating the Diagnosis As stated earlier, **a diagnosis of cervical dysplasia cannot be made by Pap smear alone. Confirmation with colposcopy and biopsy is required.**

- **Evaluating the Severity of Problems** The severity of the problem is determined histologically.
- **Managing the Patient** **A Pap smear should be obtained on every woman by age 18, or earlier if sexually active. Additionally, a Pap smear is usually obtained on pregnant women during the first trimester. The frequency of screening ranges from annually to every 3 years, depending on risk factors and previous Pap smear results.** The reliability of the Pap smear as a screening tool is based on frequent sampling, since a single test can be falsely positive or negative.

  Once an abnormality detected on a Pap smear is confirmed by colposcopy and biopsy, therapy must follow. **For low-grade lesions, cryotherapy is the preferred treatment.** Cryotherapy uses liquid carbon dioxide to cool a probe applied to the cervix. **This technique produces a cure rate of approximately 90%.**

  For higher-grade lesions, or when a tissue sample is needed, conization of the cervix can be performed. This can be accomplished using either an electrical wire loop, laser, or a cold-knife (scalpel) technique.

  The objective of management of an abnormal Pap smear during pregnancy is to rule out invasive disease. Colposcopy should be done and biopsies obtained if a high-grade lesion is suspected. The clinician should be prepared for heavier bleeding when performing biopsies on the cervix during pregnancy secondary to increased vasculature. **Only when there is a concern of invasive cancer is excisional therapy performed during pregnancy.**

Abnormal Uterine Bleeding

- **Basic Science** The hypothalamic-pituitary-ovarian axis controls ovulation and mediates the cyclic changes within the uterine endometrium that manifest as the menstrual cycle. Menses normally occurs at regular intervals (usually every 28 days) for a defined period of time (2 to 7 days), beginning at puberty and lasting until menopause. **Bleeding that is unexpected, irregular, prolonged, or excessive is called abnormal uterine bleeding (AUB) (Box 13.1). Although AUB may result from an organic etiology (e.g., a fibroid), it most commonly occurs secondary to anovulation (or dysfunctional ovulation) in the absence of identifiable pathology; such bleeding is called dysfunctional uterine bleeding (DUB).** Anovulation (or dysfunctional ovulation) occurs for a variety of reasons, but does so most often after puberty and around menopause. In all cases of uterine bleeding an organic etiology should be sought. Therefore the diagnosis of DUB is one of exclusion. Common organic causes of AUB are listed in Box 13.1.
- **History and Physical Examination** **A detailed history includes the amount and duration of current bleeding and the date of the last menstrual period. Methods of contraception and the use of medicines should be noted. The presence of pertinent diseases (e.g., blood dyscrasias, thyroid disorders) should be sought.**

  A pelvic examination should be performed to rule out

Box 13.1

COMMON ORGANIC ETIOLOGIES OF ABNORMAL UTERINE BLEEDING

In the Newborn and Child
- Placental estrogen withdrawal (in the neonate)
- Precocious puberty

In the Adult
- Bleeding disorders (e.g., leukemia, hemophilia)
- Spontaneous abortion
- Ectopic pregnancy
- Endometriosis
- Cervical cancer
- Leiomyoma
- Ovarian cysts
- Infection
- Iatrogenic (IUD, medications, and hormones)

IUD, Intrauterine device.

pregnancy, infection, cervical abnormality, adnexal mass, and uterine fibroid(s).

- Laboratory and Diagnostic Studies For excessive bleeding, a **pregnancy test, thyroid function test(s), complete blood count (CBC), coagulation profile, and Pap smear** should be ordered. The lack of ovulation may be confirmed by monitoring basal body temperature.

  A **pelvic ultrasound** may be helpful in eliminating ectopic pregnancy, fibroids, adnexal masses, or endometriosis as causes of uterine bleeding.

  Endometrial biopsy or dilation and curettage is necessary for persistent or postmenopausal bleeding to rule out neoplasia.

- Formulating the Diagnosis **DUB is the cause of AUB in approximately 85% of patients.** The differential diagnosis of AUB includes pregnancy complications, coagulation defects, thyroid or adrenal dysfunction, endometrial or cervical polyps, fibroids, adenomyosis, endometrial hyperplasia (or cancer), anovulation, endometritis, and the presence of an intrauterine device (IUD).

- Evaluating the Severity of Problems The **most severe complication** of abnormal menstruation **is hemorrhage and acute anemia.** Many patients will respond to intravenous estrogen. Invasive therapy (e.g., dilation and curettage, hypogastric artery embolization, or hysterectomy) is rarely required.

- Managing the Patient **Patients with DUB usually respond to progestins (e.g., medroxyprogesterone or norethindrone) given during the luteal phase for several cycles. Alternatively, oral contraceptive pills may be given every 6 hours for 1 week;** following this, patients can expect a heavy withdrawal bleed. **The risk of recurrence is limited if the patient is maintained on a cycled oral contraceptive pill.** Other medical management consists of progestin-containing IUDs, danazol (Danocrine), gonadotropin-releasing hormone (GnRH) agonists, or nonsteroidal anti-inflammatory drugs (which decrease menstrual flow by 20% to 50%). Patients with DUB who desire pregnancy may be treated with clomiphene (Clomid) to induce ovulation.

Patients with DUB who are older than 35 should undergo an endometrial biopsy before receiving hormonal treatment. A hysterosalpingogram (or hysteroscopy) may be necessary in women who fail to respond to hormonal treatment.

Abortion

- Basic Science **Pregnancies that fail to survive past 20 weeks are called abortions.** They may be due to chromosomal aberrations, hormonal problems, uterine structural defects, infection, environmental exposures, smoking, alcohol consumption, or spontaneous cervical dilation. **Often the cause of a pregnancy loss will not be discovered.**
- History and Physical Examination **Vaginal bleeding is usually the first symptom of an abortion.** The degree of vaginal bleeding and the loss of tissue must be ascertained by patient report. **A complaint of suprapubic cramping and vaginal bleeding usually suggests a poor prognosis for continuation of the pregnancy.**

  Vital signs and orthostatics should be determined to ensure hemodynamic stability. If the fetus is more than 10 to 12 weeks old, a fetal heart rate should be heard (120-160 beats per minute) with an ultrasound stethoscope. The abdominal examination may reveal a palpable uterine fundus. Normally the uterine fundus reaches the level of the umbilicus at 20 weeks.

  If the patient is undergoing an abortion, the speculum examination will reveal vaginal bleeding and cervical dilation. The bimanual examination should focus on uterine size (indicative of gestational age).
- Laboratory and Diagnostic Studies **A pregnancy test; complete blood count with differential, type, and screen; and a prothrombin time-partial thromboplastin time test should be ordered.**

  An ultrasound is indicated if the fetal heart rate is absent, or if the uterine size cannot be assessed. Correct estimation of uterine size and gestational age is imperative if uterine evacuation becomes necessary.
- Formulating the Diagnosis
  - **Threatened abortion**—represents **pregnancy less than 20 weeks with uterine bleeding.** The cervical **os is closed. Pain, if present, is minimal.** Half of the patients progress to abortion. Approximately 25% of all pregnancies have some degree of first-trimester bleeding.
  - **Inevitable abortion—similar to threatened abortion except that the cervix is dilated, there is heavy uterine bleeding, or the membranes are ruptured.** Menstrual cramps are present. Ultrasound may reveal a nonviable pregnancy in the process of being expelled.

• **Missed abortion—the fetus has died but it is retained in the uterus, usually for weeks.**

• **Incomplete abortion—similar to an inevitable abortion except that part of the products of conception is passed.**

• **Blighted ovum—ultrasound diagnosis of a gestational sac without fetal tissue.**

• **Complete abortion (spontaneous abortion)—all tissue has been passed through the cervix. Uterine cramping and bleeding resolve.** The uterus is smaller than the gestational age reflects. Ultrasound reveals an empty uterus.

• **Therapeutic abortion—a pregnancy that is terminated due to the patient's medical condition.**

• **Elective abortion—a patient elects to terminate a gestation.**

■ Evaluating the Severity of Problems **Maternal hemorrhage with hypovolemic shock is the most serious problem complicating abortion.** The potential for sepsis is present in women with an incomplete abortion and demands immediate recognition and treatment.

■ Managing the Patient **Avoiding alcohol and smoking during pregnancy is important in decreasing the risk of abortion.** Adequate emotional support must be provided for patients who undergo spontaneous abortion or elective termination. Patients with a history of recurrent pregnancy loss should undergo thorough evaluation to search for a possible etiology (e.g., diabetes, lupus, hypothyroidism). **All pregnant patients with suspected abortion who are Rh negative (and do not have Rh antibodies) should receive Rhogam (anti-D gamma globulin).**

Further management is as follows:

• **Nonviable pregnancy**—uterine evacuation should be considered.

• **Threatened abortion—if the patient is stable (not actively bleeding) and the fetus is viable, she may return home with precautions.**

• **Inevitable abortion—the patient may wait for the abortion to complete or may elect to complete the abortion with an obstetric procedure.**

• **Incomplete abortion—patients are admitted to the hospital for evacuation** of the remaining products of conception.

• **Missed abortion or blighted ovum** (less than 8 weeks' gestation)—**the patient may elect to pass the conceptus at home.** She must be aware of the potential for hemorrhage. **Alternatively, the patient may electively complete the abortion with an obstetric procedure.**

● **Obstetric procedures**

• **First 13 weeks—dilation of the cervix (with mechanical or osmotic dilator) and suction curettage** is commonly employed to evacuate the uterus.

• **Second trimester—the most common method of uterine evacuation is a D&E (dilation of the cervix and**

subsequent evacuation of its contents using forceps and curettage).

Advantages of an obstetric procedure to evacuate the uterus include reduced blood loss and decreased risk of infection from retained tissue. **All patients who are hemorrhaging should receive a suction completion procedure.** Between 8 and 16 weeks there is a greater risk that the entire pregnancy will not pass without a suction completion procedure.

Abruptio Placenta

■ Basic Science **Abruptio placenta is a condition that occurs primarily in the third trimester of pregnancy and is characterized by hemorrhage into the basal layer of the placenta. The hemorrhage separates the placenta from the uterus, resulting in the compromise of oxygen and nutrient delivery to the fetus. Bleeding may result in maternal hypovolemic shock.**

Blood may leak from the uteroplacental junction and pass into the vagina (external hemorrhage) or may remain trapped behind the placenta (concealed hemorrhage).

The incidence of abruptio placenta in the general population is about 1 in 150 deliveries. Perinatal mortality is approximately 20%.

■ History and Physical Examination **Several factors increase the risk of abruption, including advanced age, multiparity, hypertension (chronic or pregnancy associated), external abdominal trauma, and cocaine use.** Patients with a history of abruption also have a tenfold greater risk of a second placental abruption.

The most common sign of placental abruption is vaginal bleeding. Uterine tenderness, back pain, and increasing frequency of contractions are frequent symptoms. A nonstress test that reveals fetal distress or fetal death may be the first indication that an abruption has occurred.

A speculum examination or digital vaginal or cervical examination should not be done in a third-trimester patient with vaginal bleeding unless the location of the placenta is known. In the presence of placenta previa, a vaginal examination may worsen bleeding.

■ Laboratory and Diagnostic Studies **Because of the risk of hemorrhage and disseminated intravascular coagulation (DIC), a type and screen, hematocrit, coagulation studies, and fibrinogen level should be obtained. Urine screen for cocaine metabolites should be checked.**

Ultrasound can be helpful in ruling out placenta previa, but most placental abruptions are not visible by ultrasound. Tetanic or high-frequency contractions suggest abruptio placenta.

■ Formulating the Diagnosis The differential diagnosis of bleeding late in pregnancy includes placenta previa and abruptio placenta. Placenta previa is associated with painless vaginal bleeding, whereas abruption is described as painful vaginal bleeding. **The diagnosis, however, cannot be confirmed until after delivery.** In abruption the placenta will display a dark, adherent clot on its surface.

■ Evaluating the Severity of Problems Placental abruption can be life-threatening for both the fetus and the mother. **Maternal hemorrhage can occur, resulting in hypovolemic shock. Maternal DIC and renal failure may also occur.** The fetus may die of hypoxia since the placenta essentially functions as his or her lungs until birth.

■ Managing the Patient **Maternal vital signs (and urine output) and fetal heart tones should be closely assessed. The patient with massive abruption is managed with volume replacement, correction of coagulopathy with fresh frozen plasma, and delivery of the infant (most often by cesarean section).**

Those with slower blood loss are monitored for fetal well-being. **Delivery should occur in the event of fetal distress.**

AMENORRHEA

■ Basic Science **Normal menstruation depends on a functioning hypothalamic-pituitary-ovarian-gonadal (HPOG) axis.** It is useful to classify amenorrhea with regard to the menstrual history:

- **Primary amenorrhea—lack of menarche by age 16,** or within 2 years following the development of secondary sexual characteristics (i.e., breasts and pubic hair)
- **Secondary amenorrhea—lack of menstrual cycles for 3 months or more after menarche**

Failure at any level within the HPOG axis may result in amenorrhea. The consequence is an endometrium that fails to undergo the cyclic changes of hyperplasia and menses.

Amenorrhea is physiologic when it occurs in any of the following circumstances:

- **Before puberty**
- **During pregnancy and lactation**
- **After menopause**

If amenorrhea occurs at any other time, it is pathologic. All young women should be evaluated for medical problems if there is no evidence of puberty by age 13, or if there is no menses by age 16. Young women with true delayed puberty or delayed menses should be closely followed.

■ History and Physical Examination The physician should ask about abnormal menstrual patterns and associated symptoms. **Breast tenderness or nausea may suggest pregnancy. Bloating and mucus changes may indicate ovulation. Hot flashes and vaginal dryness may indicate menopause.**

Younger patients should be questioned about familial genetic abnormalities, stress, exercise, and dietary habits. Rigorous exercise, excessive weight loss, or stress can lead to anovulation.

Patients should be asked about **signs of increased androgen release** that may suggest an adrenal disorder. These signs manifest as **male sex characteristics** such as hirsutism, clitoromegaly, deepening voice, male pattern baldness, muscular hypertrophy, or a loss of female sexual characteristics.

The patient should be questioned about galactorrhea (expression of milk in the nonpuerperal state). **Galactorrhea or visual field deficit suggests pituitary neoplasia.**

The **medication history** is important because many medicines can inhibit ovulation (e.g., phenothiazines, monoamine oxidase inhibitors, steroids).

The patient should have a **full physical examination** (including pelvic examination). Anorexia, visual field defects, thyroid irregularities, galactorrhea, obesity, masses, and abdominal striae should be noted.

The examination in the prepubertal patient should focus on the following:

- Degree and distribution of body hair
- Degree of breast development (Tanner stage I-IV)
- Expression of breast milk
- Development of genitalia (clitoromegaly may indicate adrenal tumor or hermaphroditism)

■ **Laboratory and Diagnostic Studies** The endocrine **evaluation of patients with primary amenorrhea** may include serum determination of **follicle-stimulating hormone (FSH), luteinizing hormone (LH), prolactin (PRL), and β-hCG (human chorionic gonadotropin).** Serum **testosterone may be useful to screen for androgen insensitivity syndrome. Patients with an elevated PRL should be evaluated with a computed tomography (CT) or magnetic resonance imaging (MRI) scan of the pituitary** to evaluate for the presence of tumor. An elevated FSH may indicate ovarian dysgenesis. Karyotyping is useful to establish the chromosomal sex pattern; gonadal dysgenesis is a cause of primary amenorrhea in patients with Turner's syndrome. Less commonly an enzyme deficiency is responsible for primary amenorrhea.

The **evaluation of secondary amenorrhea** should begin with serum or urine **β-hCG. LH, FSH, and thyroid-stimulating hormone (TSH)** may be determined. **Elevated gonadotropins (FSH and LH) in women with clinically insufficient estrogen are consistent with ovarian failure; this is most commonly due to menopause. Normal or low gonadotropins are often present in women with anorexia, drug use, and severe illness or stress.** An LH:FSH ratio greater than 2 is consistent with polycystic ovarian syndrome. Alterations in TSH may reveal hypothyroidism or hyperthyroidism.

■ **Formulating the Diagnosis** The causes of amenorrhea include the following:

- Pregnancy, prepuberty, postmenopause
- Anatomical defect (e.g., uterine adhesions after D&C, intact hymen)
- Premature ovarian failure and chronic anovulation, commonly caused by:
  - Excessive exercise

- Anorexia nervosa
- Pituitary tumors
- Obesity—causes excessive estrogen production, which is converted to androgens (polycystic ovarian disease)
- Adrenal or ovarian overproduction of androgens
- Neoplasms
- Hyperthyroidism or hypothyroidism
- Medications

The history, physical examination, and diagnostic studies should point to one major diagnosis.

■ Evaluating the Severity of Problems **Amenorrhea may be the first manifestation of a pituitary, adrenal, or other tumor.**

■ Managing the Patient **Amenorrhea secondary to anatomical defects usually requires surgical intervention. Those with pituitary tumors should be referred to a neurosurgeon. Patients with premature ovarian failure should be treated with hormone replacement therapy.**

The management of anovulation (e.g., secondary to polycystic ovarian disease) **depends on the patient's desires with respect to fertility.** Ovulation can be induced to achieve pregnancy. To normalize cycles the patient may take oral contraceptives, glucocorticoids, or bromocriptine.

Benign Breast Disease

■ Basic Science **Fibrocystic disease is the most common benign breast condition. It is usually found in women between 20 and 50 years old and is characterized by proliferation of the lobular, ductal, and acinar epithelium in response to cyclic hormonal changes.**

Fibroadenomas are the second most common type of benign breast disease. They may enlarge during pregnancy and often regress during menopause. In 20% of cases they may be multiple.

Cystosarcoma phylloides are rare fibroepithelial breast tumors. They are most common in the fifth decade. One in four is malignant; only 10% metastasize.

Intraductal papilloma is often heralded by spontaneous intermittent nipple discharge in perimenopausal women. The discharge is usually from a single duct.

Fat necrosis presents as a firm, tender, indurated, ill-defined mass. The usual etiology is trauma.

■ History and Physical Examination **The classic symptom of fibrocystic breast disease is cyclic bilateral breast pain. Physical findings include engorgement, increased density, excessive nodularity, and fluctuation in the size of the cystic areas. The treatment depends on the severity of symptoms.**

Fibroadenomas are most common in adolescents and women in their twenties. They are painless and do not un-

dergo cyclic changes. They are firm, smooth, rubbery, and mobile.

All women should perform monthly breast self-examination after the age of 20. The self-examination should occur at the same time each month, within a few days after menses.

A breast examination by a physician should be performed at least annually. This examination should include inspection for contour, dimpling, symmetry, and nipple retraction or discharge. The breasts should be palpated in the supine and upright position. The axillae and supraclavicular regions should be palpated for lymphadenopathy.

■ Laboratory and Diagnostic Studies **Current American College of Obstetricians and Gynecologists and American Cancer Society guidelines recommend that women have screening mammography every 1 to 2 years from age 40 to 49, and yearly beginning at age 50. A baseline mammogram should be obtained between ages 35 and 40.** Women at high risk for breast cancer should have an annual mammogram and biannual physical examination. Additionally, mammography is indicated for all patients with a dominant mass or supraclavicular or axillary adenopathy.

In women with fibroadenoma(s), breast ultrasound is useful to distinguish solid from cystic masses. An attempt should be made to aspirate a cystic mass. If the fluid is clear or cloudy and no residual mass is palpated, follow-up 1 month later is recommended. The mass should not return within this time. **If the aspirate is bloody or the cyst recurs within 1 month, biopsy is recommended.** Aspiration of multiple macrocysts is not necessary. Fine needle aspiration cytology can be used to evaluate solid masses.

An MRI may be useful in women with complaints related to breast implants.

■ Formulating the Diagnosis The definitive diagnosis of benign breast disease remains histologic evaluation. Many women are candidates for fine needle aspiration cytology. **Absolute indications for open breast biopsy include positive fine needle aspiration cytology, cystic masses with a solid component or bloody aspirate, masses with nipple discharge, and suspicious mammographic changes without a dominant nodule or discrete thickening.**

■ Evaluating the Severity of Problems The most important differentiation in the evaluation of women with complaints related to their breasts is the differentiation between benign and cancerous breast disease. This is discussed in detail in chapter 14, Hematology/Oncology: "Breast Cancer."

■ Managing the Patient **For fibrocystic changes, support bras, premenstrual diuretics, decreased consumption of caffeine and tobacco, and oral contraceptives have been used. Danazol is the drug of choice for severe symptoms. Bromocriptine and tamoxifen may be used when danazol fails. Surgery may be indicated if the patient has intractable pain not relieved by medical therapy, multiple breast biopsies, or a precancerous lesion.**

The "cure" of fibroadenoma is achieved through surgical excision; however, if left alone, these lesions ultimately regress and calcify. Despite surgical excision, about 20% recur.

The treatment of cystosarcoma phylloides is local excision with wide margins.

The treatment of fat necrosis is excisional biopsy.

Cervical Cancer

■ Basic Science Cervical cancer accounts for one fifth of all female genital cancers. **Ninety percent of cervical cancers are squamous,** and 10% are adenocarcinomas. The **incidence of invasive cervical cancer has decreased** over the last 50 years, in large part **due to the Pap test.** The **median age for invasive cervical cancer is 45.**

Risk factors for cervical carcinoma include human papilloma virus infection, first sexual intercourse at an early age, multiple sexual partners, and smoking.

■ History and Physical Examination **Abnormal vaginal bleeding (including postcoital bleeding) is the most common presenting symptom.** Malodorous vaginal discharge and pelvic pain are also common.

While **there are no classic physical findings,** a thorough pelvic examination is essential to determine the extent of the tumor.

■ Laboratory and Diagnostic Studies **The workup of cervical cancer includes the following:**

- **Pap smear**
- **Colposcopy**
- **Biopsy of suspicious lesions.** Cervical punch biopsy or endocervical curettage are most often used to diagnose invasive cervical cancer.
- **Clinical staging of the cancer by physical examination of the pelvis.** A metastatic survey (chest x-ray examination, intravenous pyelogram, cystoscopy, or proctoscopy) is also done to determine the extent of disease progression.

■ Formulating the Diagnosis **The following is an overview of the staging classification:**

- **Stage O:** the carcinoma is in situ, intraepithelial neoplasia.
- **Stage I:** the carcinoma is strictly confined to the cervix.
- **Stage II:** the carcinoma extends beyond the cervix but not into the pelvic wall. The carcinoma involves the vagina but not the lower third.
- **Stage III:** the carcinoma has extended to the pelvic wall. On rectal examination there is not cancer-free space between the tumor and the pelvic wall.
- **Stage IV:** the carcinoma has extended beyond the true pelvis or has clinically involved the bladder or rectum.

■ Evaluating the Severity of Problems **The prognosis is proportional to the stage. Small cell carcinoma has the worst prognosis; large cell carcinoma has the best.**

■ Managing the Patient **The following are treatment options for carcinoma in situ:**

- **Excision with cold knife**
- **Laser conization**
- **Hysterectomy**

The **treatment modality depends on the patient's desire to remain fertile.** If conization is chosen, the patient should undergo a Pap smear every 3 months for the first year and at 6-month intervals thereafter.

- **Invasive squamous cell cancer (stages I to IV)**

  - **Surgery**—young women with stage I and some stage II lesions are candidates for an operation. This includes total abdominal hysterectomy and bilateral salpingo-oophorectomy. If the lesion is greater than 3 mm, then the pelvic lymph nodes are removed. Complications of surgery include bladder-vaginal fistula formation.
  - **Radiation therapy**—radiation therapy is most often used for more advanced lesions. Treatment consists of applying cesium to the cervix for a total of 70 hours. This is followed by external radiation to the pelvic wall lymphatics. Complications of radiotherapy are proctitis and cystitis.

  Squamous cell carcinoma spreads via the lymphatics. Treatment is directed at destroying the primary cancer and all affected lymph nodes while sparing healthy tissue. Radiation therapy and surgery have been used alone and together to achieve these aims. **Both radiation and surgery are equally effective** (the 5-year cure rate for stages I and II is about 85%). The 5-year survival rate for recurrent disease is less than 5%.

  All patients should have long-term follow-up. Serum carcinoembryonic antigen should be monitored, and periodic ultrasonography or CT scan should be performed.

CONTRACEPTION

■ Natural Family Planning This method **uses body temperature to predict the time of ovulation and fertility** (increased temperature indicates progesterone release). The patient should be abstinent in the first half of the cycle. Sexual intercourse can proceed from the third day after the temperature spike until the end of the menstrual cycle.

- **Advantages: low cost and lack of medication side effects**
- **Disadvantages: low effectiveness**

■ Hormonal Contraceptives

- **Estrogen/progesterone combination pill** The estrogen component of the pill inhibits FSH. The progestins are responsible for suppression of LH. Oral contraceptives also **suppress ovulation and alter the properties of cervical mucus.** Cervical mucus becomes thick, viscous, and hostile to sperm.

  Estrogenic side effects include nausea, cyclic weight gain, edema, and headache. Progestogenic side effects include depression, acne, and hirsutism.

When prescribing oral contraceptives, one must be aware of serious potential complications. **There is an increased risk of thromboembolism, myocardial infarction, and stroke. Women who smoke and are over 35 years of age on oral contraceptives are at greatest risk.** Hepatomas have also been reported, but these are rare.

Absolute contraindications of the pill include history of thrombophlebitis, cerebral vascular disease, markedly impaired liver function, known or suspected **breast cancer,** undiagnosed **vaginal bleeding, and smokers greater than 35 years old.**

Noncontraceptive benefits include **protection against endometrial cancer, ovarian cancer, and rheumatoid arthritis.** It also **decreases the risk of pelvic inflammatory disease.**

Oral contraceptives are greater than **99% effective** if taken properly.

- **Norplant** Norplant is a long-acting, low-dose, reversible, **progestin-only subdermal implant** form of contraception. The Norplant system consists of six silastic capsules. Each capsule contains 36 mg of levonorgestrel.

  The mechanism of action is most likely similar to the progestin-only pills. **Levonorgestrel suppresses the LH surge** necessary for ovulation. It also thickens cervical mucus. Norplant is usually placed just under the skin in the upper arm. It will last for 5 years.

  Absolute contraindications include thrombophlebitis, acute liver disease, and **undiagnosed vaginal bleeding. Side effects include irregular vaginal bleeding, weight gain, mastalgia, galactorrhea, and acne.**

  Norplant is greater than **99% effective.**

- **Depo-Provera** Depo-Provera is a depot **intramuscular form of medroxyprogesterone** acetate. The dose is 150 mg IM every 3 months. The mechanism of action is the same as with all progestin-only methods. **Side effects include irregular vaginal bleeding, breast tenderness, weight gain, depression, and hair loss. The incidence of irregular bleeding is 30%** within the first year. **After several injections, most women become amenorrheic.** One disadvantage is its association with infertility after discontinuing Depo-Provera. Delayed conception may persist for 9 months.

  It is considered as effective as surgical sterilization.

■ Intrauterine Devices (IUDs) **It is believed that the major mechanism of action is due to the local sterile inflammatory reaction within the uterine cavity that creates a toxic environment for sperm. Side effects include increased vaginal bleeding, perforation, infection, and displacement.** Increased flow is the primary reason women discontinue use. Perforation occurs in 1 of every 1000 insertions. Infections are believed to be secondary to contamination at the time of insertion. However, **infections occurring after 3 to 4 months of insertion** are most likely **secondary to sexually transmitted diseases. If a patient de-**

velops pelvic inflammatory disease (PID), antibiotics must be administered and the IUD must be removed. Therefore it is prudent not to insert an IUD in a nulliparous woman so as not to impair future fertility. It is also recommended not to insert one in a woman with multiple sexual partners because of the increased risk of infection.

If a pregnancy occurs with an IUD in place, spontaneous abortion will occur in approximately **50% of cases.**

The **efficacy of the IUD is 98%.**

- Barrier Methods Barrier methods of contraception include **condoms, spermicides, diaphragms, vaginal sponges, and cervical caps.** All barrier methods **protect against STDs and PID.** Barrier methods are also believed to **protect against cervical cancer.**

Diaphragms are dome shaped and made of latex rubber with a metal spring. They cover the cervix and the anterior vaginal wall and act as physical barriers to sperm. They are often coated with spermicidal creams or jellies. **Women must be instructed to leave the diaphragm in place for 6 hours after intercourse** and must apply more spermicide for each act of repeated intercourse.

The effectiveness of the diaphragm is 85%.

The **effectiveness of condoms varies from 85% to 90%** and depends on patient compliance.

Spermicides include creams, jellies, foam, and suppositories and are often used with other forms of contraception such as condoms and diaphragms. These agents damage the sperm's cell membrane. No serious side effects have been reported. A minor problem is allergy. Spermicides require application 10 to 30 minutes before intercourse and should be used with other barrier methods.

The **efficacy of spermicides is 80% if used alone.**

- Surgical Sterilization Surgical sterilization can be done via laparoscopy or mini-laparotomy. **Tubes may be ligated, electrocoagulated, or mechanically occluded.** Women undergoing surgical sterilizations require extensive counseling regarding efficacy, alternatives, safety, and complications. It is generally considered a permanent procedure, with a failure rate of 3 to 4 in 1000. **Potential complications include bleeding, infection, ectopic pregnancy, and failure.**

Vasectomies are safer, easier to perform, and less expensive than female sterilization.

Ectopic Pregnancy

- Basic Science **Ectopic pregnancy refers to any pregnancy that implants at a site other than the endometrial cavity. The most common implantation site is within a fallopian tube.** Rarely, an ectopic pregnancy may occur concurrently with an intrauterine pregnancy. This is termed a heterotopic pregnancy. The advent of assisted reproductive technology has made heterotopic pregnancy more common.

Ectopic pregnancy complicates 1% to 2% of all pregnancies. Most ectopic pregnancies occur among multiparous women (>50%); only 10% to 15% occur in primigravid women. **Ectopic pregnancies are usually symptomatic 6 to 8 weeks after conception.**

Several factors account for an increase in the diagnosis of ectopic pregnancy over the last 20 years. These include:

- **Pelvic inflammatory disease** (and sexually transmitted disease)
- **Reproductive technologies** (e.g., gamete intrafallopian transfer [GIFT], zygote intrafallopian transfer [ZIFT])
- **Tubal surgery** (e.g., tubal ligation)
- **Recurrence of ectopic pregnancy**
- **Improved diagnostic modalities**

■ History and Physical Examination **Classically, women with an ectopic pregnancy present with the following:**

- **Unilateral abdominal pain**
- **Abnormal vaginal bleeding**
- **An adnexal mass**

Only 15% of patients present with this classic triad. Many patients are asymptomatic until rupture occurs. Rupture is usually accompanied by severe abdominal pain, shock, tachycardia, and shoulder pain secondary to hemoperitoneum. Other symptoms associated with ectopic pregnancy include rectal pressure, absence of menses, dizziness, breast tenderness, nausea, and passage of a decidual cast.

The physical examination may reveal lateralized lower abdominal pain, vaginal bleeding, an adnexal mass, uterine size smaller than anticipated, cervical motion tenderness, and orthostatic vital signs (or hemorrhagic shock).

■ Laboratory and Diagnostic Studies **In patients with a suspected ectopic pregnancy, the initial blood work should include a CBC, blood type, antibody screen and crossmatch, and measurement of β-hCG.**

In the presence of a positive pregnancy test and signs and symptoms suggestive of ectopic pregnancy, an ultrasound should be performed. Ultrasound is used to identify the presence of acute bleeding or an intrauterine pregnancy, and to evaluate an adnexal mass. Either transabdominal or transvaginal ultrasound may be used.

In questionable cases, serial measurement of β-hCG titers may be useful; in a normal pregnancy the titer doubles every 48 hours, whereas ectopic pregnancies rarely display normal increases in β-hCG.

Other diagnostic modalities include measurement of progesterone, laparoscopic diagnosis, and culdocentesis. A culdocentesis is positive in 70% to 90% of ectopic pregnancies. A positive test is represented by more than 0.5 ml of nonclotting blood retrieved from the posterior cul-de-sac by needle aspiration. A hematocrit of this fluid is measured and considered positive if greater than 15%. A positive pregnancy test and positive culdocentesis predict 95% of ectopic pregnancies.

■ Formulating the Diagnosis **The diagnosis of ectopic pregnancy is suggested by a positive hCG in a woman with an empty uterine cavity and an adnexal mass.**

■ Evaluating the Severity of Problems **Intraperitoneal hemorrhage represents the most serious complication** from an ectopic pregnancy. Therefore women presenting to an emergency department with a suspected ectopic pregnancy should be evaluated urgently. Two large-bore intravenous lines should be secured early in the evaluation.

■ Managing the Patient **Multiple management strategies currently exist for ectopic pregnancy. These include:**

- **Laparotomy** with either conservative tubal surgery or salpingectomy
- **Laparoscopic surgery**
- **Medical management** (e.g., methotrexate)

Laparotomy is indicated in hemodynamically unstable patients and in patients in whom a large hemoperitoneum is suspected. During laparotomy, a salpingectomy or salpingostomy can be performed. Obviously, greater fertility rates are seen with conservative surgery, but a 15% recurrence risk of ectopic pregnancies results.

Laparoscopic surgery is now the procedure of choice in stable patients. Studies have shown no difference in conservative procedures whether done laparoscopically or by laparotomy.

The newest modality for the treatment of ectopic pregnancy is the use of methotrexate (MTX). MTX is a folic acid–inhibiting chemotherapeutic drug that attacks trophoblastic tissue. Patients are candidates for MTX if the following criteria are met:

- Ectopic size <3.5 cm
- No fetal cardiac activity seen
- Quantitative β-hCG $<15{,}000$ mIU/ml
- No evidence of rupture present
- No acute bleeding evident

Contraindications to MTX include the following:

- Poor compliance
- Fetal cardiac activity
- Abnormal liver function tests
- Active peptic ulcer disease
- White blood cell count $<3000/\mu l$
- Platelet count $<100{,}000/\mu l$

Rh-negative women with an ectopic pregnancy should receive Rhogam. Iron therapy may be necessary in anemic patients.

ENDOMETRIOSIS

■ Basic Science **Endometriosis is the presence and growth of uterine glands and stroma in an aberrant location.** Implants of endometrium may occur anywhere in the body but do so most often in the dependent portions of the pelvis. **There is a sevenfold increase in endometriosis among relatives of women with the condition.**

The three cardinal features of endometriosis are

- Ectopic endometrial glands
- Ectopic endometrial stroma
- Hemorrhage into adjacent tissue (hemosiderin-laden macrophages)

■ History and Physical Examination **Many women complain of cyclic pelvic pain beginning 2 to 7 days before menses, although one third of the women are asymptomatic.** The extent of pelvic pain is often inversely related to the amount of endometriosis in the pelvis. **Other symptoms include dyspareunia, abnormal bleeding, cyclic abdominal pain, and, rarely, urinary symptoms or rectal pain and bleeding. Infertility is often a major manifestation of the disease.**

The most prominent pelvic finding is a fixed, retroverted uterus with scarring and tenderness posterior to the uterus. The characteristic nodularity (i.e., endometriomas) of the uterosacral ligaments and cul-de-sac of Douglas may be felt on rectovaginal examination. There may also be adnexal enlargement with tenderness.

■ Laboratory and Diagnostic Studies The **definitive diagnosis of endometriosis is made by laparoscopy. Ultrasound may be helpful** since it identifies endometriomas, which characteristically have irregular edges, and internal echoes produced by chronic intracystic bleeding. CA 125 is a nonspecific blood test that may be elevated; it may be helpful in detecting and monitoring more severe cases. CT and MRI have a limited role.

■ Formulating the Diagnosis **The differential diagnosis should include pelvic inflammatory disease, ovarian neoplasia, and leiomyomas. A presumptive diagnosis may be made clinically but must be confirmed through laparoscopy or laparotomy.**

■ Evaluating the Severity of Problems When severe, endometriosis may result in disabling pain.

■ Managing the Patient **Treatment goals include preservation of fertility (when it is desired), symptom relief, and control of disease with medical therapy.** Analgesics may be required. **Medicines used in the treatment of endometriosis prevent the cyclic stimulation of ectopic endometrium.** They include the following:

- **Danazol**—suppresses menstruation; androgenic side effects
- **GnRH analogs** (e.g., nafarelin nasal spray)—inhibits ovulation; hypoestrogenic
- **Estrogen-progesterone oral contraceptives**

Surgical therapy is divided into conservative and definitive treatment. Conservative procedures (via laparoscopy) include excision, vaporization, and coagulation of endometriotic implants, excision of ovarian endometriomas, and lysis of adhesions. In patients with extensive disease (e.g., cul-de-sac obliteration and dense scarring of the ovaries to the pelvic sidewalls), laparotomy may be preferable. **Definitive therapy consists of**

total abdominal hysterectomy and bilateral oophorectomy and is indicated for patients with persistent pelvic pain after conservative management, or for patients who have completed childbearing.

Approximately one third of the women treated conservatively will have recurrent endometriosis and require additional surgery within 5 years. **Following definitive surgery (including bilateral oophorectomy), estrogen replacement therapy is recommended and may be given with little risk of reactivating residual disease. However, many physicians delay estrogen replacement therapy for 3 to 6 months.**

Genetic Counseling

■ Basic Science Approximately 2% to 3% of infants are born with a serious birth defect. **Common indications for genetic counseling include the following:**

- **Maternal age greater than 35**
- **Previous child with congenital anomalies**
- **Family history of a genetic condition** (e.g., cystic fibrosis, hemophilia, trisomy-21)
- **Family history of biochemical/metabolic conditions** (e.g., phenylketonuria deficiency)
- **Family history of chromosomal disorders** (e.g., Down syndrome, chromosomal translocations)
- **Multiple spontaneous abortions**
- **Abnormal maternal serum screening**

■ History and Physical Examination **A four-generation family history should be obtained. Known and suspected genetic conditions should be documented to identify genetic patterns.**

Ethnic and racial background should be identified. Many genetic abnormalities are associated with specific ethnic or racial groups (e.g., sickle cell anemia in African-Americans, and Tay-Sachs disease in Jews).

The physical examination of the prospective mother should include a search for obvious heritable defects.

■ Laboratory and Diagnostic Studies **Maternal serum screening between 16 and 18 weeks should be offered to all women. Standard prenatal tests include maternal serum α-fetoprotein, hCG, and unconjugated estriol.**

Open neural tube defects and abdominal wall defects have been associated with elevated levels of α-fetoprotein. Abnormally low levels of α-fetoprotein have been associated with Down syndrome. Results of this test depend on accurate gestational dating. Abnormal α-fetoprotein levels should be followed by an ultrasound and (possibly) amniocentesis.

Low levels of α-fetoprotein and unconjugated estriol, or elevated levels of hCG, are associated with chromosomal abnormalities. These tests can identify Down syndrome in about 60% of cases. An abnormal serum screen should be confirmed by amniocentesis before pregnancy termination is considered.

DNA and biochemical testing are available for many com-

mon conditions including cystic fibrosis, hemophilia, Tay-Sachs disease, and sickle cell anemia. Other genetic conditions may require family studies.

Ultrasound examination offered at 18 weeks' gestation is a noninvasive procedure that assesses fetal growth and development. Ultrasound acts as a screen for physical birth defects such as neural tube defects. Certain findings (e.g., cystic hygroma) have been associated with chromosome abnormalities. Abnormal findings should prompt other investigations (e.g., amniocentesis).

Chorionic villus sampling is available between 10 and 12 weeks' gestation. It is used in cases of suspected genetic abnormality. Chorionic villus sampling allows extremely early diagnosis of genetic diseases. The procedure involves placing a needle into the placenta to harvest chorionic cells for chromosomal, biochemical, or DNA analysis. It is an outpatient procedure done under ultrasound guidance. Chorionic villus sampling cannot detect certain birth defects (e.g., neural tube defects). Women who have this test should also have a maternal serum screen at 16 weeks' gestation.

Amniocentesis is available between 15 and 16 weeks' gestation. It is used in cases of suspected genetic abnormality. The procedure involves passing a needle into the amniotic sac to harvest amniotic fluid for chromosomal, biochemical, and DNA analysis. Alpha-fetoprotein testing of the fluid can identify the presence of a fetal neural tube defect. Amniocentesis is a safe test, with a complication rate of less than 1 per 300.

■ **Formulating the Diagnosis** The diagnosis depends on the results of the specific prenatal test(s).

■ **Evaluating the Severity of Problems** Certain genetic abnormalities are largely asymptomatic, whereas others result in intrauterine demise or death shortly after birth.

■ **Managing the Patient** Effective genetic counseling entails the following:

- Providing the family with the diagnosis
- Estimating the risk of recurrence for different family members
- Educating the family about options

The counselor must be supportive of the family throughout the decision-making process. Genetic counseling should be nondirective. The counselor should offer an unbiased and complete range of options (e.g., further testing, abortion, or carrying the fetus to term). It is important to have respect for the family's autonomy in making decisions.

Infections During Pregnancy

■ **Basic Science** This chapter covers six infections commonly encountered during pregnancy: **rubella, genital herpes, gonorrhea, chlamydia, syphilis, and acquired immunodeficiency syndrome (AIDS).**

Rubella may result in severe congenital malformations

including microcephaly, cardiac structural defects, retinopathy, and deafness.

Genital herpes infections are classified as primary or secondary based on clinical or serologic evidence of prior infection. Primary infections are usually more severe for the mother and result in increased transmission to the fetus. Congenital infections causing fetal malformations are rare. **Neonatal infections are acquired as ascending infections around the time of labor and delivery or because of direct contact during delivery.** Infection of the newborn usually causes central nervous system (CNS) disease with a 50% mortality rate. Neurologic or ophthalmic sequelae are frequent among infants who survive infections.

Neisseria gonorrhea, an aerobic gram-negative bacteria, has a prevalence rate of approximately 1% to 7% during pregnancy. **There is an association between untreated gonorrhea and spontaneous abortion, preterm delivery, chorioamnionitis, and postpartum infections.**

***Chlamydia trachomatis* is the most common sexually transmitted bacterial disease in women. Perinatal transmission is associated with neonatal conjunctivitis and pneumonia.** Vertical transmission occurs in approximately 50% of infants delivered vaginally by infected women.

Syphilis is a chronic infectious process caused by the spirochete *Treponema pallidum.* The risk of congenital syphilis depends on the stage of maternal disease. In utero infection may result in stillbirth, intrauterine growth retardation, nonimmune hydrops, and premature labor. The spirochete may cause lesions in the lungs, liver, spleen, pancreas, and bones of the newborn.

The rate of perinatal transmission of human immunodeficiency virus (HIV) is approximately 30%. Transmission is by transplacental infection or contraction of the virus at birth. There does not appear to be an increased rate of pregnancy complications such as preterm birth or restricted fetal growth in asymptomatic seropositive women.

■ History and Physical Examination **After a 2- to 3-week incubation period, women with rubella have a prodromal phase lasting 1 to 5 days. Typical symptoms include low-grade fever, headache, malaise, anorexia, pharyngitis, and cervical adenopathy. A pink macular rash then appears and spreads quickly to involve the entire body.** Fetal infection after maternal viremia leads to chronic infection, which affects organogenesis. One fourth of the maternal rubella infections are subclinical despite viremia and fetal infection.

A primary infection with herpes virus is characterized by painful vesicles on the vulva and perineum. Inguinal adenopathy and influenza-like symptoms may also occur. Recurrent infections generally have fewer lesions that shed the virus for shorter periods (2 to 5 days).

Women with gonococcal infection may display a mucopurulent cervical discharge or may be entirely asymptomatic. Infections early in pregnancy may ascend and cause endometritis and spontaneous abortion. After 10 to 12 weeks ascending infections are not seen, but adherence of gonococci to chorionic membranes may cause premature labor or premature rupture of membranes.

Chlamydial infections may be associated with several clinical syndromes that include urethritis, mucopurulent cervicitis, and salpingitis, but most pregnant women have subclinical or asymptomatic infection.

The primary infectious lesion of syphilis is the chancre. It is a painless ulcer with a raised, indurated border that appears on the perineum, vulva, or cervix 2 to 3 weeks after exposure. Between 6 weeks and several months after the appearance of the primary chancre, a systemic response is seen with fever and malaise. **Skin manifestations include genital condyloma lata and red macules on the palms or soles.** Manifestations of late syphilis usually involve the cardiovascular and neurologic systems.

Individuals with AIDS develop lymphadenopathy, night sweats, fevers, opportunistic infections, and certain neoplasms. Preliminary studies suggest that pregnancy does not appear to influence the progression of HIV infection in women.

- Laboratory and Diagnostic Studies **Maternal serum is obtained to detect IgG specific for rubella. Absence of the rubella antibody indicates susceptibility to infection. The diagnosis of perinatal rubella infection in the newborn is supported by the presence of IgM in cord blood.**

Virus isolation and tissue culture is the method used to diagnose herpes simplex virus.

Antepartum cervical sampling is the best method to detect gonorrhea and chlamydia.

The rapid plasma reagin (RPR) and Venereal Disease Research Laboratory (VDRL) serologic tests are commonly used to screen for syphilis. These tests lack specificity, and a direct treponemal test such as the fluorescent treponemal antibody absorption test (FTA-ABS) is used to confirm a positive result. Response to treatment may be followed by observing a decreasing serologic titer.

ELISA (enzyme-linked immunosorbent assay) is used as the initial screen for HIV. All positive results are followed by a Western blot test that identifies antibodies to specific viral proteins.

- Formulating the Diagnosis The diagnosis of each infection is made after considering the history, physical examination, and diagnostic test findings. Beyond diagnosing the maternal condition, specific techniques such as ultrasound, chorionic villus sampling, and cordocentesis may help to identify fetal infection.

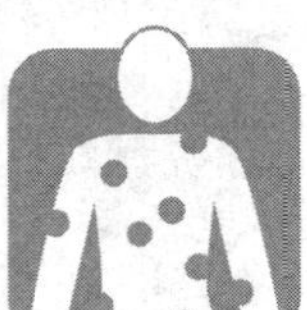

- Evaluating the Severity of Problems Each of the maternal infections (except AIDS) is usually mild and self-limited if properly treated. Fetal (and neonatal) outcomes depend on the developmental stage at the time of infection. Ultrasound may be helpful in evaluating growth restriction or congenital anomalies.

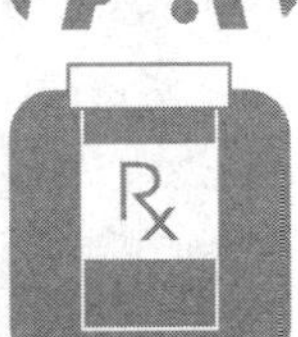

- Managing the Patient **An attenuated live virus vaccine is available to provide immunity to rubella. All women are screened**

at the start of prenatal care. Seronegative (i.e., nonimmune) pregnant women are not given the vaccine during pregnancy; rather, the rubella vaccine is administered shortly after delivery. The impact of rubella on the fetus is most severe when acquired early in gestation. Fetal manifestations range from asymptomatic to congenital rubella syndrome to intrauterine demise.

Unless previous herpes infection has been documented, all suspicious lesions should be cultured. An infant is at risk of acquiring herpes if the mother has an active lesion at the time of delivery. Maternal infection is also assumed if the woman is experiencing prodromal symptoms. If either prodromal symptoms or lesions are present, a cesarean delivery should be performed.

Cervical sampling to screen for gonorrhea and chlamydia is usually performed at the first prenatal visit. A single intramuscular dose of ceftriaxone is recommended for treating gonococcal infection. A 7 day course of oral erythromycin is the first line of therapy for chlamydial infections in pregnant women. The sexual partner of any woman with chlamydia or gonorrhea should also be treated. Patients with positive cultures should be retested later in pregnancy to identify treatment failures or reinfection.

A serologic test for syphilis is part of routine antenatal screening. Penicillin is the treatment of choice to eradicate maternal infection and to prevent congenital syphilis. Benzathine penicillin G cures early maternal infection and prevents neonatal syphilis in 98% of cases.

The pregnant woman who has AIDS or is HIV positive must be appropriately counseled regarding the risks of fetal transmission. Antepartum and intrapartum use of zidovudine (AZT) has been shown to decrease the risk of transmission from 30% to 8%. Women with AIDS should be advised against breast feeding.

INFERTILITY

■ Basic Science **Infertility is the failure of a couple to achieve a pregnancy after 12 months of intercourse.** Overall, infertility affects 10% of couples at any given time. For normal couples the probability of achieving a pregnancy during a single menstrual cycle is about 25%.

Factors responsible for infertility include the following:

- **Male factor**
- **Ovulatory factor**
- **Tubal/peritoneal factor**
- **Cervical factor**

In about 15% of cases the cause of infertility remains unknown.

Women show a significant decline in fertility after the age of 30. This relates to declining ovarian function and to other factors such as endometriosis.

■ History and Physical Examination **Important historical information** includes the **duration of infertility, the age and general health of the partners, prior pregnancies, medical problems, possible genitourinary infection, cigarette smoking, alcohol/drug use, and current medications.** Inquiry should also be made about **intercourse frequency and the use of lubricants** (which may adversely affect sperm function). Questions for men include **history of mumps orchitis, hernias, testicular injuries, and operations. Sperm function can be affected by chemotherapeutic agents, cimetidine, and calcium channel blockers.**

The physical examination should focus on the sexual characteristics and the genitalia. **Varicocele is the most common abnormality found on the male examination.**

For the female, **the menstrual cycle history is critical.** Anovulatory females may present with amenorrhea or dysfunctional uterine bleeding. **A history of hirsutism, acne, or galactorrhea may point to a specific endocrine problem. A history of IUD use, particularly if associated with a pelvic infection, may suggest a tubal factor. Progressively worsening dysmenorrhea and/or dyspareunia may suggest endometriosis.**

The physical examination should include a detailed pelvic examination. **Pelvic tenderness may suggest the presence of endometriosis, pelvic infection, or abdominal/pelvic adhesive disease.**

■ Laboratory and Diagnostic Studies

- **Male factor** The initial diagnostic test is **semen analysis.** Parameters assessed include semen volume and sperm concentration, motility, and morphology.

- **Ovulatory factor** The simplest means to reasonably confirm ovulation is to **document the basal body temperature. No increase** in basal body temperature during the midcycle **suggests anovulation.** Progesterone levels may also be measured. **If a patient does not show a midcycle progesterone increase then she is anovulatory.** If ovulatory dysfunction is suggested, hormonal testing should follow. Patients with hypothalamic disturbances have low FSH and LH concentrations. Patients with ovarian failure have high FSH concentrations. A high prolactin concentration may suggest a prolactinoma. Patients with the polycystic ovarian syndrome often have high androgen concentrations.

- **Tubal/peritoneal factor** Tubal disease is assessed by the hysterosalpingogram (HSG). HSG provides information about tubal obstruction and uterine cavity abnormalities. **Laparoscopy is done if the HSG findings are abnormal or if peritoneal disease is suspected.** Laparoscopy may reveal the presence of adhesions or demonstrate endometriosis.

- **Cervical factor** Both **cervical mucus and sperm parameters are assessed.** Abnormalities include prior cervical surgery, cervical infection, antisperm antibodies, and sperm dysfunction.

- **Formulating the Diagnosis**
  - **Male factor** The male is the cause of infertility in **30% to 40%** of cases. Most cases involve **defects in spermatogenesis.**
  - **Ovulatory factor** Ovulatory dysfunction (generally **anovulation**) is the problem in **15% to 20%** of cases.
  - **Tubal/peritoneal factor Thirty percent** of infertility cases are due to tubal/peritoneal factors. The two most common causes of infertility in this area are **endometriosis and previous PID.** *Chlamydia trachomatis*-induced PID may cause significant tubal damage in the absence of symptoms.
  - **Cervical factor** Infertility is attributed to cervical factors in **5% of cases.**
- **Evaluating the Severity of Problems** Couples affected by sterility should be provided with information about adoption.
- **Managing the Patient** **Overall, about half of couples seeking infertility care ultimately achieve a pregnancy.** Male factor cases related to defects in spermatogenesis are often not treatable, although varicocele repair may be helpful.

Anovulation is handled by treating the underlying defect. Patients with hypothalamic amenorrhea are best treated with gonadotropin preparations or pulsatile GnRH therapy. Hyperprolactinemia is treated medically with a dopamine agonist (e.g., bromocriptine). Patients with large pituitary tumors should be referred to a neurosurgeon. Patients with polycystic ovarian disease are initially treated with clomiphene citrate. If the response is unsatisfactory, gonadotropin preparations may be used.

Tubal/peritoneal factors generally require surgical therapy. Surgery ranges from laparoscopic ablation of endometriosis to tubal microsurgery. Occasionally, infertility patients with endometriosis may benefit from medical therapy such as danazol (Danocrine) or a long-acting GnRH agonist.

Couples failing to achieve pregnancy may opt to proceed with an assisted reproductive technology procedure. All procedures involve oocyte retrieval after administration of a superovulation regimen.

- **In vitro fertilization-embryo transfer** (IVF-ET). Oocytes are inseminated in vitro 4 to 6 hours after retrieval. Embryo transfer into the uterus follows in 2 to 3 days.
- **GIFT.** GIFT involves transferring both oocytes and sperm into the fallopian tube. Fertilization occurs within the fallopian tube. Women undergoing GIFT must have at least one normal, functioning tube.

Pregnancy rates are 17% for IVF-ET and 26% for GIFT.

Leiomyoma

- **Basic Science** **Leiomyomas are benign smooth muscle tumors. They are the most common pelvic tumor, affecting 20% to 25% of women of reproductive age. They are 10 times more common in black women.** In the United States, leiomyomas are the most common indication for a hysterectomy.

Leiomyomas often cause abnormal uterine bleeding. As they grow and distort the uterine cavity, they may lead to spontaneous abortions, preterm labor, and infertility. **Due to their poor vascular supply, leiomyomas may degenerate, causing severe pain and peritoneal irritation.**

Leiomyomas vary greatly in size, from microscopic growths to tumors weighing more than 100 pounds. It is common to find multiple tumors. Estrogen is thought to promote their growth.

Histologically, leiomyomas have a characteristic whorled appearance, and each myoma is surrounded by a pseudocapsule (a layer of compressed muscle cells). This pseudocapsule is helpful by providing a surgical plane during excision (myomectomy).

■ History and Physical Examination **Most women with leiomyomas are asymptomatic.** Often leiomyomas are detected as uterine masses on a routine pelvic examination. The most common symptom associated with leiomyomas is abnormal bleeding, typically abnormally heavy or prolonged menstrual bleeding. Pelvic pain, increasing in severity with menstruation, is another frequent symptom. The tumor may push on the bladder, causing urinary frequency and urgency.

Reproductive problems such as infertility; recurrent, spontaneous abortions; and preterm labor have also been associated with leiomyomas.

The pelvic examination may demonstrate an enlarged, firm, nontender, irregular uterus.

■ Laboratory and Diagnostic Studies Most leiomyomas are diagnosed through the physical examination. **An ultrasound may help confirm the diagnosis. Hysteroscopy (a scope introduced into the uterus) can be used to diagnose leiomyomas directly. When abnormal bleeding presents, an endometrial biopsy should be performed to rule out endometrial hyperplasia or cancer. Baseline laboratory tests such as a pregnancy test should be performed.**

■ Formulating the Diagnosis **Usually the diagnosis of leiomyoma is made after a pelvic examination.** In the remaining cases, an ultrasound may be needed. The differential diagnosis includes pregnancy, adenomyosis, and ovarian neoplasia.

■ Evaluating the Severity of Problems Most women with leiomyomas are asymptomatic. A hematocrit should be obtained on patients with abnormally heavy menstrual bleeding to rule out anemia. An endometrial biopsy should also be performed to rule out endometrial hyperplasia.

A uterine growth occurring after menopause is worrisome because of the potential for malignancy (e.g., leiomyosarcoma). In these circumstances, a hysterectomy should be performed.

During pregnancy, degeneration of the leiomyoma is not uncommon. The patient may present with a low-grade fever, uterine tenderness, and mild peritoneal signs. These symptoms may mimic placental abruption and chorioamnionitis.

■ Managing the Patient Most leiomyomas are asymptomatic. **Patients without symptoms are managed by periodic pelvic examina-**

tions to follow the uterine growth rate. Medical or surgical management is indicated for symptomatic leiomyomas. Age, desire for fertility, and general health determine the treatment modality.

Medical management consists of nonsteroidal antiinflammatory medications, oral contraceptives, progestins, danazol, and GnRH agonists. GnRH agonists reduce uterine volume by 50%. However, myomas regrow after GnRH agonists therapy is stopped. Because of side effects such as decreased bone density, GnRH agonists should be used for no more than 6 months.

Surgical management includes myomectomy or hysterectomy. The primary advantage of a myomectomy is that the uterus is preserved and the patient remains fertile.

Hysterectomy is the definitive treatment for uterine myomas. In women who fail conservative medical management and no longer desire fertility, a hysterectomy should be considered.

Menopause and Hormone Replacement

- **Basic Science** **Menopause refers to the last menstrual period. Menopause signals the beginning of a period in life when risk for diseases such as coronary heart disease, stroke, cancer, and osteoporosis increase.** The climacteric refers to the perimenopausal period when symptoms associated with declining ovarian function occur. **The mean age of menopause is 51 years.** The age at menopause is largely genetically determined. If menopause occurs before the age of 40, the diagnosis of premature ovarian failure is made. If menopause occurs after age 55, there is increased risk for endometrial hyperplasia or malignancy.

The basic feature of menopause is depletion of ovarian follicles. This begins in utero and continues throughout life. As follicles are depleted, cyclic hormonal patterns are lost. **Estradiol and progesterone levels decline, resulting in increases in FSH and LH.**

The loss of estrogen in postmenopausal women contributes to the development of osteoporosis and coronary heart disease (CHD). Therefore these two diseases directly benefit from hormone replacement therapy (HRT). The risk of stroke appears to be unaffected by HRT.

Long-term use of estrogen therapy may confer a slightly increased risk for breast cancer. Currently it is not recommended that HRT be used in patients who have been successfully treated for breast cancer.

In women taking unopposed estrogen there is a significantly increased risk for endometrial cancer. This risk increases with increasing dosage and duration of estrogen use. Combination therapy with progestin eliminates this increased risk.

In patients receiving HRT, there is no evidence of increased incidence of thrombophlebitis or thromboembolism. However, the use of HRT in patients with a history of deep venous thrombosis or pulmonary embolism is controversial.

There is no adverse effect of estrogen replacement therapy on blood pressure.

- **History and Physical Examination** Lack of ovarian function produces amenorrhea and sterility. **Besides manifestations related to ma-**

jor disease entities such as osteoporosis and CHD, decreasing estrogen levels may produce signs and symptoms including the following:

- **Vaginal itching, burning, discomfort, dyspareunia, and bleeding secondary to atrophy.**
- **Uterine descensus, cystocele, rectocele, and enterocele**
- **Urinary urgency, frequency, incontinence, and dysuria**
- **Hot flushes secondary to vasomotor instability**

■ Laboratory and Diagnostic Studies Serum FSH and LH will be elevated in untreated postmenopausal women. Vaginal cytology will reveal a low estrogen effect with a predominance of parabasal cells. **Pretreatment and annual mammogram, pelvic examination, and endometrial evaluation are indicated in women receiving "unopposed" estrogen replacement.**

■ Formulating the Diagnosis Menopause is the last menstrual period of a woman's life. For most women, menses occurs irregularly during the perimenopausal period. As women age, their cycles shorten. Menstrual irregularity is one feature of the climacteric. Hot flushes often precede menopause.

■ Evaluating the Severity of Problems **Any episode of vaginal bleeding needs evaluation unless the patient has had a normal evaluation within 6 months.** Treatment with combination therapy requires a baseline pelvic examination. Endometrial evaluation is indicated if:

- Vaginal bleeding occurs unexpectedly.
- Vaginal bleeding is heavy, prolonged, or frequent (for women on continuous regimens).
- Vaginal bleeding persists for longer than 10 months after beginning therapy.

■ Managing the Patient **Indications for hormone replacement therapy include** the following:

- **Treatment of vasomotor symptoms**
- **Prevention and treatment of atrophic vaginitis, atrophic urethritis, osteoporosis, and heart disease**
- **Restoration of menses for psychological reasons**

Additionally, the American College of Physicians recommends the following:

- **All women should consider hormone replacement therapy.**
- **Women who have had a hysterectomy are likely to benefit from estrogen replacement therapy alone.**
- **Women with CHD (or at increased risk of CHD) are likely to benefit from HRT.**
- **The risks of HRT may outweigh its benefits in women at increased risk for breast cancer.**

The physician needs to assess each patient individually. Hysterectomy status and the risk for CHD, breast cancer, and osteoporotic fracture should be considered. **Progestin therapy should usually be given to women with a uterus.**

The physiologic replacement dose of estrogen should be the equivalent of 0.625 mg of conjugated estrogen or 1 mg of estradiol per day. The 0.05-mg dose of transdermal estradiol provides physiologic estrogen replacement. Vaginal estrogen cream may be used to relieve atrophic symptoms, but systemic absorption is variable, so other routes are preferable for long-term use.

For combination therapy including progestins, cyclic regimens add the equivalent of 5 to 10 mg of medroxyprogesterone acetate for 10 to 14 days per month. Continuous regimens use 2.5 mg per day or the equivalent.

Ovarian Cancer

- Basic Science **Ovarian cancer occurs most often in women over 50 years of age.** One in 70 women will develop ovarian cancer during their lifetime. **About 90% of ovarian cancer originates from ovarian epithelial cells.** Five percent originates from germ cells. Germ cell tumors are most common in the first two decades of life.
- History and Physical Examination **Ovarian tumors grow insidiously and are usually large before symptoms appear. The most common symptoms of ovarian cancer are abdominal discomfort, fullness, and early satiety. Other symptoms include fatigue, increasing abdominal girth, urinary frequency, and constipation.**

  Women with infertility, low parity, delayed childbearing, or a family history of ovarian cancer are at greater risk. The use of oral contraceptives decreases the risk of epithelial ovarian cancers in women 40 to 59 years of age.

  Pelvic examination may reveal an adnexal mass. In young women this may be caused by an ovarian cyst. Follow-up examination should be performed in 6 weeks to ensure that the mass has resolved.

  The ovaries of postmenopausal women are not palpable. Therefore any palpable mass in a postmenopausal woman is an ovarian tumor until proven otherwise.
- Laboratory and Diagnostic Studies **There is currently no effective mass screening method to detect ovarian cancer. Careful pelvic examination remains the most effective screening method to date.** The use of transvaginal ultrasonography is valuable in evaluating suspected ovarian masses.

  Patients with suspected ovarian cancer should have a workup that includes colonoscopy or barium enema to detect primary or metastatic colon cancer. **Chest x-ray examination** is required to detect metastatic lung cancer. A **mammogram** should be performed to detect a primary breast cancer that may have metastasized to the ovaries. A **CT scan is used to detect liver and lymph node metastasis.**

  CA 125 is a tumor marker elevated in most patients with epithelial ovarian carcinomas, but the test is not specific and may be elevated in a variety of other disorders. CA 125 is useful for assessing the response of ovarian cancer to chemotherapy.

Alpha-fetoprotein is used to follow epithelial cell cancers. hCG is used to follow germ cell tumors.

■ Formulating the Diagnosis **Ovarian cancer is diagnosed by histological examination of the tumor.** Ovarian cancer is surgically staged as follows:

- Stage I: growth limited to the ovaries
- Stage II: growth involving one or both ovaries with pelvic extension
- Stage III: growth involving one or both ovaries with extension outside the pelvis or positive lymph nodes
- Stage IV: growth involving one or both ovaries with distant metastases

Higher stages are associated with a poor outcome. Five-year survival rates for ovarian epithelial cancer are as follows: stage I, 60% to 90%; stage II, 39% to 67%; stage III, 4% to 13%; and stage IV, 0% to 4%.

■ Evaluating the Severity of Problems Because of difficulty in detection before dissemination, ovarian cancer is the leading cause of death among women with gynecologic cancers.

■ Managing the Patient **Operative management is aimed at resection of as much tumor as possible. A total abdominal hysterectomy, bilateral salpingo-oophorectomy, node biopsy, omentectomy, and removal of all gross cancer (debulking) are performed whenever malignant disease is present.** Both pelvic and periaortic lymph nodes should be sampled.

Postoperative chemotherapy is employed in the treatment of ovarian cancer. Chemotherapy regimens often include multiple agents such as cisplatin, carboplatin, and cyclophosphamide. Tumor markers are used to follow patients for recurrence.

Pelvic Inflammatory Disease

■ Basic Science **PID is defined as infection of the fallopian tubes.** However, there may be an associated infection of the cervix, ovaries, or uterus. **Microbes are most commonly transmitted through sexual intercourse.** PID represents an ascending infection that begins inside the vagina, spreading into the uterus and then to the fallopian tubes. The fallopian tubes produce a thick exudate that leads to adhesion formation and tube occlusion.

PID is most often caused by *Chlamydia trachomatis* and *Neisseria gonorrhoea* (GC).

- *Neisseria gonorrhoea* is a gram-negative intracellular diplococcus. The organism is found within polymorphonuclear leukocytes.
- *Chlamydia trachomatis* is an obligate intracellular parasite. It is the most common cause of chronic PID and coexists with GC in up to one third of infections.

PID occurs primarily in young, sexually active women. Five percent of sexually active women in the United States have *Chla-*

mydia or GC. **Patients with an IUD are at greater risk** for the development of PID.

■ **History and Physical Examination** The following triad is classically seen:

- **Fever**
- **Increased white blood cell count**
- **Purulent cervical discharge**

PID most commonly presents with progressively **worsening lower abdominal pain beginning after menses.** Myalgia and fever may occur. **The history may reveal** any of the following: **numerous STDs, previous PID, multiple sexual partners, a new sexual partner, occasional or absent condom use, and vaginal discharge.**

The physical examination may reveal a febrile patient who walks in a stooped position, taking small, shuffling steps. Lower abdominal examination may demonstrate mild guarding and rebound since peritoneal irritation adjacent to the site of infection usually occurs.

Pelvic examination reveals tender adnexae with or without masses, cervical motion tenderness, and a purulent cervical discharge.

■ **Laboratory and Diagnostic Studies** **All patients should have a pregnancy test and urinalysis. Specimens from the cervical os should be collected and sent for culture, rapid antigen test, and Gram's stain.**

- *Chlamydia*—the definitive test for *Chlamydia* is a rapid, direct antigen test.
- GC—Gram's stain may demonstrate gram-negative intracellular diplococci indicating GC infection. The definitive diagnosis of GC infection is by culture.

Other diagnostic modalities include the following:

- *Ultrasound* is performed to rule out abscess and ectopic pregnancy (when suspected).
- *Culdocentesis* is performed if the diagnosis is uncertain. Pus indicates infection.
- *Laparoscopy* is performed if the diagnosis is uncertain. It directly visualizes the reproductive organs.
- Abdominal x-ray examination, free air series, or CT scan is performed if the diagnosis is uncertain.

■ **Formulating the Diagnosis** **To make the diagnosis of PID** the following must be present:

- **Abdominal tenderness**
- **Cervical motion tenderness**
- **Adnexal tenderness**
- **Evidence of infection or inflammation**

Beware of the wrong diagnosis. It is often difficult to differentiate PID from acute appendicitis, diverticulitis (in an older patient), endometriosis, ovarian torsion, ectopic pregnancy, pyelonephritis, and inflammatory bowel disease.

■ Evaluating the Severity of Problems Abscess (tubal, ovarian, or pelvic) may form during any stage of the disease. Perforation of an adnexal abscess may lead to peritonitis, endotoxemia, and shock (a surgical emergency).

PID may scar the fallopian tubes, resulting in ectopic pregnancy, infertility, and chronic pelvic pain.

■ Managing the Patient High-risk behaviors (e.g., numerous sexual partners, lack of barrier contraceptive use) should be identified and brought to the patient's attention. Patients with a history of any STD should be tested for *Chlamydia,* GC, syphilis, and HIV. **The patient's partner should also be tested and treated (if necessary).**

Immediately after cervical specimens are collected, the patient should receive antibiotics. It is inappropriate to wait for culture results before beginning treatment. Early antibiotic therapy will decrease the risk of infertility.

The following are recommended outpatient regimens (two antibiotics are given concurrently since the organisms often coexist):

- **Ceftriaxone** 250 mg IM (for GC) **and**
- **Doxycycline** 100 mg PO bid for 2 weeks (for *Chlamydia*), or
- **Ofloxacin** 400 mg PO for 2 weeks **and**
- Either **clindamycin** 450 mg PO qid **or metronidazole** 500 mg PO bid for 2 weeks

Patients should be seen again within 2 days. Admission to the hospital for intravenous antibiotic therapy is indicated if the diagnosis is uncertain; the patient is poorly compliant, has HIV, or fails outpatient treatment; or fallopian tube scarring or abscess has occurred.

For inpatient treatment, antibiotics are given until the patient improves. Intravenous antibiotics are then continued for an additional 48 hours. In the hospital, appropriate antibiotic regimens include intravenous cefoxitin (or cefotetan) and oral doxycycline, or intravenous clindamycin and gentamicin.

Placenta Previa

■ Basic Science **Placenta previa is a condition where the placenta covers or adjoins the internal cervical os; it is a common cause of painless third-trimester bleeding.** A total placenta previa completely covers the os, whereas a partial placenta previa covers only part of the os.

The incidence of placenta previa is about 1 in 200. **The incidence increases with age and parity. Other risk factors include previous cesarean delivery, previous therapeutic abortion, and multiple gestation.**

Placenta previa is also associated with placental abnormalities. The following terms describe abnormal growth of placental tissue:

- To the myometrium (*accreta*)
- Into the myometrium (*increta*)
- Through the myometrium (*percreta*)

■ History and Physical Examination Placenta previa classically presents with painless vaginal bleeding occurring around the 30th week. The bleeding is attributed to tearing of the placental attachments with dilation of the internal os. **Many patients are diagnosed by ultrasound before vaginal bleeding occurs.** Early ultrasound, however, is not a good predictor of placenta previa later in pregnancy.

It is extremely important that placenta previa be excluded by ultrasound before a digital vaginal examination or speculum examination is performed. This simple manipulation can cause increased bleeding.

■ Laboratory and Diagnostic Studies **The definitive diagnostic study is abdominal ultrasound** to determine placental location. In patients with bleeding, a **hematocrit** may be helpful in quantifying blood loss. A **blood type, antibody screen, and coagulation studies** should also be obtained. **Continuous fetal monitoring** should be done until the bleeding is stabilized or delivery occurs.

■ Formulating the Diagnosis **The diagnosis is established by ultrasound** that demonstrates placenta over the cervical os. Once placenta previa is excluded, other causes of bleeding late in pregnancy (e.g., placental abruption, cervical polyps or cancer, and cervical or vaginal trauma) are considered.

■ Evaluating the Severity of Problems The status of the fetus (based on fetal monitoring) and the hemodynamic stability of the mother are the main factors in evaluating the severity of problems.

■ Managing the Patient **The preterm patient with placenta previa who is not actively bleeding can often be observed at home with vaginal rest and close follow-up.** Patients may be offered phlebotomy for autologous blood banking. There is an increased likelihood of preterm delivery.

Any patient with evidence of fetal or maternal compromise should be admitted to the hospital for monitoring.

Preeclampsia and Eclampsia

■ Basic Science **Preeclampsia is a hypertensive disorder of pregnancy.** It is thought to be a vascular endothelial disease with multiorgan system involvement. It complicates 5% to 7% of all deliveries, primarily nulliparas at the extremes of reproductive age. It is more prevalent in patients with multiple gestation or molar pregnancy. **If seizure activity is involved, the diagnosis of eclampsia is made.**

The clinical presentation is mediated by arteriolar constriction and increased vascular sensitivity to pressors. Vasospasm leads to increased peripheral vascular resistance, hypertension, and vascular endothelial damage, followed by endothelial cell contraction. **This results in interendothelial leakage, in-**

terstitial edema, hemorrhage, and ultimately end-organ necrosis.

- History and Physical Examination **The hallmark symptoms reported by the patient with preeclampsia are headache, visual disturbances, rapid weight gain, increased edema, nausea, vomiting, right upper quadrant pain, and seizures (in eclampsia). The physical examination will reveal hypertension, generalized nondependent edema, hyperreflexia, right upper quadrant/epigastric tenderness (secondary to subcapsular liver hematoma), and oliguria.**
- Laboratory and Diagnostic Studies **Hemoconcentration is produced by relative intravascular volume depletion. A peripheral smear is important to rule out hemolysis. Platelet count and coagulation studies (i.e., prothrombin time, partial thromboplastin time, fibrinogen, and fibrin degradation products) may all be abnormal secondary to consumption or DIC. Serum uric acid, blood urea nitrogen, and creatinine may be elevated due to development of a prerenal state. Evaluation for proteinuria may reveal over 300 mg per 24 hours.**
- Formulating the Diagnosis **Preeclamptic/eclamptic hypertension is defined as a blood pressure (BP) >140/90, or an increase greater than 30 mm Hg in systolic BP, or greater than 15 mm Hg in diastolic BP over the baseline BP obtained in the first 20 weeks of gestation.** This elevation in BP must be distinguished from chronic hypertension. The aforementioned signs, symptoms, and laboratory results will guide the diagnosis. The differential diagnosis includes the following:
    - Cerebral vascular accident
    - Hypertensive encephalopathy
    - Pheochromocytoma
    - Space-occupying CNS lesion
    - Infection
    - Metabolic derangement
- **Evaluating the Severity of Problems** The mild preeclamptic will have a diastolic BP <100 mm Hg, proteinuria greater than 300 mg/24 hours, and/or generalized nondependent edema. Development of diastolic BP >110 mm Hg, proteinuria >5 g per 24 hours, signs of CNS irritation, right upper quadrant pain, oliguria, pulmonary edema, or the HELLP syndrome (hemolysis, elevated liver function tests, low platelets) will change the diagnosis to severe preeclampsia. Not all abnormalities must be present. Eclampsia is defined as any of the above plus seizure activity.
- Managing the Patient **Therapy for preeclampsia and eclampsia is based on the severity of disease and the gestational age of the fetus.** Ultimately, **the cure is delivery.** Outpatient management of very mild preeclampsia is successful in compliant patients. **Care consists of bed rest, daily home BP monitoring, baseline evaluation of CBC, platelet count, serum chemistries, and measurement of proteinuria. Patients are followed with**

twice-weekly tests of fetal well-being (nonstress testing and amniotic fluid index) and twice-weekly office visits.

If hospitalization is indicated, the patient is maintained on bed rest with frequent BP monitoring, daily weight, examination, and observation for symptoms. CBC, platelet count, serum chemistries, and 24-hour urine collection for creatinine clearance and total protein are followed closely. Twice-weekly tests of fetal well-being (nonstress testing and amniotic fluid index) are done. Amniocentesis for fetal lung maturity may be indicated. If the diagnosis of severe preeclampsia/eclampsia is made, management should be directed toward delivery. The patient should be medically stabilized with control of blood pressure and seizures and reversal of hypoxia and acidosis, if present. Induction of delivery should be attempted if the cervix is favorable. **Glucocorticoid therapy may be indicated to promote fetal lung maturity** if the gestational age is less than 34 weeks.

Magnesium sulfate is used in the peripartum period (labor/delivery/24 hours postpartum) for prophylaxis and treatment of seizures. Therapeutic blood levels are 4 to 8 mEq/L. A magnesium level greater than 12 mEq/L is toxic. Phenobarbital may also be used. **Antihypertensive medication has not been shown to improve outcome;** it is indicated only in the acute setting, while working toward delivery. Adrenal cortical extract inhibitors are contraindicated.

Hypertension and proteinuria usually resolve within the first week postpartum. Liver function and platelet abnormalities begin to normalize by postpartum day 4 but may persist for up to 10 days. Hypertension remains a common medical complication of pregnancy. It accounts for a significant proportion of all maternal and neonatal morbidity and mortality.

Prenatal Testing

- Nonstress Test **The nonstress test (NST) is the most widely used technique for antepartum assessment. It is based on the observation that fetuses normally display oscillations and fluctuations of their baseline heart rate;** when these are absent, there is an increased risk of perinatal morbidity and mortality.

A reactive test is one in which the fetal heart rate demonstrates at least two accelerations during 20 minutes of monitoring. An acceleration is defined as at least 15 beats per minute above baseline lasting at least 15 seconds.

If the criteria for reactivity are not met, then the test is considered nonreactive. The most common cause for a nonreactive test is a period of quiet sleep or inactivity. Vibroacoustic stimulation may be utilized to change fetal state from quiet to active sleep. The predictive accuracy of a nonreactive NST is not nearly as high as that of a reactive test.

The NST is utilized in high-risk pregnancies (e.g., diabetes mellitus, intrauterine growth retardation (IUGR), hypertension, postdates pregnancies) **and in situations of perceived decreased fetal movement.** Fetal compromise in postdates pregnancy is more often due to umbilical cord compression (secondary to decreased amniotic fluid volume) and not to uteroplacental insufficiency. For this reason, it is recommended that NST be performed twice weekly along with weekly ultrasound assessment of amniotic fluid volume.

- Ultrasonic Assessment Real-time ultrasonography is used to assess fetal well-being. Specifically, **fetal movement, fetal breathing, and amniotic fluid are assessed. Normally, there are at least three fetal movements and 30 fetal chest wall movements within 10 minutes. The largest pocket of amniotic fluid should measure at least 2 cm. Additionally, the amniotic fluid is measured in each of the four quadrants of the gestational sac; the sum of these linear measurements should be between 5 and 20 cm.**

  A normal ultrasonic assessment and a reactive NST together represent a normal biophysical profile.

- Contraction Stress Test **The contraction stress test (CST) attempts to detect uteroplacental insufficiency.** Baseline fetal heart rate and uterine tone are first recorded for 10 to 20 minutes. **An adequate CST requires uterine contractions of moderate intensity lasting approximately 40 to 60 seconds, with a frequency of three in 10 minutes.** If uterine activity is absent or inadequate, nipple stimulation is used (to initiate contractions) or intravenous oxytocin is begun.

  The CST is interpreted as negative if no late decelerations appear on the tracing when adequate contractions are present. The CST is positive if there are late decelerations present with the majority of contractions without uterine hyperstimulation. The test is also considered positive when persistent late decelerations are present before adequate contractions. A suspicious test is one in which there are inconsistent late decelerations.

  A negative CST has been associated with a good fetal outcome. A positive test usually indicates the need for delivery.

Preterm Labor

- Basic Science **Preterm labor (PTL) is defined as labor beginning before the 37th week of gestation.** PTL occurs in about 10% of pregnancies. Preterm delivery accounts for over 60% of perinatal morbidity and mortality.

  In most cases, the cause of PTL is unknown. Prostaglandins, estrogen, progesterone, and oxytocin all play a role in the initiation of labor. **Conditions associated with PTL include premature rupture of membranes, genital tract infection, pyelonephritis, urinary tract infection, overdistension of the uterus (e.g., multiple gestation), and bicornuate uterus.**

- History and Physical Examination The history should focus on known causes of PTL, particularly infectious etiologies. **A history of PTL is a significant risk factor for recurrence.** Smoking and cocaine abuse also increases the risk of PTL. Common symptoms of PTL include change in cervical discharge, uterine cramps, lower back and pelvic pressure, and bloody show. **Patients with PROM (prema-**

ture rupture of membranes) often describe a sudden gush of fluid followed by continued leakage.

If the membranes are not ruptured, and there is no history of placenta previa, a digital examination should be done to determine if the cervix is dilated or effaced.

- Laboratory and Diagnostic Studies **Cultures for gonorrhea, chlamydia, and group B streptococci should be obtained. Urinalysis should be performed to detect asymptomatic urinary tract infections.** If chorioamnionitis is suspected, amniotic fluid should be sent for Gram's stain and culture.

Amniocentesis may be indicated to determine fetal lung maturity. **A lecithin:sphingomyelin ratio greater than 2:1 signifies fetal lung maturity.** In cases of PROM, a sample of amniotic fluid from the vagina can be tested for phosphatidylglycerol. Its presence is an indicator of pulmonary maturity. **There are two rapid bedside tests that can confirm the diagnosis of PROM:**

- Vaginal fluid that changes the color of **nitrazine paper** (from yellow-green to dark blue) indicates an amniotic fluid leak and confirms PROM.
- The second test is the **fern test.** Amniotic fluid is air-dried on a glass slide and then examined under the microscope. A characteristic ferning pattern is seen if amniotic fluid is present.

- Formulating the Diagnosis **True labor is characterized by regular uterine contractions** (six to eight per hour, or four contractions within 20 minutes) **associated with cervical change** (either dilation or effacement). False labor is characterized by irregular uterine contractions without cervical change. Uterine contractions alone may suggest false labor or threatened PTL.
- Evaluating the Severity of Problems Maternal temperature and fetal heart rate should be monitored. Elevation in either vital sign suggests chorioamnionitis. The presence of palpable contractions and uterine tenderness may indicate abruptio placentae.
- Managing the Patient **Gestational age determination is crucial in the management of PTL.** The fetus is not viable at less than 24 weeks. Fetal survival rises with increasing gestational age.

Pharmacologic tocolysis (i.e., terminating labor with medication) is not used in pregnancies over 35 weeks. Between 33 and 35 weeks tocolytics should be used only in cases of documented lung immaturity. Not all cases of PTL should be stopped. It must be decided whether the fetus would do better in utero or birthed. In cases of chorioamnionitis, antibiotics (ampicillin and gentamicin) should be started and labor should ensue.

The first-line treatment to stop labor is bed rest and IV hydration. The goal of this therapy is to increase uterine blood flow and thereby "quiet" the uterus.

If contractions continue, then a tocolytic agent should be used. Most studies show that the tocolytics are successful only in the short term by delaying labor up to 48 hours. **Commonly used medications include magnesium sulfate and the β-mimetics: ritodrine and terbutaline.** Magnesium acts as a calcium antagonist, competing with it for intracellular smooth muscle uptake. It is administered intravenously as a 4-g bolus. Deep tendon reflexes should be tested frequently since they are a reliable index of high blood magnesium concentrations. **Loss of the patellar reflex is the earliest sign of magnesium toxicity.** Respiratory depression begins at 10 to 15 mg/dl, and cardiac arrest occurs at 30 mg/dl. Calcium gluconate is the antidote to magnesium toxicity.

The β-mimetics act on β_2 receptors (of the myometrium) to cause relaxation. Side effects (e.g., tachycardia, tremor) occur since these agents have both β_1 and β_2 effects. They are contraindicated in patients with cardiac disease or hyperthyroidism.

Because of the devastating effects of group B streptococcus on the newborn, prophylactic ampicillin is given in the setting of PTL. Glucocorticoid administered to the mother will enhance fetal lung maturity before 33 weeks' gestation.

Uterine Cancer

■ **Basic Science** Uterine cancer is the most common female genital tract malignancy in the United States. Most patients with endometrial cancer are greater than 40 years old. **Risk factors for endometrial cancer include the following:**

- **Obesity** (conversion of androstenedione to estrone in adipose tissue)
- **Nulliparity**
- **Estrogen replacement therapy**
- **Early menarche or late menopause**
- **Diabetes mellitus**
- **Polycystic ovarian syndrome**

The most significant risk factor for the development of uterine adenocarcinoma is endometrial hyperplasia; this results from the excessive endometrial estrogen stimulation found in many of the aforementioned conditions.

Hormonal contraceptives (e.g., birth control pills and parenteral progesterones) **are associated with up to a 50% reduction in endometrial cancer.** Smoking may modestly decrease the risk of an estrogen-dependent endometrial cancer by lowering circulating estrogen levels.

■ **History and Physical Examination** The initial history should focus on age of menarche and menopause, number of pregnancies, and the use of steroid contraceptives or hormone replacement therapy. **The most common symptom of uterine cancer is abnormal vaginal bleeding.** Therefore a complaint of postmenopausal bleeding always warrants investigation.

Most cancers are discovered early; thus manifestations of distant metastases to lung, brain, and bone are rare. Inguinal lymphadenopathy and an enlarged uterus are occasionally present. The adnexae must be examined for evidence of neoplasia.

■ **Laboratory and Diagnostic Studies** **Any bleeding following 6 months of postmenopausal amenorrhea deserves evaluation. This should include a speculum examination (with biopsy**

of any suspicious lesion), Pap smear, and endometrial biopsy.

The Pap smear is not a screening test for any neoplasm except squamous cell lesions of the cervix. Nevertheless, endometrial cells found on a postmenopausal woman's cytology report should be investigated with a follow-up endometrial biopsy since many of these patients will have an endometrial cancer.

The only metastatic survey routinely indicated is a chest x-ray examination. In special circumstances, pelvic ultrasound may be helpful to evaluate the degree of uterine invasion. However, ultrasound, CT, and MRI are not part of the routine evaluation.

- **Formulating the Diagnosis** **Thirty percent of postmenopausal women with vaginal bleeding will have endometrial cancer.** No therapy should be prescribed until a histologic diagnosis is obtained.

  Approximately 95% of uterine neoplasms are adenocarcinomas. Up to 40% of patients will progress to cancer from hyperplasia with atypia if they remain untreated.

- **Evaluating the Severity of Problems** **Occasionally a patient with uterine cancer may present with significant vaginal bleeding.** Management options to control hemorrhage include emergency hysterectomy and arterial embolization.

- **Managing the Patient** Endometrial hyperplasia (with or without atypia) can be managed in a premenopausal woman who desires to remain fertile with long-term progesterone (e.g., 12 weeks) and periodic endometrial biopsies.

  Stage I cancer is confined to the uterus and should be adequately treated with hysterectomy and salpingo-oophorectomy. During the hysterectomy, **lymph node biopsies should be obtained** from the pelvis and periaortic areas to identify those patients who will benefit from postoperative adjuvant therapy. **Patients at high risk for recurrence or those with extrauterine disease should receive radiation treatment to the pelvis and vaginal apex. Patients with evidence of cancer in the upper abdomen may be offered systemic chemotherapy or whole abdomen radiation treatment.**

  Most patients with uterine cancer are cured. Therefore **long-term follow-up is needed** to detect rare recurrences seen after treatment of early stage disease. Although patients with advanced disease are at high risk for recurrence, up to 50% may eventually be cured if the recurrence is detected early. **A surveillance regimen should begin with a pelvic and speculum examination plus Pap smear every 3 to 4 months for the first 3 years following the completion of therapy.**

  Endometrial cancer is a terminal disease if untreated. It can cause life-threatening vaginal bleeding and symptoms attributable to metastatic disease. Stage I disease has a 90% cure rate, but even the most advanced cases can benefit from palliative chemotherapy or radiation treatment. Besides stage, prognosis is related to histology and grade.

Multiple Choice Review Questions

1. A 44-year-old white married woman, G3P3, presents in your office. Menses are regular at 28-day intervals × 4 to 5 days. The patient is S/P tubal ligation. Approximately 5 days after her last period, the patient notices a firmness in the left breast. There is slight tenderness and no corresponding area in the other breast. There is no history of previous breast lumps and a negative family history for breast cancer. Mammography and ultrasound were done and gave a presumptive diagnosis of fibroadenoma. Which of the following is a physical characteristic of a fibroadenoma?

a. Pain
b. Indistinct borders
c. Freely mobile lump
d. Bloody discharge

2. With a working diagnosis of fibroadenoma, which of the following is the correct treatment?

a. Repeat ultrasound in 3 to 6 months to confirm the diagnosis.
b. Fine-needle aspiration to confirm the diagnosis.
c. Treat with Danazol for 6 months and repeat the ultrasound.
d. Wide excision of the area.

3. A 29-year-old white single woman, G0P0, presents with complaints of serous vaginal discharge that is bloody after intercourse. The patient has been sexually active since the age of 16 and has been treated in the past for venereal warts. Examination reveals a friable verrucous lesion of the cervix. Which of the following is not an immediate appropriate test for the patient?

a. Cervical cultures
b. Hysteroscopy
c. Pap smear
d. Cervical biopsy

4. Which of the following statements about cervical cancer is false?

a. HPV is a risk factor.
b. 90% are adenocarcinoma.
c. Carcinoma in situ can be treated effectively with cone biopsy.
d. Median age for invasive cervical cancer is 45.

5. A 39-year-old woman, gravida one, presents to the office with general malaise, nausea, vomiting, and diarrhea for 2 days. Examination reveals BP 130/84, P 94, R 18, T 37.5°, weight 165 lbs (up 5 lbs in 10 days). You note generalized 1+ edema and epigastric tenderness. The fundal height is 32 cm (consistent with gestational age of 32 weeks) and fetal heart rate is 146. Reflexes are 3+/4+, no clonus. You review her chart and note her BP was 124/64 at 10 weeks' gestation. The appropriate next step in management should be to:

a. Hospitalize her for bed rest and hydration.
b. Diagnose preeclampsia and make arrangements for delivery.
c. Evaluate for proteinuria; obtain serum chemistries, CBC, platelet count, and NST.
d. Diagnose viral gastroenteritis, review comfort measures, and send her home to rest.

6. Further evaluation reveals repeat BP 136/82, hemocrit 35%, platelet count 182,000, and uric acid 6.9. BUN/creatinine/AST/LDH are all within normal limits. 24-hr urine reveals 350 mg protein. NST is reactive. The most appropriate diagnosis is:

a. Hyperemesis gravidarum.
b. Mild preeclampsia.
c. Severe preeclampsia.
d. Viral gastroenteritis.

Chapter 14

Hematology and Oncology

Acute Leukemia

■ Basic Science **Leukemia is a malignant neoplasm of blood-forming organs characterized by diffuse replacement of bone marrow with proliferating leukocyte precursors. The disease presents with abnormal numbers and forms of immature white cells in the circulation that infiltrate the spleen, lymph nodes, liver, and other sites.**

Approximately 20,000 new cases in adults and 2500 in children are diagnosed each year in the United States. Leukemia has been linked to ionizing radiation, benzene, and other chemicals that are toxic to the bone marrow. The risk of the disease is increased in Down syndrome. Acute lymphoblastic leukemia **(ALL) is more common in children,** and acute myelogenous leukemia **(AML) is more common in adults.**

■ History and Physical Examination **Patients with acute leukemia are almost always acutely ill. Symptoms arise because of pancytopenia. Patients have severe anemia leading to fatigue, pallor, and dyspnea. Thrombocytopenia causes bleeding and petechiae. The nonfunctional leukocytes make patients susceptible to overwhelming infection; patients may present with fever or even life-threatening sepsis.**

Tissue infiltration by the leukemic cells may occur in tissues such as the gums and meninges, leading to gingival inflammation and cranial nerve deficits, respectively. Massive numbers of white blood cells may occlude small blood vessels and cause priapism and central nervous system symptoms (e.g., headache, blurring of vision, and stroke). The heart, kidneys, and lungs may also be diseased.

The physical examination will often reveal an ill-appearing, pale, febrile patient. The patient may be hemodynamically unstable because of sepsis or blood loss. The skin may have petechiae or leukemic infiltrates (i.e., raised red lesions of various sizes). The lungs should be examined carefully for evidence of pneumonia or effusion.

■ Laboratory and Diagnostic Studies **The complete blood count will usually demonstrate anemia, various degrees of thrombocytopenia, and low, normal, or elevated white cell counts.** The peripheral smear in most patients with acute leukemia will have a small number of normal maturing white cells. **Bone marrow aspiration and biopsy should be performed for cell marker studies to determine the origin of the malignant cell.**

If the patient is hypotensive or febrile, blood cultures should be drawn. Blood should also be sent for type and cross-match in the event that transfusion is necessary.

- Formulating the Diagnosis **The diagnosis of acute leukemia is made after demonstration of blasts in the peripheral blood smear** in a patient with a history and physical examination suggestive of acute leukemia. However, **approximately 10% of patients with acute leukemia will not have peripheral blasts, and the diagnosis is made only when bone marrow examination reveals an increase in blasts.**

  Immature cells may be seen in a variety of other conditions, including sepsis, bone marrow infection with fungus or mycobacteria, or a nonhematologic malignancy that has metastasized to the marrow.

- Evaluating the Severity of Problems **Infection, bleeding, and leukostasis are the most important concerns in the initial evaluation.** If the neutrophil count is abnormally low, or if the patient appears septic, broad-spectrum parenteral antibiotics should be given immediately. Platelet transfusions should be given if there is active bleeding or if the platelet count is less than 20,000/µl. If a patient with AML has a high peripheral blast count (greater than 100,000/µl), leukapheresis should be performed. Often chemotherapy needs to be started within a day or two after presentation.

- Managing the Patient **In addition to treatment of infection and transfusion of blood products, patients with acute leukemia are most often treated with myeloablative chemotherapy.** Usually patients require a prolonged hospital stay for chemotherapy and for treatment of infectious and bleeding complications until their bone marrow recovers after chemotherapy. **Patients who go into remission are treated with additional courses of chemotherapy. Bone marrow transplantation is used in patients with both AML and ALL.**

  Elderly patients or patients with complicating medical problems may elect not to undergo treatment with high-dose chemotherapy and may choose supportive care, involving treatment of infections, transfusion, and oral agents such as hydroxyurea to help control the abnormal cell proliferation.

Chronic Myelogenous Leukemia

- Basic Science **Chronic myelogenous leukemia (CML) is a myeloproliferative disorder of the hematopoietic stem cell. Unlike the other myeloproliferative disorders** (polycythemia vera, essential thrombocytosis, and myelofibrosis), **CML always progresses to acute leukemia. Most patients with CML have the Philadelphia chromosome, an abnormal chromosome 22 that arises by translocation of material from chromosome 9.** The resulting gene encodes a protein with tyrosine kinase activity that may make the CML cells abnormally sensitive to growth-stimulating factors and resistant to programmed cell death.

- History and Physical Examination **The course of CML is divided into three phases:**

  - **Chronic phase**

- **Accelerated phase**
- **Blast phase**

Many patients with CML are asymptomatic at the time of diagnosis. In these patients, the finding of an abnormally high white blood cell count is the first evidence of the disease. **If symptoms are present, they include fatigue, modest weight loss, fevers, and night sweats.** Moderate splenomegaly may cause left upper quadrant fullness and is usually the only abnormal physical finding. Anemia, if present, is mild and rarely symptomatic.

As the disease progresses into the accelerated and then the blast phase, patients develop symptoms of worsening fatigue, fever, and weight loss. The spleen often markedly increases in size, adenopathy and bony pain develop, and anemia worsens.

Patients in the full-blown blast phase will have the same complications of cytopenia and infection as those with acute leukemia. Patients with high numbers of myeloblasts may require leukapheresis. Although most CML patients transform into AML, one third will transform into acute lymphoblastic leukemia.

■ Laboratory and Diagnostic Studies **The complete blood count of the patient with CML shows an increase in the total number of white blood cells** (as high as 1 million per microliter). **The platelet count is often increased. Anemia, if present, is mild. The urate and lactate dehydrogenase may be elevated** in the patient with CML because of the tumor cell burden.

In contrast to the patient with acute leukemia, the patient with CML will have the full spectrum of myeloid maturation (that is, an increase in immature forms but no corresponding decrease in the numbers of segmented and band neutrophils). An increase in eosinophils and basophils is also characteristic of the peripheral blood in CML.

Bone marrow biopsy and aspirate are done for morphologic diagnosis and chromosomal analysis (looking for the Philadelphia chromosome).

■ Formulating the Diagnosis The diagnosis is based on the laboratory results described above.

■ Evaluating the Severity of Problems Patients in blast phase are acutely ill, suffering from the same complications of cytopenia and infection as those with acute leukemia.

■ Managing the Patient

- **Chronic phase** Patients with CML in the chronic phase may be followed off therapy, but nearly all require treatment to control the white cell count. **Hydroxyurea is highly effective in controlling both white cell and platelet proliferation. Alpha-interferon may also be used in the chronic phase and appears to reduce the number of cells bearing the Philadelphia chromosome.**

  The only treatment shown to alter the natural course of CML is allogeneic bone marrow transplantation. Sixty percent of patients who undergo transplantation will be cured. Therefore, patients who are otherwise in good health and have a matched sibling donor should be considered for transplantation.

- **Accelerated and blast phase Patients with progressive disease are more difficult to treat.** In fact, difficulty controlling the white cell count with the patient's former medical regimen is a hallmark of progression to the accelerated phase. Intensive chemotherapy, such as that used in acute leukemia, usually fails to put the patient into a true remission. **Allogeneic bone marrow transplantation is also much less effective once the patient is in the advanced stages of the disease.**

Chronic Lymphocytic Leukemia

■ Basic Science **Chronic lymphocytic leukemia (CLL) is a cancer that leads to the accumulation of a functionally abnormal monoclonal population of B lymphocytes.** These lymphocytes are found in the peripheral blood, bone marrow, lymph nodes, and spleen. **CLL is the most common leukemia in adults.**

Immunologic abnormalities (e.g., hypogammaglobulinemia, impaired cellular immunity, and autoimmune hemolytic anemia and thrombocytopenia) are common in patients with CLL and account for many clinical problems.

■ History and Physical Examination The diagnosis of CLL is often made in the asymptomatic patient who is found to have an elevated white blood cell count (often during evaluation for an unrelated medical problem).

When symptoms are present, they include fatigue, malaise, and weight loss. Evidence of splenomegaly (left upper quadrant fullness and early satiety) **and symptoms of thrombocytopenia and anemia can occur.**

Patients with CLL have reduced levels of immunoglobulins and impaired cellular immunity, making them susceptible to infection. Because fever and night sweats are uncommon in CLL, a history of fever in a patient suspected of having CLL should prompt an evaluation for possible infection.

The physical examination in a patient with CLL is often entirely normal. A careful lymph node and spleen examination must be done because the presence or absence of lymphadenopathy and splenomegaly determines the stage to which the patient is assigned.

■ Laboratory and Diagnostic Studies **The diagnosis of CLL requires a lymphocyte count over 5000/μl.** The complete blood count may also reveal anemia and/or thrombocytopenia.

Because cell marker studies can be done on the peripheral blood to demonstrate monoclonality, a bone marrow biopsy is not required to make the diagnosis but may help evaluate the patient's reserve of normal bone marrow. As described above, patients with CLL often have hypogammaglobulinemia.

■ Formulating the Diagnosis **Lymphocytosis may be seen in conditions other than CLL.** A variety of infections (e.g., infectious mononucleosis, cytomegalovirus, and *Toxoplasma gondii*) can cause high lymphocyte counts. **In these situations, however, the population of lymphocytes is polyclonal,** and the patients are generally younger than the typical patient with CLL.

- **Evaluating the Severity of Problems** The presence of lymphadenopathy, splenomegaly, and nonimmune anemia and/or thrombocytopenia indicate advancing disease. Median survival in those with advanced disease is less than 2 years. Infectious complications may occur in any stage of the disease.

- **Managing the Patient** **Early asymptomatic CLL does not require specific treatment. Intravenous immunoglobulins (IV Ig) may decrease the incidence of recurrent infections in the patient with known hypogammaglobulinemia, but it has not been shown to prolong survival.**

  Advanced stage disease (symptomatic lymphadenopathy, splenomegaly, anemia, or thrombocytopenia) is generally treated on an outpatient basis. Steroids and chlorambucil are the most commonly used drugs, but combination chemotherapy with newer agents has also been successful.

  Autoimmune hemolytic anemia and thrombocytopenia usually respond to steroids. Infections are treated with standard antibiotic therapy in addition to IV Ig when indicated.

NON-HODGKIN'S LYMPHOMA

- **Basic Science** While **lymphoma most often arises in lymph nodes,** it may also originate in extranodal sites such as the gastrointestinal tract, skin, bone, central nervous system, nasopharynx, thyroid, and testes. **Most lymphomas are B cell in origin.** Epstein-Barr virus and human T cell leukemia virus have been linked to lymphoma development.

  Immunodeficient patients (e.g., those on immunosuppressive medications after organ transplantation and **those with acquired immunodeficiency syndrome [AIDS]) are at increased risk for the development of lymphoma.**

- **History and Physical Examination** **Most patients with lymphoma present with painless adenopathy in the neck, inguinal regions, or axillae.**

  Mediastinal lymphoma may cause a persistent nonproductive cough. Patients may come to medical attention after being treated with antibiotics for presumed bronchitis. Massive mediastinal disease may also present with superior vena cava syndrome (swollen face and arms and prominent venous pattern on the chest) due to obstruction of the superior vena cava.

  Patients with intraabdominal lymphoma may have abdominal or back pain, bowel obstruction, extrinsic compression of urinary tract, or gastrointestinal bleeding.

  Lymphoma of the central nervous system (CNS) may present with headache, cranial nerve palsies, or symptoms of spinal cord compression. The diagnosis of CNS lymphoma should be considered in any AIDS patient with neurologic symptoms.

  So-called B symptoms (fever, chills, night sweats, and

weight loss) may be present at diagnosis and usually suggest advanced disease.

Physical examination may reveal enlargement of the lymph nodes, liver, and spleen. Rapidly growing lymph nodes may be tender to palpation.

■ Laboratory and Diagnostic Studies **Tissue biopsy is mandatory. Patients should undergo surgical biopsy of the most accessible lymph node. Because lymph node architecture is important in defining the type of lymphoma, needle aspiration and frozen section of the lymph node are considered inadequate in making a precise diagnosis of lymphoma.**

Cell marker studies may be helpful in making the diagnosis if they confirm monoclonality of the cell population.

In order to rule out the diagnoses of tuberculosis and fungal infection, tissue should also be sent for microbiologic studies, including acid-fast bacillus and fungal stains.

Staging involves imaging the neck, chest, abdomen, and pelvis by CT scanning.

■ Formulating the Diagnosis **The most important feature in determining outcome and treatment of the non-Hodgkin's lymphomas is the pathologic grade.** The working formulation classifies the non-Hodgkin's lymphomas into three major groups **based on microscopic appearance alone:**

- **High-grade lymphoma**
- **Intermediate-grade lymphoma**
- **Low-grade (indolent) lymphoma**

■ Evaluating the Severity of Problems Emergent medical care is indicated in only a few situations: severe pain, spinal cord compression, increased intracranial pressure or cranial nerve palsies, superior vena cava syndrome, ureteral obstruction or obstruction of another vital organ, and metabolic abnormalities such as tumor lysis syndrome. In most other circumstances, treatment and staging workups can be done on an outpatient basis.

Tumor lysis syndrome is a condition of rapid cell turnover and death that leads to release of intracellular potassium, phosphate, and nucleic acids. Irreversible renal damage may be the end result of overwhelming tumor lysis.

■ Managing the Patient

● **Low-grade lymphoma** The low-grade lymphomas often present with advanced stage disease (for example, with bone marrow, splenic, or liver involvement). The median survival, however, is generally greater than 5 years. **Low-grade lymphomas are highly treatable but are usually incurable with conventional therapy.**

Treatment of the indolent lymphomas is usually undertaken only when the patient develops symptomatic lymphadenopathy, cytopenia, or compression of a vital organ.

Commonly used agents to treat low-grade lymphoma include the following:

- **Chlorambucil**

- **Cyclophosphamide** (used alone or with prednisone)
- **Combination chemotherapy** with regimens such as cyclophosphamide, vincristine, and prednisone

- **Intermediate- and high-grade lymphoma The intermediate- and high-grade lymphomas are characterized by rapid disease progression but higher rates of remission and cure.**

  Combination chemotherapeutic regimens such as CHOP (cyclophosphamide, doxorubicin, vincristine, and prednisone) **effect a complete remission in many patients.** Multiple other regimens are also commonly used. **Combination radiation therapy and chemotherapy are used and improve complete remission rates in patients with localized disease.**

Hodgkin's Disease

■ **Basic Science** Hodgkin's lymphomas comprise approximately 25% of the lymphomas. **The malignant cell is thought to be the Reed-Sternberg cell,** but controversy persists regarding the normal counterpart of this cell.

Hodgkin's disease tends to spread along contiguous lymph node regions. The staging of the disease, rather than the histopathology, is the most important prognostic feature. The Ann Arbor system (also used for the non-Hodgkin's lymphomas) classifies the stage of the disease according to the number and distribution of lymph nodes involved.

- **Stage I** is disease in a **single lymph node region** on one side of the diaphragm.
- **Stage II** is disease in **two or more lymph node regions on one side of the diaphragm.**
- **Stage III** is disease in lymph nodes on **both sides of the diaphragm.**
- **Stage IV** is **disseminated disease** in one or more extranodal organs (such as lung, liver, or bone marrow).

Patients who have bulky mediastinal disease (measuring greater than one third of the transthoracic diameter) are treated like those with stage IIB disease.

Histopathologically, the Rye classification scheme identifies four subtypes of Hodgkin's lymphoma: lymphocyte predominant, nodular sclerosing (most common), mixed cellularity, and lymphocyte depleted.

■ **History and Physical Examination** The history and physical examination are the same as those described for non-Hodgkin's lymphomas.

■ **Laboratory and Diagnostic Studies** The laboratory and radiographic workup are the same for the patient with Hodgkin's lymphoma and non-Hodgkin's lymphoma.

■ **Formulating the Diagnosis** As in all malignancies, tissue is required to make the diagnosis of Hodgkin's disease.

■ **Evaluating the Severity of Problems** The prognosis correlates with the stage of the disease at diagnosis. Another important prognostic feature is the responsiveness of the tumor to treatment.

- Managing the Patient **Treatment of the patient with Hodgkin's disease is usually curative. Stages I and IIA disease may be cured with radiation therapy; stage IIB requires combination therapy with radiation and chemotherapy, and stages III and IV are treated with chemotherapy alone.**

Anemia

- Basic Science Anemia is a reduction in the oxygen-carrying capacity of the blood due to a decrease in hemoglobin. Different processes can lead to anemia. The finding of a low hemoglobin should always prompt a workup. Anemia may be approached in two ways, the mechanistic or morphologic approach:
  - **Mechanism (underproduction, destruction or loss, or abnormal maturation)** The reticulocyte count is helpful in learning the mechanism of the anemia.
  - **Morphology of the red blood cells (microcytic, normocytic, or macrocytic)** The mean corpuscular volume (MCV) and peripheral smear are evaluated in defining the morphology.

    The morphologic approach will be described here.

    Microcytic anemia (low MCV) may be due to iron deficiency, thalassemia, lead poisoning, sideroblastic anemia, and chronic disease. Iron deficiency is suggested by a history of blood loss and the finding of a microcytic anemia.

    Normocytic anemia (normal MCV) may be due to blood loss (acute), chronic disease, renal insufficiency, hemolytic anemia, overwhelming involvement of the bone marrow with tumor or infection, multiple myeloma, or aplastic anemia.

    Macrocytic anemia (high MCV) may be due to megaloblastic anemias that arise as the result of vitamin B_{12} or folate deficiency, liver disease, alcoholism, myelodysplastic syndrome, and hypothyroidism.
- History and Physical Examination **Patients with mild anemia are usually asymptomatic.** Pallor may be evident on physical examination.

  Moderate anemia may present with fatigue and decreased exercise tolerance. **Severe anemia may lead to fatigue, dyspnea, chest pain suggestive of angina, syncope or dizziness, and headache.**

  Chronic slow blood loss, particularly from the gastrointestinal tract, may not be apparent to the patient. **All patients with anemia should have their stool tested for occult blood.**
  - **Iron deficiency** Patients with severe iron deficiency may have glossitis, complaining of tongue soreness or the urge to eat ice. **Iron deficiency should never be attributed to poor dietary intake of iron.**
  - **Thalassemia A family history of anemia raises the possibility of thalassemia,** particularly in those of Mediterranean or Asian descent.

- **Hemolytic anemia Patients with hemolytic anemia may be jaundiced but do not necessarily have dark urine** because the indirect bilirubin released in excess from red blood cell lysis is insoluble in urine. Urinary hemosiderin, if present, may color the urine.
- **B_{12} and folate deficiency Symptoms and signs of advanced vitamin B_{12} deficiency, which almost always occurs because of pernicious anemia, include mental status changes (e.g., memory changes or dementia), ataxia, and peripheral neuropathy.** Folate deficiency should be suspected in those with poor diets, malabsorption syndromes, or increased folate requirements (e.g., during pregnancy).

■ Laboratory and Diagnostic Studies **Three laboratory tests are essential in working up anemia:**

- **The mean corpuscular volume (MCV)**
- **The peripheral blood smear**
- **The reticulocyte count**

Reticulocytes are immature red blood cells released from normally functioning bone marrow in response to a decrease in hematocrit. **If the reticulocyte count is inappropriately low for the hematocrit, a primary bone marrow disorder** (such as red cell aplasia or myelodysplasia) **or deficiency of one of the "building blocks" of red blood cells** (iron, folate, or vitamin B_{12}) **may be responsible. A very high reticulocyte count suggests red blood cell hemolysis or blood loss.**

- **Iron deficiency Low serum iron, low ferritin, and high total iron binding capacity (TIBC) support the diagnosis of iron deficiency in the patient with microcytosis.**
- **Anemia of chronic disease In anemia of chronic disease, which arises because of ineffective iron use, both the iron and the TIBC will be low.**
- **Hemolytic anemias Hemolysis may cause an elevation of the bilirubin and lactate dehydrogenase (LDH) and may lead to fragmentation** (microangiopathic process) **or spherocytosis** (autoimmune process). If autoimmune hemolysis is suspected, a Coombs' test should be ordered.
- **Folate and vitamin B_{12} deficiency Folate and vitamin B_{12} levels should be checked in the patient with macrocytosis. The neutrophils are often polysegmented in folate or B_{12} deficiency states.**

■ Formulating the Diagnosis **The cause of the patient's anemia can usually be determined with the evaluation just described.** A bone marrow aspirate and biopsy may eventually need to be done in the case of elusive anemia. **A bone marrow study should also be done in patients suspected of having primary bone marrow failure, pancytopenia, or infiltrative disease of the marrow.**

■ Evaluating the Severity of Problems Orthostasis, syncope, angina, and dyspnea all suggest severe anemia and should prompt immediate evaluation and treatment, usually with transfusion of red blood cells.

- Managing the Patient **Treatment of anemia should be focused toward the underlying cause.** Transfusion of red cells is generally indicated for significant acute and ongoing blood loss or for marrow failure.
  - **Iron deficiency Iron deficiency can be treated with once- or twice-daily oral iron in most cases;** the more important intervention in the iron-deficient patient is discovering and treating the cause of the iron deficiency.
  - **Folate deficiency Folate is easily replaced with oral or intravenous folate.**
  - **B_{12} deficiency Patients with B_{12} deficiency are treated with intramuscular vitamin B_{12}.**
  - **Renal insufficiency Patients with anemia due to chronic renal insufficiency respond to intravenous or subcutaneous erythropoietin.**
  - **Hemolytic anemias Many autoimmune hemolytic anemias respond to treatment with steroids.**

Hemophilia

- Basic Science **The hemophilias are inherited defects that result in abnormal amounts of certain plasma clotting proteins. The two common hemophilias are hemophilia A (factor VIII deficiency)** and **hemophilia B (factor IX deficiency).**

  Both hemophilias result in joint and soft tissue hemorrhage as well as exsanguinating hemorrhage after minor trauma.

  Hemophilia is classified as follows:

  - Severe: < 1% coagulant activity
  - Moderate: 1% to 5% coagulant activity
  - Mild: > 5% coagulant activity

  As many as 20% of patients with hemophilia A have antibodies to factor VIII, making their management extremely challenging.

- History and Physical Examination **The diagnosis of hemophilia is usually made by family history. In the typical family, male patients are affected and female patients are carriers.** Severe hemophilia usually leads to spontaneous hemorrhage early in life. Milder forms of the disease may not be apparent until after trauma or surgical procedures.

  Hemarthrosis is the most common spontaneous bleeding complication in the patient with hemophilia. Recurrent bleeding into the joints leads to pain and deformity and often requires fusion of the joint or total joint replacement. Muscle hematomas may occur, causing compartment syndrome. Ecchymoses, retroperitoneal hematomas, oral bleeding, and gastrointestinal bleeding are all common in severe hemophilia. Intracranial bleeding is the most common cause of hemorrhagic death.

  Seventy-five percent of patients with hemophilia are infected with transfusion-acquired hepatitis B or C. The resul-

tant cirrhosis and complications of portal hypertension contribute significantly to morbidity and mortality. Patients who received blood products in the mid-1980s have a high rate of transfusion-acquired HIV infection.

- Laboratory and Diagnostic Studies **Most patients with hemophilia have an elevated partial thromboplastin time (PTT). The prothrombin time (PT), bleeding time, and fibrinogen level are all normal.** Patients with mild hemophilia may have a normal PTT. Direct assays of both factor VIII and factor IX levels can be done to measure the coagulant activity in the patient's blood.
- Formulating the Diagnosis The diagnosis of hemophilia is usually made in a bleeding patient with a positive family history. The results of assays of specific factors will confirm the clinical suspicion. Patients with mild hemophilia may go undiagnosed until significant bleeding occurs.
- Evaluating the Severity of Problems **All bleeding in a patient with hemophilia must be considered life-threatening.** Physical examination of the hemophiliac may not reflect the severity of the bleeding.
- Managing the Patient Symptomatic patients should be evaluated immediately. Consultation with a hematologist should be sought promptly.

  Factor VIII and factor IX concentrates are the mainstays of treatment. The aim is to replace the diminished factor to 50% to 100% of normal coagulant activity.

  Hemarthroses should be managed with orthopedic consultation. The suspicion of compartment syndrome in a limb should lead to prompt surgical evaluation.

  Perioperative management involves replacing the abnormal factor until the levels are 50% to 100% of normal. Treatment with factor continues for up to 7 days postoperatively.

  Patients with inhibitors (i.e., antibodies to factor VIII) are exceedingly difficult to treat and should be managed together with a hematologist.

von Willebrand's Disease

- Basic Science **von Willebrand's disease refers to a group of hereditary coagulation disorders that arise because of abnormal synthesis or release of von Willebrand factor (vWF) from endothelial cells.** von Willebrand factor is a large protein required for platelet adhesion to subendothelial surfaces and for stabilization of factor VIII. The most common form of the disease, type I, or classic von Willebrand's disease, is inherited as an **autosomal dominant** trait. The trait is **seen in both males and females.**
- History and Physical Examination **The most common type of bleeding in von Willebrand's disease is mucocutaneous bleeding. Examples of such bleeding are epistaxis, gingival bleeding, menorrhagia, and easy bruising.** Variability of bleeding is characteristic of the disease. Therefore, although bleeding after surgery, childbirth, and dental extractions is common, the patient may not have a consistent bleeding history.

The family history may be helpful.

- Laboratory and Diagnostic Studies The **bleeding time,** a measure of platelet function, is often **prolonged** in patients with von Willebrand's disease. The **PTT is often but not always elevated. The ristocetin cofactor assay is a sensitive test for von Willebrand's disease.**

- Formulating the Diagnosis **The history, physical examination, and the laboratory tests mentioned will help make the above diagnosis of von Willebrand's disease.** Because of the variability of the disease, however, one set of normal blood results does not rule out the diagnosis.

  Mild hemophilia may simulate this disease.

- Evaluating the Severity of Problems The extent of bleeding in the past is not necessarily a reliable indicator of disease severity. **The level of factor VIII and the bleeding time are probably accurate measures of the status of the disease at the time of evaluation.** Fatal bleeding from von Willebrand's disease is rare.

- Managing the Patient **von Willebrand's disease can be treated prophylactically (or during bleeding episodes) with cryoprecipitate or factor VIII preparations that are "contaminated" with von Willebrand factor.** Both must be given approximately every 12 hours.

  Patients with type I von Willebrand's disease, in which the vWF is synthesized normally but secreted abnormally, may respond to treatment with ddAVP (i.e., desmopressin, an analog of vasopressin). This hormone, given intravenously, stimulates release of vWF from endothelial cells.

Sickle Cell Disease

- Basic Science **Sickle cell disease results from a single base mutation in DNA that leads to an amino acid substitution in the hemoglobin molecule. The abnormal hemoglobin (Hb S) polymerizes in the deoxygenated state and causes red blood cell sickling.** The sickled cells in the microvasculature cause vasoocclusion. Hemolysis of the sickled cells leads to anemia.

  An individual must be homozygous to develop sickle cell disease. Heterozygosity leads to the carrier state; such patients are said to have sickle trait.

- History and Physical Examination **Patients with sickle cell anemia experience both acute crises and chronic medical problems. Most problems are due to vasoocclusion rather than anemia. Dehydration, infection, pregnancy, and fever can precipitate vasoocclusion.** Vasoocclusion affects the bones, joints, mesentery, and lungs. This causes painful crises, bone infarction, and stroke.

  Acute chest syndrome is a vasoocclusive process characterized by fever, chest pain, dyspnea, hypoxia, and pulmonary infiltrates.

  Chronic sickling and splenic infarction lead to functional asplenia. Sickle cell patients are particularly susceptible to infection with encapsulated pathogens (e.g., pneumococcus) and *Salmonella*.

Chronic medical problems also contribute to morbidity and mortality. Growth retardation, chronic bone disease, leg ulcers, cardiac disease due to chronic anemia and iron overload (from hemolysis and blood transfusions), renal disease, proliferative retinopathy, and chronic pulmonary disease develop in the adolescent and adult patient. The chronic hemolysis leads to pigment gallstones, frequently demanding cholecystectomy.

■ Laboratory and Diagnostic Studies **Widespread screening of infants in the United States detects both sickle trait and sickle cell disease.** Early detection helps in management and decreases morbidity and mortality.

Quantitative hemoglobin electrophoresis confirms the diagnosis by showing a predominance of hemoglobin S. Heterozygotes have both hemoglobin A and hemoglobin S.

The patient with sickle cell disease who presents with a painful crisis should have a complete blood count, chemistry panel (including LDH and bilirubin as measures of hemolysis), and a reticulocyte count. Blood cultures and type and cross are usually obtained during the initial evaluation as well. Magnetic resonance imaging scans may be helpful in differentiating between infarcts and osteomyelitis.

In the patient with pulmonary symptoms, a chest radiograph and arterial blood gas should also be done.

Abdominal ultrasound may be indicated in the patient with possible cholelithiasis or cholecystitis.

■ Formulating the Diagnosis **The screening of infants provides early diagnosis.** In the rare patient who eludes diagnosis in infancy, a painful crisis coupled with sickled cells on the peripheral smear will lead to the diagnosis. The finding of osteomyelitis caused by *Salmonella* should also raise the suspicion of sickle cell disease.

■ Evaluating the Severity of Problems **The patient's pain is the most important indicator of crisis severity.** The degree of anemia often does not correlate with the severity of the painful crisis, particularly as dehydrated patients may have a falsely elevated hematocrit upon presentation. **The reticulocyte count will correlate with the degree of hemolysis and can be followed during the crisis.**

■ Managing the Patient

- **Pneumococcal vaccination** is indicated for all sickle cell patients.
- Patients should avoid dehydration, cold, and high altitude to prevent vasoocclusive crises.
- **Early treatment with penicillin in the febrile pediatric patient reduces morbidity from meningitis.**
- **Sickle cell crisis treatment includes IV hydration, narcotic analgesics, folate supplements, antibiotics if an infectious process is suspected, and red blood cell transfusions as indicated.**
- Acute chest syndrome is treated with oxygen, fluids, antibiotics, and exchange transfusion.

- Patients with iron overload can be treated with deferoxamine, an iron chelator.

THALASSEMIA

■ Basic Science **The thalassemias are a group of genetic disorders that affect the globin protein of hemoglobin.** Defects are due to mutations in the DNA that code for these globin chains. **Each normal adult has four α-globin genes and two β-globin genes.**

■ History and Physical Examination The clinical manifestations of thalassemia depend on the location of the DNA mutation and the number of involved genes.

■ Laboratory and Diagnostic Studies

- **α-Thalassemia There are four α-thalassemia syndromes, most of which arise because of α-globin gene deletions:**

  - The ***silent carrier state,*** in which one of the four α-chain loci is absent, has **no clinical or laboratory implications** for the carrier.
  - ***α-thalassemia trait,*** in which two of the four loci are abnormal or absent, will produce **microcytosis, with or without mild anemia.** Hemoglobin electrophoresis is normal. Patients are rarely symptomatic unless they have a second cause of anemia.
  - ***Hemoglobin H disease*** arises from deletions of three of the four α-chain genes and **causes moderate to severe microcytic anemia with red blood cell changes, including target cells, stippling, and hypochromia. Besides anemia, affected patients have jaundice and hepatosplenomegaly as early as 1 year of age.**
  - ***Hydrops fetalis,*** caused by deletions of all four α-chain loci, is **incompatible with life,** and affected fetuses are stillborn.

- **β-thalassemia The β-thalassemias are the result of mutations within the genes encoding the β-globin chains. There are two β-thalassemia syndromes:**

  - ***β-thalassemia minor*** (also called β-thalassemia trait) which occurs in those who are heterozygous for the β-thalassemia gene. **Findings include microcytosis, hypochromia, target cells, cigar-shaped cells, basophilic stippling, and hypochromia. Anemia, if present, is usually mild.** Hemoglobin electrophoresis demonstrates increased amounts of hemoglobin A_2. **Because of the microcytosis, many patients with β-thalassemia trait have been treated with iron for what was thought to be iron-deficiency anemia.**
  - ***β-thalassemia major*** (Cooley's anemia) arises from the homozygous state of the β-thalassemia gene. **Affected patients are severely anemic and have signs of hemolysis, iron overload, and skeletal abnormalities (due to expansion of the marrow).** The peripheral smear demonstrates marked shape changes and often normoblasts. Hemoglobin electrophoresis

reveals an increased amount of fetal hemoglobin (Hb F). **Patients are transfusion dependent and have a decreased life expectancy. Folate supplementation is appropriate. Splenectomy may be helpful in increasing the red blood cell survival. Bone marrow transplantation is potentially curative in the patient with a human lymphocyte antigen-compatible sibling. Many patients require iron chelation therapy to avoid the complications of iron overload.**

- Formulating the Diagnosis **The diagnosis of α-thalassemia trait or β-thalassemia minor is suspected in the patient with microcytosis (with or without anemia) whose iron studies are normal.** A family history of mild anemia also suggests thalassemia. As just described, **hemoglobin electrophoresis will help** distinguish these two syndromes, being normal in α-thalassemia trait and showing increased amounts of hemoglobin A_2 in β-thalassemia minor.

  Hemoglobin H disease and β-thalassemia major are generally diagnosed because of the unique clinical features described in the preceding paragraphs.
- Evaluating the Severity of Problems The degree of anemia, the transfusion requirement, and the severity of hemolysis are important factors in determining the severity of the disease.
- Managing the Patient **Transfusion support is often needed.** Patients do not need to be transfused to a normal hematocrit; most are asymptomatic with hematocrits in the low 20% range.

  Because of the hazards of iron overload from chronic transfusion therapy, most patients should be considered for treatment with chelation therapy. The most common chelator is deferoxamine, a drug that can be given via subcutaneous injection through a home pump.

Lung Cancer

- Basic Science In the United States **lung cancer is the most common cancer** and the leading cause of cancer deaths. The incidence of death due to lung cancer continues to increase in both men and women.

  Cigarette smoking is responsible for 80% to 90% of lung cancer cases. The risk of developing lung cancer increases with both the number of cigarettes smoked per day and the number of years of smoking. After a patient stops smoking, the risk of developing lung cancer declines over 2 to 15 years. Other potential carcinogens include asbestos, radiation, chromates, nickel, arsenic, hydrocarbons, vinyl chloride, and mustard gas.

  Although early localized lung cancer may be cured with surgical resection, more advanced disease is incurable, and the prognosis for affected patients is dismal.

  In the discussion of tumor behavior and treatment strategies, lung cancer is broadly divided into two groups: small cell and non-small cell (based on the histopathology of the tumor). Adenocarcinoma and squamous cell carcinomas are included in the discussion of non–small cell cancer.
- History and Physical Examination **Patients with lung cancer may present with or develop cough, dyspnea, hemoptysis, wheezing, stridor, chest wall pain, hoarseness, pleural effusion,**

dysphagia, and weight loss. Not uncommonly, patients will present with manifestations of metastatic disease, such as seizures from intracranial metastases or pain from bony metastases. Paraneoplastic syndromes (e.g., Eaton-Lambert syndrome) are seen most commonly with small cell lung cancer.

Large lung masses can cause superior vena cava (SVC) syndrome because of extrinsic compression (by tumor or lymph nodes) of the SVC. Patients have swelling of the face, neck, and arms and a prominent thoracic venous pattern. Thrombosis of the SVC may also occur.

A Pancoast's tumor is a lung cancer located in the superior pulmonary sulcus. Patients present with pain in the shoulder or medial scapula, radicular pain in the distribution of the ulnar nerve (with or without weakness), **and Horner's syndrome** (ptosis, myosis, and hemifacial anhidrosis).

■ Laboratory and Diagnostic Studies **Chest radiographs are reviewed carefully in all patients suspected of having lung cancer. Computed tomography (CT) scans will define the location, size, and nature of a parenchymal lung lesion and will aid in pursuing tissue.** Additionally, the mediastinum is more readily evaluated with CT scan than with x-ray film. The scan should include views of the liver and adrenals as part of the staging workup.

In addition to chest and abdominal scans, staging usually includes a bone scan and, if the patient has neurologic symptoms, a contrast CT scan of the head. Patients who have resectable non–small cell carcinoma do not need bone scans in most situations.

Routine chemistries may raise suspicion for bone or liver metastases.

If the cancer is thought to be resectable, the patient needs to have an assessment of the expected postoperative pulmonary reserve.

■ Formulating the Diagnosis **As with all malignancies, a confident diagnosis requires the microscopic examination of tissue. Transbronchial biopsy, transthoracic needle aspiration, thoracentesis, and thoracotomy/mediastinoscopy are techniques used to obtain tissue from primary lung lesions.** Patients with metastatic disease often have lesions that are more easily sampled; examples include cervical lymph nodes, bone lesions, and liver metastases. Sputum cytology may occasionally reveal malignant cells.

■ Evaluating the Severity of Problems The prognosis is clearly correlated with the stage of the disease. In general, patients with significant weight loss, chest wall pain, large primary tumors, and metastatic disease have poor prognoses. If the tumor is considered resectable, the patient's prognosis is markedly improved, and cure can be expected in a significant proportion of these patients.

■ Managing the Patient

- **Non–small cell lung cancer In non–small cell lung cancer, potentially curative surgical resection is indicated when a complete resection is anticipated, even in those with more advanced disease. Patients with nonresectable disease are treated primarily with radiation therapy; chemotherapy has effected a modest improvement in sur-**

vival in patients who are active and have minimal symptoms. The treatment of metastatic disease is aimed toward palliation of symptoms such as pain, dyspnea, and hemoptysis. Chemotherapy, in hope of obtaining a response, can be offered to patients with good performance status.

- **Small cell lung cancer Limited stage small cell lung cancer is treated with combination radiation and chemotherapy. Extensive disease is treated with chemotherapy. In contrast to non-small cell lung cancer, small cell lung cancer does respond to chemotherapy.** Survival is increased from 2 to 3 months without chemotherapy to 8 to 14 months with chemotherapy. Long-term survival (>5 years) is highly unlikely (<1%).

Breast Cancer

■ Basic Science **Breast cancer is the most common cancer among women, and the second leading cause of cancer death among women.** Breast cancers arise from the ductal and glandular structure of the breast. Some data suggest that endocrine, environmental, and genetic factors may be involved in the initiation and the promotion of breast cancer.

■ History and Physical Examination Cancers of the breast are usually first discovered by the patient or the physician as a solitary painless mass in the breast. Nipple discharge is an unusual presenting symptom. Breast cancer may also be detected on mammography in an asymptomatic patient. On occasion, the patient will present with symptoms of metastatic disease, commonly in the bone.

Risk factors for breast cancer include the following:

- **Family history** of breast cancer in first-degree relatives
- Reproductive history of **early menarche, late age at menopause, nulliparity, or first pregnancy over age 30**
- **Prolonged use of estrogen replacement therapy**
- **Diet history of high lipid and alcohol consumption** (possibly increase risk)
- **Radiation exposure** (e.g., in patients treated for lymphoma)

Despite these known risks, however, most women with breast cancer do not have an obvious risk factor. The recently discovered "breast cancer gene" is seen in only 5% to 10% of breast cancers.

Physical examination begins with the breasts and axillary regions. **The size, consistency, and mobility of any breast mass should be noted. Examination of the skin for erythema, edema, dimpling, or ulceration and the nipples for retraction, discoloration, or discharge should be done.**

■ Laboratory and Diagnostic Studies **Screening mammography leads to earlier detection of breast cancer. A baseline mammogram is recommended for all women between the ages of 35 and 40. Although controversy exists regarding the use of mammography in women between the ages of 40 and 50,**

it is recommended that women in this age group have an annual or biannual mammogram. It is recommended that women over the age of 50 have an annual mammogram.

Mammographic abnormalities of a malignant breast mass include clustered microcalcifications and stellate densities.

Biopsy or removal of a breast lesion is mandatory if there are suspicious mammographic abnormalities or a dominant breast mass. Breast biopsy provides histologic information on tumor type and grade. Tissue is also sent for detection of estrogen and progesterone receptors. **Axillary lymph node dissection is necessary for staging the disease.**

- Formulating the Diagnosis **Microscopic examination of the breast biopsy sample will reveal the histopathology of the cancer;** infiltrating lobular and ductal carcinomas carry a worse prognosis. The axillary nodes are also examined carefully for the presence of malignancy.
- Evaluating the Severity of Problems **Staging of breast cancer is based on tumor size, the involvement of regional lymph nodes, and the presence of distant metastases.** As in most malignancies, the stage of the disease correlates with the prognosis. Other **prognostic features include the presence of estrogen and progesterone receptors (their presence portends an improved prognosis)** and the histology of the cancer. Bone scans and chest radiograph are indicated in most patients with a breast cancer greater than 2 cm in diameter.
- Managing the Patient **Treatment of the breast cancer patient will depend on the stage of breast cancer. Options include local and systemic therapy:**
  - **Local therapy**
    - Surgical resection lumpectomy or mastectomy
    - Radiotherapy given after lumpectomy and in the treatment of bony metastases
  - **Systemic therapy**
    - Chemotherapy is beneficial in premenopausal patients with positive nodes. The most commonly used regimens include cyclophosphamide, Adriamycin or methotrexate, and 5-fluorouracil (5-FU).
    - Hormone manipulation with the antiestrogen tamoxifen is used following local therapy and adjuvant chemotherapy, particularly in patients with estrogen receptor–positive tumors. Other hormonal agents such as megesterol acetate and aminoglutethimide are used in selected patients with refractory metastatic disease.

Colon Cancer

- Basic Science Colon cancer implies a carcinoma arising from intestinal mucosa. **Most colon cancers arise from adenomatous polyps, protrusions of the mucosal surface. Villous polyps are more likely to be malignant than tubular polyps.** Epidemiologic evidence implicates a diet low in fiber and high in fat and cholesterol as a possible etiologic factor. **Other risk factors include** a history of inflammatory bowel disease (in particular, **ulcerative colitis**), **a family history of colon cancer, and autosomal dominant familial colonic polyposis.**

- History and Physical Examination **Symptoms that suggest the presence of colon cancer include a change in bowel habits, narrowing of stool caliber, rectal bleeding, or symptoms of anemia due to chronic insidious blood loss.** Occasionally, obstruction or perforation results from a large neoplasm. **A history of weight loss is particularly worrisome for a malignant colonic process.**

  Physical examination may be unremarkable. The abdominal examination rarely reveals an abdominal mass. Digital rectal examination may provide evidence of a rectal mass. **The presence of heme-positive stool demands further evaluation.**

- Laboratory and Diagnostic Studies Laboratory data may reveal only a hypochromic, microcytic anemia suggesting iron deficiency. Liver function test abnormalities should raise the concern of liver metastases.

  Current recommendations for patients without risk factors include:

  - **A digital rectal examination at age 40**
  - **A fecal Hemoccult test and/or flexible sigmoidoscopy at age 50 and every 3 to 5 years thereafter**

  A positive Hemoccult test (which tests for the presence of microscopic blood) **should prompt further medical evaluation with sigmoidoscopy, barium enema, and/or colonoscopy. However, a negative Hemoccult test does not rule out colon cancer.**

  A lesion identified by sigmoidoscopy, colonoscopy, or barium enema should be biopsied to make a definitive diagnosis. All colonic polyps larger than 0.5 cm should be removed and examined by a pathologist.

- Formulating the Diagnosis **The diagnosis is often made because of routine screening. The most important information in determining the diagnosis is the biopsy result.** Adenocarcinoma accounts for most colon cancers.

  Staging involves evaluating the extent of the tumor in the pathologic specimen. The Dukes' staging system stages tumors on the extent of both local and distant involvement and correlates with survival. **The metastatic workup includes a chest radiograph, bone scan** (in the patient with bony pain), **and abdominal imaging** to rule out liver metastases (if the liver function tests are abnormal).

- Evaluating the Severity of Problems The earlier the stage, the more favorable the prognosis. Metastatic disease carries a very poor prognosis (5-year survival less than 5%). Younger patients tend to have more advanced cancers. Rectal hemorrhage generally leads to early detection and may therefore be a favorable prognostic sign.

- Managing the Patient **The only curative therapy for colon cancer is surgical resection; cure is a reasonable goal if the cancer is localized.** Patients with more advanced disease may also benefit from surgery if they have impending obstruction, significant pain, or solitary liver metastases. **Radiation therapy can reduce local recurrence and is particularly appropriate in patients with rectal cancer. Combination chemotherapy with 5-FU and leucovorin is used for the treatment of advanced local disease and metastatic disease.** For metastatic disease, patients are treated when they develop symptoms.

Multiple Choice Review Questions

1. An 8-year-old boy with severe hemophilia A comes to the Emergency Department complaining of pain in the right forearm that began 2 hours ago, after he fell at school. The arm is visibly swollen, and he is holding it across his chest. He is unable to make a fist and reports decreased sensation in the fingers. The arterial pulse is strong. Your first step in the management of this patient is:
   - **a.** Transfusion of red blood cells.
   - **b.** Factor VIII infusion.
   - **c.** Emergent venogram.
   - **d.** Radiograph of the arm.
2. Which of the following is likely to be seen in this patient's lab work?
   - **a.** Prothrombin time (PT) of 17.0 seconds
   - **b.** Prolonged bleeding time
   - **c.** Partial thromboplastin time (PTT) of 70 seconds
   - **d.** PTT of 28 seconds
3. A 30-year-old woman with sickle cell disease who had a cholecystectomy at the age of 18 presents with severe abdominal pain that has been increasing in severity for the last 3 days. She reports a week-long history of rhinitis and sore throat. Her oral analgesics are no longer controlling her pain. She denies fever and chills but does report some nausea. On examination she is afebrile, has dry mucous membranes, mild scleral icterus, no axillary sweat, and diffuse abdominal tenderness without evidence of peritonitis. Her hematocrit, usually 24%, is 18%, and her bilirubin is 2.8 mg/dL. All of the following are appropriate measures *except:*
   - **a.** Pregnancy test.
   - **b.** Broad-spectrum antibiotics.
   - **c.** Parenteral hydration.
   - **d.** Transfusion.
   - **e.** Parenteral narcotic analgesics.
4. A 67-year-old woman presents to your office for a follow-up physical examination. She describes fatigue but no other symptoms. The physical examination is remarkable only for trace guaiac positive stools and a palpable liver edge. She has a mild anemia and a slightly elevated alkaline phosphatase. All of the following procedures would be appropriate *except:*
   - **a.** Liver biopsy.
   - **b.** Colonoscopy.
   - **c.** Sigmoidoscopy and barium enema.
   - **d.** Iron studies.
   - **e.** CT scan of the abdomen.

Chapter 15

Internal Medicine

Adult Vaccinations

■ **Influenza Vaccine** Usually this vaccine contains three subtypes: influenza A H_1N_1, Influenza A H_3N_2, and one subtype of influenza B virus. The specific strains are determined each year based on the previous year's isolated strains. The following people should receive this vaccination each year:

- Persons 65 years or older
- Adults with chronic illnesses:

  1. Pulmonary disease (e.g., asthma, COPD)
  2. Cardiac disease (e.g., ischemic coronary artery disease)
  3. Diabetes mellitus
  4. Renal disease
  5. Hemoglobinopathy

- Immunosuppressed adults (including HIV disease and malignancy induced)
- Health care personnel, home care providers, and household members of patients with the aforementioned conditions

A hypersensitivity to eggs is a contraindication for this vaccine.

■ **Pneumococcal Vaccine** The current vaccine includes purified capsular polysaccharides of 23 serotypes of pneumococcus. Most healthy adults respond well to this vaccine, and antibodies persist for at least five years after immunization. The vaccine should be administered to the following people every 10 years:

- Persons 65 years or older
- Adults with chronic illnesses:

  1. Pulmonary diseases as mentioned earlier
  2. Cardiac diseases as mentioned earlier
  3. Alcoholism
  4. Cirrhosis
  5. Hodgkin's lymphoma
  6. Asplenia (including functional asplenics, e.g., sickle cell patients)
  7. Multiple myeloma
  8. Nephrotic syndrome
  9. Diabetes mellitus

- Immunosuppressed adults (including HIV disease and malignancy induced)

There is essentially no contraindication to the pneumococcal vaccine.

- **Tetanus Toxoid Vaccine** Covered as a separate topic later in this chapter.
- **Hepatitis B Vaccine** Hepatitis B vaccine is a series of three injections and should be given to
  - Adult health care workers
  - People who work in institutions such as schools and day care centers
  - Immigrants from Asia and Africa
  - Intravenous drug users
  - Homeless persons
  - Household contacts or sexual partners of patients who carry the hepatitis B virus chronically
- **Varicella Vaccine** Varicella vaccine for adults is indicated for patients who have no varicella titer but risk exposure, such as health care workers, college students, day care workers, and parents of young children.

AIDS

- **Basic Science** **The acquired immunodeficiency syndrome has a complex definition: HIV-infected patients with CD4 T lymphocyte cell counts <200/µl alone, or patients with CD4 T lymphocyte cell counts higher in the presence of certain disease (e.g., pulmonary tuberculosis, *Pneumocystis carinii* pneumonia, Kaposi's sarcoma, non-Hodgkin's lymphoma).** However, clinicians are encouraged to consider HIV infection as a syndrome that advances from an asymptomatic phase to severely advanced disease.

  Human immunodeficiency virus is a member of the human retrovirus family and lentivirus subfamily. Both HIV-1 and HIV-2 (recently reported in the United States) have been described. The HIV binds to the gp120 protein of the CD4 molecule on the surface of T helper lymphocytes. The virus is integrated into the T cell's DNA, leading to dysfunction and destruction of the T helper cells. B lymphocytes and macrophages are affected to a lesser extent.

  HIV is transmitted by body fluid contact. The commonly documented modes of transmission are sexual contact (both homosexual and heterosexual), **blood exposure** (as with intravenous drug abusers and health care workers stuck with contaminated needles, via blood products), **and from mother to infant** (either intrapartum or through breast milk). Homosexual men represent the largest percentage of patients, but the percentages of heterosexual women and children are increasing. Currently, the most common mode of transmission is between heterosexual partners.

- **History and Physical Examination** **Acute infection with HIV is characterized by a syndrome of lymphadenopathy, fever, pharyngitis, anorexia, and malaise. After this acute syndrome, a period of clinical latency occurs and is variable in duration. However, CD4 T cells continue to be affected and eventually fall to low levels. As the CD4 T cell count falls below 500/µl, symptoms such as herpes zoster reactivation,**

lymphadenopathy, and thrush occur. Eventually, more opportunistic infections and neoplasms (e.g., Kaposi's sarcoma, lymphoma) may develop, and patients may develop AIDS dementia.

- Laboratory and Diagnostic Studies **Demonstration of antibodies to the HIV makes the diagnosis.** Most laboratories use an enzyme-linked immunosorbent assay (ELISA), which has >99% sensitivity. If the ELISA is positive or inconclusive, a more specific assay, the Western blot, is done for confirmation.
- Formulating the Diagnosis As described earlier, the diagnosis of AIDS is made when the CD4 count falls below 200 cells/μl or when an "AIDS-defining opportunistic illness" occurs.

  The more common opportunistic infections in AIDS patients are *Pneumocystis carinii* pneumonia, tuberculosis, cytomegalovirus, *Mycobacterium avium,* and toxoplasmosis. AIDS patients are seemingly more susceptible to lymphoma, Kaposi's sarcoma, and other neoplasms.
- Evaluating the Severity of Problems A CD4 count of <50 cells/μl portends a poor prognosis and a significantly increased risk of death.
- Managing the Patient **Antiretroviral drugs should be the first step in therapy; currently, patients with CD4 counts below 300 cells/μl are candidates for zidovudine.** Very recent research has indicated that CD_4 count <500 should receive antiretroviral therapy and that 2 drugs (AZT and ddI) combined prolong the asymptomatic period. Drugs such as zidovudine (AZT), didanosine (ddI), or zalcitabine (ddC) are available and have been documented to slow the progression of HIV disease. A new class of drugs called protease inhibitors shows promise by halting the assembly of infectious virions. **CD4 counts should be initially monitored every 3 months to assess the efficacy of therapy.**

  Once the CD4 T cell count has dropped below 200/μl, prophylactic antibiotics for *Pneumocystis carinii* should be started. The best choice is trimethoprim/sulfamethoxazole, but dapsone and inhaled pentamidine can be used in sulfa-allergic patients. When the CD4 T cells fall below 100/μL, some clinicians favor starting rifabutin for *Mycobacterium avium* complex prophylaxis. **HIV patients should receive the pneumococcal vaccine** since they are also more susceptible to the usual bacterial diseases. **The influenza vaccine should be given each fall.** HIV-infected patients who are HBsAb-negative should receive the hepatitis B vaccine. A PPD and anergy panel should be documented.

  The treatment of common opportunistic infections is beyond the scope of this text; however, *Pneumocystis carinii* pneumonia is covered separately in this section.

Deep Venous Thrombosis

- Basic Science **Deep venous thrombosis (DVT) refers to intravascular clotting in the deep veins of the upper (only 10%) and lower extremities (90%). Virchow defined three factors that promote DVT formation:**

- Venous stasis
- Abnormalities of the vessel wall
- Alterations of the coagulation system

Deep venous thrombi usually develop in the region of a venous valve. Platelets aggregate, forming a nidus (i.e., a white thrombus), followed by development of a large fibrin (or red) thrombus. This process may occur rapidly; large thrombi can develop within minutes.

Resolution occurs by fibrinolysis and organization. Fibrinolysis can result in complete resolution of the thrombus in a few hours to days. Any remaining thrombus will undergo organization, leaving a fibrotic zone. This process reaches a stable state in approximately 7 to 10 days. **It is during the first few days after formation that embolic risk is highest in patients with DVTs.**

■ **History and Physical Examination** The clinical diagnosis of a DVT is often inaccurate. **A DVT may be present in the absence of any clinical signs. When present, the clinical signs of a DVT are nonspecific but may include pain, warmth, tenderness, and swelling or color changes in the area of the thrombus, or a palpable "cord."**

■ **Laboratory and Diagnostic Studies** **The "gold standard" for evaluation of a lower extremity DVT is contrast venography. This is an invasive test and can be painful. The hazards of giving IV contrast also exist. For these reasons, some physicians start with a noninvasive test such as duplex ultrasonography and/or impedance plethysmography.** Ultrasonography can be obtained quickly and has good sensitivity for detecting proximal thrombi. Impedance plethysmography has poor sensitivity for calf-vein thrombi and nonobstructive proximal thrombi. False positives have occurred in patients with peripheral vascular disease and postoperative lower extremity swelling.

■ **Formulating the Diagnosis** Clinical suspicion based on the present risk factors is a good start in making the diagnosis. **The risk factors for a DVT are as follows:**

- **Any surgical procedure requiring 30 minutes or more of general anesthesia**
- **Prolonged immobilization (e.g., in the head-injured patient)**
- **Fractures or injuries involving the lower extremities**
- **Use of estrogens or birth control pills**
- **Cancer**
- **Obesity**
- **History of thromboembolic disease or chronic venous insufficiency of the legs**
- **Hematological disorders (e.g., lupus anticoagulant or protein C deficiency)**
- **Age greater than 60**

■ **Evaluating the Severity of Problems** **The most obvious serious con-**

sequence of a DVT is a pulmonary embolus. This topic is covered separately as "Pulmonary Embolism," in Chapter 4.

- Managing the Patient **DVT prophylaxis includes the use of low-dose heparin, (i.e., 5000 units sq bid) or pneumatic stockings with at least one compression per minute.** Alternatively, low molecular weight heparin may be used for prophylaxis and requires no monitoring of the partial thromboplastin time (PTT).

  A patient with a documented DVT requires anticoagulation. Initially, continuous IV heparin is used. Coumadin is begun simultaneously. Once the patient is on coumadin, the protime, specifically the international normalized ratio (INR), is monitored until a maintenance dose is determined. When the coumadin reaches a therapeutic range (an INR of two to three is desirable), the heparin is discontinued. The exact duration of therapy depends on the etiology of the DVT. A patient with a hypercoagulable state will likely require prolonged anticoagulation.

HYPERKALEMIA

- Basic Science **Hyperkalemia is an elevation of serum potassium above 5.5 mEq/L. The etiologies of hyperkalemia can be divided into three categories: increased intake, increased movement from cells into the extracellular space, and decreased excretion.**

  The body maintains several protective measures against potassium loads. Insulin, catecholamines, alkalosis, and passive uptake all promote potassium movement into the intracellular space in response to increased serum concentrations. In addition, **aldosterone secretion promotes urinary excretion of potassium.** Although mechanisms exist to compensate for variations in intake, an acute ingestion or infusion of as little as 160 mEq can transiently overwhelm homeostatic mechanisms and produce cardiac arrhythmia. Other sources of significant amounts of potassium include penicillin VK and stored whole blood.

  Potassium leaves cells in response to a variety of stimuli, including acidosis, insulin deficiency, and β-adrenergic blockade. Cell lysis (as in tumor lysis, trauma, or massive hemolysis) releases large amounts of intracellular potassium. Drugs such as digoxin, ACE inhibitors, and succinylcholine may result in hyperkalemia. Pseudohyperkalemia is an artificially elevated serum potassium, usually secondary to hemolysis during venipuncture.

  The most significant source of potassium elimination is aldosterone-mediated renal excretion. Decreases in glomerular filtration rate, whether from intrinsic renal failure or a prerenal state, can impair the kidney's ability to handle a normal potassium intake. Similarly, adrenal insufficiency can result in hyperkalemia secondary to decreased aldosterone.

- History and Physical Examination **The patient should be questioned regarding dietary intake, medications, history of**

renal disease, previous hyperkalemia, and diabetes mellitus. An estimate of volume status and the presence of muscle weakness should also be sought.

- Laboratory and Diagnostic Studies **The diagnosis of hyperkalemia is made by the measurement of serum potassium from an unhemolyzed specimen. Measurement of serum blood urea nitrogen and creatinine (to estimate renal function), serum pH (to evaluate for acidosis), and glucose level (to evaluate for hyperglycemia and possible diabetic ketoacidosis) can all be useful in discovering the etiology of hyperkalemia.** Urinary potassium levels are often unhelpful.

  If hypoaldosteronism is suspected, an investigation of its stimulation, secretion, and effect is warranted. To test aldosterone, all medications previously listed as contributing to hyperkalemia must be discontinued. After an overnight dehydration (furosemide is given in the evening to ensure volume depletion) a morning renin, aldosterone, and cortisol are drawn. The results can be used to evaluate the function of renin release, adrenal function, and renal response to aldosterone.

  An immediate electrocardiogram is indicated if the measured potassium is greater than 6.0 mEq/L. Characteristic electrocardiographic changes (e.g., peaked T waves, shortened QT interval, loss of P waves, and eventual QRS widening) can be seen as the serum potassium level rises. ECG changes can occur very quickly and lead to ventricular fibrillation.

- Formulating the Diagnosis **Increased intake, intravenous fluids, or medication-induced hyperkalemia can be elicited from the history. Metabolic disturbances causing potassium shifts to the extracellular space and renal failure are often apparent on routine labs. If no apparent cause is found, or dramatic hyperkalemia is present with a normal electrocardiogram, then pseudohyperkalemia should be suspected. After this is eliminated, a workup of aldosterone stimulation, release, and activity is warranted.**

- Evaluating the Severity of Problems Any symptomatic elevation of potassium (particularly with electrocardiographic changes) requires emergent intervention. Patients with chronic hyperkalemia (e.g., those with end-stage renal failure) tolerate higher potassium levels than those in whom the disturbance is acute.

- Managing the Patient **Symptomatic hyperkalemia requires cardiac monitoring and urgent correction. The treatment falls into three categories: membrane stabilization, driving potassium into cells, and removing potassium from the body. One to two ampules of 10% calcium gluconate can stabilize cardiac cell membranes and temporarily reverse electrocardiographic changes. Ten to fifteen units of insulin (accompanied by glucose if the patient is not extremely hyperglycemic) will quickly drive much of the potassium into the intracellular space. Sodium bicarbonate, particularly in an acidotic patient, accomplishes the same function. Forced diuresis with furosemide and normal saline, cation-exchange resins**

(e.g., Kayexalate, which binds potassium in the gastrointestinal tract), and dialysis are less immediate in their action but often necessary to definitively correct the potassium overload.

Chronic hyperkalemia can be controlled with use of loop diuretics, cation-exchange resins, or exogenous aldosterone if correction of the underlying problem is impossible.

Hypernatremia

- Basic Science **Hypernatremia is an elevation of the serum sodium >145 mEq/L. The etiologies of hypernatremia fall into two broad categories: water loss and sodium retention. Vigorous exercise with profuse sweating, fever, or burns can result in increased skin losses. Hyperventilation from pulmonary infection, acidosis, or other irritants increases respiratory water loss. Osmotic diarrhea or normal GI losses that go unreplaced due to depressed mental status or physical disability can also cause dehydration.**

  Renal water losses resulting in hyperosmotic states are more complicated. The term *diabetes insipidus* refers to a deficiency in ADH secretion or function resulting in renal free water losses and subsequent hyperosmolality. Central diabetes insipidus results from a process impairing the normal production or release of ADH. Dysfunction of this pathway may result from neurosurgical procedures, trauma, tumor invasion, hypoxia secondary to shock or cardiac arrest, infiltration from sarcoidosis or histiocytosis, or anorexia nervosa.

  Patients with nephrogenic diabetes insipidus have normal production and release of ADH but impaired or absent renal response to the hormone. The many causes of nephrogenic diabetes insipidus include impaired concentrating ability from renal failure, hypercalcemia, hypokalemia, loop diuretic use, sickle cell anemia, or osmotic loads (e.g., hyperglycemia). Drugs (e.g., lithium, demeclocycline), amyloidosis, Sjögren's syndrome, and rare congenital disorders may render the collecting tubule relatively impermeable to water.

  Sodium retention is most commonly seen in hospitalized patients receiving hypertonic intravenous fluids or feedings. In addition, hypothalamic dysfunction may lead to alterations in the thirst mechanism, resulting in inadequate water ingestion.

- History and Physical Examination **Neurologic features may be the first and only symptoms of hypernatremia. Lethargy, weakness, agitation, seizures, coma, and eventually death may occur as the hyperosmolality worsens.** Clues to the diagnosis may come from signs of total body volume depletion. **Patients with diabetes insipidus will complain of polyuria, polydipsia, and nocturia.**

- Laboratory and Diagnostic Studies **The diagnosis is usually made by routine serum sodium measurement. Urine osmolality (U_{osmol}) and the kidney's response to exogenous ADH help to determine the etiology.**

- Formulating the Diagnosis **Suspicion of a hyperosmolar state is confirmed by basic serum chemistries.** The plasma osmolality can be calculated or directly measured. Patients with impaired mental status most likely have inadequate intake. Evidence for increased skin, respiratory, or gastrointestinal losses should be clear from the history and physical exam. **If no obvious source of water loss is found, the urine osmolality should be measured. If the U_{osmol} is >400 mOsm/kg, then the urine is (appropriately) maximally concentrated, and a source of water loss or decreased intake should be sought. If the U_{osmol} is <250 mOsm/kg, then at least a partial defect in ADH release or activity is present.** Before remeasurement of the U_{osmol}, the patient should be given five units of aqueous vasopressin. If the U_{osmol} increases in response, then a defect in ADH secretion is present. **Failure of the U_{osmol} to increase in response to vasopressin is consistent with nephrogenic diabetes insipidus.**
- Evaluating the Severity of Problems The presence of any neurologic symptoms requires urgent attention.
- Managing the Patient **The treatment of hypernatremia should be aimed at its cause. The type of fluid for rehydration depends on the patient's volume status. Hypovolemic patients may initially require normal saline, whereas euvolemic or hypervolemic patients are often better treated with 0.45 NS or D_5W. As with hyponatremia, too rapid correction (>12 mEq/L/day) may result in worsening neurologic status.**

  The chronic treatment of patients with central diabetes insipidus can be achieved most conveniently with DDAVP (vasopressin) nasal spray every 12 hours. Drugs that lower urine output and a low-sodium/low-protein diet minimize polyuria. Drugs such as chlorpropamide and carbamazepine may increase the secretion of ADH in some patients. **In nephrogenic diabetes insipidus the defect is usually reversible. If it persists, the polyuria can be decreased by thiazides and a low-sodium/low-protein diet as in central diabetes insipidus.** Other measures that may be beneficial include the diuretic amiloride and NSAIDs to decrease renal prostaglandins.

Hypokalemia

- Basic Science **Hypokalemia is a serum potassium <3.5 mEq/L. Hypokalemia can be divided into three broad categories: decreased intake, increased losses (mostly renal), and increased entry into cells.** Potassium is ubiquitous in a normal diet. **Therefore hypokalemia is rarely seen solely because of dietary deficiency.**

  Increased intracellular flux of potassium can cause hypokalemia even though total body potassium remains normal. **Alkalemia, insulin, hypothermia, β-adrenergic agonists, and new cell production** (in patients receiving vitamin B_{12}, folate, or cellular growth factors) **promote the intracellular movement of potassium.**

  Net potassium losses can be grouped as extrarenal or

renal. Extrarenal losses include profuse sweating, dialysis, and gastrointestinal losses. Diarrhea contains high concentrations of potassium, resulting in significant losses, whereas vomiting results in hypokalemia secondary to volume depletion and metabolic alkalosis.

Urinary potassium loss is mediated by several mechanisms. Causes of primary mineralocorticoid excess, such as adrenal adenomas or carcinomas, Cushing's disease, ectopic ACTH (e.g., from oat cell carcinoma), or exogenous substances (e.g., glycyrrhizic acid in licorice), can all lead to hypokalemia. High renin states (as in decreased effective circulating volume in congestive heart failure and cirrhosis) or renal artery stenosis can lead to a secondary increase in aldosterone levels. **Thiazide and loop diuretics are very common causes of hypokalemia.** Mild diuretic-induced hypovolemia (resulting in increased aldosterone levels) may enhance potassium loss. Other syndromes of potassium wasting exist due to primary or secondary tubular defects.

■ History and Physical Examination Patients with hypokalemia should have their medications reviewed and be questioned regarding intake and sources of potassium loss. **In the undiagnosed patient, symptoms of muscle weakness, arrhythmias, glucose intolerance, polydipsia, and polyuria all warrant a serum potassium measurement. Blood pressure measurement may help to differentiate hyperaldosteronism (i.e., in the hypertensive patient) from other causes of potassium loss.**

■ Laboratory and Diagnostic Studies **The diagnosis of hypokalemia is confirmed by serum potassium measurement. Electrocardiographic changes of hypokalemia include increased ectopy (especially in patients on digoxin therapy), decreased T wave amplitude, increased P wave amplitude, and eventually QRS widening.**

The differential diagnosis of hypokalemia can be narrowed by urinary potassium measurement. In a euvolemic patient off diuretic therapy, **potassium excretion of <25 mEq/day suggests appropriate potassium conservation and an extrarenal or exogenous cause of hypokalemia. Conversely, values of >30 mEq/day imply renal wasting and either mineralocorticoid excess or a tubular defect.** Random urine potassium concentrations are useful but are less accurate than 24-hour collections. In severe hypokalemia (<2.0 mEq/L) the urine potassium concentration may be low regardless of the cause. The urinary chloride concentration may be helpful since low levels suggest a gastrointestinal source of potassium loss. **In perplexing cases, surreptitious diuretic use should be considered.**

If hyperaldosteronism is suspected, the lack of appropriate plasma renin increase in response to dehydration and upright posture, and lack of aldosterone suppression in response to volume load and recumbency, both suggest mineralocorticoid excess. Once the diagnosis has been established, CT or MRI scanning as well as adrenal vein aldosterone levels can be used to distinguish unilateral adenoma or carcinoma from bilateral adrenal hyperplasia.

- **Formulating the Diagnosis** Once identified by laboratory assay, the etiology of hypokalemia can be discovered through a stepwise approach. **Diuretic use and gastrointestinal losses constitute the vast majority of cases and may be identified through the history.**
- **Evaluating the Severity of Problems** **While the potassium should be repleted to the normal range, symptomatic hypokalemia, particularly with electrocardiographic changes, should be corrected urgently. Uncorrected hypomagnesemia makes correction of hypokalemia difficult.**
- **Managing the Patient** Since most potassium is intracellular, the total body deficit can be estimated only from the serum measurement. **An approximation of the potassium deficit is 100 mEq for levels <3.5 mEq/L, 200 mEq for <3.0 mEq/L, 300 to 400 mEq for a level <2.5 mEq/L, and 400 to 600 mEq for <2.0 mEq/L.** Potassium is most commonly combined with chloride in replacement preparations. Occasionally potassium bicarbonate or potassium phosphate may be preferable in metabolic acidosis or phosphate deficiency, respectively.

  Oral potassium is the preferable route of repletion. Due to gastric upset, a maximum of 40 mEq per dose, and 160 mEq/day, is all that is commonly tolerated. An elixir can be used for rapid absorption, whereas long-acting tablets are better tolerated for chronic use. Many commercially available salt substitutes consist of potassium chloride and may be significantly less expensive than prescription preparations. **Intravenous repletion is limited by the rate of infusion and the volume of the fluid. A maximum rate of 10 mEq/hr in a peripheral vein and 20 mEq/hr through central venous access is standard at most institutions** (although greater concentrations can be administered in life-threatening circumstances). Too rapid repletion may precipitate cardiac arrhythmias.

  Treatment of the underlying cause, from correction of diarrhea to surgical removal of adenoma or change in medication, should always be attempted.

Hyponatremia

- **Basic Science** **Hyponatremia is a plasma sodium concentration <135 mEq/L.** This results from net sodium loss or free water retention. **The normal manifestation of excess water is a lowering of plasma osmolality (P_{osmol}), which leads to a decrease in antidiuretic hormone** (ADH). This decrease in ADH renders the collecting tubules of the kidney relatively impermeable to water reabsorption, resulting in a dilute urine and elimination of excess water.

  With normal renal function, states of pure volume overload are uncommon but may be seen in hospitalized patients receiving hypotonic intravenous fluids or in patients with primary polydipsia.

  Two basic defects impair the body's ability to maximally excrete free water. The first is hypovolemia. This may represent a truly decreased circulating volume secondary to GI losses (e.g., vomiting, diarrhea), renal losses (e.g., diuretic use), or skin losses (e.g., burns, vigorous exercise); or hypovolemia may represent an effective volume depletion that results in a lack of free water available to the distal nephron as seen in CHF, ascites, or nephrosis. **Volume depletion, whether true or effective, results in a decreased GFR and an increase in the proximal tubule reabsorption of water.**

The second major factor responsible for hyponatremia is an increased permeability of the collecting tubules to water, usually mediated by increased ADH. Excessive ADH occurs with hypovolemia (discussed earlier), the syndrome of inappropriate ADH secretion (SIADH), adrenal insufficiency, and hypothyroidism.

- History and Physical Examination **The possibility of hyponatremia should be sought in patients with a history of volume loss** (e.g., profuse vomiting, diarrhea, fever, diuretic use) **or edematous states such as CHF, ascites, or nephrosis.** The history should also include a review of the common causes of SIADH.

  The physical examination should include an assessment of volume status and observation for symptomatic hyponatremia (which presents as lethargy, encephalopathy, obtundation, seizure, or coma). Unfortunately, the more severe neurologic manifestations of hyponatremia may be irreversible despite correction of the serum sodium.

- Laboratory and Diagnostic Studies **The diagnosis is usually obtained by screening serum chemistry. A measurement of P_{osmol} and urine sodium should be obtained in hyponatremic patients.**

- Formulating the Diagnosis **The P_{osmol} will delineate true hyponatremia with a decreased P_{osmol} from pseudohyponatremia in which another solute (e.g., glucose) is diluting the sodium concentration, resulting in a normal or increased P_{osmol}. The urine sodium is often helpful in determining the etiology. A urine sodium <25 mEq/L suggests that the body is appropriately conserving sodium. Conversely, a urine sodium >40 mEq/L reveals an inappropriate loss of sodium and warrants an investigation for SIADH, diuretic use, renal failure, or adrenal insufficiency.**

- Evaluating the Severity of Problems The presence of any neurologic symptoms requires urgent attention.

- Managing the Patient **There are two basic principles to the correction of hyponatremia: correct the plasma sodium at a safe rate and treat the underlying cause.** If the patient is truly volume depleted then normal saline can be given; conversely, the volume-overloaded patient requires water restriction.

  The sodium deficit can be estimated by multiplying the volume of distribution of sodium in the body by the sodium deficit per liter: Na deficit = 0.6 (weight in kg) × (140 − plasma Na).

  Too rapid correction can lead to central demyelinating lesions. It is generally assumed safe to correct the sodium at a rate less than 1.0 mEq/L/hr, and not greater than 12 mEq/L

over the first 24 hours. Correction at a rate of 0.5 mEq/L/hr over the first 24 hours seems appropriately safe in a patient without severe neurologic symptoms. If obtundation, seizures, or coma is present then more rapid correction is indicated. Whatever the intended rate of correction, frequent checks of sodium and potassium are advisable during the initial treatment period.

In SIADH and edematous states, water restriction is the first line of therapy. In severe or refractory hyponatremia, the combination of normal saline and a loop diuretic may be cautiously employed to achieve a net loss of free water. Patients with SIADH should have prolonged free water restriction if the underlying process cannot be corrected. **The antibiotic demeclocycline inhibits the action of ADH and can be used when water restriction alone is insufficient.**

Infectious Mononucleosis

- Basic Science **Infectious mononucleosis is a clinical syndrome of pharyngitis, fever, and lymphadenopathy caused by the Epstein-Barr virus.** The virus replicates in oropharyngeal epithelial cells and B lymphocytes. There is an incubation period of approximately 4 to 8 weeks.
- History and Physical Examination **The patient experiences a prodrome of malaise, anorexia, headaches (occasionally), and chills for about 1 week before the pharyngitis, fever, and lymphadenopathy occur.** Abdominal pain is present only rarely, unless splenic involvement has occurred.

  Up to 90% of patients have either posterior or anterior cervical adenopathy and fevers. The fever can be as high as 40° C. **The pharyngitis can be exudative, and palatal petechiae are present in some patients. Approximately one half of the patients will have splenomegaly as well. The clinical syndrome lasts about 3 to 4 weeks.**
- Laboratory and Diagnostic Tests **Heterophil antibody testing is done, but 10% to 15% of patients have a negative test in the first week of illness. The monospot test is more specific and sensitive.** On the complete blood cell count, **a lymphocytosis is present in 75% of patients, and atypical lymphocytes (i.e., activated T cells) are present.** The patient may also develop neutropenia and thrombocytopenia. Immunoglobulins M, G, and A are elevated.
- Formulating the Diagnosis **Infectious mononucleosis is diagnosed clinically, based on the presence of fever, pharyngitis, lymphadenopathy, and the aforementioned laboratory findings.** Some patients with exudative pharyngitis will have β-hemolytic streptococcal pharyngitis as well.
- Evaluating the Severity of Problems Rare complications include autoimmune hemolytic anemia, splenic rupture, encephalitis, mild neutropenia, thrombocytopenia, and airway obstruction (usually in younger patients).
- Managing the Patient **Supportive care, rest, and avoiding con-**

tact sports (for several weeks) are the mainstays of treatment. Airway obstruction can occur and should be treated with glucocorticoids. There are no documented benefits to antiviral therapy.

INFLUENZA

■ Basic Science **Influenza refers to the acute respiratory illnesses caused by influenza viruses** A, B, and less often C. These RNA viruses belong to the Orthomyxoviridae family. Influenza outbreaks can be local, pandemic or epidemic. The **pandemics and epidemics are caused mostly by influenza A virus since it can change its genome (antigenic shift).** Influenza B virus causes less severe disease than influenza A. Influenza C virus is rarely associated with human disease.

Transmission of the virus commonly occurs secondary to contact with aerosolized respiratory secretions from an infected individual, as occurs with coughing and sneezing, or by direct hand-to-hand contact. Viral replication in infected cells occurs in 4 to 6 hours, and spreads to adjacent cells. **The incubation period of illness can be from 18 to 72 hours.**

■ History and Physical Examination **Most patients with influenza describe an abrupt onset of symptoms: fever, headache, sore throat, myalgia, and cough.** There is a wide spectrum of disease severity. Most patients have fever, ranging from 38 to 40° C for the first 3 to 4 days. Myalgia is a prominent symptom, usually localized to the legs and lumbosacral area. Anorexia and arthralgias may also be present.

The physical examination is often unproductive in mild cases. Vascular injection of the oropharyngeal mucous membranes is often present, as is mild cervical lymphadenopathy. The lung examination is clear in uncomplicated cases, but rhonchi, rales, and wheezes can be present. Mottled skin is present in some elderly patients as well.

■ Laboratory and Diagnostic Studies **An attempt is made to isolate the virus via throat or nasopharyngeal swabs.** The type of influenza virus can be determined by immunofluorescence. Other laboratory tests are not usually helpful in making the diagnosis.

■ Formulating the Diagnosis **The diagnosis of influenza should be made with caution unless a known outbreak is present. Other diseases, such as that caused by *Mycoplasma pneumoniae* or streptococcal pharyngitis, can present similarly to influenza.** Findings such as a markedly elevated leukocyte count, purulent sputum, or exudative pharyngitis suggest a bacterial etiology. In a known outbreak, however, the diagnosis of influenza is more certain if the characteristic symptoms are present.

■ Evaluating the Severity of Problems The complications from influenza infection include pneumonia (either primary viral, secondary bacterial, or mixed viral and bacterial), Reye's syndrome in children, myositis, myocarditis, encephalitis, transverse myelitis, and the Guillain-Barré syndrome.

■ Managing the Patient **Rimantadine and amantadine are two an-**

tiviral agents with activity against influenza A virus. These drugs interfere with the replication cycle of the influenza A virus but not the influenza B virus. They are used prophylactically during influenza A epidemics and are approximately 70% to 90% effective. These two agents are also effective in reducing the severity and duration of symptoms when given to patients with influenza A infection within 48 hours after the onset of disease. Rimantadine and amantadine doses need to be adjusted in patients with liver disease and renal insufficiency.

Acetaminophen (not aspirin) is the preferred analgesic and antipyretic in patients with influenza due to the risk of Reye's syndrome with aspirin.

Lyme Disease

- Basic Science **Lyme disease is a tick-borne *(Ixodes sp.)* inflammatory disorder caused by the spirochete *Borrelia burgdorferi*.** In 1975 a group of children in Lyme, Connecticut, developed symptoms similar to juvenile rheumatoid arthritis. This unusual geographic grouping lead to an investigation and the eventual discovery of Lyme disease. **The disease begins mostly in the late spring or early summer. Shortly after the tick bite, a characteristic skin lesion develops (i.e., erythema migrans) and can be accompanied by mild influenza-like symptoms (early Lyme disease). The disease can worsen several weeks later with cardiac, neurologic, and arthritic changes (late Lyme disease).** Lyme disease is among the most common vector-borne illnesses in the United States and is most prevalent in the Northeast (from Maryland to Massachusetts), the Northern Midwest, and in the Pacific Northwest. It has also been reported in Europe and Asia.
- History and Physical Examination **Recent travel or residence in an endemic area with a reported tick bite is the most important part of the initial history. Patients usually present with the characteristic skin lesion, erythema migrans. This is a well-demarcated, circular, expanding, erythematous rash occurring at the site of the tick bite. The lesion must be larger than 5 cm in diameter, pruritic, but not painful. Patients may also complain of myalgias, headache, arthralgias, and general malaise in the early stages. Patients can be quite ill with systemic spread of the infection. They may appear toxic and have high fevers, a stiff neck, and severe headaches.** Secondary skin lesions may also occur at this time. Multiple organs may be affected, (e.g., the heart, spleen, liver, and nervous system). Other organs, such as the kidneys, skeletal muscle, and gut, are generally unaffected.
- Laboratory and Diagnostic Studies **Three serologic tests are available: an immunofluorescent assay, an enzyme-linked immunosorbent assay (ELISA), and Western blot. Early localized disease usually will not produce enough antibodies to be detected in any of these tests. Patients with disseminated Lyme disease and with late Lyme disease will typically have**

a positive ELISA. False positives and negatives do occur, and confirmation with a Western blot test is usually indicated if the ELISA result remains in question. Cerebrospinal fluid will have a positive serology if CNS Lyme disease is present. **Serologic testing cannot, however, distinguish between active and previous infection.**

- Formulating the Diagnosis **The diagnosis of early Lyme disease is easiest if the patient (who lives in an endemic area) presents with a tick bite followed by the typical rash. However, the classic presentation is atypical.** Early, disseminated, and late Lyme disease are usually diagnosed based on clinical criteria. **The laboratory data is used as supportive evidence of *Borrelia* infection.**

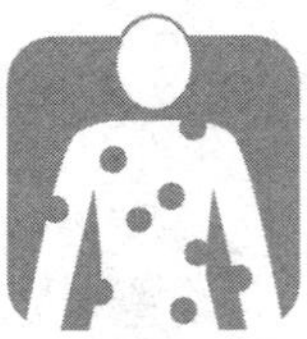

- Evaluating the Severity of Problems **The more serious aspects of Lyme disease are meningitis, cranial neuropathies (e.g., Bell's palsy), myopericarditis, and associated conduction abnormalities.** Long-term sequelae include significant arthritis, usually involving the knee, which can result in permanent joint destruction. Occasionally, late Lyme disease can manifest itself with central nervous system symptoms. Patients can have cognitive dysfunction and upper motor neuron disease. This is called tertiary neuroborreliosis.

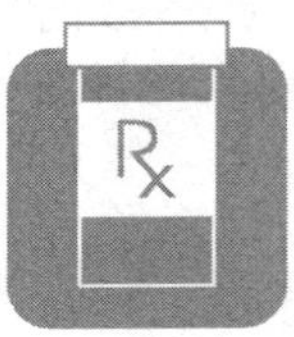

- Managing the Patient **Treatment for early, disseminated, and late Lyme disease is amoxicillin, tetracycline, or doxycycline for 3 to 4 weeks.** CNS involvement requires IV ceftriaxone or penicillin G for at least 28 days. Some patients with Lyme arthritis respond better to IV antibiotics. **The most frequent reason for treatment failure is an incorrect diagnosis. The role of prophylaxis following a tick bite remains a controversial and unresolved issue.**

Medical Ethics

Over the past few decades, patient autonomy has become a major focus for medical ethics. That which is an acceptable treatment or life for one patient may be considered unacceptable by another. The physician must realize that the patient has the right to choose a path that may be different from what the physician has recommended. **For the physician to act ethically, he or she must follow the following principles:**

- **Do not cause the patient harm, including overtreating and giving false hope.**
- **Provide care in the best interest of the patient medically, balancing potential benefits and burdens to arrive at the most net benefit.**
- **Informed decision making by the patient is possible only if the physician provides all relevant information concerning the patient's diagnosis, prognosis, and treatment options. This information must be in terms that the patient can understand; the physician must refrain from coercion.**
- **The physician must provide fair options and treatment for each patient regardless of socioeconomic factors, race, sex, religion, and so on. Avoid personal and financial conflicts of interest.**

- **The physician must be honest and maintain professional confidentiality.**
- **The physician must respect the patient's dignity and decisions. Physicians should recognize that the patient's religious and cultural beliefs are not always the same as their own, and therefore decisions about treatment options may differ. For example, a patient who is a Jehovah's witness may refuse a blood transfusion even if death is likely without it. It is unethical for the physician to impose this treatment on the patient even if the physician is, in his or her mind, acting in the patient's best interest. The right to choose treatment is not limited to the right to agree with the physician. The patient must, however, understand the consequences of the decision; otherwise, a surrogate decision maker should be involved.**
- **One must always base treatment decisions on what is best for the patient, in consideration of patient autonomy. This becomes particularly important in treating pediatric, incompetent, or elderly patients.**

Nutritional Deficiencies

■ Water-Soluble Vitamin Deficiencies: Vitamins B, C, and Folate

- **Thiamine (vitamin B_1) deficiency** Thiamine serves as a coenzyme in carbohydrate oxidation and plays a role in peripheral nerve conduction. The clinical features of thiamine deficiency are seen most commonly in the cardiovascular and/or nervous system of alcoholics (secondary to malnutrition and decreased absorption). Classically, the features of thiamine deficiency are "wet beriberi" (CHF with cardiomegaly secondary to high output heart failure) and "dry beriberi" (nervous system involvement). Central nervous system involvement may manifest as the Wernicke-Korsakoff syndrome. **Wernicke's encephalopathy presents with nystagmus, ophthalmoplegia, truncal ataxia, and confusion, whereas Korsakoff's syndrome is characterized by anorexia, confabulation, and impaired learning.**

  Although biochemical assays are available to assess thiamine deficiency, the clinical response to empiric treatment with thiamine is most often used. The most common form of replacement therapy for thiamine deficiency is thiamine 50 to 100 mg per day.
- **Riboflavin (vitamin B_2) deficiency** Riboflavin is an important cofactor in oxidation-reduction reactions. Riboflavin deficiency is commonly seen when accompanied by deficiencies of other water-soluble vitamins in malnourished or alcoholic patients. **Riboflavin deficiency manifests as cheilosis, angular stomatitis, seborrheic dermatitis, anemia, and weakness.** When the diagnosis is suspected, it is usually treated empirically with riboflavin, 5 to 15 mg per day.
- **Niacin deficiency** Niacin is an important cofactor in oxidation-reduction reactions and is sometimes used in the treatment of hyperlipidemia. As in other water-soluble vitamin deficiencies, de-

ficiency of niacin is seen in alcoholics and malnourished patients. **Severe deficiency results in the classic triad of pellagra: diarrhea, dermatitis, and dementia.** The diagnosis of pellagra is usually made on clinical grounds and treated with replacement doses of niacin ranging from 10 to 150 mg per day.

- **Pyridoxine (vitamin B_6) deficiency** Pyridoxine is involved in heme production and amino acid metabolism. Most commonly, deficiency of this vitamin arises secondary to drug interactions (e.g., with contraceptives, isoniazid, and other drugs), or from alcoholism. **The manifestations of deficiency are similar to other B vitamin deficiencies (e.g., glossitis, cheilosis).** The diagnosis can be confirmed by measurement of pyridoxal phosphate in the blood. Treatment consists of oral supplementation. Patients taking isoniazid should be instructed to supplement their diet with pyridoxine.

- **Folate and B_{12} deficiency** Vitamin B_{12} and folate are important cofactors in DNA synthesis. Vitamin B_{12} also serves as a coenzyme in the formation of succinyl CoA (necessary for myelin metabolism).

  Folate deficiency may be secondary to poor dietary intake, decreased absorption (e.g., tropical sprue), drugs (e.g., phenytoin), increased requirements (e.g., chronic hemolytic anemia, pregnancy), or increased loss (e.g., hemodialysis). Vitamin B_{12} deficiency can be secondary to decreased intake seen in strict vegetarians, decreased intrinsic factor (pernicious anemia or gastrectomy), competitive use from a parasite *(Diphyllobothrium latum),* bacterial overgrowth, or ileal disease.

  Folate or B_{12} deficiency presents as megaloblastic anemia characterized by macrocytosis, macroovalocytosis, and hypersegmented neutrophils on a peripheral smear. There are also mucosal cell changes resulting in glossitis, anorexia, and diarrhea. Additionally, vitamin B_{12} deficiency results in paresthesias and posterior column symptoms (i.e., impaired balance and decreased proprioception).

  The diagnosis is based on the history and physical examination and is supported by laboratory findings, which include macrocytic anemia, low reticulocyte counts, hypersegmented neutrophils and (possibly) increased LDH and bilirubin (secondary to dysfunctional erythropoiesis). Low levels of red blood cell folate or serum B_{12} represent more specific findings. The Schilling test is used to identify patients with inadequate B_{12} absorption.

  The treatment consists of repletion of folate or B_{12} deficiency. Patients with B_{12} deficiency who are mistakenly treated with folate may have improvement in their anemia; however, the neurologic manifestations persist.

- **Vitamin C (ascorbic acid) deficiency** Vitamin C is involved in collagen metabolism. Defects in collagen synthesis secondary to lack of vitamin C are the basis for scurvy. **Patients with vitamin C deficiency present with bleeding, large areas of ecchymosis on the extremities, and perifollicular hemorrhage on the lower extremities and trunk. Less fre-**

quently, patients present with hemarthroses or poor wound healing.

Scurvy is seen in patients who are malnourished (e.g., alcoholics) and chronically ill or debilitated patients. Although vitamin C levels can be measured, it is usually unnecessary since empiric treatment with vitamin C is inexpensive.

■ Fat-Soluble Vitamin Deficiencies (Vitamins A, D, E, and K)

- **Vitamin A deficiency** Vitamin A is important in epithelial cell growth and differentiation. Therefore this nutrient is particularly important in wound healing. The antioxidant role of vitamin A in cancer prevention is assuming a more prominent role.

  Worldwide, deficiency of vitamin A is not uncommon; it is frequently associated with blindness. In the United States, deficiency is most often secondary to fat malabsorption syndromes, laxative abuse, or alcoholism.

  The earliest symptom of deficiency is night blindness. Xerosis (i.e., dry skin) and corneal ulceration are other common manifestations. Although serum levels may be measured, it is important to begin empiric therapy in patients with suspected deficiency.

- **Vitamin D deficiency** Vitamin D deficiency is important in bone mineralization. In children, severe deficiency results in rickets, whereas osteomalacia occurs in adults. **The symptoms of vitamin D deficiency are those of hypocalcemia.** For a more detailed review of vitamin D, see the topic, "Disorders of Calcium Balance," in Chapter 11.

- **Vitamin E deficiency** Vitamin E, like vitamins A and C, functions as an antioxidant. The most common reason for deficiency is malabsorption. **The manifestations of deficiency include gait disturbances, ophthalmoplegia, hyporeflexia or areflexia, and decreased vibration and position sense.** The treatment is supplementation.

- **Vitamin K deficiency** Vitamin K is used in the γ-carboxylation of clotting factors II, VII, IX, X, protein C, and protein S. Predisposing conditions to vitamin K deficiency are fat malabsorption (e.g., secondary to biliary disease and celiac sprue), long-term antibiotic treatment (which alters gut flora, a source of vitamin K production), and warfarin, which competitively inhibits the effect of vitamin K on the carboxylation of the clotting factors just mentioned.

  Signs and symptoms at presentation are bleeding from any site, especially after venipuncture. The diagnosis of vitamin K deficiency is supported by a prolonged prothrombin time and a low vitamin K level. The treatment includes supplementation with parenteral vitamin K.

Pneumocystis Carinii Pneumonia

■ Basic Science ***Pneumocystis carinii*** **is an organism that** asymptomatically infects most individuals but **commonly becomes symptomatic in immunosuppressed patients,** most commonly those with AIDS. Unless patients with AIDS receive prophylaxis to prevent pneumocystosis, up to 80% will become infected; in this population it is a major source of mortality. Infection of extrapulmonary sites occurs but is much less common. **A CD4 count of less than 200/µl is a significant risk factor.**

- History and Physical Examination **Early symptoms** may be mild but usually include **a nonproductive cough and,** depending on the duration of illness, **dyspnea.** AIDS patients should be questioned about the use of prophylactic measures and recent CD4 counts; **fatigue, fever, and weight loss may all accompany the infection.**

Vital signs may reveal fever and tachypnea. Patients with severe systemic illness often appear wasted, with intercostal and supraclavicular retractions. **The lung examination may reveal bibasilar crackles or be unremarkable.**

- Laboratory and Diagnostic Studies Routine labs, including a complete blood count and electrolytes, should be obtained. **An elevated lactate dehydrogenase (LDH) is characteristic.** Often the total white cell count is low, with a markedly decreased CD4 count.

Culture and antigen tests of pneumocystis are unreliable. **Usually, the organism is identified on silver or Giemsa's stain of an induced sputum sample.**

Examination of the chest x-ray typically reveals a diffuse interstitial pattern, though a small minority of patients has an unremarkable chest film. Other findings less often seen include a consolidation, nodules, cavitation, and cystic changes.

The arterial blood gas may reveal significant hypoxia (and decreased CO_2), despite supplemental oxygen.

- Evaluating the Severity of Problems Patients with this infection may progress to respiratory distress despite aggressive treatment. **Intubation and mechanical ventilation may be necessary if hypoxia or the work of breathing become overwhelming.**

- Formulating the Diagnosis **The combination of fever, dyspnea, a nonproductive cough, and a bilateral interstitial infiltrate suggests *Pneumocystis carinii* pneumonia (PCP). The diagnosis is confirmed by finding the organism within lung tissue.**

- Managing the Patient In patients at risk for PCP, **prophylaxis should be administered.** When the CD4 count has fallen below 200/µl, trimethoprim/sulfamethoxazole should be offered; however, dapsone and inhaled pentamidine can be used in sulfa-allergic patients.

The two most commonly administered drugs used to treat PCP are trimethoprim/sulfamethoxazole and pentamidine. Either drug should be administered for 14 to 21 days to treat a confirmed infection. Treatment-related side effects are common and occur in up to 50% of patients. Atovaquone is an alternate drug used for patients who cannot tolerate either trimethoprim/sulfamethoxazole or pentamidine. **Besides antibiotics, supplemental oxygen is often required** to maintain adequate oxygenation. **The role of steroids in patients with PCP is controversial, but some studies have demonstrated benefit** of steroids when administered to patients with moderate to severe disease.

Pneumonia

- **Basic Science** **Pneumonia is an infection of the lung parenchyma. Pneumonia can be caused by bacteria, viruses, parasites, and fungi.** The environment in which the patient acquired the infection and the patient's health may predict the infectious agent. For example, **community-acquired pneumonia in a healthy patient is often caused by *Streptococcus pneumoniae, Haemophilus influenzae, Chlamydia,* or *Mycoplasma.* Up to one third of the cases can be caused by viruses. Patients in the hospital are at risk of developing a pneumonia from enteric gram-negative bacilli, *Pseudomonas aeruginosa,* and *Staphylococcus aureus.***

  Additionally, the time of the year and the area of the country may suggest the etiologic agent. **In the wintertime, bacterial pneumonia may occur secondary to influenza A or B infection. These postinfluenza pneumonias are most often caused by *S. pneumoniae, S. aureus,* and *H. influenzae.***

  Finally, **the age of the patient also affects which infectious agent is likely. Infants less than 6 months old are susceptible to respiratory syncytial virus and *Chlamydia.* Young adults tend to be infected by *Mycoplasma* and *Chlamydia.* Elderly patients with chronic lung disease are more likely to be infected with *H. influenzae, Legionella pneumophila,* and *Moraxella catarrhalis.***

  Mortality from pneumonia can reach up to 30%, especially in the elderly. Patients with chronic diseases such as renal failure, emphysema, cardiac disease, malnutrition, and alcoholism are most at risk, as are diabetics and the immunocompromised. **Patients who lack a spleen are at risk for infection from encapsulated organisms (primarily *S. pneumoniae,* but *H. influenzae* as well).**

- **History and Physical Examination** The presenting symptoms of pneumonia vary with the infecting agent. **The "typical" pneumonia has an acute onset with fever, chills, productive cough, and pleuritic chest pain. The organisms involved are usually *S. pneumoniae, H. influenzae, Klebsiella pneumoniae, S. aureus,* anaerobes, and gram-negative bacilli.**

  The "atypical" pneumonias, usually due to viruses, *Mycoplasma pneumoniae,* and *L. pneumophila* are insidious in onset. Lower fevers, nonproductive cough, headaches, myalgias and the absence of organisms on Gram's stain of sputum characterize these infections.

  Viral pneumonia is often due to influenza A and B, but can also be due to adenovirus, parainfluenza, and respiratory syncytial virus. Bacterial pneumonia with *Legionella* is acquired by the aerosol route and often has extrapulmonary symptoms: confusion, ataxia, hyponatremia, diarrhea, and increased lactate dehydrogenase. Aspiration pneumonia is usually seen in patients with impaired mental status or tracheoesophageal fistula and is often due to mouth organisms (i.e., anaerobes).

- Laboratory and Diagnostic Studies **All patients with pneumonia should submit sputum for Gram's stain and culture and sensitivity.** Patients with suspected tuberculosis, pneumocystis (both covered elsewhere), or other less common agents (e.g., *Histoplasma*) should have appropriate special stains made. **To be interpretable, the sputum Gram's stain should have greater than 25 polymorphonuclear leukocytes and less than 10 epithelial cells per low power field.**

  Blood cultures may be obtained on febrile patients in whom the diagnosis is suspect or bacteremia is suspected. **Serologic tests are not usually helpful initially but are done when *Legionella* or viruses are suspected. Cold agglutinins can be checked for *Mycoplasma* infection.**

  A chest x-ray is commonly obtained. In patients with the "typical" pneumonias, a lobar infiltrate is usually present. Patients with an "atypical" pneumonia more commonly present with a diffuse interstitial pattern.

- Formulating the Diagnosis **Patients who present with a productive cough, fever, chills, and pleuritic chest pain may be assumed to have pneumonia. A chest x-ray will confirm the diagnosis.**

- Evaluating the Severity of Problems Patients with underlying pulmonary or cardiac disorders may become dyspneic with even a mild pneumonia. **Arterial blood gas measurement may identify those patients in need of supplemental oxygen or more aggressive support during treatment.**

- Managing the Patient **The keys to determine appropriate therapy are clinical suspicion, patient characteristics (e.g., smoker, immunocompromised, diabetic), and the sputum Gram's stain. Most authorities recommend starting with a broad-spectrum antibiotic, such as a third-generation cephalosporin or ampicillin/sulbactam. The antibiotic can be changed once the sputum culture and sensitivity are reported. If *Legionella* is suspected, high-dose erythromycin should be started. *Mycoplasma* can be treated with lower-dose erythromycin. The elderly should be treated with erythromycin and a second- or third-generation cephalosporin.**

Rabies

- Basic Science **Rabies is a viral encephalitis caused by infection with a rhabdovirus usually following an animal bite. The pathognomonic lesion on brain histology is the Negri body,** although this may be absent in 20% of cases.

  The virus travels to the brain via nerve axons and then migrates along afferent nerves to the salivary glands. The incubation period is dependent in part on the distance the virus must travel to reach the brain and may vary from 10 days to several years but averages 3 to 7 weeks. **Animal vectors vary geographically in the United States; skunks in the Midwest, Southwest, and California; raccoons in the East; coyotes in Texas; and foxes in the**

Southwest, New England, and Alaska. Dogs and cats may be infected in developing countries, whereas rodents and rabbits are very unlikely to be infected. Scratches from rabid animals when compared to bites are about 50 times less likely to cause rabies.

- History and Physical Examination **With the development of the disease, pain is followed by paresthesias at the site of the bite; hydrophobia results as attempts at drinking cause extremely painful laryngeal spasm. The patient is restless and behaves in an unusual manner. Extreme excitability, convulsions, and paralysis occur. The production of thick tenacious saliva is characteristic.**

- Laboratory and Diagnostic Studies **To decide whether to provide prophylaxis after a bite, the animal should be kept under observation for 7 to 10 days. Alternatively, the brains of sick or dead animals should be examined for evidence of rabies. If the animal cannot be examined (e.g., skunks, bats, coyotes, foxes, and raccoons), they should be assumed to be rabid.**

  In unimmunized individuals, a fourfold rise in the neutralizing antibody titer is considered diagnostic. Examination of the CSF will reveal antibody titers between 1:200 and 1:160,000, whereas postexposure prophylaxis will reveal titers of less than 1:64, if at all.

- Formulating the Diagnosis The involvement of the brainstem distinguishes rabies from other viral encephalitides, though in practice this distinction is difficult. **The history of exposure is the most important clue.** It is difficult to distinguish rabies from Guillain-Barré syndrome, poliomyelitis, and rabies vaccine–induced allergic encephalomyelitis.

- Evaluating the Severity of Problems Untreated rabies infection uniformly results in death.

- Managing the Patient As the disease is almost always fatal, prevention is the most important consideration. In particular, immunization of household cats, dogs, and persons with a significant exposure risk should receive the vaccine.

  The wound should be thoroughly cleaned and repeatedly rinsed with soap and water. If rabies immune globulin or antiserum is used, half is infiltrated locally and the remainder is given intramuscularly (IM).

  Human rabies immune globulin (20 IU/kg) is the optimal passive immunization dose, with up to 50% infiltrated locally and the rest IM. Equine rabies antiserum (40 IU/kg) is an alternative. **An inactivated human diploid cell rabies vaccine should also be given on days 0, 3, 7, 14, and 28 after exposure.** The vaccine and immune globulin should be administered at different sites in separate syringes. Local reactions to the vaccine occur in up to 25% with mild systemic reactions (e.g., headaches, myalgias, and nausea) in 20% of recipients; true allergic reactions are rare. For those with previous preexposure or postexposure vaccination, rabies immune globulin should not be given, but rather vaccine 1 ml IM on days 0 and 3.

  Once symptoms have appeared, death generally follows within 7 days, usually from respiratory failure.

Statistics in Medicine

Test Characteristics and Use

- **Screening tests should be**

  Inexpensive

  Sensitive and specific

  Painless

  Able to detect disease before symptoms occur, and early enough to affect outcome

- **Prevalence/prior probability: all patients with the disease divided by all patients tested**

- **Sensitivity and specificity** (does not depend on prevalence but does depend on the spectrum of patients in whom the test is being evaluated)

  Sensitivity = percentage of patients with the disease in whom the test is positive (Table 15.1):

$$\frac{a}{a + c}$$

  Specificity = percentage of patients without the disease in whom the test is negative:

$$\frac{d}{b + d}$$

- **False positive and false negative**

  False positive = percentage of patients with a positive test who do not have the disease:

$$\frac{b}{b + d}$$

  False negative = percentage of patients with a negative test result who actually have the disease (see Table 15.1):

$$\frac{c}{a + c}$$

- **Predictive value**

  Positive predictive value (PPV) = percentage of patients with a positive test result who actually have the disease:

$$\frac{a}{a + b}$$

  Negative predictive value = percentage of patients with a negative test result who really do not have the disease:

$$\frac{d}{c + d}$$

Table 15.1 *2 x 2 Table*

	Disease Present	Disease Absent	
Test positive	*a*	*b*	*a* + *b*
Test negative	*c*	*d*	*c* + *d*
	a + *c*	*b* + *d*	*a* + *b* + *c* + *d*

Example 1
Sensitivity = 90%
Specificity = 95%

Interpretation of Test Results When Pretest Probability Is 10%

	Disease Present	Disease Absent	
Test positive	9,000	4,500	13,500
Test negative	1,000	85,500	86,500
	10,000	90,000	100,000

Positive predictive value = 9,000/13,500 = 67%
Negative predictive value = 85,500/86,500 = 99%
False negative rate = 1,000/10,000 = 10%
False positive rate = 4,500/90,000 = 5%

Example 2
Sensitivity = 90%
Specificity = 95%

Interpretation of Test Results When Pretest Probability Is 50%

	Disease Present	Disease Absent	
Test positive	45,000	2,500	47,500
Test negative	5,000	47,500	52,500
	50,000	50,000	100,000

Positive predictive value = 45,000/47,500 = 95%
Negative predictive value = 47,500/52,500 = 99%
False negative rate = 5,000/50,000 = 10%
False positive rate = 2,500/50,000 = 5%

— ***These values are dependent on the prevalence of the disease in the tested population. A test with a particular sensitivity and specificity has different PPV when used in populations that have different prevalences.***

Population Description

- **Mean: the arithmetic average of a series of numbers (total of numbers/number in series)**
- **Median: the number that 50% of numbers are above and 50% are below**

- **Mode: the most frequently occurring number in a series**
- **For example, here is a series of hospital salaries:**
  - 1 Chairman: $45,000
  - 1 Full professor: $15,000
  - 2 Assistant professors: $10,000 each
  - 1 Fellow: $5,700 ← **Mean** (arithmetical average)
  - 3 Chiefs: $5,000 each
  - 4 Associate residents: $3,700 each
  - 1 Assistant resident: $3,000 ← **Median** (12 workers above and 12 below)
  - 12 Interns: $2,000 each ← **Mode** (occurs most frequently)

■ Basic Comparisons

- **Student's t-test:** used to calculate the probability with which two sets of measurements would have come from normal distributions with the same mean, assuming that the standard distributions are identical
- **Chi square:** used to test for dependence of outcome on treatment when the outcomes to be measured are divided into discrete outcomes or variables. This is a two-tailed hypothesis in that it checks for both negative and positive correlation; therefore data must be inspected carefully.
- **P value:** the probability of obtaining a given result (difference between two samples); usually set at $p < .05$, meaning that there is a less than 5% probability that the observed difference occurred by chance alone
- **Correlation coefficient:** a number that indicates the degree of correlation between two sets of data or random variables. It is usually denoted as r on a graph, and the range is usually −1 to 1; the closer to zero, the less correlation. The closer to 1 or −1 indicates a positive or negative correlation, respectively.

■ Error Types and Power

- **Type I** occurs when one rejects the null hypothesis when it is in fact true. It is also called an α error or false positive.
- **Type II** occurs when one fails to reject the null hypothesis when it is in fact false. This can lead to the error of failure to detect a true treatment effect. Also called a β error or false negative.
- **Power (1 − β error):** the probability of rejecting the null hypothesis when the alternative is true (i.e., when it should be rejected)

TETANUS

■ Basic Science **Tetanus is caused by tetanospasmin, a neurotoxin elaborated by *Clostridium tetani*, a motile anaerobic gram-positive sporulating bacteria. The disease is characterized by increased muscle tone and spasm.** *C. tetani* is ubiquitous in soil and is found in animal feces and occasionally in human feces. Although the bacteria is easily inactivated and is sensitive to several antibiotics, the spores are resistant to some disinfectants and can survive for long periods in the environment and up to 20 minutes in boiling water. The toxin mediates its effect by preventing the neuronal release of the inhibitory neurotransmitter, gamma amino butyric acid (GABA).

In the United States most cases occur in unvaccinated individuals. At-risk groups include the elderly, newborns, migrant workers, and IV drug users. The incubation period is between 5 days and 15 weeks, averaging 8 to 12 days.

■ History and Physical Examination **Potential sufferers may present with a history of an acute injury, such as a puncture wound, laceration, or abrasion, possibly acquired during farming or gardening;** the injury may have been quite trivial. Tetanus may also follow burns, frostbite, ear infections, surgery, abortion, childbirth, or drug abuse (i.e., "skin popping") and may be associated with chronic conditions such as skin ulcers, abscesses, or gangrene.

The first sign may be pain and swelling at the site of inoculation and possibly spasticity of nearby muscles. More commonly the patient presents with stiffness of the jaw and neck, dysphagia, and irritability. This is later followed by hyperreflexia; spasm of the jaw (i.e., trismus, or lockjaw) and facial muscles (i.e., risus sardonicus); and rigidity and spasms of the abdomen, neck, and back. Painful tonic convulsions can be precipitated by minor stimuli, and acute asphyxia may result from spasms in the respiratory muscles and glottis. The patient is normally awake and alert throughout the illness, and the temperature is normal or only slightly elevated.

■ Laboratory and Diagnostic Studies **Wound cultures should be done; however, a positive culture for *C. tetani* provides confirmation only. The organism may be present in the absence of tetanus and is frequently not cultured even during active disease.** The WBC may be elevated; however, the CSF is normal. **An electromyogram may show continuous discharge of motor units and shortening or absence of the normal silent period after the action potential.**

■ Formulating the Diagnosis **The diagnosis is made clinically with historical queues. The differential diagnosis includes various acute central nervous system infections such as meningitis, encephalitis, or rabies, as well as acute intraabdominal processes that produce a rigid abdomen. Trismus may occur with acute local processes such an alveolar abscess, phenothiazine or metoclopramide use, strychnine poisoning, and hypocalcemia.**

A marked increase in the tone of central musculature with superimposed generalized spasms involving the neck and face (sparing the hands and feet) strongly suggests tetanus.

■ Evaluating the Severity of Problems The most common scenario is the assessment of risk after a potential inoculation. The decision of how to provide secondary prophylaxis is discussed later.

With the development of disease, the prognosis is worst in the elderly, neonates, and those with a short incubation period, a short period of onset, or rapid progression.

- **Managing the Patient** Management should be conducted in an intensive care unit with minimal environmental stimulation. **The goals are to eliminate the source of the toxin, neutralize unbound toxin, prevent muscle spasm, and provide supportive measures (particularly respiratory) until recovery occurs.**

Initially, it is important to adequately clean and debride the wound. Intravenous penicillin (10 to 12 million units per day for 10 days) is given to destroy the vegetative cells. For patients with penicillin allergy, clindamycin, metronidazole, or erythromycin may be used. **Tetanus immune globulin (TIG), 3000 to 6000 units, is given to neutralize unbound toxin and should be given before manipulating the wound. Vaccination (with Td 0.5 cc intramuscularly) should also be initiated, as the small amounts of toxin-producing disease are not generally enough to induce immunity.**

Muscle spasms may be controlled with diazepam or, in severe cases, vecuronium (a neuromuscular blocking agent). Mechanical ventilation will be necessary if paralytic agents are used. When trismus, dysphagia, or laryngospasm occur, intubation or tracheostomy may be required to protect the airway and prevent aspiration. **Labetalol may be used to control autonomic dysregulation (i.e., sympathetic overactivity).** With its shorter half-life, esmolol may be safer if bradycardia is likely.

The course of the illness generally extends over 4 to 6 weeks. Increased tone and minor spasms can last for months, but recovery is usually complete. The overall mortality in the United States has been reported as 2%.

The disease is completely preventable by active immunization. Active immunization for children is accomplished by using the DTP vaccine at 2, 4, 6, and 18 months, and again at 4 to 6 years of age with boosters. For previously unimmunized adults, tetanus toxoid (Td 0.5 cc IM) is administered in two doses 4 to 6 weeks apart, followed by a third dose 6 to 12 months later. Booster doses should then be given every 10 years or at the time of a major injury if it occurs more than 5 years after a dose.

Wound prophylaxis consists of combined adult absorbed tetanus and diphtheria toxoid (Td) and should be considered in nonimmunized, incompletely immunized (less than three doses of vaccine) **greater than 10 years** (>5 years for a major wound) **after a booster, or in those with uncertain status** whenever a wound may contain contaminated material or devitalized tissue. **Tetanus immune globulin (TIG 250 units IM) is indicated for wounds (other than clean or minor wounds) if the patient is not fully immunized. Active immunization with tetanus toxoid should be started concurrently, using separate syringes and sites.**

Multiple Choice Review Questions

1. A 79-year-old man comes to the emergency room with a myocardial infarction. He has had two previous myocardial infarctions and has made himself DNR (do not resuscitate) with the help of his longtime primary care physician. The cardiologist that admits him to the CCU wants to take him for percutaneous transluminal coronary angioplasty (PTCA). There is a chance that the patient will arrest or have significant cardiac arrhythmias while undergoing the procedure. The cardiologist reverses the DNR order (i.e., he makes the patient a full code) for the procedure without talking to the patient. When asked about this by the patient's nurse, the cardiologist says that the PTCA is a temporary intervention and often resuscitation efforts are necessary even in younger, healthier patients. The nurse calls the ethics committee to help, as the patient has clearly stated to many people his wishes to be DNR. The most appropriate viewpoint is:
   - **a.** The cardiologist is correct; that DNR orders can be reversed by physicians who "know better" than the patient.
   - **b.** The cardiologist is correct; the DNR can be reversed without the patient's approval if it is only for one procedure
   - **c.** The cardiologist is incorrect; a 79-year-old should not have a PTCA anyway because he has already outlived his life expectancy and should be DNR.
   - **d.** The cardiologist is incorrect; the patient's wishes must be respected at all times, providing that the patient has a clear understanding of the risks and benefits of treatment and DNR orders.
   - **e.** The issue needs further investigation as to the cost-effectiveness of PTCA in elderly patients.

2. A patient on chronic hemodialysis has missed the last three treatments for unclear reasons and comes to the emergency room for vague complaints of fatigue and malaise. His BUN is 110 and potassium is 6.9. What is the best management of this patient?
   - **a.** Recheck the potassium and schedule hemodialysis because that is the best method to remove the extra potassium. No other intervention is necessary.
   - **b.** Give 100 mg of furosemide IV and, if no urine is made in ten minutes, give 200 mg of furosemide IV. Schedule hemodialysis if no urine is made after an hour.
   - **c.** Give intravenous D_{50} and insulin, and admit the patient to the medical floor for further management.
   - **d.** Give the patient intravenous D_{50} and insulin, followed by intravenous calcium to stabilize the cardiac membranes; call in the dialysis nurse stat and check an ECG.
   - **e.** Call the hemodialysis nurse and place the patient on telemetry.

3. What are the likely electrocardiographic changes associated with hyperkalemia in the aforementioned patient?
   - **a.** T wave inversion in lateral leads
   - **b.** Atrial flutter
   - **c.** Initial shortening of the QRS interval and then widening as potassium levels increase
   - **d.** Left bundle branch block
   - **e.** ST elevations in inferior leads

4. A 30-year-old male physician from Baltimore comes to your office with complaints of arthralgias and general fatigue from suspected Lyme disease. He does not recall a skin rash or fever symptoms. He is an avid golfer and often has to get his golf ball out of a wooded area. He gave himself a 4-week trial of amoxicillin for presumed Lyme disease but has had no improvement of symptoms. He did not have any blood tests done before now. What is your next step?
   - **a.** Repeat the 4 weeks of antibiotics using doxycycline.
   - **b.** Check the ELISA, because it will be positive if Lyme disease is present for this duration in this patient.
   - **c.** Admit him to a hospital for 4 weeks of IV penicillin.

d. Tell him to avoid wooded areas.
e. Initiate therapy with tetracycline for 4 weeks.

5. You read an article in a recent journal where the authors stated that they found a difference between groups A and B where the p value was <0.01. What does that mean?

a. There is no real difference between groups A and B; the observed difference could have been due to chance alone.
b. There is a <1% chance that the difference observed between groups A and B was due to chance alone.
c. There is a <0.01% chance that the difference observed between groups A and B was due to chance alone.
d. The difference between the groups is likely to be accounted for by nonrandom sampling.
e. The difference between the groups was due to chance 99% of the time.

Chapter 16

Emergency Medicine

Advanced Cardiac Life Support

■ Basic Science **Advanced cardiac life support (ACLS) is a collection of guidelines for resuscitation of respiratory failure,** hemodynamic instability **(shock), dysrhythmias,** and **cardiac arrest.**

- **Respiratory distress** is **impaired breathing due to inadequate airway, ventilation, or oxygenation.**
- **Respiratory arrest (apnea)** is the **cessation of spontaneous breathing,** causing tissue anoxia and acidosis. Cardiac arrest follows within minutes.
- **Shock** is a syndrome caused by decreased perfusion and oxygen delivery to tissues. Causes include hypoxia, hypovolemia, vasodilation, cardiac pump failure or obstruction, and dysrhythmias.
- **Dysrhythmias** are **abnormal cardiac rhythms that,** if sustained, may **impair cardiac output.** See the topic, "Arrhythmias and Conduction Disorders," in Chapter 5.
- **Cardiac arrest** is **cessation of organized cardiac contractions.** Blood flow is halted, and anoxic damage to brain and heart occurs within minutes.
- **Sudden cardiac death** is cardiac arrest from a dysrhythmia, usually **ventricular fibrillation (V-fib).**

■ History and Physical Examination

- **Shock—mental status changes, syncope, lightheadedness, dyspnea, skin pallor** or flushing, **diaphoresis** and **oliguria**
- Hypotension—a late sign
- **Dysrhythmias—altered pulse** rate, regularity, and strength
- **Respiratory distress**—appears as **dyspnea, tachypnea,** hypoventilation, stridor, wheezing, cyanosis, poor chest excursion, or accessory chest muscle use
- **Respiratory arrest**—presents with unresponsiveness and **apnea**
- **Cardiac arrest**—the triad of **unresponsiveness, apnea,** and **no** palpable **carotid pulse**

■ Laboratory and Diagnostic Studies **The following should be obtained:**

- **Electrocardiogram (ECG)**—crucial to identify dysrhythmias; also helps to identify myocardial ischemia, infarction, or other cardiac diseases
- **Pulse oximetry**—monitors oxygen saturation but it is not accurate during hypotension or arrest

- **Arterial blood gases**—detect acid-base disorders, hypoxemia, or abnormal CO_2 exchange
- **Urine output**—by foley catheter; reflects tissue perfusion

■ Formulating the Diagnosis Clinical presentation identifies shock or arrest. However, the exact etiology is often obscure, since features of different shock states may overlap simultaneously.

• **Hypotension** on examination **is due to shock until proven otherwise.**

• Pulse rate and regularity, or absence of pulse, are assigned a rhythm diagnosis by cardiac monitor.

• Chest pain from myocardial infarction is usually associated with ECG abnormalities. Chest pain with **cardiogenic shock is associated with pulmonary edema, jugular venous distention, murmurs, or rubs.**

• Pump obstruction may cause hypotension, absent breath sounds and tracheal deviation (tension pneumothorax), or muffled heart sounds, distended neck veins, and electrical alternans on ECG (pericardial tamponade).

■ Evaluating the Severity of Problems

• **Shock** and **cardiopulmonary arrest are obvious emergencies requiring immediate resuscitation**

• Respiratory distress with stridor, agitation, exhaustion, poor chest movement, or cyanosis (despite high-flow oxygen) indicates severe hypoxia and/or impending respiratory arrest.

• Supine hypotension, with systolic blood pressure (BP) <90 mm Hg in adults, is a late sign of severe shock.

• An orthostatic BP drop >20 mm Hg corresponds with significant blood volume loss or vasodilation.

• Cardiogenic shock, with pulmonary edema and hypotension signifies infarction of ≥40% of the left ventricular myocardium and carries an in-hospital mortality >80%.

■ Managing the Patient **All *conscious* patients with signs of shock, acute mental status changes, or abnormal vital signs immediately require the following:**

- **Assessment of airway and ventilation**
- **High-flow oxygen**
- **IV access**
- **Cardiac monitoring**
- **ECG**

If the patient is unconscious, assess responsiveness (to voice or to pain), breathing, and carotid pulse.

- **Respiratory distress** High-flow oxygen may be attempted, but **endotracheal intubation** is **immediately indicated** for severe, refractory stridor; cyanosis; hypoxia; hypoventilation; chest wall exhaustion; mental status depression with vomiting; or development of apnea.

  In the unresponsive patient, maintain an open airway. Hypoventilation requires bag-valve-mask ventilation with oxygen. If a blood glucose chemstrip is <80 mg/dl, give 50% dextrose (50 cc D_{50} IV). Also give naloxone (0.2-0.4 mg IV push) and thiamine (100 mg IV push) if narcotic or alcohol use is suspected. **Persistent coma, apnea, hypoxia, or hypoventilation requires immediate endotracheal intubation and assisted ventilation.**

- **Unresponsive, pulse absent (cardiac arrest)**

— ***V-fib or ventricular tachycardia (V-tach)*** **Immediately defibrillate with 200, 300, and 360 joules,** as needed. If V-fib or V-tach persist, **cardiopulmonary resuscitation (CPR), endotracheal intubation,** oxygen and IV access are initiated.

The drug of choice for any cardiac arrest is epinephrine (1 mg IV push **every 3 minutes).**

Refractory V-fib or V-tach is further treated with:

- Lidocaine (1.5 mg/kg IV push, every 3 minutes, up to 3 mg/kg)
- **Bretylium** (5 mg/kg IV push, then 10 mg/kg every 5 minutes, up to 35 mg/kg)
- **Magnesium sulfate** (2 g IV over 1-2 minutes).

These drugs decrease ventricular automaticity by different mechanisms.

After each drug dose, CPR is continued for 30 seconds, followed by defibrillation with 360 joules. CPR, ventilation, and oxygen are maintained between shocks. Rapidly alternating drugs with shocks is critical to successful resuscitation.

— ***Asystole or pulseless electrical activity (PEA)*** **For asystole or pulseless organized rhythm (electromechanical dissociation), electrical therapy is not effective.**

CPR, intubation, ventilation, oxygen, and IV saline should be started.

The drugs of choice are epinephrine (1 mg, 1:10,000 solution, IV push every 3 minutes) and **atropine** (1 mg IV push every 3 minutes, up to 3 mg in adults), to offset vagal inhibition of the heart rate.

— ***Reversible causes of asystole and PEA***

- **Hypoxemia is treated with oxygen,** hypovolemia with IV saline, and acidosis with hyperventilation and CPR.
- **Hyperkalemia is treated with 10% calcium chloride** (1 g IV push) and **sodium bicarbonate** (1 mEq/kg IV push).
- Cardiotoxic drugs require specific treatments (e.g., biocarbonate for tricyclic antidepressants).
- **Hypothermia** (rectal temperature <30°C) requires rewarming of the core circulation **(warmed IV fluid, oxygen, and,** if necessary, **peritoneal lavage).**
- **Absent breath sounds indicate misplaced endotracheal tube or pneumothorax.**

- Suspected **tension pneumothorax requires immediate needle thoracostomy.**
- Suspected **pericardial tamponade indicates pericardiocentesis.**

- **Tachycardia with a pulse**
  - ***Hypotension or refractory tachyarrhythmias*** (other than sinus tachycardia) are treated with urgent **synchronized cardioversion.**

    Oxygen and IV sedation are given, and the defibrillator is "synchronized" to deliver shock during the QRS complex. This avoids a shock during the T wave (the "relative" refractory period), which could provoke ventricular fibrillation. The procedure is otherwise the same as defibrillation, starting at 100 joules and increased as needed.
  - ***Hemodynamically stable tachyarrhythmias***
    - **Sinus tachycardia** is managed by treating the cause (pain, sympathomimetic drugs, etc); **treated with beta-blockers during acute myocardial infarction.**
    - **Paroxysmal supraventricular tachycardia (PSVT)** is treated with **adenosine** (6 mg rapid IV bolus, followed in 1-2 minutes with a 12-mg rapid IV bolus if unsuccessful).
    - **Use verapamil if adenosine does not extinguish the PSVT** (5 mg IV over 3 minutes, and repeated at 10 mg in 15 minutes, if necessary). However, it may cause vasodilation, hypotension, and acceleration of a wide-QRS, accessory pathway SVT (as in the Wolff-Parkinson-White syndrome).
    - **Atrial flutter** and **fibrillation with a rapid ventricular response may be treated with IV diltiazem, beta-blockers, or digoxin.**
    - **Wide-QRS tachycardias** (QRS ≥0.12 seconds) may be ventricular or supraventricular with aberrant ventricular conduction. **Assume ventricular tachycardia,** since this is more unstable. Treatment includes **lidocaine** (1.5 mg/kg IV push, and 0.75 mg/kg every 5 minutes, up to 3 mg/kg and **procainamide** (20 mg/minute IV infusion, up to 17 mg/kg). If either drug leads to conversion to a normal rhythm, a 2 mg/minute IV infusion is started.
    - **Torsades de pointes** is sinusoidal oscillation of the QRS amplitude and a rotational shift in QRS axis. It is **treated with defibrillation, magnesium sulfate** (2 g IV push), and an overdrive pacemaker; or isoproterenol (2-10 µg/min IV). **Low serum potassium, magnesium and calcium aggravate this rhythm** and should be repleted. Drugs prolonging the QT interval are contraindicated (e.g., quinidine, procainamide, tricyclic antidepressants).

- **Bradycardia with pulse** Symptomatic bradycardia is treated with **oxygen** and **atropine** (0.5-1.0 mg IV push, every 5 minutes, up to 3 mg in adults).

  Refractory bradycardia, or **high-grade heart blocks** (2° Type II and complete AV blocks), are not responsive to atropine.

These require oxygen, IV sedation, and **transcutaneous (external) pacemaker,** followed by central IV access and insertion of a **transvenous pacemaker.**

- **Hypotension (systolic BP <90 mm Hg)**
  - Dysrhythmias are rapidly treated, as described earlier.
  - **Hypovolemia** is treated with IV boluses of **isotonic saline.** With hemorrhage, packed red cell transfusions are also indicated.
  - **If** IV saline fails to raise the BP, **dopamine** (5-20 μg/kg/minute IV) or **epinephrine** (2-10 μg/minute IV) is used.
  - For **cardiogenic shock, dobutamine** (5-20 μg/kg/minute IV) increases contractility and causes a reflex drop in afterload and preload, without aggravating tachycardia or myocardial ischemia.
- **Pregnancy**
  - **Shock and cardiac arrest in pregnancy are treated with standard ACLS.**

  The best way to save an unborn baby is to rapidly stabilize the mother. For the third-trimester mother in shock or arrest, **the right pelvis must be propped up to displace the uterus off the vena cava.**

  Immediate emergent cesarean section may be indicated.
- **Termination of efforts Terminate efforts if the patient is found to have a "do not resuscitate" status or if all indicated ACLS measures have failed after 10 to 15 minutes.** Prognosis is dismal beyond this time, unless unusual circumstances (e.g., hypothermia) make more prolonged efforts potentially successful.

Advanced Trauma Life Support

■ Basic Science Emergency medicine physicians and surgeons manage trauma utilizing a standard protocol called **advanced trauma life support (ATLS).**

ATLS prioritizes treatment according to the potential lethality of injury. The physician addresses the patient according to the alphabetic algorithm ABCDE:

- **Airway (with cervical spine [c-spine] control)**
- **Breathing**
- **Circulation**
- **Disability (or neurologic status)**
- **Exposure** (undress) with temperature control
- **Secondary survey**

■ History and Physical Examination

- **Airway** The airway is **assessed by**
  - Looking for agitation or obtundation (signs of hypercarbia), checking for peripheral cyanosis (a sign of hypoxemia), or retraction

- Listening for **abnormal sounds such as stridor, snoring, or hoarseness**
- Feeling for air movement

- **C-spine** The **patient's neck should be immobilized. C-spine injury is suggested by c-spine pain or deformity; definitive c-spine assessment is by c-spine radiographs with clinical correlation.**
- **Breathing** Ventilation is **assessed by**
  - Visualizing the chest wall for motion abnormalities
  - **Listening to the lung fields to ensure air exchange**
  - Percussion (to assess air or blood in the lung field)
- **Circulation** Hemodynamics are **assessed by**
  - Level of consciousness (decreased blood volume leads to decreased consciousness)
  - **Pulse (thready, rapid pulse is a sign of hypovolemia)**
  - Skin color (gray or white skin signifies hypovolemia)
- **Disability or neurologic status Assessment is by** mini-neurologic examination:
  - **Level of consciousness**
  - Pupillary reactions
  - Lateralizing extremity weakness
  - Assessment of the **Glasgow coma scale**
- **Exposure (undress) with temperature control The patient should be rolled** in order for the physician to visualize their back and **to check for injuries** and localized tenderness.
- **Secondary survey** Once the primary ABCDE survey is finished and stabilizations initiated the physician begins a head-to-toe physical examination of the patient. **The following is a list of the common injuries found on secondary physical examination:**
  - *Head:* palpation will reveal depressed skull fractures; subcutaneous emphysema or pain may indicate facial fractures.
  - *Ears:* blood behind the tympanic membrane indicates basilar skull fracture.
  - *Mouth:* broken teeth
  - ***Abdomen:* tenderness over the liver indicates possible liver laceration. Tenderness over the spleen indicates possible splenic laceration.**
  - ***Pelvis:* movement with pushing on the pelvis indicates pelvis fracture.**
  - ***Rectal examination:*** inability to feel the prostate indicates possible urethral damage. **Blood in the rectal vault indicates damage to the intestine.**
  - ***Extremities:*** pain with palpation indicates fracture. **Pulseless extremities should be reduced immediately.**

■ Laboratory and Diagnostic Studies

- Blood should be drawn for type and cross-match, complete blood count, electrolytes, and pregnancy test (when appropriate).
- **ECG should be performed (cardiac contusion and arrhythmias are common in trauma).**
- Frequent blood pressure measurements should be performed.
- The patient should be monitored with pulse oximetry.
- X-ray films, including portable c-spine, chest and pelvic x-ray films, should be taken.
- **If head trauma is suspected, a head computed tomography (CT) scan should be obtained.**
- **If the patient** is hemodynamically stable and **has abdominal pain,** either **diagnostic peritoneal lavage or abdominal CT scan with contrast should be performed.**

■ Formulating the Diagnosis Many injuries in trauma are obvious, such as a patient with a broken arm. However, the nature of trauma is such that obvious injuries distract from more subtle, often life-threatening injuries.

■ Evaluating the Severity of Problems **The most common causes of death in trauma patients brought to hospital are subdural and epidural hemotomas, pneumothorax, hemothorax, ruptured spleen, lacerated liver, pelvic fractures, or major blood loss.**

■ Managing the Patient

- **Airway** The airway must be unobstructed for the patient to breathe. The **most common reason for airway obstruction is** the relaxed **tongue blocking** the **airway.**

  This situation is **treated by** performing either of two procedures, **chin lift or jaw thrust.**

  In an unconscious patient, an **oropharyngeal airway may** be placed to **prevent** the **tongue from obstructing** the **airway.**

  The **patient should be intubated if** he or she is **apneic, to prevent aspiration,** or **for potentially unstable airway (e.g.,** after severe burns with **inhalation of smoke). If** an **intubation is impossible,** then **a surgical airway must be placed (e.g., cricothyroidotomy).**

- **Cervical spine** The patient's **neck should be immobilized at all times until c-spine injury is ruled out.** The c-spine is assessed radiographically by a cross-table lateral x-ray film. While the lateral view of the c-spine reveals most serious neck injuries, additional views (including CT scanning) are often necessary.

  If the cervical collar must be removed, the patient's neck should be stabilized by inline immobilization.

- **Breathing**
  - **Patients** should be **given supplementary oxygen.**
  - Tension pneumothorax should be treated immediately by decompressing the chest with a needle. The needle should be

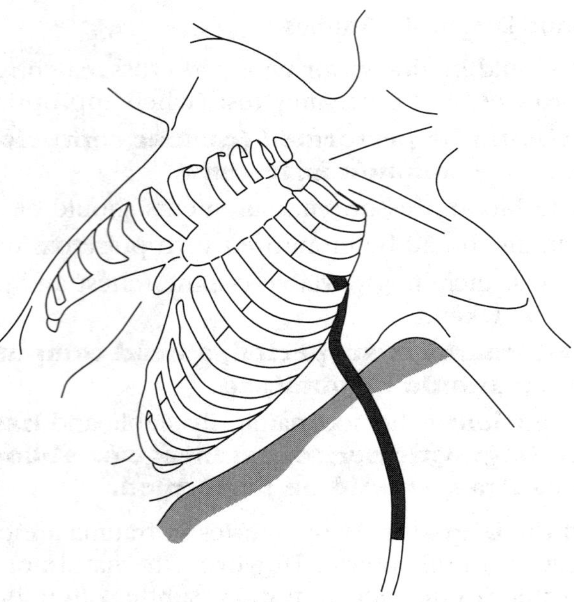

Fig. 16.1 Improper placement of a chest tube outside the ribs and beneath the scapula will not be functional even though an anteroposterior chest film may be normal.

placed in the second intercostal space in the midclavicular line.

- **Absent breath sounds indicate a possible hemothorax or pneumothorax. These conditions are treated by immediate placement of a chest tube (Fig. 16.1).**

- **Circulation A rapid thready pulse with low blood pressure indicates severe bleeding.** The **patient must receive fluid immediately. Bring the patient to the operating room immediately after stabilizing.**
  - **Two large bore IVs should be placed.**
  - **Ringer's lactate or normal saline** should be initially used for volume resuscitation **(2 to 3 L).**
  - **Type-specific blood or type O negative blood should be given if the response to IV fluid is inadequate.**
  - **Hypovolemic shock should not be treated with vasopressors or sodium bicarbonate.**
  - IV placement in infants and children is often difficult. **In children, intraosseous lines are an excellent alternative if IV access is unattainable.**
- **Disability or neurologic status** If the patient **presents with any of the following findings, a head CT scan should be obtained:**
  - **Altered mental state**
  - **Extended loss of consciousness**
  - **Appears intoxicated**
  - **Skull or facial injury on physical examination**

Head CT scan will demonstrate whether subdural, epidural, or other intracranial bleeding has occurred. **If the CT scan does demonstrate blood in the cranium a neurosurgical evaluation should be immediately obtained.**

Immediate treatment includes giving **mannitol** to decrease intracranial pressure, **elevating the** patient's **bed** to 15 degrees, and **hyperventilating the patient** to reduce intracranial pressure. In extreme cases a ventriculostomy can be performed to lower the intracranial pressure.

- **Exposure (undress) with temperature control** The **patient should be covered with blankets and warmed** to prevent hypothermia.

Altered Mental Status

- **Basic Science** The term *altered mental status* (AMS) represents a **continuum of disease** that **ranges from confusional** states **to coma.** Although there are **many etiologies** of AMS **(e.g., toxic-metabolic, infectious, cerebrovascular, or traumatic),** each manifests through **disruption** of normal **neuronal function** of either the **reticular activating system or bilateral cerebral cortices.**

- **History and Physical Examination** **Patients with AMS are not reliable historians;** therefore, **whatever history is available is** usually **obtained from medic alert tags, paramedics, friends, or family members.**

  The **physical examination begins with assessment of** the **ABCs** (i.e., airway, breathing, and circulation). **Vital signs (including core temperature and the Glasgow coma scale score) should be recorded. A cervical spine injury should be assumed to be present in all patients with AMS; therefore neck immobilization should be maintained until cervical injury can be excluded.** The **physical examination** may reveal several **important clues: track marks** in the intravenous drug abuser, **skull fracture** in the traumatized patient, **fruity breath** in the patient with diabetic ketoacidosis, and so on. A **complete neurologic examination should be done.**

- **Laboratory and Diagnostic Studies** Among the first studies in the patient with AMS are **arterial blood gas** and **glucose determination.** These will identify two common metabolic derangements (i.e., hypoxemia and hypoglycemia or hyperglycemia). Meanwhile, **blood** should be sent **for electrolytes, calcium, toxicology,** and **complete blood count.**

  Patients in whom the **diagnosis** is **not** readily **apparent** may **require head CT scan. When** the **CT scan is nondiagnostic or an infectious etiology is suspected, a lumbar puncture should be performed.** C-spine films should be obtained to exclude radiographically evident abnormalities.

- **Formulating the Diagnosis** Once the ABCs have been assessed (and corrected when necessary), **the history** (in the stable patient) **and physical examination are used to identify obvious causes.**

Thiamine, glucose, oxygen, and naloxone are usually **administered immediately** and **may be both diagnostic and therapeutic.**

- **Toxic-metabolic causes include opiate overdose** (pinpoint pupils, hypoventilation), **sedative-hypnotic intoxication, alcohol intoxication, hypoxia, hepatic encephalopathy, derangements of glucose and electrolytes, and uremia.**
- **Infectious causes include meningitis** (aseptic, bacterial), **encephalitis, and septic shock.**
- **Cerebrovascular causes include stroke** (ischemic, hemorrhagic).
- **Traumatic causes include head injury** (resulting in hemorrhage, contusion, or cerebral edema).
- **Other causes include seizure, psychiatric (e.g., pseudocoma), cerebral neoplasms, and hypothermia.**

■ Evaluating the Severity of Problems Certain causes of AMS are not readily correctable and portend a poor prognosis (e.g., pontine hemorrhage and other inoperable lesions).

■ Managing the Patient Until the diagnosis can be established, the **patient is treated presumptively. First-line agents include thiamine, naloxone, glucose, and oxygen.** Laboratory and diagnostic studies are usually helpful in establishing a diagnosis when it is not clinically evident.

The **treatment** is **directed toward** the **cause.** For example, the diabetic who has taken too much insulin may require only supplemental glucose, observation, and an adjustment of insulin dosage.

Anaphylaxis

■ Basic Science **Anaphylaxis** is a **life-threatening condition.** It is an **IgE-mediated immune reaction that occurs upon reexposure to a stimulus to which a person has been previously sensitized.** The release of chemical mediators such as histamine, prostaglandins, and leukotrienes results in manifestations that range from a benign urticarial rash to laryngeal edema, bronchospasm, and shock. **Substances reported to cause anaphylaxis** are manifold but **include drugs (most commonly penicillin), proteins (e.g., insect venoms, foods),** and, in certain patients, exercise. The onset of anaphylaxis **is** usually **abrupt** (i.e., within 30 minutes).

■ History and Physical Examination **Patients may present with early complaints consistent with anaphylaxis (e.g., a lump in the throat, a pruritic rash, nausea, vomiting, abdominal cramps, facial swelling) or may be moribund.** Although a thorough history is desirable, in many cases the time consequences of obtaining it are prohibitive. An **attempt** should be **made to identify the causative agent or factor, so that future exposure may be avoided.**

The **physical examination** of the patient with anaphylaxis may be **remarkable for hypotension, tachycardia, angioedema** (evident as facial swelling, particularly of the eyelids and lips), **wheezing, stridor, and an urticarial rash.**

- **Laboratory and Diagnostic Studies** The diagnosis of anaphylaxis is made independently of laboratory investigation.
- **Formulating the Diagnosis** When the clinical course of events includes **exposure to an etiologic agent of anaphylaxis followed by the classic manifestations** as noted earlier, the **diagnosis is unmistakable.**

  However, certain features of anaphylaxis may be present in other disorders. For example, hereditary angioedema (a disorder characterized by the absence of a component in the complement cascade) may result in facial and upper airway swelling, but the urticarial rash of anaphylaxis is absent and the treatment is different.
- **Evaluating the Severity of Problems** **The two most lethal complications of anaphylaxis are laryngeal edema and refractory hypotension.**
- **Managing the Patient** Initial efforts should be directed toward evaluating the airway. **Patients with any evidence of airway difficulties (e.g., angioedema, hoarseness) should receive 0.3 cc of epinephrine (1:1000 dilution) administered subcutaneously.** Drastic measures (e.g., intravenous epinephrine, intubation, or cricothyroidotomy) may be necessary in the patient with impending airway obstruction. **The patient's ability to exchange air may be impaired but should improve with epinephrine and supplemental oxygen. Hypotension should initially be treated with a crystalloid fluid bolus (e.g., 1 or 2 L of normal saline, 20 ml/kg in children).** Refractory hypotension may be responsive to an epinephrine drip; other cardioactive agents, such as dopamine, may be added as necessary to maintain the blood pressure.

  The offending agent, (e.g., a bee stinger) should be removed as soon as possible. An antihistamine (e.g., diphenhydramine) is routinely administered. Steroids (e.g., methylprednisolone), although not immediately effective, mitigate late reactions.

  Any patient with a life-threatening manifestation of anaphylaxis requires hospital admission. Patients who are discharged should be given a short course of an oral steroid and an antihistamine. Follow-up with an allergist should be arranged.

Head Injury

- **Basic Science** Head injury is a common consequence of motor vehicle accidents, assaults, and falls. The injury may be relatively benign (e.g., a concussion) or severe (e.g., intracranial bleeding with resultant herniation and death).

  Concussion describes a syndrome of **temporary neurologic dysfunction following head trauma.** Commonly, this **includes confusion, dizziness, nausea, vomiting, and headache, often following a brief period of loss of consciousness.** When these symptoms persist for weeks or longer, the patient is said to have postconcussion syndrome.

  Intracranial hemorrhage is **classified according to the location of the blood** and **includes epidural, subdural, intrace-**

rebral, and subarachnoid hemorrhage. Contusions are often seen on CT scans and clinically associated with longer periods of unconsciousness.

- History and Physical Examination **A history of loss of consciousness** or changes in the level of alertness **should be obtained from the patient or witnesses.** Additionally, the **presence of seizure activity, incontinence, and ability to move all extremities should be sought.** Questions should be asked about events that may have precipitated the injury (e.g., alcohol intoxication) as well as the event itself.

Classically, the history of a patient with an **epidural hemorrhage** includes a **head injury that precedes a brief period of loss of consciousness, followed by a lucid interval. Minutes to hours later, the patient demonstrates signs of neurologic deterioration.**

Patients with an **acute subdural hemorrhage** often have **loss of consciousness following head trauma, usually with return of consciousness, though with some level of cognitive dysfunction.**

The **vital signs and ABCs should be addressed first.** Patients should be assumed to have a cervical spine injury until proven otherwise. A complete **neurologic examination must be performed;** the **most important part of** the **neurologic examination is** the **level of consciousness. Frequent evaluation of** the **head-injured patient** is **necessary,** primarily to detect evidence of clinical deterioration, readily discovered through changes in the level of consciousness. The Glasgow coma scale is appropriate for monitoring level of consciousness and gross neurologic function.

- Laboratory and Diagnostic Studies **Patients presenting to the emergency department following a significant loss of consciousness or with neurologic deficits** should **undergo a brain CT scan.** Patients with a suspected subarachnoid hemorrhage that is not identified on CT scanning should undergo a lumbar puncture to look for blood.

An epidural hemorrhage appears biconvex on a head CT scan, whereas a subdural hemorrhage appears crescent shaped. Intracerebral hemorrhages most commonly occur in the frontal or anterior portion of the temporal lobes.

- Formulating the Diagnosis **The diagnosis is made after a complete history, physical examination, and appropriate studies are interpreted, as described earlier.**

The Cushing response to markedly increased intracranial pressure consists of bradycardia and hypertension. Hypotensive patients should be assumed to be hypovolemic since this finding in a patient with head injury means severe, mostly fatal injury. If hypovolemia is uncorrected, mild cerebral injury can become severe.

- Evaluating the Severity of Problems The **mortality rate for most forms of severe head trauma is high;** patients with an acute subdural hemorrhage have a mortality rate of approximately 70%. Subdural hematomas generally have significant underlying damage, whereas epidural hematomas generally have less underlying damage and better prognosis, if treated promptly. **Signs of posturing (decerebrate or decorticate) or herniation (unilaterally dilated pupil with hemiparesis, central hyperventilation) indicate need for immediate decompression.**

■ Managing the Patient **All head injury patients** should **receive supplemental oxygen** and **c-spine immobilization** (until the c-spine can be radiographically and clinically cleared of injury).

In **patients without** a **focal examination or loss of consciousness, a period of observation** (several hours) may be **adequate. Patients with a demonstrable head injury are at risk for seizure;** accordingly, an agent such as **phenytoin can be used prophylactically** to raise the seizure threshold.

Severe head injury may result in increased intracranial pressure. Increased intracranial pressure is **treated with mannitol (0.5 g/kg up to 50 g IV). Intracranial pressure monitors** may be **used to direct therapy (e.g., diuretics, hyperventilation)** to maintain this pressure within the normal range. Intracranial pressure can be further elevated during intubation (blunted somewhat by lidocaine), hypoventilation, and lying flat (raise the head 15%). The injured brain is particularly sensitive to fever, hypoglycemia, hypovolemia, and hypoxia.

Patients with an epidural or subdural hemorrhage usually require neurosurgical evacuation of the extravasated blood, whereas a subarachnoid hemorrhage is typically not amenable to operative intervention. **Patients with** extravasated **intracranial blood** should **receive neurosurgical consultation.**

Hemorrhagic Shock

■ Basic Science **Hemorrhagic shock is defined as acute blood loss leading to impaired tissue perfusion and hypotension. This condition** is most often **seen** in association **with trauma** but **also occurs with GI and reproductive system bleeding. Aneurysms** and **dissections may also result in hemorrhagic shock.** The physiologic responses to acute blood loss involve an increase in cardiac rate and contractility (secondary to an increase in norepinephrine and decreased vagal tone), blood flow redistribution, retention of sodium and water (primarily mediated by aldosterone and antidiuretic hormone), and local control of blood loss (primarily through activation of the coagulation system).

■ History and Physical Examination In patients with hemorrhagic shock questions should be directed toward assessing the source and volume of blood loss. **The source of hemorrhagic shock in trauma patients may be obvious** (e.g., a bleeding stab wound) **or obscure (e.g., a retroperitoneal hematoma).** Similarly, patients with gastrointestinal blood loss include those with bleeding arteriovenous malformations, peptic ulcers, and diverticuli. The female reproductive system can be a source of blood loss as well; sources include abnormal uterine bleeding and ruptured ectopic pregnancy.

Physical examination should first establish the presence of an airway, followed by adequate respiration and circulation.

• **Small amounts of blood loss** (e.g., <20% of circulating blood volume) **produce cool, clammy skin; a narrowed pulse pressure; and tachycardia.**

• **Moderate amounts of blood loss** (e.g., 20% to 40%) **produce the above findings with tachypnea, orthostatic changes in blood pressure, and possibly confusion or agitation.**

• **Large amounts of blood loss** (e.g., >40%) **produce hypotension (with weak pulses), obtundation, and oliguria.**

■ Laboratory and Diagnostic Studies The initial treatment of hemorrhagic shock does not depend on the results of laboratory tests; however, **baseline labs including a type and crossmatch, complete blood count, prothrombin time, partial thromboplastin time, and pregnancy test (in women of childbearing age) should be obtained.** Changes in hemoglobin and hematocrit are often not apparent in patients with acute blood loss.

■ Formulating the Diagnosis The diagnosis of hemorrhagic shock is made in patients with a history of blood loss who have impaired perfusion as the result of hypotension.

■ Evaluating the Severity of Problems **Acute massive blood loss may lead to irreversible changes at the cellular level, precluding successful resuscitation, despite volume repletion.** Blood loss of 15% (750 cc in an adult) leads to mild tachypnea and possibly orthostatic changes in blood pressure. Blood loss of 30% (1500 cc) leads to tachycardia and a decreased blood pressure. Blood loss of 40% (2000 cc) leads to marked tachycardia and decreased blood pressure. Above this amount severe hypotension and anuria result. Patients are generally confused.

■ Managing the Patient The **first priority** in hemorrhagic shock **is providing** an **adequate airway** and **ventilation** (supplemental oxygen or intubation).

The **next** priority is to **identify** and **stop the blood loss.** If the patient is **bleeding externally, direct pressure may stop blood loss. Patients with suspected internal bleeding should be taken to the operating room for immediate exploration** (stabilization procedures should continue).

Two large-bore intravenous lines should be placed. Patients should **receive** a 20-40 ml/kg **crystalloid fluid bolus** (e.g., about **2 L of normal saline or lactated Ringer's solution** in the average patient). **If the patient does not improve** clinically **(i.e., improved color, urine output, mental status, or vital signs), packed red blood cells should be given** (blood transfusions should be type specific, but if there is no time to obtain a crossmatch type O blood may be used).

Evidence to support the widespread use of medical antishock trousers (MAST) is lacking; however, they do benefit patients with unstable pelvic fractures, many of whom have retroperitoneal hematomas.

Poisoning

■ Basic Science Poisonings result from an overdose of medications (e.g., acetaminophen, digoxin) or street drugs (e.g., cocaine, heroin) or exposure to a variety of nonpharmaceutical products (e.g., carbon monoxide, drain cleaners, gasoline, pesticides, mushrooms). Ingestions account for 75% of exposures; inhalation, parenteral, and dermal are other important routes of exposures. Poisonings in young children are usually accidental and benign. Intentional overdoses (adolescents and adults) are more serious.

The most common serious pharmaceutical poisonings result from acute overdose of acetaminophen, salicylates, iron, and tricyclic antidepressants (TCA). Chronic poisoning from salicylates is also a major concern. Overall, more deaths occur from carbon monoxide (CO) poisoning than any other toxin.

- Acetaminophen overdose may result in fulminant hepatic failure toxic metabolite formation.
- Salicylism may cause a respiratory alkalosis, metabolic acidosis, hyperpyrexia, and cerebral edema.
- Iron is a gastrointestinal corrosive but may also cause multisystem injury.
- Tricyclic antidepressants have a pronounced cardiac and neurologic effect manifest by wide complex tachydysrhythmias and hypotension and seizures.
- CO acts as a chemical asphyxiant, impairing the transport and utilization of oxygen.

■ **History and Physical Examination** Since many overdose patients present to the emergency department with altered mental status, obtaining a history from family, friends, paramedics, pill bottles, and old medical records is critically important. In the awake patient, the history may be incomplete or even misleading. The local regional poison control center can be quite helpful in properly identifying potential toxins and providing management recommendations.

Vital signs including a rectal temperature should be obtained immediately and closely followed. Physical examination should search for features of various toxidromes.

- **Miosis, coma, and hypoventilation suggest opiate overdose** (e.g., heroin, methadone).
- **Tachycardia, mydriasis, agitation, hyperpyrexia, diminished bowel sounds, and dry skin suggest anticholinergic overdose (e.g., antihistamines).**
- **Salivation, lacrimation, bronchorrhea, incontinence, and muscle weakness suggest cholinergic excess (e.g., organophosphates).**

A careful search for evidence of concomitant traumatic injury should also be performed.

■ **Laboratory and Diagnostic Studies** The laboratory evaluation may provide essential information that facilitates management of these patients.

- **Serum acetaminophen levels should be obtained on all intentional overdose patients.**
- **Serum salicylate levels or urinary ferric chloride tests should be performed on all patients whose presentations are**

Box 16.1

COMMON POISONINGS AND THEIR TREATMENT

Poison	*Treatment*
Theophylline and phenobarbital for tricyclic antidepressants	Repetitive dose activated charcoal
Bicarbonate for cardiac abnormalities	Serum alkalinization with sodium conduction
Carbon monoxide	Hyperbaric oxygen

consistent with possible acute or chronic salicylism (e.g., altered mental status, hyperventilation).

- A carboxyhemoglobin test should be ordered on patients with a history of CO exposure or who present with symptoms (e.g., headache, fatigue) that might be attributable to CO.
- A high anion gap (>16) metabolic acidosis may be the result of salicylate, methanol, ethylene glycol, iron, or isoniazid ingestion.
- **Toxicology screens are helpful but do not detect many important drugs.**
- **The ECG and continuous monitoring are essential in the assessment of TCA overdose.**

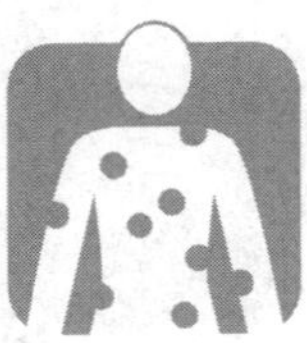

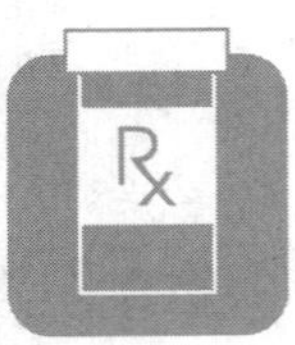

■ Evaluating the Severity of Problems Patients with abnormal vital signs often need emergent treatment. Agitated patients require proper sedation to avoid complications of rhabdomyolysis and hyperthermia. Comatose patients may need an airway. **A widened QRS interval >0.10 seconds in the setting of TCA overdose is associated with potential for arrhythmias and seizures.** There are nomograms for severity of ingestions for several toxins (e.g., acetaminophen, salicylates).

■ Managing the Patient **The regional poison center should be notified about all poisons. They will provide the latest information** about treatment and antidotes.

- **Activated charcoal** is the best treatment modality to limit absorption and **should be used in the treatment of most overdoses.**
- **Gastric lavage should be restricted to early presentations (<1 hour) of life-threatening ingestions.**
- **Syrup of ipecac to induce vomiting is limited to home use** in patients where the level of consciousness is not likely to change.
- For the majority of overdoses, good supportive care including airway control and cardiac monitoring is all that is required once decontamination is completed.
- **Hemodialysis is seldom required** in the management of poisoning; its use is generally limited to severe cases of salicylate, methanol, ethylene glycol, lithium, and theophylline toxicity.

A few specific treatments exist (Box 16.1).
Antidotes exist for few common poisonings (Box 16.2).

Box 16.2

ANTIDOTES

Poison	*Antidote*
Acetaminophen	*n*-acetylcysteine
Opiates	Naloxone
Benzodiazepines	Flumazenil
Iron	Defexoxamine
Digoxin	Digibind

Septic Shock

- Basic Science **Septic shock is a form of shock that occurs when endotoxins are released from certain bacteria in the bloodstream.** The endotoxins **cause a complex immunologic reaction that results in the release of numerous vasoactive substances. This results in decreased vascular resistance and a drastic fall in blood pressure.**

Septic shock is usually preceded by signs of severe infection, often of the genitourinary or gastrointestinal (GI) systems. The **causative organism is most frequently gram-negative (e.g., *E. coli, Pseudomonas, Klebsiella, Haemophilus*),** followed by a gram-positive organism (e.g., *Staphylococcus, Streptococcus*), and, less often, a fungus (especially in patients with acquired immunodeficiency syndrome [AIDS]) and other microorganisms. **This syndrome most often occurs in hospitalized patients, especially those who are diabetic or immunosuppressed.**

Patients often develop acute renal failure, adult respiratory distress syndrome, and multiple organ failure. **Death occurs in approximately one quarter of cases.**

- History and Physical Examination
  - **History Age, recent illnesses,** and factors predisposing to infection (e.g., **immunosuppression, recent hospitalization**) should be noted.
  - **Physical examination Sepsis presents with tachypnea, tachycardia, fever, chills, and oligouria. Neonates, alcoholics,** and the **elderly may have a normal or low temperature. Skin is warm. Confusion and disorientation** are **common.**

    The physician should closely examine potential sites of infection such as the skin, GI tract, intravenous sites, and the lungs.

- Laboratory and Diagnostic Studies
  - **Complete blood count usually reveals a leukocytosis.**
  - **Routine chemistries may demonstrate hyperglycemia.**
  - **Urinalysis may indicate signs of infection.**
  - **Serum lactate levels are elevated.**
  - **Arterial blood gas reveals hypoxemia and a metabolic acidosis.**
  - **Coagulation studies should be obtained;** patients with

septic shock are at risk for thrombocytopenia and disseminated intravascular coagulation.

- **Blood, urine, and sputum should be sent for culture and sensitivity.**
- **A chest x-ray** film may identify pneumonia as a causative factor.
- **Suspected sites of infection** (e.g., an abscess, or indwelling catheter) **should be Gram's stained and cultured.**

■ Formulating the Diagnosis **A definitive diagnosis requires isolation of the microorganism.** However, this is often not possible.

The differential diagnosis should include other causes of shock (e.g., anaphylaxis, occult hemorrhage).

Occasionally the history and physical examination may suggest an etiologic agent. For example, pneumococcal septicemia should be suspected in splenectomized individuals; purpuric lesions may suggest *Neisseria;* and erythroderma with exfoliation suggests gram-positive organisms responsible for toxic shock syndrome (e.g., *Staphylococcus*).

■ Evaluating the Severity of Problems **Septic shock is fatal in approximately one quarter of cases.** The presence of underlying disease and extremes of age worsen prognosis.

■ Managing the Patient **Patients with septic shock should be admitted to the intensive care unit. All patients with septic shock should receive supplemental oxygen.**

Swan-Ganz catheters are often placed to measure the central venous pressure (CVP) or pulmonary artery pressure. Patients are hydrated with normal saline until their wedge pressure is between 12 and 15 mm Hg. If the patient remains hypotensive despite IV fluids, vasopressors (e.g., dopamine) should be started.

The antibiotic treatment for septic shock should be tailored to the source of infection. **Cultures should be obtained before the administration of antibiotics if this will not delay therapy. Empiric antibiotics should include coverage against both gram-negative and gram-positive organisms.**

- **A cephalosporin (e.g., ceftazidime) and an aminoglycoside (e.g., gentamicin) represent an acceptable empiric regimen.**
- **When gram-positive organisms are strongly suspected (e.g., in the intravenous drug abuser), either vancomycin or a penicillinase-resistant penicillin should be added (e.g., nafcillin).**
- **Anaerobic coverage (with either clindamycin or metronidazole) should be included in patients in whom an intraabdominal source is suspected.**

Syncope

■ Basic Science Syncope is defined as a brief lapse in consciousness caused by transient cerebral hypoxia. **Syncope is caused by one of the following:**

- **Cardiac causes**
  - **Reduced cardiac output—e.g., myocardial infarction, massive pulmonary embolism, aortic stenosis, hypertrophic cardiomyopathy,** atrial myxoma, pericardial tamponade, aortic dissection, and congenital heart disease
  - **Arrhythmias—e.g., atrioventricular block, ventricular tachyarrhythmias,** supraventricular tachycardia, pacemaker malfunction
- **Noncardiac causes**
  - **Vasovagal—the "common faint"; classically precipitated by fear, excitement, pain, fatigue or stress, especially in a warm environment; preceded by premonitory symptoms such as nausea, diaphoresis, or lightheadedness;** especially common in those patients **with hypovolemia or those receiving antihypertensive or vasodilating medications**
  - Inadequate peripheral vasoconstriction
  - **Hypovolemia**
  - **Postural hypotension—may cause syncope after a positional change from lying to sitting**
  - **Vertebrobasilar insufficiency—frequently accompanied by diplopia, vertigo, dysarthria, hemiparesis,** or other symptoms of brainstem ischemia
  - Situational syncope—syncope while coughing, defecating

History and Physical Examination

- **History**
  - **Cardiac syncope develops abruptly with minimal or absent warning symptoms and is followed by rapid recovery.**
  - Vasovagal syncope is often heralded by nausea, sweating, and yawning. It develops as a result of disturbing emotional or physical stimuli.
  - Orthostatic syncope usually appears in the elderly after prolonged sitting or lying.
  - Patients with seizures will sometimes experience an aura prior to losing consciousness and have a more prolonged recovery period.

  Patients should be closely questioned about recent changes in medications or the presence of recent head trauma.

- **Physical examination**
  - Vital signs, orthostatic vital signs, and pulse quality should be noted.
  - Particular attention should be paid to heart rate and rhythm.
  - **Complete neurologic examination should be done.**
  - Maneuvers that reproduce symptoms assist in diagnosis. For example, cartoid sinus massage or the Valsalva maneuver may reveal the source of syncope.

■ **Laboratory and Diagnostic Studies** An ECG should be performed. Young, healthy patients (under 40 years) with normal ECGs do not usually require extensive testing.

A hematocrit and serum glucose may identify factors contributing to the syncopal episode (e.g., hypoglycemia, anemia).

Ambulatory cardiac monitoring (Holter), electrophysiologic studies, tilt-table testing, echocardiography, and electroencephalography aid in further patient workup.

In young women a pregnancy test should be done.

Patients with localizing nuerological symptoms (double vision, dysarthria, etc.) should have a CT scan.

■ **Formulating the Diagnosis** **History, physical examination, and results of laboratory tests will usually point to a diagnosis. If arrhythmias or life-threatening causes of syncope have not been ruled out, the patient should be admitted to the hospital for further diagnostic testing and observation (e.g., Holter monitoring).**

Conditions such as hypoxemia, hyperventilation, and anxiety attacks can mimic syncope.

■ **Evaluating the Severity of Problems** **A cardiac etiology for syncope should be suspected in elderly patients or patients with a history of ischemic heart disease, valvular disease, or an abnormal ECG. These patients are at high risk for sudden death and significant morbidity from falls.**

■ **Managing the Patient**

- **Life-threatening causes of syncope** **Patients with potentially life-threatening causes of syncope as just listed should be stabilized, started on oxygen, placed on continuous cardiac monitoring, and admitted for further evaluation.** Any disease-specific treatment should be started immediately (e.g., thrombolytics for myocardial infarction).

  Pacemakers may be effective for those with profound bradycardia.

- **Non-life-threatening causes of syncope**
  - **Inadequate peripheral vasoconstriction—increased fluid intake and compression stockings; sudden changes in position should be avoided.**
  - **Hypovolemia**—patients should be rehydrated with normal saline. Patients with poor states of health should be admitted to the hospital.
  - **Vasovagal—the "common faint";** patients should be advised to rest and avoid the situations precipitating syncope.
  - Situational syncope—**patients with situational syncope should be instructed to modify their habits** (e.g., sit while urinating, use stool softeners, try cough suppressants, quit smoking).

Multiple Choice Review Questions

1. A 56-year-old man in a grease fire sustains burns to his right hand, arm, and neck. Realizing that this is a serious injury, you prepare him for transfer to the nearest burn center 70 miles away. Which of the following is contraindicated?

a. Liberal doses of morphine and ibuprofen
b. Immediate cleaning and débridement of blisters
c. Tetanus toxoid
d. Cold dressings and cold packs for relief of pain
e. I.V. lactated Ringer's solution

2. A 6-year-old child is seen because a dime is lodged in his throat. He reports that his throat feels funny but is otherwise asymptomatic. X-ray shows that the coin lies in the sagittal plain with the flat edge visible on a lateral photograph. The next step is to:

a. Cautiously observe, offer the child slow sips of water, and consider midazolam or glucagon to enhance passage of the coin to the stomach.
b. Place the child on oxygen, monitor, and call for appropriate surgical help to remove the coin.
c. Sedate the child and attempt removal via an inflated catheter.
d. Send the child to an otolaryngologist for removal of the coin.
e. Perform the Heimlich maneuver (abdominal thrust).

3. A man attempting to obtain "discount gasoline" via a siphon tube presents after accidentally ingesting a swallow. Your first step should be to:

a. Administer ipecac immediately.
b. Decontaminate via gastric lavage.
c. Administer charcoal.
d. Continue with your history and physical, including chest x-ray and pulse oximetry.
e. Nothing; gasoline is relatively nontoxic to the GI tract.

4. An inexperienced volunteer calls in on the radio because a woman has just delivered a child in an ambulance. The child is spontaneously crying. Which of the following actions is *not* indicated?

a. Suctioning the airway
b. Tying the umbilical cord
c. Cutting the umbilical cord
d. Drying and wrapping the child
e. Leaving the placenta in the uterus and continuing to the hospital

5. A 6-year-old falls from the monkey bars, landing on her buttocks and her outstretched left arm. She is holding her flexed left arm to her side and appears comfortable if left alone. A superficial examination shows isolated left elbow tenderness. The next step is to:

a. Grasp her elbow in one hand, her wrist in the other, and via extension, supination, and flexion, reduce the "nursemaid's elbow."
b. Rule out snuff box tenderness, neurovascular compromise, and send to an orthopedic physician for further evaluation.
c. X-ray the elbow, then continue the examination.
d. Do a thorough examination to evaluate motor and sensory ability. If a fracture is suspected, confirm with an x-ray and refer to an orthopedic physician.
e. Splint the arm in extended position.

Chapter 17

Pediatrics

BRONCHIOLITIS

- Basic Science **Bronchiolitis is an acute infectious illness of the small airways occurring most commonly in infants 2 to 12 months of age. Respiratory syncytial virus (RSV) is the most common agent causing bronchiolitis** and encompasses nearly 75% of all cases of bronchiolitis. Parainfluenza, influenza, and adenoviruses comprise the rest of the agents responsible for this clinical picture.
- History and Physical Examination Bronchiolitis begins as an upper respiratory infection with clear nasal discharge and variable degrees of fever. Over 3 to 7 days there is a progression to **lower respiratory tract infection** that manifests as **tachypnea, wheezing, and hyperinflation of the lungs.**
- Laboratory and Diagnostic Studies **Pulse oximetry** is useful to assess oxygen saturation in infants with bronchiolitis. More severe cases may require serial arterial blood gas monitoring. **Chest x-ray** film is useful in severe cases to determine the degree of air trapping.

  A fluorescent antibody test and an ELISA are available for the rapid diagnosis of RSV.
- Formulating the Diagnosis **The diagnosis is usually clear from the history and physical examination.** Recurrent episodes of wheezing in a young child may suggest foreign body aspiration, reactive airway disease, or cystic fibrosis.
- Evaluating the Severity of Problems Infants with significant respiratory distress manifested by increased work of breathing, tachypnea, and cyanosis are considered to have severe disease. **Infants with persistently increasing oxygen requirements or rising P_{CO_2} levels may require intubation.**
- Managing the Patient **Otherwise healthy infants with mild to moderate symptoms can be managed at home. Hospitalization is required for children with severe respiratory distress or hypoxemia, which can lead to apnea in infants. These patients require supplemental oxygen.** Initial and subsequent monitoring of oxygenation by pulse oximetry facilitates the appropriate management of hypoxia.

  Ribavirin is an antiviral agent that is active against RSV and is recommended for use in children with severe bronchiolitis or with underlying heart or lung disease. Some children may benefit from the use of aerosolized bronchodilators (e.g., albuterol).

Child Physical Abuse

■ Basic Science **Child abuse often presents to the physician with a chief complaint of "accidental injury."** Thus knowledge of accidental versus deliberate injury patterns, and the factors that put families at risk for child abuse, are essential to identify and manage this problem.

■ History and Physical Examination **Child abuse must be suspected when**

- The history of an injury is vague.
- **History is incompatible** with physical signs.
- **History changes** when the story is retold.
- **Parents give different histories.**
- History is incompatible with the child's developmental stage.
- **Child gives a different history from the parent.**

  Note that often the child will not reveal that he or she was deliberately injured.

Other **red flags** include the following:

- Significant **delay in seeking care**
- A detached or inattentive parent
- Unrealistic parental expectations (e.g., demanding a 1-year-old to sit still)

The physical examination includes checking for the following:

- Head swelling (possible skull fracture)
- Abdominal injury
- Genital trauma
- Skin bruising

Injuries are classified as follows.

- **Bruises** Accidental bruises tend to occur over bony prominences, such as the forehead, chin, and elbows. Inflicted bruises are usually found on soft tissue areas such as the cheeks and buttocks. Belts and cords often leave suspicious marks.
- **Head injury** Children rarely sustain significant head injury or skull fractures due to falling from a couch or bed, or running and falling onto the ground. Evidence of head trauma includes vomiting, altered mental status, and swollen areas on the scalp.
- **Shaken baby syndrome** Shaken baby syndrome occurs when an infant is shaken so violently that cerebral hemorrhage occurs. The infant usually presents with signs of increased intracranial pressure. These include **apnea, alternating lethargy and irritability, altered level of consciousness, a full fontanelle, retinal hemorrhage(s), or poor or increased muscle tone.** These children often suffer severe neurologic injury with permanent deficits.
- **Thoracic injuries** Thoracic and abdominal injuries result from the child's being hit, kicked, thrown, or struck with an object. Often these battered children present with a story lacking a history

of trauma. Symptoms of thoracic trauma include respiratory distress, chest pain, or collapse from severe injury.

- **Abdominal injuries** Abdominal injury often presents with abdominal pain, sometimes accompanied by distention and signs of obstruction (e.g., vomiting). The major injuries include duodenal hematoma, mesenteric tears, pancreatic and renal contusions, and liver and splenic laceration/hematoma.
- **Skeletal injuries Accidental fractures rarely occur in children under 18 months of age.** Fractures highly associated with abuse are
    - Skull fractures with a history of minor trauma
    - Metaphyseal fractures
    - Spiral fractures of the humerus or femur
    - Posterior rib fractures

  Old fractures in various stages of healing suggest child abuse.
- **Burns** Children are burned by objects such as cigarettes, irons, and hot liquids. Accidental burns generally occur on the hands. **Accidental burns are rarely full thickness** because the child quickly withdraws from pain. Inflicted burns are often full thickness and can imprint the source of the burn.

  Inflicted burns are often caused by immersing the child in hot water. Immersion burns often leave a sharp line of demarcation between the normal and the burned skin. The imprints are symmetrical with no splash marks. Inflicted splash burns, on the other hand, are very difficult to discern from accidental causes. **The association between burns and child abuse is high.**

Laboratory and Diagnostic Studies

- **Bruising** A CBC, platelet count, and coagulation screen should be performed to rule out medical causes of abnormal bruising such as leukemia.
- **Shaken baby syndrome** The child should have an eye examination to check for retinal hemorrhages and head CT scan to look for intracranial hemorrhage.
- **Thoracic injuries** Diagnosis of the injuries is by plain radiographs or chest CT scan.
- **Abdominal injuries** Plain radiographs may show dilated loops of bowel. An abdominal CT scan may be required for definitive diagnosis.
- **Fractures** Medical causes of unusual fractures, such as osteogenesis imperfecta, should be ruled out. **Children younger than 5 with any unusual fracture warrant an entire skeletal survey to detect old fractures.**

Formulating the Diagnosis If the physician becomes suspicious of child abuse a comprehensive investigation is necessary. A social worker generally examines the psychosocial factors that place families at risk for child abuse. Some of these factors include financial problems, unemployment, social isolation, domestic violence, young or single parents, use of corporal punishment, poor impulse control, depression, and substance abuse.

After the history, physical examination, x-ray films, and social work analyses are complete, a determination is made regarding whether child abuse is suspected.

- Evaluating the Severity of Symptoms The psychologic sequelae of abuse and neglect are often the most permanent and devastating injury suffered by the child.
- Managing the Patient **If child abuse is highly suspected,**
    - **The child should not be allowed to return home until any risk of injury is ruled out.**
    - **The physician must report findings immediately to the proper child protective agency.**
    - **All medical problems should be treated.**

Child Sexual Abuse

- Basic Science The annual number of cases of child sexual abuse reported in the United States is approximately 250,000 to 350,000. **Perpetrators are most often male, relatives, or are well known to the victim.**
- History and Physical Examination **Often the victim does not reveal a history of any sexual abuse.** When a child does disclose sexual abuse, a history should be taken by someone skilled in interviewing children.

  Children may have symptoms that suggest sexual abuse. These include difficulty walking or sitting; abdominal pain; constipation; dysuria; vaginal, urethral, or rectal discharge or bleeding; change in behavior; and depression. Children may also have recurrent UTIs or may present with a sexually transmitted disease.

  Approximately 50% of abused children have no physical findings.
- Laboratory and Diagnostic Studies Sexually transmitted disease testing should be done. **This includes appropriate specimens for gonorrhea, clamydia, syphilis, hepatitis, and HIV.**
- Formulating the Diagnosis The diagnosis is made when there is adequate suspicion, despite the absence of physical findings.
- Evaluating the Severity of Problems Even without significant physical injury, sexual abuse victims develop severe emotional problems.
- Managing the Patient **Once it has been determined that a child has been sexually abused, management includes treatment of medical problems, protection from the perpetrator, and counseling for the child and family.**

Colic

- Basic Science **Colic is defined as excessive, unexplained crying, lasting more than 3 hours per day, and more than 3 days per week, in an otherwise healthy and well-fed infant.**

At 2 weeks of age, infants normally cry about 2 hours per day. Crying tends to increase to an average of 3 hours per day at 6 weeks and then decreases to 1 hour per day by 3 months of age.

Ten percent of all infants have colic. Crying episodes typically begin in the first 3 weeks of life. Persistent crying episodes often start abruptly and may last for several hours. Most often, crying occurs in the evening hours. Infants often draw their knees up and pass flatus. The infant with colic is difficult to console with usual measures.

The etiology of colic is poorly understood and probably multifactorial. Proposed theories have included immaturity of the infant's gastrointestinal tract, trapped intestinal gas, hyperperistalsis, allergy, and maternal anxiety.

■ History and Physical Examination **The history and physical examination should rule out other causes of excessive infant crying.** The history should review the timing, duration, and character of the crying, as well as the parent's responses. Vomiting, diarrhea, or blood in the stools may indicate a sensitivity to cow's milk protein. A social history should consider potential child abuse.

Physical examination must include a complete neurologic examination looking for evidence of infection or shaken baby syndrome. Otitis media should be ruled out. Abdominal and genitourinary examination should look for evidence of trauma, intussusception, incarcerated hernia, or testicular torsion. The extremities should be examined for trauma and digits examined for strangulation by an encircling hair. A fluorescein eye exam will rule out corneal abrasion as the cause.

■ Laboratory and Diagnostic Studies **Laboratory studies are not indicated** unless the history or physical examination suggests an etiology other than colic.

■ Formulating the Diagnosis The diagnosis of colic is one of exclusion.

■ Evaluating the Severity of Problems **Colic is benign.** The infant is generally healthy and will outgrow the behavior.

■ Managing the Patient **The natural history of colic is improvement over time, with resolution by 3 to 4 months of age regardless of the management strategy used. Reassurance, support, and a review of behavioral tactics should be the mainstays of management.** Parents should be reassured that their infant is healthy and the crying will not hurt the baby but will improve with time. The physician should acknowledge parental feelings of frustration toward their baby.

Behavioral tactics that may be beneficial include swaddling, walking, rocking, and other comforting measures. To avoid overstimulation, parents should not use more than one or two comforting methods per 20-minute period.

There is no drug proven effective and safe for colic. Parents should be discouraged from switching formulas unless there is evidence of dietary intolerance such as diarrhea, vomiting, or blood in stools.

Croup

■ Basic Science **Croup,** or laryngotracheobronchitis, **is an acute, viral infection of the larynx and surrounding areas.** The vocal cords become inflamed and edematous, leading to stridor, hoarseness, and a barking cough. Croup is usually a mild, self-limited process; severe cases with significant upper airway obstruction may be life-threatening.

Parainfluenza virus is the most common pathogen; influenza A, adenovirus, respiratory syncytial virus (RSV), and other viruses have also been implicated. Croup occurs most frequently in the winter months.

■ History and Physical Examination Croup most commonly affects children ages 6 months to 3 years. Fever and upper respiratory symptoms are usually present for 1 to 2 days before the onset of **hoarseness; stridor; and a characteristic, barking cough.** Wheezing and retractions may be present. Dehydration is common. Children do not appear toxic. Severe cases may result in tachypnea with impaired air exchange, cyanosis, and altered mentation.

■ Laboratory and Diagnostic Studies **There is little role for lab testing in croup.** Airway radiographs, if obtained, show subglottic narrowing and are most useful to rule out retropharyngeal abscess or airway foreign bodies.

■ Formulating the Diagnosis **Croup is diagnosed clinically.** Epiglottitis, foreign body aspiration, and retropharyngeal abscess must be considered in the differential diagnosis.

Spasmodic croup is a variant consisting of the precipitous, nighttime onset of barking cough, stridor, and dyspnea in an otherwise apparently well child. Symptoms resolve quickly with mist or cool air but often recur on several successive nights.

■ Evaluating the Severity of Problems Children with severe retractions, stridor at rest, tachypnea, hypoxia, cyanosis, agitation, or lethargy may progress to respiratory failure and require close observation. Respiratory difficulty may interfere with drinking, leading to dehydration.

■ Managing the Patient **Home management consisting of the use of a humidifier, placing the child in a steamy bathroom, or exposure to cool night air is often effective. In the hospital, humidified air or nebulized saline is helpful.** As with epiglottitis, invasive procedures should be kept to a minimum. **Children with moderate to severe disease usually benefit from parenteral corticosteroids,** often given as a single IM injection of dexamethasone. **Administration of nebulized racemic epinephrine is an effective way to rapidly reduce airway inflammation in children with severe croup.** Since rebound airway edema may occur, children should be observed for several hours before being discharged home.

Dehydration

■ Basic Science **Maintenance fluid requirements for children can be calculated based on their body weight according to the following chart:**

First 10 kg body weight	100 cc/kg/24 hr
Second 10 kg body weight	50 cc/kg/24 hr
Remaining body weight	20 cc/kg/24 hr

Sources of increased fluid loss, such as fever, vomiting, diarrhea, or decreased fluid intake, can lead to dehydration.

■ History and Physical Examination **The history may reveal sources of increased fluid loss (e.g., fever, vomiting, diarrhea, increased respiratory rate) or decreased intake. There is usually a history of decreased urine output as well.**

Moderate dehydration presents with dry skin, eyes, and mucous membranes, sunken eyes, sunken fontanelle, and tachycardia. Indications of severe dehydration include altered mental status, skin tenting, delayed capillary refill, or decreased blood pressure. Severe dehydration is a medical emergency.

■ Laboratory and Diagnostic Studies **The CBC may show an elevated WBC with a left shift due to stress. The hematocrit may be falsely elevated due to hemoconcentration. Electrolytes are usually abnormal** and vary based on the amount and source of fluid loss(es).

Urinalysis shows an elevated specific gravity, often accompanied by the presence of ketones. Other studies may be necessary depending on the underlying etiology of the dehydration.

■ Formulating the Diagnosis **The diagnosis is based on the patient's history. Particular attention should be paid to oral intake and urine output.**

■ Evaluating the Severity of Problems Children with decreased urine output, dry skin, dry mucous membranes, mild tachycardia, and a normal mental status are mildly to moderately dehydrated and can often be orally rehydrated. Children with no urine output for 24 hours, delayed capillary refill, decreased blood pressure, or altered mental status require intravenous fluid resuscitation.

■ Managing the Patient **Children with mild to moderate dehydration can be orally rehydrated with electrolyte solutions such as Infalyte and Pedialyte. Children with moderate to severe dehydration (or those with persistent vomiting) require intravenous fluid resuscitation. This usually consists of an initial 20 cc/kg bolus of isotonic fluid without glucose, such as lactated Ringer's solution or normal saline. This bolus can be repeated twice for patients with severe dehydration. The estimated fluid deficit should be replaced over 24 hours in the following manner:**

- **Hours 1-8: maintenance + ½ deficit**
- **Hours 16-24: maintenance + ½ deficit**

Patients with electrolyte abnormalities such as DKA or hypernatremia require special management.

ENURESIS

■ Basic Science **Enuresis is defined as the involuntary passage of urine. It usually refers to bedwetting at night.** It can also occur during the day. Enuresis is a common occurrence. **Most children develop bladder control between the ages of 3 and 6,** but as many as 15% of 8-year-olds still wet the bed.

Although **the etiology of enuresis is still unclear,** it is believed that some children experience persistent bedwetting because of a small bladder capacity or the inability to control their urge to urinate. Others drink too much before bed and do not wake up to empty their bladder. Some may make large amounts of urine because they do not have a nighttime rise in antidiuretic hormone.

A few patients (< 5%) have enuresis as a symptom of an underlying condition, most commonly a urinary tract infection or diabetes mellitus. Other neurologic or genitourinary tract abnormalities should be considered. Some children who have recently been toilet trained may regress and have secondary nocturnal enuresis under stress such as illness, a family move, or a new baby.

■ History and Physical Examination **Most children with enuresis not caused by an underlying condition will have a normal history and physical examination.** Important details in the history include birth history, growth, development, previous urinary tract infections, urinary symptoms, and stream. The family history often reveals childhood enuresis in one or both parents.

The physical examination is usually normal. The examiner should check closely for elevated blood pressure and abdominal, genital, and neurologic abnormalities.

■ Laboratory and Diagnostic Studies **All patients should have a urinalysis and urine culture.** The decision to obtain other studies should be based on findings in the history and physical examination.

■ Formulating the Diagnosis **A normal history, physical examination, urinalysis, and urine culture rules out most medical and surgical etiologies of enuresis of concern.**

■ Evaluating the Severity of Problems Successful treatment of underlying disorders that cause enuresis will usually result in resolution of the enuresis.

■ Managing the Patient Once a pathologic etiology has been excluded, treatment can include behavior modification, bladder-stretching exercises, and pharmacologic intervention when appropriate. Parents should be reassured that enuresis is common among school-age children and that spontaneous resolution occurs in 15% of children each year. **Children usually do not wet the bed purposefully and should not be punished.** Instead, they should be encouraged to take responsibility when bedwetting does occur; for example, they should be involved in changing sheets and doing laundry. **They should be actively involved in their treatment plan and receive positive reinforcement, such as gold stars on dry days.**

Initial attempts to manage enuresis should begin around 4 years of age with establishing a day and night toileting routine. Daytime urination should be postponed to increase bladder capacity. Daytime fluid intake should be encouraged while nighttime fluids discouraged. Children who have secondary nocturnal enuresis as a symptom of stress or regressive behavior can be managed with time and patience.

If enuresis persists, children can use bedwetting alarms at night that sense moisture and waken them before fully voiding. Alternatively, a bedside alarm can be set as a reminder to urinate in the middle of the night.

Older children who continue to have enuresis may benefit from intranasal desmopressin (an analog of antidiuretic hormone) **or imipramine** (a tricyclic antidepressant). Both drugs have been effective, but when stopped the enuresis often recurs. It may be helpful to use these drugs with other therapies such as the alarm and bladder-stretching exercises.

Epiglottitis

- Basic Science **Epiglottitis is a life-threatening acute infection of the supraglottic region of the upper airway. As the arytenoids, aryepiglottic folds, and the epiglottis swell, the upper airway becomes obstructed.**

  Epiglottitis is usually caused by infection with *Haemophilus influenzae* type b (Hib). The incidence of epiglottitis has decreased considerably since the introduction of Hib vaccination. *Streptococcus pneumoniae* and other bacteria have also been implicated and should be considered in a fully immunized child.

- History and Physical Examination The diagnosis of epiglottitis is usually readily apparent. **The affected child is usually 2 to 5 years of age,** although adult and infant cases have been reported. **The child will have had a sudden onset of fever and sore throat, followed in a few hours by drooling, muffled voice, and a very ill appearance. Children assume a characteristic posture: sitting quietly while leaning forward with the neck extended, mouth open, and an anxious expression.** Children with advanced airway obstruction may have decreased air exchange, altered level of consciousness, and gray color. **Since agitation of the child may lead to laryngospasm, no attempt should be made to separate a conscious child from his parents, force the patient to lie down, or to view his pharynx with a tongue blade.**

- Laboratory and Diagnostic Studies **Although an enlarged epiglottis seen on a lateral neck radiograph will confirm the diagnosis, this is not advisable given the tenuous nature of the patient's airway.**

- Formulating the Diagnosis **Epiglottitis is a clinical diagnosis made on the basis of the history and physical examination.** Chemical or thermal injury to the upper airway may cause an epiglottitis-like picture. Foreign body aspiration may also mimic epiglottitis. Retropharyngeal abscess and croup must be considered in the differential diagnosis; however, the onset of retropharyngeal abscess is usually less abrupt than epiglottitis. **Barking cough, absent in epiglottitis, suggests croup.**

- Evaluating the Severity of Problems Epiglottitis is a medical emergency. Complete airway obstruction is the most important complication.

Table 17.1 ***Commonly Isolated Bacterial Agents in the Pediatric Population***

Newborn (0-6 Weeks)	6-8 Weeks to 4-5 Years	Over 5 Years
Group B streptococcus (*Streptococcus agalactiae*)	*Haemophilus influenzae* type B	*Streptococcus pneumoniae*
Gram-negative enterics, mostly *E. coli*	*Streptococcus pneumoniae*	*Neisseria meningitidis*
Listeria	*Neisseria meningitidis*	*Staphylococcus aureus*

■ **Managing the Patient** Every emergency department should have a plan for management of the epiglottitis patient. **Oxygen should be administered and monitors applied as tolerated. Equipment and medications for intubation and resuscitation should be at the child's bedside, an operating room rapidly prepared, and a team skilled at endotracheal intubation and tracheostomy assembled.** In the operating room, the trachea is intubated, intravenous access obtained, and broad-spectrum antibiotics administered. **Should endotracheal intubation fail, tracheostomy is performed.**

Once the airway is secured, the patient is admitted to an intensive care unit. Intubation is usually necessary for 48 hours; as bacteremia is often present, **intravenous antibiotics are usually administered for 10 to 14 days** (third-generation cephalosporin or chloramphenicol).

Venipuncture and IV placement should be performed after the airway is secured; a blood culture should then be drawn since the patient is often bacteremic. A swabbed culture of the epiglottis may reveal the infecting organism. CSF should be collected if meningitis is suspected.

If transfer of the child with epiglottitis is necessary, he or she should be accompanied by someone skilled in airway management.

Fever Without Localizing Signs

■ **Basic Science** Acute febrile illnesses constitute the majority of patient visits to a pediatric office or emergency room. Many infectious diseases present with classic features that make the diagnosis obvious. **Occasionally, infectious agents may cause fever without localizing signs,** presenting the physician with a diagnostic challenge. Although most infectious illnesses are benign and self-limited, **high fever in a child (particularly a child under 2 years of age) should prompt an investigation for more serious bacterial disease.**

At different ages, children have different exposure and susceptibility to bacterial agents (Table 17.1). Newborns usually become infected with organisms acquired by passage through the birth canal or by ascending infection of the lower female genital tract. Older infants and children are exposed to pathogens through close contact by caregivers, pets, day care, and school attendance.

- History and Physical Examination **High fever (38.5° C or higher) in infants less than 2 years of age should be considered as a separate group because of the frequency of severe illnesses, particularly for infants younger than 6 months of age.** Several subjective features are often cited to distinguish those infants who are more ill from those who have a benign, self-limited febrile disease. Among large studies of febrile infants, the degree of playfulness, alertness, consolability, use of eye contact, and other age-appropriate behaviors are all favorable clinical observations. However, proper interpretation of these features requires the experience of the examiner. **A thorough physical examination should be performed to attempt to identify the source of the fever.**

- Laboratory and Diagnostic Studies Adjuncts to the clinical observation of the infant include a **complete white count and differential, cultures of blood and urine for bacteria, and consideration of a lumbar puncture for culture, cell count, and chemistries.**

- Formulating the Diagnosis **A white count of greater than 15,000/µl in a child with a fever of 38.5° C or greater, combined with a bandemia and a suggestive clinical picture, identified a significant number of infants who subsequently were found to have bacteremia.**

- Evaluating the Severity of Problems Young infants are poorly able to localize infection and may rapidly disseminate bacteria. Clinical and laboratory evaluations should be thorough but selective, with an emphasis on attempting to identify those infants who are at high risk of bacterial disease. This is one of the most critical and difficult decisions in the care of children.

- Managing the Patient **In the past, many infants under the age of 3 months with fever were automatically admitted to the hospital for several days of empiric antibiotic therapy pending the results of blood, urine, and spinal fluid cultures. With the advent of broad-spectrum antibiotics that could be administered by the intramuscular route, some physicians now favor the outpatient use of IM antibiotics if a febrile infant is at low risk for bacterial infection. However, this should be done only by experienced clinicians who can clinically evaluate the child. In addition, adequate follow-up must be ensured, and the child must have parents willing and capable of assessing their child for changes in status. If all these criteria are not met, the child should be hospitalized pending results of cultures.**

  Since the bacterial agents causing sepsis are different in different pediatric age groups, the empiric antibiotic therapy will vary with the age of the child. **Ampicillin and gentamicin or ampicillin and ceftriaxone are often used in the newborn period and can be useful until about 2 or 3 months of age. These two regimens will cover group B streptococci, gram-negative enterics, and listeria. Third-generation cephalosporins alone are not appropriate for this age group because *listeria* are resistent to third-generation cephalosporins. Third-generation cephalo-**

Table 17.2 *Common Bacterial and Viral Pathogens in Diarrhea*

Pathogen	Pathogenesis	Blood	Mucus	Treatment
Bacterial				
Campylobacter	Invasion and enterotoxin	Yes	Variable	Supportive; erythromycin if severe
Cholera	Invasion and enterotoxin	No	No	Bactrim
E. coli	Invasion and enterotoxin; cytotoxin	Variable	Variable	Supportive, or Bactrim, or ampicillin
Salmonella	Invasion and enterotoxin	Variable	Moderate	None unless severe, ampicillin, chloramphenicol, Bactrim
Shigella	Invasion and enterotoxin	Yes	Yes	Bactrim
Yersinia	Invasion and enterotoxin	Yes	Yes	None, or Bactrim
Viral				
Rotavirus	Mucosal destruction	No	No	Supportive

sporins alone are used at many centers for infants greater than 2 to 3 months of age. An alternative regimen of ampicillin and chloramphenicol is also acceptable in this age group. High percentages of *Haemophilus influenzae* type b are ampicillin resistant and therefore prompt the use of third-generation cephalosporins or chloramphenicol until culture results are known. **The emergence of penicillin-resistant pneumococci and other resistant organisms may alter the recommendations for empiric antibiotic therapy in the near future. At all ages, antibiotics should be directed at the most likely pathogen(s); if an organism is isolated, therapy should be modified to cover only the isolated organism.** If bacterial cultures are negative within 48 to 72 hours and the child is clinically well without any development of new signs or symptoms suggesting bacterial disease, antibiotics can often be stopped.

GASTROENTERITIS

- Basic Science **Gastroenteritis is a syndrome of diarrhea (≥5 loose stools/day) usually accompanied by emesis. In children, it is most often of viral etiology.** During the **summer months enteroviruses** are the most common agents, whereas in the late **winter and early spring, rotavirus** is the usual culprit. However, many other viruses can cause gastroenteritis.

  The most common bacterial causes of gastroenteritis include *Salmonella, Shigella, Yersinia, Campylobacter,* and enteroinvasive *E. coli*. Less common are *Staphylococcus, Clostridium difficile* and the parasites, *Entamoeba histolytica,* and *Giardia lamblia.* The specific characteristics of selected pathogens are listed in Table 17.2.

- History and Physical Examination **In general, viral disorders of the gastrointestinal tract produce a gastroenteritis with vomiting preceeding or concurrent with the diarrhea. In contrast, most (but not all) bacterial pathogens produce enteritis with little or no emesis. The exception to this is staphylococcal food poisoning, which produces both vomit-**

ing and diarrhea. In addition, bacterial diarrhea is much more likely to produce bloody or mucusy stools. There may be history of contact with ill persons, animals, contaminated water, undercooked food, or travel outside the area of residence.

On physical examination, the presence of fever should be sought. The examination may be normal but is often remarkable for hyperactive bowel sounds, a scaphoid abdomen, and the clinical signs of dehydration.

■ Laboratory and Diagnostic Studies **Examination of a freshly voided stool is the mainstay of diagnosis** for both viral and bacterial pathogens. **A positive test for occult blood and a Wright's stain of stool revealing many polymorphonuclear cells, often in sheets, are characteristic of stools positive for bacterial pathogens.** Viral diarrhea usually does not contain blood but may produce blood-streaked stool if the child develops an anal fissure; an irritated, bleeding hemorrhoid; or a severe diaper rash secondary to the passage of copious stools. If the history and stool examination suggests a bacterial pathogen, a stool culture and susceptibility test should be done. If the child's age and season of the year suggest viral diarrhea, a rotazyme (an ELISA specific for rotavirus), an ELISA for enteric adenovirus, or a stool culture for enterovirus may be appropriate considerations. **Depending on the severity of illness, ancillary tests including electrolytes, complete blood count with differential, and blood cultures may be warranted.**

■ Formulating the Diagnosis **Appropriate history and examination of the stool will aid in the identification of most cases of gastroenteritis.** Occasionally an acute abdominal process or a UTI may present with similar history and physical findings. Examination of the urine or abdominal films may be warranted in children who do not have a clearly identified case of gastroenteritis, particularly if the child is acutely ill.

■ Evaluating the Severity of Problems **Questions regarding the quantity of stool passed, ability of the child to retain oral fluids, and assessment of the child's hydration status by physical and electrolyte evaluations will allow the clinician to determine the severity of the child's condition.** Any child unable to retain oral fluids or with continued GI fluid losses should be admitted for IV fluids and observation. In addition, any child with a toxic appearance should be admitted and given antibiotics pending the results of cultures.

■ Managing the Patient Viral gastroenteritis requires only supportive therapy; there is no specific antiviral therapy. Most infants and children can be managed with oral rehydration consisting of a balanced salt solution with a simple sugar (Pedialyte, Lytren, or the WHO oral rehydration solution). Bland meals such as the BRAT diet (bananas, rice, applesauce, and toast) are often well tolerated. If emesis is protracted or severe, an infant or young child can become significantly dehydrated, necessitating intravenous rehydration.

Bacterial diarrhea is also often treated symptomatically, without the use of antimicrobial agents. The exceptions to this are *Shigella* (treated to prevent spread of infection) and

other bacterial diarrhea that is particularly severe and not resolving spontaneously. The choice of antibiotic for each pathogen is listed in the table. In general, the use of Lomotil (diphenoxylate) or Imodium (loperamide) to decrease gastrointestinal motility is not recommended for children.

Gastroesophageal Reflux

- Basic Science **Gastroesophageal reflux occurs during the first 6 months of life due to poor lower esophageal sphincter tone.**
- History and Physical Examination **There is a history of projectile, nonbilious vomiting after feeding.** The infant is afebrile, with normal appetite and activity. The infant should be vigorous and nontoxic appearing, with a normal physical examination.
- Laboratory and Diagnostic Studies **Laboratory studies such as CBC and electrolyte analyses are usually not required but should be normal. Barium swallow or pH probe studies can confirm the diagnosis.**
- Formulating the Diagnosis **The diagnosis can often be suspected based on the history and physical examination. It may be confirmed by one of the previously mentioned studies.**
- Evaluating the Severity of Problems Gastroesophageal reflux rarely causes dehydration or failure to thrive. However, the infant must be assessed for these conditions. The presence of these, or an ill-appearing infant, warrants consideration of other possible diagnoses.
- Managing the Patient **Treatment consists of giving the infant thickened feeds and placing him or her in an upright posture after feeding. On occasion, cisapride or metoclopramide is given to enhance gastric emptying. Gastroesophageal reflux usually resolves by the time the infant reaches 6 months of age.**

Immunizations

- Basic Science In "active" immunization, a form of the infectious agent is introduced via injection or ingestion, which in turn stimulates the immune system to develop antibodies. **Most routine childhood immunizations, such as DTP** (i.e., diphtheria, tetanus, and pertussis), **MMR** (i.e., measles, mumps, and rubella), **and OPV** (oral polio vaccine), **are forms of active immunization.**

  Passive immunization involves using antibodies produced by other animals to treat a patient already exposed to a disease. Examples include hepatitis B immune globulin (HBIG) **and tetanus immune globulin.** Although passive immunization takes effect immediately, it is short-lived. It is often combined with active immunization to result in long-term disease prevention.
- Specific Immunizations
  - **MMR (measles, mumps, and rubella) vaccine The first dose of MMR is given at 12 to 15 months. The second**

dose of MMR should be given at entry to junior high school. Current measles and mumps vaccine (also yellow fever and influenzae vaccines) contain small amounts of egg proteins. **Skin testing with the vaccine is recommended before administration in individuals with a history of systemic anaphylactic symptoms after egg ingestion.**

- **Polio virus vaccine Two types of vaccine are available:**
  - **Live oral** polio virus vaccine (OPV)
  - **Inactivated intravenous** polio virus vaccine (IPV)

  Either vaccine is given at ages 2, 4, and at 6 or 18 months.

  Live (bacterial, viral) vaccines are contraindicated in patients with immune system compromise (HIV, chemotherapy). Children with immunologically compromised housemates should not receive the oral polio virus vaccine (OPV). This virus is shed in the stool and can be transmitted to immunocompromised persons. **In such cases IPV should be given. Note that live MMR may be given because virus shedding does not occur.**

- **Hepatitis B vaccine (HBV) HBV is given at birth, 1 month, and 6 months, although the vaccine series may be started at any age. When the mother is hepatitis B positive, the infant should receive HBIG and HBV immediately after birth. This provides passive and active protection.**

- **DTP (diphtheria, tetanus, and pertussis) vaccine The immunization series consists of five doses of vaccine given at 2, 4, 6, 15 to 18 months, and before school. The DTP vaccine efficacy is so high that diphtheria and tetanus are literally absent in immunized individuals. However, the pertussis vaccine is less effective, and this disease occasionally occurs.** A long list of side effects includes erythema, pain at the injection site, fever, and irritability. **Children with the following conditions should not receive further doses of pertussis vaccine:**
  - **Seizure within 3 days of receiving the vaccine**
  - **Inconsolable crying for 3 or more hours within 48 hours of the vaccination**
  - **Collapse or shock within 48 hours of vaccination**
  - **High fever within 48 hours of vaccination**

  For children over 7, diphtheria immunization consists of Td (the tetanus and diphtheria toxoids without pertussis). This vaccine is less likely to produce adverse reactions in adults and older children.

- *Haemophilus influenza* type B (Hib) Now rarely seen due to immunization, this organism was the most common cause of bacterial meningitis in infants and young children and the most frequent cause of epiglottitis and septic arthritis in young children. Generally, **it is administered at 2, 4, and 6 months.**

- **Varicella vaccine A single subcutaneous dose of this live attenuated vaccine is recommended for all healthy children between 1 and 12 years of age. It is contraindicated in immunocompromised children and in those who have had varicella infection.** Its efficacy is good, although experience with this vaccine in the United States is limited.

■ Site and Route of Immunization Injectable vaccines should be administered in the anterolateral aspect of the upper thigh or the deltoid area of the upper arm. The upper, outer aspect of the buttocks in infants should not be used for injections since these injections can damage the sciatic nerve.

■ Lapsed Immunizations **A lapse in an immunization schedule does not require one to repeat the entire schedule. If a dose of DTP, OVP, Hib, or HBV is missed, immunization should occur on the next visit as if the usual interval had elapsed.**

In a patient who is lacking at least three doses of tetanus vaccine and has a wound that requires tetanus prophylaxis, both active and passive immunization are necessary. Such a patient would initially receive tetanus immune globulin (TIG) and Td or DTP at the time of injury, and then he or she would be scheduled to complete the Td or DTP series in 2-month intervals.

If the patient has previously received the series of four tetanus vaccines, he or she would require Td if it has been over 5 years since the last vaccine.

■ Misconceptions Regarding Vaccine Contraindications Some physicians inappropriately defer or delay immunizations. In the following cases immunizations *should* be given:

- Mild acute illness with low-grade fever or mild diarrhea in an otherwise well child
- Child taking antibiotics
- Reaction to previous DTP involving only soreness, redness, or swelling at vaccination site or mild temperature
- *Prematurity:* the appropriate age for initiating immunizations in a preemie is the chronological age. Vaccine doses should *not* be reduced for preterm infants.
- Pregnant household contacts
- Recent exposure to infectious disease
- Allergies to penicillin or any other antibiotic except anaphylactic reactions to neomycin or streptomycin (IPV contains trace amounts of streptomycin and neomycin)
- Family history of SIDS
- Family history of seizures
- Malnutrition

INTUSSUSCEPTION

■ Basic Science **Intussusception is the prolapse of one segment of the bowel into the lumen of another segment.** This kind of intestinal obstruction may involve segments of the small intestine, the colon, or the terminal ileum and cecum. **Intussusception is the most common cause of bowel obstruction in the 3- to 12-month age group.**

- History and Physical Examination **Intermittent abdominal pain is common** (manifesting as crying episodes in the preverbal child). **Anorexia and vomiting usually have occurred. Microscopic blood in the stool is common. Currant jelly stool is a late finding.**

  Physical examination demonstrates lethargy or irritability. The abdomen may be tender. Sometimes the intussusception can be palpated. **It is commonly described as a sausage-like mass, most often in the child's right upper quadrant.**

- Laboratory and Diagnostic Studies **The contrast enema is the best study for diagnosing and reducing intussusception.** Plain film findings are variable. The radiograph may be normal or may demonstrate signs of bowel obstruction, such as distention and air fluid levels. Laboratory studies, such as a complete blood count (CBC), may reflect a stress response or, in the case of perforation, may be markedly elevated with a left shift. Electrolytes are usually normal.

- Formulating the Diagnosis **The diagnosis is based on the history, physical, and contrast enema** (when appropriate, see below) **findings.**

- Evaluating the Severity of Problems Protracted intussusception can lead to gangrenous bowel or perforation, which are surgical emergencies. **Any child with suspected intussusception, who is ill-appearing with altered mental status** (i.e., lethargy to the point of not responding to parents); **signs consistent with shock,** such as delayed capillary refill or low blood pressure, **peritonitis, or perforation** (free air on plain film); **requires immediate surgical consultation and surgical repair. A contrast enema would not be safe in such a patient.**

- Managing the Patient **Contrast enema is the procedure of choice to reverse intussusception. Any child with suspected intussusception should be made NPO and placed on intravenous fluids. A free air series should be done to search for signs of obstruction and perforation. Surgical consultation should be obtained before a contrast enema** for the following reasons:

  - There is a risk of perforation during contrast enema.
  - The child may have evidence of gangrenous bowel or perforation, which are contraindications to contrast enema.
  - Even in cases of successful reduction of the intussusception with contrast enema, there is a small recurrence rate.

  If the contrast enema reduces the intussusception, the child should remain in the hospital for 24 hours to observe for signs of recurrence. If the enema is not successful, surgical repair is necessary.

MEASLES

- Basic Science **Measles, or rubeola, is a paramyxovirus transmitted by contact with respiratory secretions.** Patients are contagious from 24 to 48 hours before the onset of symptoms and remain so for 4 days after the appearance of the rash. The incubation period is from 8 to 12 days.

- **History and Physical Examination** Measles is characterized by the three Cs, **cough, conjunctivitis, and coryza, as well as fever and morbilliform rash. Koplik spots** (small white spots that appear inside the buccal and labial mucosa early in the disease) are the pathognomonic enanthem of measles. The rash usually appears 3 to 5 days after the initial symptoms. **Red maculopapular lesions begin on the head and progress to cover the trunk and extremities, becoming confluent.** During the next 6 to 7 days, the rash turns a brownish color and then fades. Fever is common in the first 3 to 4 days of illness.
- **Laboratory and Diagnostic Studies** The WBC is usually low in measles, although this is a nonspecific finding. The diagnosis can be confirmed by immunologic testing (usually ELISA) for antibodies.
- **Formulating the Diagnosis** **The diagnosis is made based on the clinical findings, history, and exposure to measles in an unvaccinated child.**
- **Evaluating the Severity of Problems** Most children with measles experience fever, pharyngitis, and rash. Lymphadenopathy, splenomegaly, vomiting, and diarrhea are also common symptoms but can be usually managed on an outpatient basis. Other **complications include pneumonia, both viral and bacterial, which may require hospitalization for oxygen and parenteral antibiotics.**

  Neurologic complications include encephalitis, which occurs in 1 per 1000 to 2000 cases of typical measles and has a 15% to 24% morbidity and mortality rate, **and subacute sclerosing panencephalitis (SSPE),** which occurs in about 1 per 100,000 cases. This is a degenerative nervous system disease caused by persistent measles infection of the central nervous system. It causes intellectual deterioration, progressing to dementia and rigidity, and is usually fatal.
- **Managing the Patient** **Typical measles in a normal host is managed at home with antipyretics, fluids, and rest.** Measles pneumonia may be viral or have a secondary bacterial component and often requires hospitalization.

  Atypical measles, worth mentioning here, is an infection in an immunized host whose antibody titer has dropped. It is most often seen in adults who were immunized between 1963 and 1967 with the killed virus vaccine, which is no longer in use. It begins with fever, headache, myalgia, weakness, abdominal pain, and a cough. The rash is atypical, usually starts on the distal extremities, and may be maculopapular or purpuric. Most patients get pulmonary involvement, with chest radiograph findings of hilar adenopathy and pleural effusions, and hospitalization is often necessary.

MUMPS

- **Basic Science** **Mumps is a paramyxovirus infection of the salivary glands (predominantly the parotid glands). It is transmitted by exposure to oral secretions** from an infected patient.

The incubation period from exposure to glandular swelling is approximately 18 days. Patients with mumps are infectious from 1 to 2 days before the onset of glandular swelling and remain so for 8 to 10 days. Currently the disease may be avoided by using a vaccine.

- History and Physical Examination **Mumps may range from an asymptomatic infection (up to 40%) to the classic illness, which is characterized by fever, malaise, headache, abdominal pain, and glandular swelling.**
- Laboratory and Diagnostic Studies Although laboratory studies are usually not necessary to make the diagnosis, the serum amylase, if measured, may be elevated. The diagnosis can be confirmed by an ELISA.
- Formulating the Diagnosis **The diagnosis of mumps is based on the clinical presentation of generalized symptoms in combination with parotid swelling in unimmunized, previously uninfected individuals. Usually the parotid swelling begins in one gland and then becomes bilateral,** although some patients have only unilateral involvement. Before vaccination, the most common age of infection was 5 to 9 years; presently, however, in states with relaxed vaccination policies, the peak incidence occurs in 10- to 14-year-old children.
- Evaluating the Severity of Problems **Mumps is usually a mild, self-limited disease that requires no specific treatment,** but it can be complicated by **orchitis** in males, which **occurs in approximately one third of postpubertal males** who contract mumps. Far less frequent complications include meningoencephalitis; pancreatitis; thyroiditis; oophoritis; mastitis; neuritis, especially of the facial and auditory nerves; myelitis; arthritis; and myocarditis. Note that parotitis may also be caused by other viruses and bacteria, such as *Staphylococcus aureus.* In these cases, there is usually erythema and tenderness of the skin overlying the involved gland, and antibiotic therapy is indicated.
- Managing the Patient **Most patients with mumps require only symptomatic treatment, such as acetaminophen, fluids, and rest until symptoms resolve (in 7 to 10 days).** Orchitis is also treated with rest and avoidance of contact sports until it is completely resolved. **If patients have high fever, headache, and vomiting, they should be assessed for pancreatitis and meningoencephalitis.**

  The other less common complications should be considered if suggested by the patient's symptoms.

Normal Newborn Behavior

- Feeding **Newborns eat every 2 to 4 hours, between 2 and 4 ounces per feeding, or 5 to 10 minutes on each breast.**
- Sleeping **Newborns sleep about 16 hours daily.** This number decreases in the first year of life.
- Voiding/Stooling **Newborns urinate up to 20 times daily and stool between one and seven times daily. The stools are often soft, yellow-green, and "seedy" in appearance.**

- Jaundice **The newborn liver is slow to conjugate bilirubin, and mild *indirect* hyperbilirubinemia is common. It is more prevalent in breast-fed than formula-fed infants, but it rarely requires treatment in the healthy, term infant.**
- Crying Crying is the only way a newborn can communicate his or her needs such as hunger, cold, and fatigue. **Please refer to the "Colic" section of this chapter for details.**
- Temperature Control/Color **The absence of fever in a newborn does not eliminate the possibility of severe bacterial infection.**

  Full-term newborns often have undeveloped distal circulation, which may result in a purple appearance to their hands and feet when they are cold or crying. If they remain pink centrally (lips, tongue) and "pink up" distally when warmed and calmed down, this condition is not a concern in the absence of other symptoms.

PHARYNGITIS

- Basic Science Many viruses can cause pharyngitis, particularly the group of viruses that cause other respiratory symptoms such as influenza viruses, parainfluenza viruses, adenoviruses, and rhinoviruses. Streptococci, usually group A (but also group C and group G), may cause bacterial pharyngitis. *Mycoplasma pneumoniae* has also been implicated as a cause of acute pharyngitis. Epstein-Barr virus can cause pharyngitis as part of the clinical picture of infectious mononucleosis. Diphtheria, although rare in this country, can cause membranous pharyngitis and should be considered in an unimmunized patient.

  For convenience one can classify the appearance of the tonsils or pharynx into one of three classifications: exudative, ulcerative, or membranous. Exudative pharyngitis is defined by the presence of a white or gray film on the surface of the tonsils or pharynx that resembles skim milk and could be readily wiped off without bleeding. White, hard material occasionally seen in the tonsillar crypts is more appropriately called *crypt debris* and is not an exudate. **Exudative pharyngitis is the most common manifestation of group A streptococcal pharyngitis.** Ulcerative pharyngitis is characterized by the presence of circular or shallow ulcers on the soft palate, tonsillar area, or posterior pharynx. ***Herpangina* is an older term that is sometimes used to describe ulcerative pharyngitis (usually caused by coxsackie viruses). Herpes simplex viruses may also cause an ulcerative pharyngitis.** Membranous pharyngitis is the presence of a grayish-white membrane adherent to the surface of the pharynx and tonsils. When removed, there is bleeding from the surface underneath. In the United States, **most cases of membranous pharyngitis are caused by Epstein-Barr virus (EBV) mononucleosis, particularly in teenagers and young adults. Pharyngitis coupled with fever, fatigue, generalized nontender adenopathy, and enlarged liver or spleen should make one consider the possibility of EBV mononucleosis.** In unimmunized children, diphtheria is a more common cause of membranous pharyngitis.

- **History and Physical Examination** Pharyngitis may occur alone or in combination with other symptoms in many pediatric infectious diseases. When evaluating complaints of a sore throat it is important to get a thorough look at the pharynx using a good light and whatever method of restraint is necessary to adequately visualize the area. **One should closely examine the oral and buccal mucosa for associated redness or lesions, examine the cervical lymph nodes for lymphadenopathy or adenitis, and palpate the abdomen for the presence of an enlarged liver or spleen.**

- **Laboratory and Diagnostic Studies** **The most useful adjunct to the diagnosis of pharyngitis is the use of a throat culture for identification of group A streptococcus.** Not every patient with a complaint of sore throat should receive a throat culture. This should be reserved for those patients with history of exposure to streptococci; streptococcal infections in the past; or the clinical features that suggest streptococcal pharyngitis, including exudative or membranous pharyngitis, enlarged tonsils with tender cervical adenopathy, and fever. The recovery of β-hemolytic colonies on throat culture does not prove the presence of a streptococcal infection, but in the presence of suggestive clinical features, a positive throat swab for group A streptococcus is sufficient to warrant therapy. Culturing for other agents of pharyngitis is possible but usually unnecessary.

  Those patients who have features suggesting EBV infection should undergo a complete blood count, differential, and a monospot test. Often one will find an elevation of the total white count with a predominance of atypical lymphocytes and monocytes from the peripheral smear. Eighty percent of patients with EBV mononucleosis can be diagnosed by a monospot test. However, **in children under 5 years of age, a monospot test will not necessarily be positive,** even in cases of proven EBV infection. **The preferred method of diagnosis in a young child is the evaluation of IgG and IgM antibody titers** against EBV viral capsid antigen (EBV-VCA). The IgM antibody is positive during the acute phase of illness, whereas IgG antibodies develop during convalescence. These may also be useful in monospot negative EBV infections to clarify the diagnosis.

- **Formulating the Diagnosis** **The diagnosis of pharyngitis is based on the clinical presentation and appropriate physical findings.** Laboratory tests for group A streptococcus, viruses, or Epstein-Barr virus (mono) will not always be necessary but can help confirm the specific diagnosis.

- **Evaluating the Severity of Problems** Pharyngitis is rarely a serious problem. Occasionally a patient with severe infectious mononucleosis or diphtheria will have obstructive symptoms requiring medical intervention.

- **Managing the Patient** **Most viral pharyngitis is managed with symptomatic therapy. The treatment for group A streptococcal pharyngitis is a 10-day course of oral penicillin. Children are considered noninfectious after 24 hours of therapy and can return to school then if they are well enough. For the penicillin-allergic patient, erythromycin may be used.** The rationale for treating patients with group A β-hemolytic streptococci is to prevent rheumatic heart disease. In contrast, the incidence and severity of glomerulonephritis are not altered by treatment of streptococcal pharyngitis.

Treatment of EBV infection is mostly supportive. Children with splenomegaly should not be allowed to participate in contact sports until their spleen has returned to normal to prevent splenic rupture. Rarely, steroids may be necessary for cases of EBV infection where the tonsillar enlargement is severe enough to cause compromise of the airway, but the routine use of steroids for uncomplicated EBV infection is unwarranted and may prolong the course of infection.

Pneumonia

- Basic Science **In all ages of pediatric patients, pneumonia is most commonly caused by viruses.** Respiratory syncytial virus, influenzae, parainfluenza, and adenoviruses comprise the majority of viral agents responsible for pediatric viral pneumonia. **Primary bacterial pneumonia is less common in pediatric patients but still occurs.** Children with poor control of their oral secretions are particularly prone to aspiration pneumonia. **Newborn infants may develop group B streptococcal pneumonia *(Streptococcus agalactiae)* or, less commonly, gram-negative pneumonia due to aspiration of secretions contaminated by maternal vaginal flora at the time of delivery. *Chlamydia trachomatis* pneumonia may also present in the first few weeks of life** in infants exposed to maternal genital tract chlamydia. **Older infants and children are more likely to develop pneumococcal pneumonia.** *Streptococcus pyogenes* or *Staphyloccus aureus* may also cause pneumonia in children, particularly following a viral illness such as measles or influenza. ***Mycoplasma pneumoniae* and *Chlamydia pneumoniae* may cause an atypical pneumonia syndrome in school-aged children and young adults.**

- History and Physical Examination **In the newborn or young infant, pneumonia may present with extreme respiratory distress, hypoxia, hypercapnia, and cyanosis.** Many infants will be tachypneic and have varying degrees of respiratory difficulties manifested by nasal flaring, substernal or intercostal and supraclavicular retractions, and occasionally abdominal breathing patterns. Some infants may even be apneic, particularly those infants with respiratory syncytial virus. Cough and audible wheezing may also be clinical findings in infants with pneumonia. Posttussive emesis may make feeding difficult for the small infant and predispose to dehydration if severe. **During physical examination, an examiner may have difficulty hearing crackles or wheezes in a young infant with pneumonia. In older children, decreased breath sounds, crackles, and rhonchi are all characteristics of pneumonia. *Chlamydia trachomatis*** pneumonia is usually an afebrile pneumonia and may be preceded by conjunctivitis.

- Laboratory and Diagnostic Studies **Arterial blood gas sampling is suggested for children with severe respiratory distress, but**

measurement of pulse oximetry may be sufficient for less severe cases. **A CBC and differential may reveal an elevated white count with a left shift in cases of bacterial pneumonia.** For infants with chlamydia pneumonia, a peripheral eosinophilia may be a clue to the diagnosis. **Viral pneumonia can often be diagnosed by obtaining a viral culture** of respiratory secretions either obtained through an endotracheal tube, by a throat swab, or by a nasopharyngeal wash. Rapid diagnostic tests for RSV, parainfluenza, and influenza can also be applied to these specimens. Nasal secretions can also be obtained for culture or ELISA against chlamydia. **A febrile child with pneumonia should also have a blood culture since there may be an associated bacteremia.**

A chest x-ray can be a useful adjunct to the physical examination of a patient with suspected pneumonia. Viral pneumonia often shows patchy alveolar infiltrates or interstitial disease. *Mycoplasma* and *Chlamydia pneumoniae* pneumonia may have mild chest symptoms and, when viewed by radiography, demonstrate a fairly extensive interstitial pneumonia. Bacterial pneumonia is much more likely to produce a consolidation with an opacified area of lung. Chest x-rays can also be useful in cases where a pleural effusion is present. If there is sufficient pleural fluid, a pleural tap should be performed for diagnostic and therapeutic purposes. The fluid should be analyzed for protein, LDH, cell count, Gram's stain, and bacterial cultures.

- Formulating the Diagnosis **The presence of rales, rhonchi, and absent or decreased breath sounds in a febrile child suggests pneumonia. A chest x-ray confirms the diagnosis,** but often children are initially treated based on clinical findings alone.

- Evaluating the Severity of Problems Significant respiratory distress, cyanosis, or labored breathing signify significant lung pathology and warrant prompt medical intervention.

- Managing the Patient Most mild viral pneumonia can be managed by symptomatic support. **Supplemental oxygen should be administered to any child with pneumonia who has a significant degree of hypoxemia. Intubation and assisted ventilation may be necessary for some severe cases of pneumonia, particularly in a very small infant. Severe influenza pneumonia can be treated with amantadine. RSV pneumonia can be treated with ribavirin, if necessary. Pneumococcal pneumonia can be successfully treated with intravenous or oral penicillin. A mild aspiration pneumonia may be successfully treated with ampicillin, but more severe aspiration pneumonia, where anaerobes are suspected, may require a combination therapy of clindamycin and gentamicin. *Chlamydia trachomatis* pneumonia can be treated with erythromycin ethyl succinate, orally. Erythromycin or tetracyclines (in a child older than 9 years) is acceptable therapy for *Mycoplasma* or *Chlamydia pneumoniae* pneumonia.**

Pleural effusions will often resolve spontaneously, but pleural empyemas associated with pneumonia may require longer courses of antibiotics before resolution (up to 3 weeks). Early removal of a pleural empyema with serial taps or a chest tube may hasten clinical improvement. Open thoracotomy to remove empyema fluid is rarely necessary in pediatric patients. Unlike adults, children do not develop restrictive lung disease from empyemas. Follow-up chest x-rays in children with empyema demonstrate complete or nearly complete resolution within 6 months.

Pyloric Stenosis

- Basic Science **Pyloric stenosis is hypertrophy of the pyloric sphincter that prevents passage of gastric contents into the duodenum. It generally occurs in male infants under the age of 2 months.**
- History and Physical Examination **The history consists of projectile, nonbilious vomiting directly after feeding.** The infant does not have fever, diarrhea, or loss of appetite. **The hypertrophied pylorus may be palpated as an "olive" in the epigastric area.**
- Laboratory and Diagnostic Studies **Electrolytes are remarkable for hypochloremic alkalosis. Barium swallow will show delayed gastric emptying and a positive "string sign."**
- Formulating the Diagnosis **The diagnosis can be made based on the history and physical examination when an olive is felt by an experienced clinician; however, an olive is not always palpable, and in these cases barium swallow (or in some institutions ultrasound) can confirm the diagnosis.**
- Evaluating the Severity of Problems Initially infants with pyloric stenosis are vigorous with a normal to increased appetite; however, they may become **dehydrated** over time and should be assessed for this (see the section on "Dehydration").
- Managing the Patient Once the diagnosis has been made, treatment consists of **correction of dehydration and electrolyte abnormalities with intravenous fluids in preparation for definitive treatment, which is pyloromyotomy.** The operation involves surgical incision of the hypertrophied pylorus and is curative.

Retropharyngeal Abscess

- Basic Science **Infection can fill the potential space between the esophagus and the cervical vertebrae, leading to stridor and airway compromise. Group A streptococci, anaerobes, and *Staphylococcus aureus* are the most common pathogens.**
- History and Physical Examination These infrequent infections occur most commonly before age 6. A history of trauma to the neck or pharynx or recent infection should be sought. **A child with a retropharyngeal abscess usually complains of a stiff neck, preceded by nasal congestion or sore throat. High fever and toxic appearance are common. Drooling and stridor are often present. A midline mass may be visible on inspection of the posterior pharynx.**

- **Laboratory and Diagnostic Studies** **A lateral radiograph of the soft tissues of the neck usually demonstrates enlargement of the retropharyngeal space. Blood cultures** may identify the infecting organism. As fever and stiff neck may simulate meningitis, lumbar puncture for CSF collection may be necessary. **Gram's stain and culture should be performed on abscess material.**
- **Formulating the Diagnosis** The diagnosis is made both clinically and after results of a neck x-ray. **Other common causes of upper airway obstruction include peritonsillar abscess, epiglottitis, croup, and infected foreign bodies.**
- **Evaluating the Severity of Problems** Major complications include complete airway obstruction and aspiration. The abscess may spontaneously rupture or erode into other major structures, such as the mediastinum or lung.
- **Managing the Patient** **All children with retropharyngeal abscess require hospitalization and surgical abscess drainage.** Broad-spectrum intravenous antibiotics and fluids should be administered. Endotracheal intubation or, rarely, tracheostomy may be necessary to secure the airway.

Reye's Syndrome

- **Basic Science** Reye's syndrome is a clinical and pathologic entity of **unknown etiology.** Reye's syndrome is **characterized by a rapidly progressive encephalopathy and fatty degeneration of the liver.**

  The syndrome is **preceded by a viral illness** (influenza A and B, varicella, Epstein-Barr virus, enteroviruses). A **35-fold increase in the incidence of Reye's syndrome has been reported when aspirin or other salicylate products are used during an acute viral illness.** Because of this, the use of **aspirin in children is contraindicated** during "flu" season or chicken pox infection.
- **History and Physical Examination** Patients most commonly present with a history of a **viral infection** (upper respiratory tract infection). The patient may also report **aspirin usage** during the illness. About a week after the start of symptoms **severe vomiting** begins, followed by **mental status changes.** Mental status changes range in severity from a mild amnesia to irritability, confusion, disorientation, delirium, and coma.

  Children are usually not febrile and not jaundiced (although hepatomegaly does occur in about half of the patients). Meningeal signs and focal neurologic signs are absent. As the illness worsens the patient may manifest **decorticate or decerebrate posturing. Seizure and respiratory arrest may also occur.**
- **Laboratory and Diagnostic Studies** **The definitive diagnosis is made by liver biopsy.** A presumptive diagnosis can be made if **blood ammonia levels** are increased and **liver function tests are elevated** (AST and ALT are greater than three times normal, and the prothrombin time is elevated). Hypoglycemia is an early laboratory finding in more severe cases.

- **Formulating the Diagnosis** Reye's syndrome should be suspected in any child with a history of viral illness and/or aspirin usage in combination with mental status change and vomiting.

  Serum chemistry tests that demonstrate high levels of ammonia and liver dysfunction support the diagnosis.

- **Evaluating the Severity of Problems** **The prognosis is related to the degree of neurologic deficit.** The overall mortality rate is 21%; it is higher in patients who develop seizures and respiratory arrest.

  Neurologic deficits (mental retardation, epilepsy, and the like) are commonplace among those patients who develop decerebrate and decorticate posturing or seizures while ill.

  Recurrence is uncommon, and most survivors have a good prognosis.

- **Managing the Patient** The brain is the principal organ affected in Reye's syndrome. **Cerebral edema is the major factor contributing to morbidity and mortality.** Therefore **intracranial monitoring and reduction of increased intracranial pressure with mannitol and dexamethasone is essential. Hypoglycemia must also be corrected.**

  Vitamin K and fresh frozen plasma may help to reverse the patient's coagulopathy. Lactulose and neomycin are given to decrease ammonia levels. Serum electrolytes should be monitored frequently.

Rubella

- **Basic Science** **Rubella, also known as "German measles," is a viral infection that has its greatest clinical impact on fetal development during the first and second trimester of pregnancy** (see "Infections During Pregnancy," in Chapter 13). The virus is a togavirus acquired by exposure to oral droplets or via the placenta in the fetus. The incubation period is 14 to 21 days, with a 1- to 5-day prodrome of cough, conjunctivitis, coryza, lymphadenopathy, and low-grade fever. An erythematous maculopapular rash follows and lasts about 3 days. The patient is contagious the week before and after onset of this rash.

- **History and Physical Examination** **The history is one of low-grade fever, respiratory symptoms, and lymphadenopathy for about 5 days before the onset of the rash.** The most commonly affected lymph nodes are the cervical, posterior auricular, and suboccipital. **Adolescents and adults (females more than males) may also have arthritis and arthralgia of multiple joints, both small and large.** This usually begins during or after the rash. **As in rubeola (i.e., measles), the rash of rubella begins as discrete, red, maculopapular lesions, first on the head, progressing down to the trunk and extremities. However, it does not coalesce or turn brown and has a shorter duration than the rubeola rash.** The symptoms usually resolve within 2 weeks.

- **Laboratory and Diagnostic Studies** The diagnosis can be confirmed by obtaining acute and convalescent titers 10 to 14 days apart.

- **Formulating the Diagnosis** Rubella's symptoms and rash are nonspecific and difficult to distinguish from those associated with enterovirus, scarlet fever, mycoplasma, or human papovavirus 19 infection. **The history of missed MMR immunization is important in making the diagnosis, as is serologic testing.**
- **Evaluating the Severity of Problems** **Rubella ranges from asymptomatic illness (25% of patients) to arthritis requiring aspirin or NSAID therapy, but it is usually a mild, self-limited infection, managed at home with supportive care.**

  Congenital rubella, however, causes anomalies in 30% to 60% of infants infected during the first month of pregnancy. These include hepatosplenomegaly, lymphadenopathy, anemia, thrombocytopenia, meningoencephalitis, myocarditis, nephritis, cataracts, glaucoma, microphthalmia, sensory neural deafness, patent ductus arteriosus, and pulmonary artery stenosis. Infants infected in the second trimester have an increased risk of ocular, hearing, and developmental abnormalities.
- **Managing the Patient** **The prime directive of management is to prevent exposure of nonimmunized pregnant women to infected patients.** The infection in the nonpregnant patient is usually mild and requires only rest and analgesics. **In pregnant women, the infection during the first or second trimester is managed by counseling women about the risk to the fetus. Immunization after exposure is not helpful and is actually contraindicated during pregnancy.** (See the section in this chapter, "Immunizations.")

Urinary Tract Infection

- **Basic Science** Urinary tract infections (UTI) are common in pediatrics. The incidence of true urinary tract infection in the newborn and infant periods has been estimated to be less than 1% of the population at risk. **In the first 3 months of life, greater than 75% of all diagnosed urinary tract infections are in males. However, outside infancy, both symptomatic and asymptomatic urinary tract infections are much more common in females,** because of the relatively short urethra and close proximity to the vaginal and anal areas.

  The organism most frequently involved in UTIs is *E. coli*. *Proteus mirabilis* has been isolated more commonly from boys and from patients with renal stones. *Staphylococcus aureus* is responsible for occasional cases of hematogenous dissemination, and for renal parenchymal infections including pyelonephritis. Sexually active females may have a higher incidence of *Staphylococcus saprophyticus,* in addition to the common gram-negatives such as *E. coli.*
- **History and Physical Examination** **History suggestive of UTI in an infant may include foul-smelling urine, poor feeding or growth, and recurrent episodes of unexplained fever. Older children may complain of frequency, urgency, or dysuria. Some children may present with nighttime wetting or the re-**

appearance of frequent accidents following successful toilet training. Physical examination is often unhelpful in an infant, but **in an older child suprapubic tenderness or flank tenderness are useful physical findings.**

- Laboratory and Diagnostic Studies The diagnosis of urinary tract infection is not based on pyuria alone. White cells in the urine should be interpreted cautiously since they may arise from other structures, such as the vagina. Fever, glomerulonephritis, renal stones, appendicitis, or any extrinsic ureteral irritation may produce white cells in the urine without bacterial infection. A freshly voided, unspun urine sample should be examined by Gram's stain for the presence of bacteria. If present, this suggests at least 10^5 colony forming units (CFU)/ml of bacteria. **In general, suprapubic aspiration is the most reliable method for the clear diagnosis of urinary tract infection, but it cannot be performed in infants after approximately age 1 year since the bladder is no longer an abdominal organ.** In catheterized specimens, infection is highly likely if there are greater than 10^5 organisms of a single type isolated. For less bacteria isolated, infection is suspicious and repeat catheterization is usually recommended. For a clean voided specimen in a boy, greater than 10^4 organisms make the diagnosis of UTI likely, but for girls the issue is more complicated. Often more than one properly obtained, clean voided specimen is necessary to make the diagnosis of a UTI in a girl, and again **greater than 10^5 CFU of a single organism make the diagnosis likely.**

  Many studies have been conducted in an attempt to differentiate upper tract infection from lower tract infection (i.e., to differentiate a pyelonephritis from cystitis). However, in most acutely ill patients, the clinical symptoms and signs of infection will be the most important source of information. The use of radionuclide studies such as technetium glucoheptinate or technetium dimercaptosuccinic acid (^{99m}Tc-DMSA) are helpful in severe pyelonephritis to identify areas of scarring due to infection.

- Formulating the Diagnosis **The definitive diagnosis of urinary tract infection is based on the finding of significant bacteria in the urine obtained by catheterization, suprapubic aspiration, or clean voided specimen in a patient with an age-appropriate constellation of symptoms.**

- Evaluating the Severity of Problems The febrile infant, particularly if less than 3 months of age, should be admitted to the hospital for intravenous antibiotic therapy.

- Managing the Patient **The management of urinary tract infections is influenced by several factors, particularly the age of the child. In the febrile infant less than 3 months of age, the initial choice of antibiotics should include a synthetic penicillin such as ampicillin for gram-positive coverage and an aminoglycoside or second- or third-generation cephalosporin for gram-negative coverage.** A repeat urine culture is recommended after 2 or 3 days to ensure clearance of the organism and at the completion of antibiotic treatment. Intravenous antibiotics are continued for at least 5 days or until the patient has resolution of fever and other symptoms. Then the child can often be placed on an oral antibiotic therapy to complete a 10- to 14-day course. **In children older than 6 months of age who have symptoms and signs of a UTI, outpatient treatment with oral antibiotics for 7 to 10 days is recommended unless patients have high fever or other symptoms that imply upper tract disease. Although successful in adults, there are little data currently to support the use of short-course antibiotic therapy for uncomplicated UTIs in children. Useful oral antibiotics in this age group include ampicillin, amoxicillin, or trimethoprim/sulfamethoxazole (Bactrim, Septra).** Antibiotic therapy should be modified according to the antibiotic susceptibility pattern of the isolated bacteria.

Following all first-time UTIs, children should have an evaluation of their urinary tract, including a voiding cystourethrogram (VCUG) and a renal ultrasound. These studies allow for the diagnosis of vesicoureteral reflux or anatomic abnormalities that predispose to repeated urinary tract infections.

VARICELLA (CHICKENPOX)

- Basic Science **Chickenpox represents infection with the varicella-zoster virus, a herpes virus. It is highly contagious and strikes over 90% of the population at some point** during the first two decades of life. Zoster (or shingles, covered in Chapter 10), the latent form of this virus, becomes reactivated later in life, usually during illness or stress. At this stage the infection consists of burning pain, followed by a vesicular eruption in areas usually limited to one, but occasionally up to three, sensory dermatomes.
- History and Physical Examination In healthy children, chickenpox is routinely a mild, self-limited disease. **The primary infection often starts with a day or two of fever and mild systemic symptoms (e.g., malaise and headache). Then follows the characteristic rash, consisting of groups of vesicles, usually beginning on the trunk or face and spreading distally. The rash has the initial appearance of "dewdrops on a rose petal" (i.e., small, clear vesicles on an erythematous base), which may spread over a 5- to 7-day course, later crusting and healing.**
- Laboratory and Diagnostic Studies **The diagnosis is largely clinical,** but it can be confirmed by two methods: the Tzanck preparation and immunofluorescent assays.
- Formulating the Diagnosis **The diagnosis is commonly based on recognition of the rash, especially in a patient who has recently been exposed to varicella. (The incubation period is 7 to 21 days.)** The diagnosis can be confirmed by the Tzanck preparation or immunofluorescent assays; however, this is usually not necessary.
- Evaluating the Severity of Problems **Encephalitis and Reye's syndrome are rare, serious sequelae to varicella infection.**
- Managing the Patient **Chickenpox is a mild, self-limited illness in healthy children. It is treated with rest and medications**

to control itching (diphenhydramine, Aveeno baths) and fever (acetaminophen). Aspirin is contraindicated because of the increased risk of Reye's syndrome.

A complication afflicting healthy children is cerebellar ataxia, which occurs approximately 7 to 21 days after the development of the varicella rash.

Adolescents and adults are often more seriously afflicted during varicella infection and may develop mild hepatitis.

The elderly and immunocompromised are at increased risk for severe varicella complications, such as pneumonia, pancreatitis, hepatitis, and central nervous system problems (transverse myelitis, encephalitis, aseptic meningitis, and peripheral neuritis, including Guillain-Barré syndrome).

A single varicella infection generally gives lifelong immunity. However, immunocompromised persons who have been exposed to varicella should receive VZIG (i.e., varicella-zoster immune globulin) **to prevent infection;** they are very susceptible to severe illness and complications.

Immunocompromised patients should be treated with acyclovir. A live-attenuated vaccine is now available (and is discussed in the section "Immunizations").

MUTIPLE CHOICE REVIEW QUESTIONS

1. A 1-month-old infant is brought to the emergency room with a history of crying inconsolably over the past 4 hours. The infant's mother appears to be about 17 years old and seems very distraught. Past medical history, including birth history, is negative. Appropriate actions at this time would be:

a. A CBC, blood culture, and spinal tap to rule out medical causes of irritability.
b. Reassurance that this is colic and will pass with time.
c. A careful physical examination looking for causes of irritability.
d. Report to child protective services because this mother is high risk because of her age.

2. A 2-month-old infant is in your office for a well-child visit. She has a family history of seizures; however, her own past history is negative. Developmental history and neurologic examination are normal.

a. This baby should not receive DTP because of the family history.
b. This baby may receive vaccination for diphtheria and tetanus but should not receive the pertussis vaccine because of the family history.
c. This baby should receive IPV instead of OPV because of the family history.
d. This baby may receive the usual age-appropriate vaccines in spite of the family history.

3. You have a child with leukemia on chemotherapy in your practice. He has a 4-month-old sister who is now in your office for a well-child visit. The immunizations she should receive are:

a. DTP, OPV, and HIB.
b. DTP, IPV, and HIB.
c. DTP, OPV, and HBV.
d. DTP, MMR, and IPV.

4. You are working as the ship doctor on a cruise in the South Pacific. You are asked to see the 2-year-old child of one of the kitchen workers. The family is from Western Samoa, the child's immunization status is uncertain. The child has had a URI with cough and conjunctivitis. Now she has high fever, and when you examine her, you find small white spots inside the labial and buccal mucosa of her mouth. Besides conjunctival injection and rhinorrhea, her physical examination is unremarkable. Her temperature is 39°, but she does not appear toxic. Her parents note that her cousin had a rash, high fever, and pneumonia a little over a week ago, and the children were together just before his illness. Appropriate steps would be:

a. To transport her to the mainland immediately for work up for meningococcemia.
b. To start broad-spectrum antibiotics.
c. Supportive care with close observation and respiratory isolation.
d. A throat culture and treatment with appropriate antibiotics for strep.

5. A 3-month-old infant is brought to your urgent care center for vomiting. He has been seen several times over the past month for this problem. His mother has tried clear liquids in addition to numerous formula changes without success. On physical examination, the infant is alert and smiling. He is afebrile with normal vital signs. His weight is at the 75th percentile for his age. His physical examination is normal, and he seems hungry. He takes 4 oz of formula eagerly but, while being burped, he spits up about 15 ml. Appropriate action at this time would be:

a. Referral to a pediatric emergency facility for evaluation of the emesis.
b. Referral to a pediatric surgeon for evaluation for pyloric stenosis.
c. Adding rice cereal to thicken feedings and keeping the baby upright for 30 minutes after feedings.
d. Switching to a different formula for 1 to 2 weeks.

Chapter 18

Surgery

Abdominal Aortic Aneurysms

■ Basic Science An abdominal aortic aneurysm **(AAA)** is a **localized dilation of** the **abdominal aorta.** It is usually **secondary to atherosclerosis** and **hypertension,** which cause local aortic wall weakness and dilatation. Less frequently the aneurysm can be the result of trauma, infection, or congenital weakness (e.g., Marfan's syndrome). Over 90% of AAAs begin below the level of the renal arteries.

The natural history of AAAs is related to their size.

- **Aneurysms less than 5 cm are unlikely to rupture.**
- **Aneurysms over 6 cm** are at **significant risk for rupture.**

Thrombus may line the wall of the aneurysm.

■ History and Physical Examination

- **History** AAAs present in one of the following ways:
  - **Asymptomatic AAAs—no historical findings;** usually **discovered incidentally** by physical examination or radiographically.
  - **Symptomatic AAAs—low back or flank pain.** Pain may radiate into the testis or leg. Pain may be described as **severe, deep, boring;** it may be confused with renal colic or benign low back pain.
  - **Ruptured AAA**—acute onset of **severe abdominal pain;** may be accompanied by **hypotension or shock or myocardial event.**
- **Physical examination** Greater than 90% of AAAs present as a ***pulsating mass.*** AAAs may be difficult to feel in the obese. **Unstable aneurysms are** frequently **tender.**

■ Laboratory and Diagnostic Studies Depending on the severity of presentation, preoperative laboratories may be drawn.

- **Plain x-rays** often **pick up AAAs incidentally** because of calcifications in the aortic wall.
- **Computed tomography (CT) with IV contrast** is the **study of choice for defining the size, anatomy, and the presence or absense of rupture of AAA.**
- Ultrasound may also be used to define the size of a AAA and is often the screening test of choice.
- Aortography with contrast is indicated when the aneurysm extends above the renal arteries.

- Formulating the Diagnosis **Clinical support and physical examination is the key to diagnosing AAA.** Diagnostic testing will delineate the extent of the AAA.
- Evaluating the Severity of Problems **Rupture or potential rupture is** a **surgical emergency.** Vascular surgical consultation should be sought immediately.
- Managing the Patient
  - **Asymptomatic AAA** Patients with asymptomatic AAA should be referred for **vascular surgical evaluation.**
    - **AAA less than 5 cm may be monitored as** an **outpatient** for increase in size.
    - **AAA greater than 5 cm should be electively repaired.** Repair involves **excision of the aneurysm** and **replacement of vessel with a synthetic graft.**

    Because patients almost always have atherosclerosis, the patient's **cardiovascular status should be assessed prior to surgery.**
  - **Symptomatic AAA** Patients should be **admitted** to the hospital **for immediate definition of size and status of AAA.** Patients with symptoms clearly referable to the aneurysm will undergo operation.
  - **Ruptured AAA**
    - **Stabilization—IV fluid and packed red blood cells**
    - **Immediate operation**

Acute Appendicitis

- Basic Science The appendix arises from the cecum. It is generally about 6 inches long and lies directly behind the cecum.

  Acute appendicitis results from **obstruction of the appendiceal lumen.**

  Obstuction may be **caused by fecaliths, fibrous bands, foreign bodies, or lymphoid hypeplasia.**

  Appendiceal inflammation and edema cause increased wall tension and ultimately disrupts blood flow to the appendix. This leads to ischemia and ultimately to appendiceal perforation. The time from obstruction to perforation may be **as little as 24 hours.**
- History and Physical Examination The **four classic symptoms of appendicitis are as follows:**
  - **Right lower quadrant (RLQ) abdominal pain**
  - **Anorexia**
  - **Nausea and vomiting**
  - **Fever**
  - **History** Appendicitis usually **begins as nonspecific abdominal discomfort in the periumbilical region of the abdomen.** This is followed by **anorexia, nausea,** and often **vomiting.** Several hours after symptoms begin the **pain usually moves to the right lower abdominal quadrant.** Right lower

quadrant **pain** is due to peritoneal irritation, which is **made worse by movement** or change in the intraabdominal pressure (e.g., **sneezing or coughing**).

In the case of a retrocecal or pelvic location of the appendix, the pain may remain poorly localized, making the diagnosis difficult.

- **Physical examination** The patient may experience **low-grade fever of 100° to 101° F** (high temperatures are rare in uncomplicated appendicitis). The patient will exhibit **tenderness to palpation in the right lower quadrant, most commonly over McBurney's point.**

  Peritoneal signs **(guarding and rebound tenderness) appear with peritoneal irritation. Other helpful signs** of appendicitis include the following:

  - **Psoas sign**—pain from passive extension of the right hip
  - **Internal obturator sign**—pain on passive internal rotation of flexed thigh
  - **Rovsing's sign**—RLQ pain during palpation of the left lower quadrant

  There is a **great deal of variation in the signs and symptoms of acute appendicitis. Many** patients with acute appendicitis **report pain only with rectal examination.** This is **particulary true of children** where pain is often not localized. Because of this variation in presentation a **rectal examination is mandatory in all cases of suspected appendicitis.** Rectal examination may reveal tenderness in the right lower quadrant, a periappendiceal mass, or an alternative diagnosis.

  A pelvic examination is **required in young women. Pelvic inflammatory disease** may **mimic** the **symptoms of appendicitis.**

  Note that less than one half of patients have typical signs and symptoms of appendicitis.

■ **Laboratory and Diagnostic Studies** The CBC will often reveal a **leukocytosis (12,000-15,000/µL) and bandemia.** One third of patients, particularly the elderly, will have a normal WBC in the presence of acute appendicitis.

■ **Formulating the Diagnosis** In the classic case, the history should be diagnostic and readily supported by the physical examination and ancillary tests. The diagnosis is usually more difficult in the young female patient, in which the differential diagnosis includes gynecologic causes of abdominal pain.

■ **Evaluating the Severity of Problems** **Appendicitis is a life-threatening disease.** As stated earlier, the time from obstruction to perforation may be as little as 24 hours. A ruptured appendix may result in peritonitis and death.

■ **Managing the Patient** The **treatment** of **acute appendicitis is prompt surgical removal of the appendix** (note that the appendix may rupture within 24 hours of the onset of symptoms).

Removal of the appendix, either through the traditional open technique or laparoscopically, should be carried out after the patient is adequately fluid resuscitated and prepared for surgery. (Antibiotics are given before and for 2 days after surgery in uncomplicated cases.)

Anal Fissure

- **Basic Science** An anal fissure is a **tear or ulcer in the anal canal.** Most anal fissures result from **trauma due to passage of large, hard stool.**

  Other less common reasons for anal fissure include anatomical abnormalities, infections (sexually transmitted diseases), and inflammatory bowel disease.

  If the fissure rests on the **internal sphincter,** it **may** cause it to **spasm.** This results in **avoidance of defecation and development of hard, constipated stools.**

- **History and Physical Examination** Symptoms include **pain during defecation and bleeding upon wiping.** Pain may persist for hours after defecation.

  Anal fissures may be **associated with** a **skin tag protruding from the anus.**

- **Laboratory and Diagnostic Studies** None.

- **Formulating the Diagnosis** The **diagnosis** should be **apparent after history and physical examination.**

  If the fissure does not heal, the fissure can be tested for syphilis or TB. Biopsy may also be considered.

- **Evaluating Severity of Problems** Nonhealing fissures may be associated with Crohn's disease, TB, syphilis, or cancer.

- **Managing the Patient** Primary therapy includes **local anesthetic, glycerin suppositories, stool softeners, laxatives (psyllium), high-fiber diet, and frequent sitz baths.**

  Fissures **refractory to** 3-4 weeks of **conservative therapy** should be **considered for surgery.**

Complications of Anesthesia

Anesthesia complications usually occur during or immediately after operation. The most **common or serious complications are as follows:**

- **Nausea and Vomiting** Nausea and vomiting are the **most common complications** of anesthesia.

  Narcotic analgesics are a frequent cause of postoperative nausea (40% of patients experience nausea with meperidine or morphine). Postoperative nausea is also attributable to **induction and inhalational anesthetics.** Induction agents implicated include propofol (the least), barbiturates (intermediate), ketamine, and etomidate (the worst).

  Other causes of nausea and vomiting **include hypotension, hypoxemia, hypoglycemia, increased intracranial pressure,** gastrointestinal bleeding, gastric distension, and ileus.

  Postoperative nausea can be prevented by administering an antiemetic. Droperidol, metoclopramide, transdermal scopolamine, and Compazine and **Phenergan** are all effective.

- **Persistent Sedation** Persistent sedation most **commonly results from overuse of sedatives** (pentobarbital, droperidol, lorazepam, and diazepam).

  Sedation linked to inhalational agents is **characterized by shallow respirations.** Persistent inhalational anesthetic sedation is most likely to occur after relatively long periods of use of the inhalational agent and is **worse in obese patients** because of the high lipid solubility of the agents.

  Oversedation secondary to **narcotic overdose** is **characterized by pinpoint pupils and slow respiratory rates.** If narcotic sedation is deep enough to interfere with respiratory drive, it may be reversed with the use of naloxone.

  In the cases of persistent sedation secondary to **benzodiazepine overdosage, reversal with flumazenil** (a benzodiazepine antagonist) **is appropriate.**

  Persistent paralysis may be mistaken for oversedation and should be excluded by observing the patient's ventilation, active movements, and reflex activity.

  Metabolic derangements such as hypercapnia, hypoxia, hypoglycemia, hyperglycemia/hyperosmotic coma, and hyponatremia must be excluded. Hypothermia must also be considered.

- **Bronchospasm** **Patients with asthma are six times more likely to develop bronchoconstriction in response to intubation.**

  Bronchoconstriction is **treated with β-agonists (albuterol nebulizers** or metered dose inhalers). Administration of intravenous corticosteroids will help decrease bronchospasm.

- **Hypothermia** **Perioperative hypothermia** is an extremely **common problem. Trauma and burn patients,** patients with a large blood loss, and **patients at the extremes of age are** all at risk for hypothermia.

  Temperatures less than 33° C cause platelet and **coagulation abnormalities, dysrhythmias,** and increased blood viscosity. Intraoperative normothermia is proven to reduce the postoperative infection rate and length of hospital stay.

  Warmed air heating blankets and **warmed IV solutions** are commonly **used to raise** the patient's core **body temperature.** Additionally, low doses of meperidine (Demerol) are used to control excessive shivering and decrease oxygen consumption.

- **Malignant Hyperthermia** Malignant hyperthermia is a **autosomal dominant genetic trait.** The condition is **characterized by often fatal hyperthermia with rigidity of muscles** occurring **in** affected **patients exposed to certain anesthetic agents,** particularly **halothane, succinylcholine, and methoxyflurane.** It is thought that the triggering agent causes the patient's cells to lose the ability to control intracellular levels of calcium.

  Signs of malignant hyperthermia include **tachypnea, tachycardia, unstable blood pressure, cyanosis, rigidity, sweating, and fever.** Laboratory abnormalities include **respiratory and metabolic acidosis, electrolyte abnormalities** (especially hypercalcemia and hyperkalemia), **rhabdomyolysis, and myoglobinuria.**

Treatment of malignant hyperthermia involves the following:

- **Discontinuing all possible agents that could have triggered the attack**
- **Hyperventilation** with 100% oxygen
- **Dantrolene (IV)**
- **Cool the patient**
- **Correction of acidosis and hyperkalemia**
- Transfer to ICU

Patients with **malignant hyperthermia** should be told that malignant hyperthermia is an autosomal dominant **genetic trait and therefore likely to affect one half of their children and parents.**

Patients with known malignant hyperthermia **are able to undergo future anesthesia as long as triggering agents** that caused the previous attack **are** completely **avoided.**

Complications of Surgery

■ Surgical Complications **Any postoperative occurrence that adversely affects a patient's morbidity and/or mortality is termed a complication.** Complications may be specifically related to the operation performed. Others are more general and can occur after any surgical procedure.

A very **important aspect of surgical care is to anticipate potential problems and take preventive measures to minimize their incidence.** Recognized early, problems may be treated so as to lessen the severity and prevent more serious outcomes. This ultimately has an impact on decreasing a patient's morbidity and mortality.

■ History and Physical Examination Often the **risk** of any given patient's **suffering postoperative complications is** directly **related to that patient's general health.** A thorough history and physical examination will alert the physician to an individual patient's risk for complications. Preventive measures and surgical care can be tailored to minimize postoperative problems. Specific information helpful that can be obtained by preoperative evaluation includes the following:

- Allergies
- Medications
- Smoking and alcohol consumption
- Chronic medical conditions (e.g., coronary artery disease [CAD], chronic obstructive pulmonary disease [COPD], diabetes mellitus [DM], etc.)
- Past surgical history, including history of complication
- Review of indicated preoperative tests (e.g., ECG)

The most common nonspecific postoperative complications are discussed in the following sections:

■ Respiratory Respiratory complications of surgery are extremely common. **Patients with underlying lung disease and smokers** are at significantly **increased risk of postsurgical respiratory com-**

plications. Two common surgical respiratory complications are atelectasis and aspiration pneumonia.

***Atelectasis* is the collapse and failure to reexpand of small airways and alveoli.** Clinical **signs include decreased breath sounds and rales. Fever and tachycardia** can also occur. Chest x-ray demonstrates platelike opacities. **Atelectasis** may be **prevented by encouraging** the **patient to cough and take** frequent, **deep breaths. Untreated atelectasis may lead to pneumonia. Inhalation of oropharyngeal and/or gastric contents** can **cause aspiration pneumonitis.**

■ Infection **Factors increasing the risk for infection include** presence of **necrotic and devitalized tissue, hematomas or seromas, foreign bodies, use of steroids, existence of underlying chronic disease, and malnourishment.**

Preventive measures include meticulous attention to operative detail, hemostasis, maintaining sterile technique, and occasionally preoperative antibiotics. Early recognition of infection is important so that serious complications (e.g., dehiscence and/or incisional hernia) can be minimized.

The cardinal signs of the nonspecific inflammatory response are as follows:

- **Calor (fever, warmth)**
- **Dolor (pain)**
- **Rubor (erythema)**
- **Tumor (edema and swelling)**

The four Ws represent the **most common sites of infection during the postoperative period.**

- **Wind (respiratory)**
- **Walk (phlebitis)**
- **Wound (wound infection)**
- **Water (urinary tract infection)**

■ Thrombosis/Pulmonary Embolism Virchow's triad indicates **factors that put patients at risk for thrombosis.**

- **Trauma (surgery)**
- **Venous congestion (stasis)**
- **Hypercoagulability**

Advanced age, obesity, birth control pills, underlying malignancy, orthopedic and pelvic procedures all **increase** the patients **risk of venous thrombosis.**

Measures used to prevent DVTs should be tailored to the individual patient. These may include use of **elastic stockings, pneumatic compression stockings, subcutaneous heparin, and low-dose oral anticoagulants.**

Classic **signs of deep vein thrombosis include a swollen, tender calf, fever,** and **Homan's sign** (calf pain with dorsiflexion of the foot).

Diagnostic **studies to confirm a DVT include Doppler ultrasound,** plethysmography, and venography. **DVTs** are **treated with heparin** (bolus and then constant infusion).

Pulmonary embolus results from **migration of a DVT to the lungs. Signs** and symptoms of pulmonary embolus **include tachypnea, tachycardia, and decreased O_2 saturation. An A-a gradient that does not improve with O_2** administration is a classic finding. **Ventilation/perfusion scans indicate the likelihood that the patient has a pulmonary embolus.**

Treatment for pulmonary embolus is anticoagulation with heparin (bolus and then at a constant infusion rate); if anticoagulation is contraindicated, Greenfield filter placement should be considered. Thrombolytic agents and surgical treatment are also options.

Decubitus Ulcer

- **Basic Science** A **decubitus ulcer** is also known as a bed or pressure sore. It is defined as the **necrosis of tissue** and subsequent ulceration **in areas with bony prominences.** Decubitus ulcers are a **complication of prolonged immobilization.** They are a direct **result of localized pressure.** Decubitus ulcers most frequently **occur in debilitated patients** that are either bedridden, wheelchair bound, or left in the same position for prolonged periods of time. Many of the affected patients have decreased or absent sensation and are unable to feel pain associated with ischemia. Tissues overlying the occiput, scapulae, sacrum, ischia, greater trochanters, calcanei, and lateral malleoli are the most frequent sites of pressure necrosis.

  Prolonged pressure over a period greater than 2 hours will lead to decreased circulation of blood, tissue ischemia that results in tissue necrosis. The necrosis can involve any tissue that is located between the external pressure source and the bony prominence. Skin, subcutaneous tissue, and muscle can be involved.

  Factors that **increase** a patient's **predisposition** to the development of a decubitus ulcer **include contact with perspiration, urine,** and **feces.** Poor nutritional status of the patient, anemia, as well as infection predispose the patient to formation of decubitus ulcers.

  A decubitus ulcer is a complication of immobility.

- **History and Physical Examination** The history is of a patient who is paralyzed, insensate, or debilitated in some other way, and who **spends prolonged periods of time in the same position.** Most of the patients are either bedridden or wheelchair bound. Additional factors can include **poor nutritional status, anemia, diabetes, incontinence, or concomitant infection.**

  Physical examination will demonstrate an ulcer in one of the areas just mentioned. The depth of the ulcer determines it severity.

- **Laboratory and Diagnostic Studies** None.

- **Evaluating the Severity of Problems**

  - Stage 1 consists of an area of reddened skin. There is no underlying tissue involvement.
  - Stage 2 is characterized by reddened, edematous skin with induration and dermal necrosis.
  - Stage 3 involves exposed fat or muscle tissue.
  - Stage 4 involves destruction of soft tissue and the involvement of bone, which may lead to osteomyelitis.

- **Formulating the Diagnosis** Diagnosis is evident from history and physical examination.

- **Managing the Patient** The **best treatment for decubitus ulcer is its prevention.** This entails **relief of pressure over the areas of bony prominence.** Special devices are available that distribute the patients weight evenly and alleviate increased pressure over the points of prominence. Additional padding should be placed over areas that are prone to the development of decubiti. It is also imperative that the **patient's position be changed every 2 hours** in order to prevent the development of ischemic changes that will ultimately lead to necrosis. Patients that are wheelchair bound are at risk for the same complications and should be able to change their position frequently. In the case of braces or casts, windows may have to be cut at potential pressure sites in order to decrease the risk.

  Patients at risk must be carefully monitored for any cutaneous changes. This entails careful inspection of pressure points, at least on a daily basis. **Patients** have to be **kept clean and dry.** Especially in cases of incontinence it is imperative that the patient be checked frequently in order to avoid prolonged exposure to moisture and soilage.

  Patients' **nutritional status must be optimized,** and any deficits, such as anemia, corrected in order to facilitate the healing process.

  The early stage of the decubitus requires aggressive measures of prevention and optimization of the patient nutritional status. All pressure should be relieved from the area of the potential decubitus, and it should be kept dry and exposed. Unless progression or infection occurs at this site, the early stage of a decubitus ulcer should heal spontaneously.

  The later stages of a decubitus ulcer are manifested by the presence of tissue necrosis. On the discovery of such changes an aggressive approach must be instituted in order to minimize the progression of necrosis. **Surgical treatment is required for any area that is full thickness and greater than 2 cm in diameter. Any necrotic tissue must be debrided,** and the underlying tissues protected from desiccation with, for example, wet to dry dressings.

  The **likelihood of infection of the open wound is high.** Uncontrolled **local infection may lead to bacteremia and death.** For bacterial counts greater than 10^3 organisms, topical antibiotics should be instituted. Once the progression of necrosis has been stopped and healing has started, closure of the defect may be undertaken. In the case of small defects, less than 2 cm in diameter, they may heal spontaneously once granulation tissue develops. **Larger defects may have to be closed surgically,** either by simple closure, split thickness skin grafts, or by the utilization of flaps.

  In the case of chronically debilitated patients or in the case of significant bone involvement, removal of the affected bone may be necessary.

Hemorrhoids

- **Basic Science** **Hemorrhoids are varicose veins of the anorectal area.**

  Hemorrhoids are thought to be **caused by increased pressure in the anorectal vascular network.** This pressure may occur secondary to **straining while defecating, congestive heart failure, pregnancy, or portal hypertension.**

- History and Physical Examination
  - **History** Common **symptoms** include the following:
    - **Rectal bleeding—**blood on toilet paper, blood on stool; bleeding is **usually painless and bright red.**
    - **Pain—caused by thrombosis of** an external **hemorrhoid**
    - **Rectal itch**
    - **Prolapse—**protrusion of the hemorrhoid; hemorrhoids may reduce spontaneously or may need to be reduced manually.
  - **Physical examination** of the rectum may demonstrate tender **bluish swellings** or a reducible (or unreducible) **mass.**
- Laboratory and Diagnostic Studies **Anoscopy** is utilized to evaluate painless rectal bleeding.
- Formulating the Diagnosis History and physical examination, with possible anoscopy, makes the diagnosis of hemorrhoids. External hemorrhoids involve sensate skin and are thereafter painful; internal hemorrhoids are painless.
- Evaluating the Severity of Problems Other more serious conditions should be ruled out before rectal bleeding is attributed to hemorrhoids.
- Managing the Patient **Asymptomatic hemorrhoids need no treatment.**

  Initial **therapy for symptomatic hemorrhoids** consists of **sitz baths, stool softeners (psyllium), increased fluid intake, bran, avoiding prolonged sitting, and topical pain and itching preparations (witch hazel, hydrocortisone cream).**

  Surgical therapy should be considered **if there is no response to** 1-2 weeks of **conservative measures.**

  Surgical treatments include the following:

  - **Injection sclerotherapy—**sclerosing agent injected into protruding vein
  - **Rubber band ligation—**protruding hemorrhoid ligated with rubber band, causing necrosis and sloughing
  - **Hemorrhoidectomy—**rarely performed

HERNIA

- Basic Science A **hernia is a protrusion of intraabdominal tissue through a defect in the abdominal wall.**

  The **majority of hernias are inguinal hernias.** Other less common hernias include ventral (i.e., incisional or umbilical) and femoral hernias.

  Inguinal hernias are **more common in men than women.** Inguinal hernias are classified as direct, indirect, and femoral.

  A hernia **may remain unrecognized until a change in intraabdominal pressure causes the defect to increase in size.** The most **common causes** of increased in intraabdominal pressure are **obesity, pregnancy, excess straining, chronic cough, and ascites.**

- **History and Physical Examination**
  - **History** Most inguinal hernias are asymptomatic and are found on routine physical examination. On occasion a patient may present with a bulge in the groin that causes aching on exertion.
  - **Physical examination** Palpation of an inguinal hernia is performed by placing one's finger at the superior border of the scrotum in men and invaginating the skin into the inguinal canal toward the superficial ring. A **mass or bulge structure can be felt when the patient coughs or strains.** Differentiating between indirect versus direct hernia is of limited clinical significance and often impossible on physical examination.
- **Laboratory and Diagnostic Studies** None.
- **Evaluating the Severity of Problems** Occasionally a **hernia may become incarcerated.** This is a more serious presentation, as the **hernia contents can become strangulated** and eventually **infarcted.** The **patient will complain of a mass that does not reduce and that is tender to palpation.** Often with gentle pressure and occasionally sedation the mass can be reduced into its proper place. If this is not possible, emergent surgery is required.
- **Managing the Patient** Patients with hernias should be referred to a surgeon. Almost **all hernias will require surgical repair** because the defect will only increase in size, resulting in the possibility of bowel incarceration.

  For inguinal hernias, various repairs of the hernia have been proposed, but they all have one goal in common, which is to fortify the inguinal floor and tighten the internal ring. Similar principles are used for the other types of hernia where closure of the defect is the desired result.

  If the protrusion is not reducible, immediate surgical consultation should be sought.

Peripheral Vascular Disease

- **Basic Science** **Atherosclerosis** is the **most common cause of peripheral arterial insufficiency. Other causes** include **thrombosis, embolism,** and **trauma.**

  Inadequate blood flow manifests as ischemic pain in the affected area. **In the lower extremity** this phenomenon is **called *intermittent claudication*** and is **relieved by cessation of exercise. If** the **obstruction worsens,** flow becomes insufficient even at rest, and **pain is present at all times (rest pain).** The end stage of arterial insufficiency to the extremity is tissue loss due to insufficient flow.
- **History and Physical Examination** Patients are often **older men with risk factors for atherosclerosis such as smoking, hypertension, diabetes, lack of exercise, and elevated cholesterol.** They will often give a history of atherosclerotic disease in other areas such as the **carotid or coronary circulation.** Patients will give a **history of claudication** and can often describe how many feet they walk before pain ensues. Patients with aortoiliac disease may present with buttock pain and impotence. Those with femoral disease will

often have calf claudication. Rest pain and tissue loss signify advanced disease.

Physical findings are classically described by **the five P's:**

- **Pain**
- **Pallor**
- **Pulseless**
- **Paresthesia**
- **Paralysis**

The physical examination may also **reveal cool skin and hair loss, as well as some muscle wasting in the area that is hypoperfused. Pulses may be decreased or absent in the affected limb.** Patients with severe disease have lower extremity blood flow that is dependent on gravity. They may complain of symptoms when lying flat that are relieved by hanging their legs over the side of the bed.

■ Laboratory and Diagnostic Studies The ankle-brachial index (ABI) is defined as the ratio of systolic BP at the ankle to SBP in the arm (normal ABI is greater than or equal to 1). In peripheral vascular disease the ABI will be decreased to less than 1. Patients with an ABI is less than 0.5 generally have severe peripheral vascular disease.

Doppler ultrasound studies may demonstrate a dampened flow pattern caused by the obstruction proximal to the measured point.

The **gold standard for evaluating the peripheral vasculature is angiography.** This will demonstrate areas of stenosis, collateral circulation, adequacy of vessels distal to obstruction, and also the presence of anomalies or aneurysms.

■ Formulating the Diagnosis Diagnosis is made by history and physical examination and from the results of the diagnostic tests just listed.

■ Evaluating the Severity of Problems **Severity of problems can be measured clinically** (i.e., rest pain in the involved extremity) **or with the use of diagnostic studies** as outlined earlier.

■ Managing the Patient Most **patients who present with claudication will improve with cessation of smoking, a graduated exercise program, and improved diet.**

Those **patients with rest pain or tissue loss often require revascularization. This is done with autologous grafts (the patient's own saphenous vein).** Other materials that may be used for bypass include prosthetic materials such as polytetrafluoroethylene (PTFE).

Patients whose limbs are not salvageable with bypass require amputation.

Mutiple Choice Review Questions

1. A 17-year-old white male has crampy abdominal pain, poor appetite, and diarrhea for the past two days. He presents to the ED with a temperature of 38° C. On physical examination he exhibits mild RLQ tenderness. Which of the following would not be helpful in making the diagnosis?

a. Rovsing's sign
b. Psoas sign
c. Murphy's sign
d. Rectal examination
e. A plain abdominal X-ray showing a gas-filled appendix

2. A 67-year-old white male with a past medical history significant for hypertension, atherosclerosis, and a 3.5 cm AAA presents to your office complaining of vague lower abdominal and flank pain. Appropriate management of this patient includes:

a. Outpatient workup of abdominal pain including monitoring of AAA for increase in size.
b. Admission to the hospital with CT scan with IV contrast of the abdomen.
c. Elective repair of the AAA.
d. Transfer to ED for aortography.
e. Immediate outpatient CT scan with IV contrast to diagnose an expanding AAA.

3. The most common expected physical finding in this patient is:

a. A pulsating abdominal mass.
b. Absent femoral pulses.
c. Severe abdominal tenderness to palpation.
d. Altered mental status.
e. Fever.

4. A 48-year-old woman presents with painless rectal bleeding associated with bowel movements. Upon examination she is found to have hemorrhoids. No other pathology is noted. These hemorrhoids are most likely:

a. External
b. Internal
c. Not hemorrhoids
d. Premalignant
e. Thrombosed

5. Asymptomatic external hemorrhoids require:

a. Injection sclerotherapy to prevent bleeding.
b. Rubber band ligation.
c. No treatment.
d. Topical steroids.
e. Hemorrhoidectomy.

Chapter 19

Radiology

This chapter is designed to show common radiologic presentations of disease. Each image is accompanied by a history and pertinent description. Following is a short discussion including clinical information and differential diagnosis.

- Head CT, Fig. 19.1 Sudden onset of severe **headache** and effortless **vomiting** that has progressed to **obtundation.**
  - **Description** Axial computed tomography (CT) image at the level of the pons. **Blood** is present **throughout** the **basal cisterns,** even outlining the basilar artery. The blood extends into the Sylvian fissures bilaterally and is also present in the fourth ventricle. No mass effect is demonstrated.
  - **Discussion** This is a classic presentation and appearance of a **subarachnoid hemorrhage** (SAH). The most common causes are trauma, ruptured aneurysm or arteriovenous malformation, and hypertension. Complications include obstructive or communicating hydrocephalus, cerebral infarction due to vasospasm (maximal at 48-72 hours), and cerebral herniation.

    The initial radiologic study for a patient with suspected SAH is a noncontrasted head CT scan. Acute blood has a high attenuation, which progressively falls to below that of the brain over the next 2 weeks. Subarachnoid blood classically fills the basilar cisterns and sulci. An intraventricular component may be present. The blood is often distributed asymmetrically. Cerebral angiography may be performed on an urgent basis to identify the bleeding source prior to surgery.
- Head CT, Fig. 19.2 **Motor vehicle accident with head trauma.** Patient is unconscious.
  - **Description** Axial CT image at the level of the thalami. **Blood is present in** an extraaxial crescentic distribution in the **left frontal and right parietal regions** as well as the posterior interhemispheric fissure. **No mass effect** is demonstrated at this level. Foci of calcification are present in the choroid plexus.
  - **Discussion** This patient has **bilateral acute subdural hematomas** (SDH). These are a neurosurgical emergency requiring immediate evacuation. Despite appropriate care, acute SDH carries a poor long-term prognosis due to underlying brain injury. The most common location for SDH is overlying the cerebral convexities. They also occur over the occipital poles, adjacent to the tentorium, and in the interhemispheric region. The latter is very characteristic of child abuse (shaken baby syndrome).

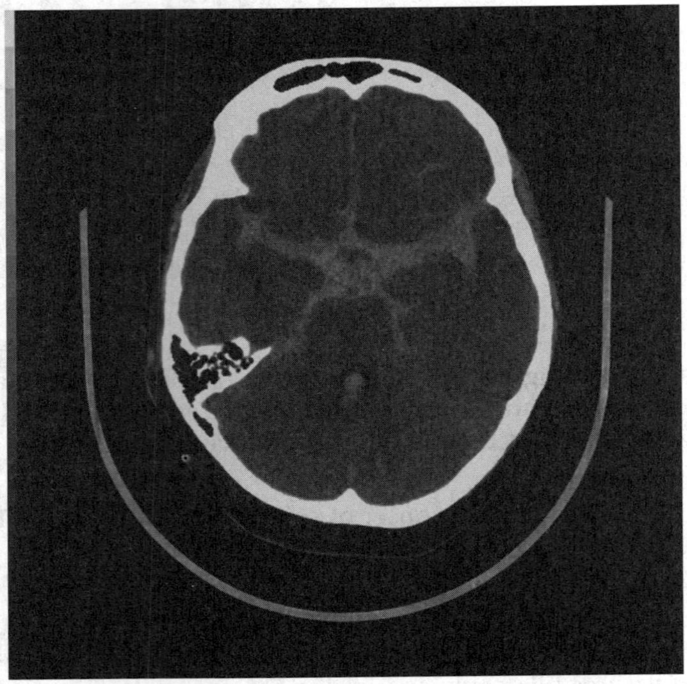

Fig. 19.1 Sudden onset of severe headache and effortless vomiting that has progressed to obtundation.

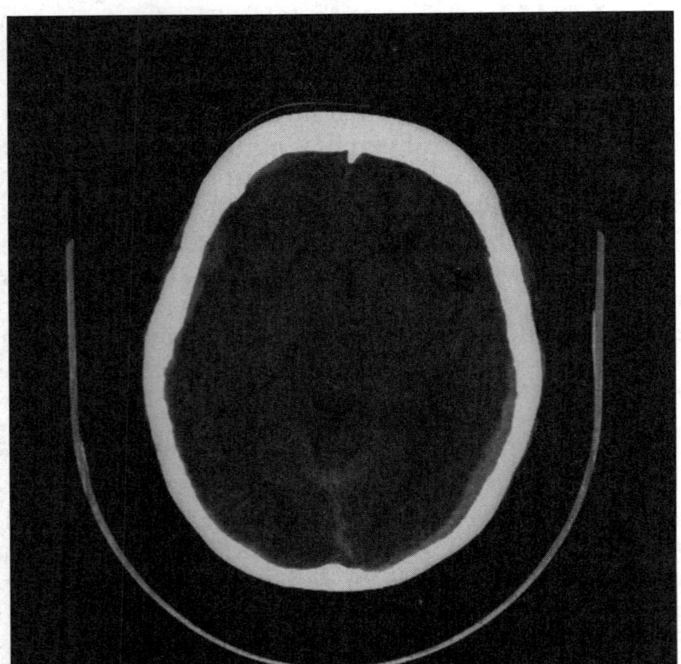

Fig. 19.2 Motor vehicle accident with head trauma. Patient is unconscious.

Chronic subdural hematomas (not shown) are often due to minor or forgotten trauma at a remote time from clinical presentation. Although the shape seen on CT scan is the same as for acute SDH, they are often of low or mixed attenuation due to blood breakdown or rebleeding from a neovascular membrane. Fluid-

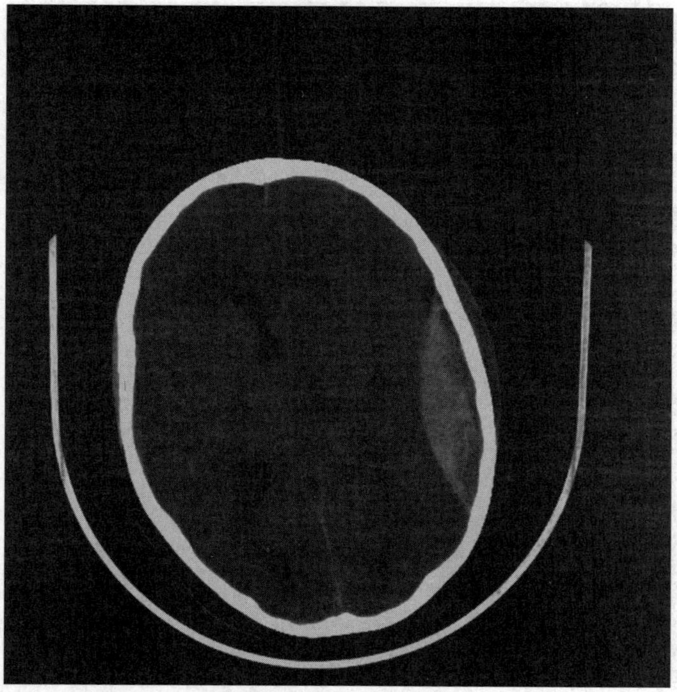

Fig. 19.3 Head trauma with transient loss of consciousness.

fluid levels may be seen. Clinical signs and symptoms of chronic SDH are vague and unreliable. They are not a surgical emergency.

- Head CT, Fig. 19.3 **Head trauma** with transient loss of consciousness.
  - **Description** Axial image from a head CT scan at the level of the cerebral hemispheres. **Blood in the left frontoparietal region exerts mass effect** (convex surface) on the underlying brain. The ipsilateral lateral ventricle is effaced and subfalcine herniation is demonstrated by a **shift of the midline structures to the right.**
  - **Discussion** This patient has an acute **epidural hematoma,** a neurosurgical emergency. They are usually caused by laceration of a meningeal artery or vein, commonly with associated skull fracture. The majority occur in the temporoparietal region. Less common locations include the frontal region, the posterior fossa, or between the occipital poles. The latter is differentiated from a subdural hematoma by crossing of the falx. An epidural hematoma will not cross a suture.
- Head CT, Fig. 19.4 **Child with a new onset generalized seizure** and now **unresponsive.**
  - **Description** Axial image from a noncontrasted head CT scan at the level of the cerebral hemispheres. A lobulated **mass (hematoma)** fills the lateral ventricle and exerts significant mass effect on surrounding structures. A low attenuation halo indicates **brain edema.**
  - **Discussion** This patient has a large **intraparenchymal hematoma.** Patients are invariably symptomatic with altered mental status, headache and/or effortless vomiting. Neurologic signs are

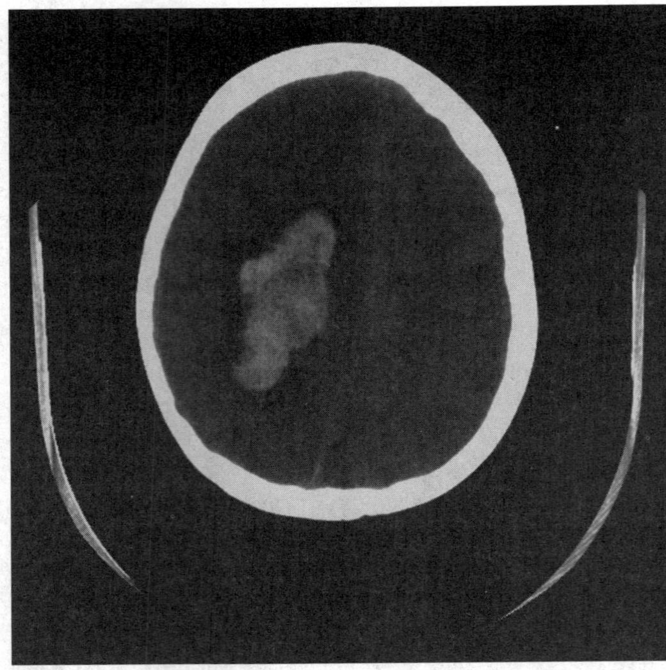

Fig. 19.4 Child with a new onset generalized seizure and now unresponsive.

present as well. Common etiologies include ruptured aneurysms, hypertension, arteriovenous malformations, trauma, tumors, bleeding diathesis, and hemorrhagic infarction. The distribution and associated findings will often give clues as to the cause. Hypertensive hemorrhages classically occur in the basal ganglia, brainstem, or cerebellum. Trauma will often have associated extraaxial blood or skull fractures. Complications and sequelae include encephalomalacia, rebleed, and abscess formation.

- Ultrasound, Fig. 19.5 **Intermittent right upper quadrant pain.**
  - **Description** A longitudinal ultrasound image of the gall bladder demonstrates hyperechoic material layering dependently with foci of obstruction of the sound beam (shadowing).
  - **Discussion** This patient has **multiple gall stones** lying in "sludge" (inspissated bile). Stones present a strong echo with farfield shadowing. Gall stones can be seen to move during a real-time examination. Ultrasound is over 95% sensitive in detecting gall stones.

    Acute cholecystitis is diagnosed when **stones** are present, the **gall bladder wall is thickened,** and a **sonographic Murphy's sign** is present. The latter is maximal pain elicited by pressing the transducer directly over the gall bladder. Associated findings may be biliary dilatation and ductal stones, or pericholecystic fluid.
- Abdominal CT, Fig. 19.6 **Pulsatile abdominal mass.**
  - **Description** Single axial CT image of the abdomen at the level of the transverse duodenum. The patient received oral and intravenous contrast. The **aorta is grossly enlarged with a peripheral rim of calcification.** A large eccentric mural thrombus is

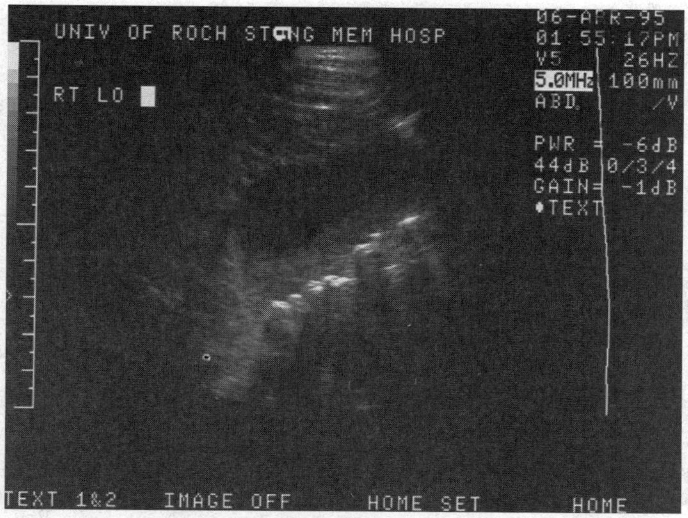

Fig. 19.5 Intermittent right upper quadrant pain.

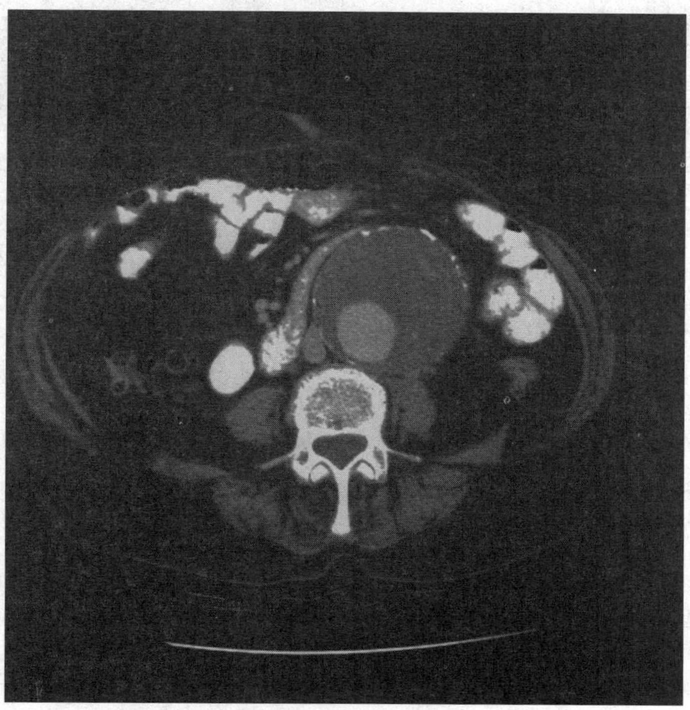

Fig. 19.6 Pulsatile abdominal mass.

present with a near-normal diameter true lumen. A high attenuation slit at one o'clock in the aortic cross section may be a normal vascular structure or intramural dissection.

- **Discussion** This patient has an **abdominal aortic aneurysm (AAA).** Atherosclerotic aneurysms are a manifestation of generalized vascular disease. Risk factors include smoking, diabetes mellitus, high cholesterol, and sedentary lifestyle. Aneurysms may rupture, dislodge emboli to the periphery, become infected, or

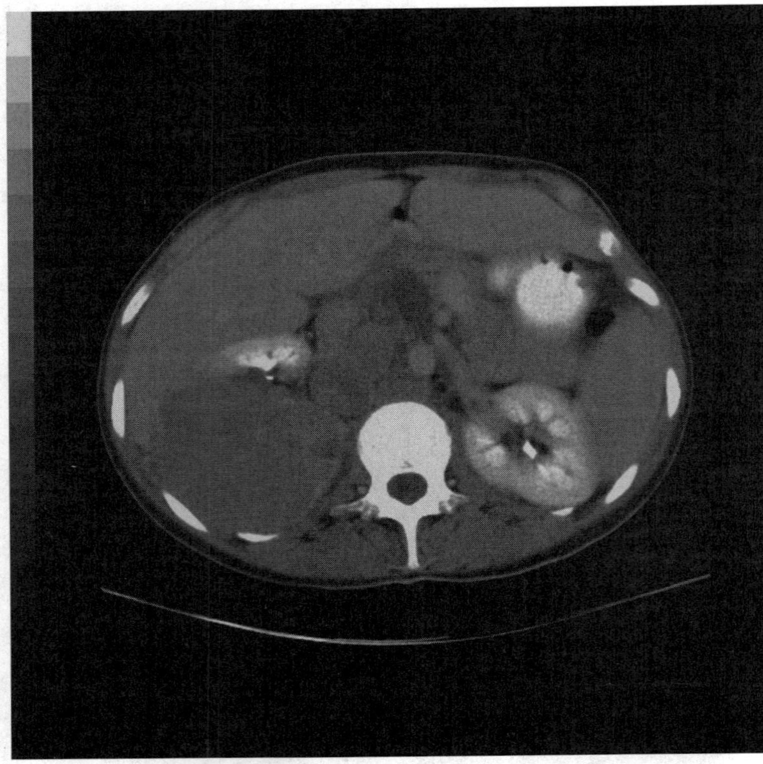

Fig. 19.7 Right flank pain and microhematuria.

spontaneously occlude. Signs of rupture are usually clinically apparent and include acute onset of abdominal or back pain and hypotension. Radiologic evaluation is most commonly by CT scan and ultrasound, although angiography may play a role. AAA is diagnosed with aortic diameter greater than 3 cm. The risk of rupture increases greatly at 6 cm or more in size.

- Abdominal CT, Fig. 19.7 **Right flank pain and microhematuria.**
  - **Description** Axial CT image of the abdomen at the level of the kidneys. The patient received oral and intravenous contrast. A large mixed attenuation **mass** arises **from the right kidney,** filling the perinephric space. The pararenal space is compressed but not invaded. Bulky adenopathy displaces the inferior vena cava anteriorly.
  - **Discussion** This patient has a large **renal cell carcinoma** with associated retroperitoneal adenopathy. These tumors occur later in life, especially in smokers, dialysis patients, and those with von Hippel-Lindau disease. Staging is based on spread into contiguous structures and to distant structures. This patient's tumor is limited to the perinephric space, and there is no invasion of the renal vein (not shown).
- Hand Radiograph, Fig. 19.8 **Pain in hand after a fight.**
  - **Description** Two views of the hand show a simple **transverse fracture of the distal fifth metacarpal shaft** with volar angulation of the distal fragment (or dorsal angulation of the fracture site).

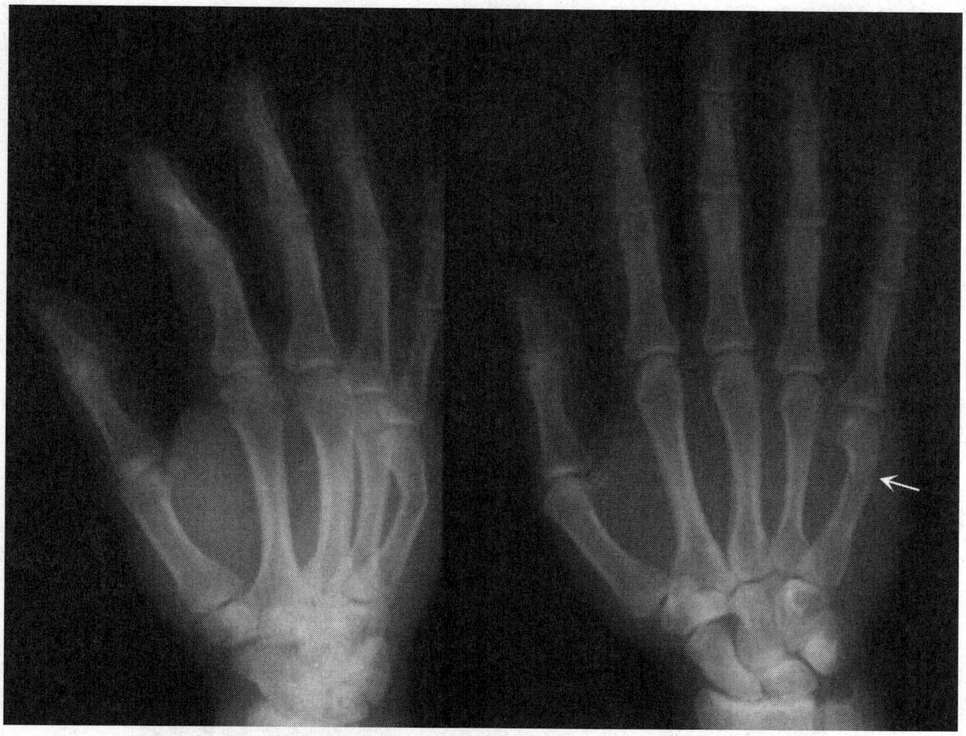

Fig. 19.8 Pain in hand after a fight.

- **Discussion** This is a form of **boxer's fracture.** It is caused primarily by longitudinal forces on the metacarpal head. Key concerns are rotation, excessive angulation causing protrusion of the metacarpal head through the palm pad, and associated lacerations caused by human teeth.

■ Pelvis Radiograph, Fig. 19.9 **Elderly male found on floor** at home.

- **Description** There is a simple **extraarticular fracture through the greater trochanter** extending to the basicervical region medially. Varus angulation is present.
- **Discussion Hip fractures** are actually fractures through the femoral neck or trochanteric region. They occur most commonly in elderly or debilitated patients. Osteoporosis is a risk factor. The final position seen on the plain radiograph depends on the number of fragments and residual muscular forces. The prime concern is that intraarticular fractures may compromise the blood supply to the femoral head. Treatment is with pins or screws for fractures less likely to lead to osteonecrosis. More severe fractures are treated by femoral head resection and replacement. Preexisting medical conditions contribute to morbidity and mortality.

  Did you notice the large right inguinal hernia containing stool-filled colon? A smaller hernia is present on the left.

■ Barium Enema, Fig. 19.10 **Weight loss and guaiac positive stools.**

- **Description** Lateral rectal view from an air contrast barium enema. An **apple-core lesion** is present at the rectosigmoid junction. The presacral space is **widened.**

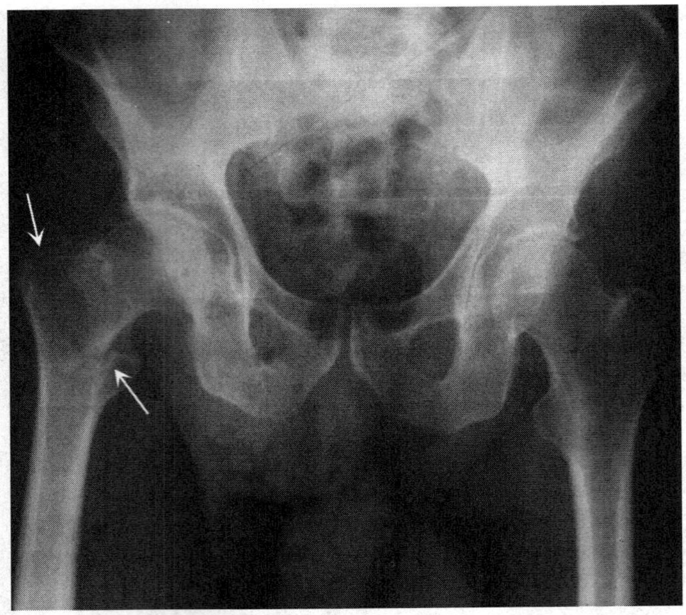

Fig. 19.9 Elderly male found on floor at home.

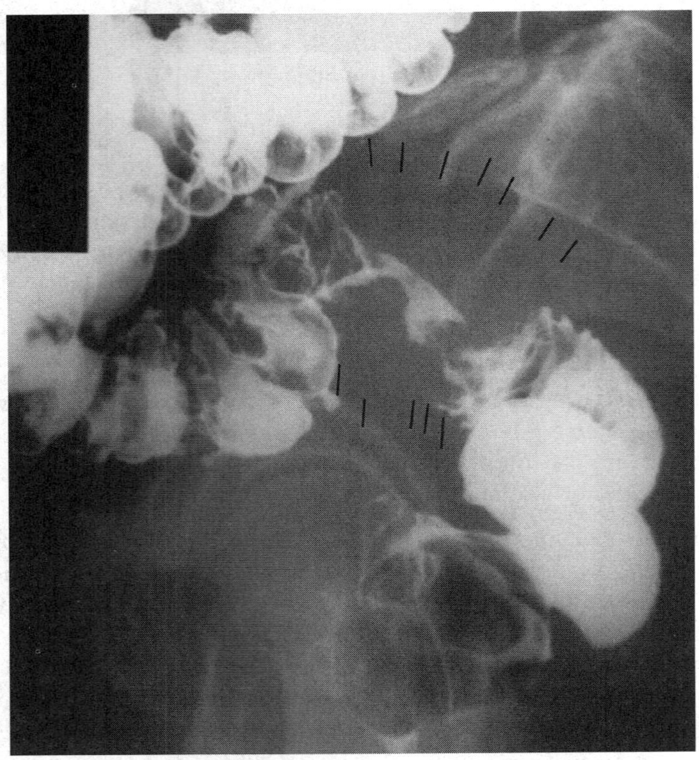

Fig. 19.10 Weight loss and guaiac positive stools.

- **Discussion** This patient has a **colonic adenocarcinoma.** Colon cancer is the third most common cancer in the United States and is the second most common cause of cancer deaths. Presentation usually involves anemia due to chronic occult blood loss, change in bowel habits, or weight loss. Risk factors include family history, low fiber/high fat diet, inflammatory bowel disease,

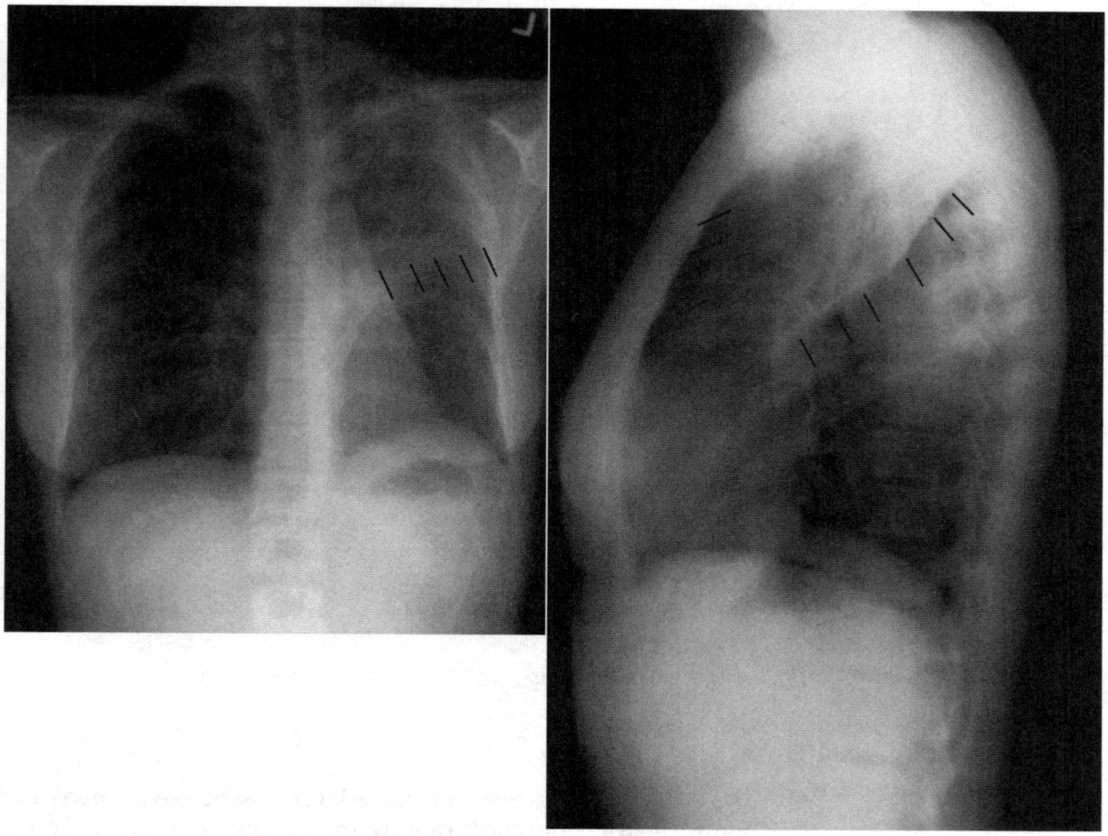

Fig. 19.11 Fever and productive cough.

and various polyposis syndromes. Diagnosis is made by barium enema or endoscopy. Staging is based on depth of penetration of the bowel wall and involvement of lymph nodes and distant organs. Metastases are primarily to the liver and are well seen by CT scan or ultrasound.

- Chest Radiograph PA, LAT, Fig. 19.11 **Fever and productive cough.**
  - **Description** Posteroanterior and lateral radiographs of the chest demonstrate **confluent density and volume loss in** the apical segments of the **left upper lobe** with **air bronchograms** throughout. A portion of the aortic knob is silhouetted.
  - **Discussion** This patient has a **left upper lobe pneumonia.** Bacterial pneumonias tend to involve contiguous segments of a lung rather than the diffuse disease of viral pneumonia. Etiologic division into community- and hospital-acquired disease directs treatment plan.

    The differential diagnosis of the radiographic pattern includes atelectasis (such as that due to mucus plugging or endobronchial tumor), localized pulmonary hemorrhage, or pulmonary infarction. Rarely a bronchoalveolar cell carcinoma will present this way.
- Chest Radiograph, Fig. 19.12 **Shortness of breath.**
  - **Description** Posteroanterior radiograph of the chest demonstrates a **shrunken right lung with absence of pulmonary vessels** (lung markings) beyond a thin white pleural line.

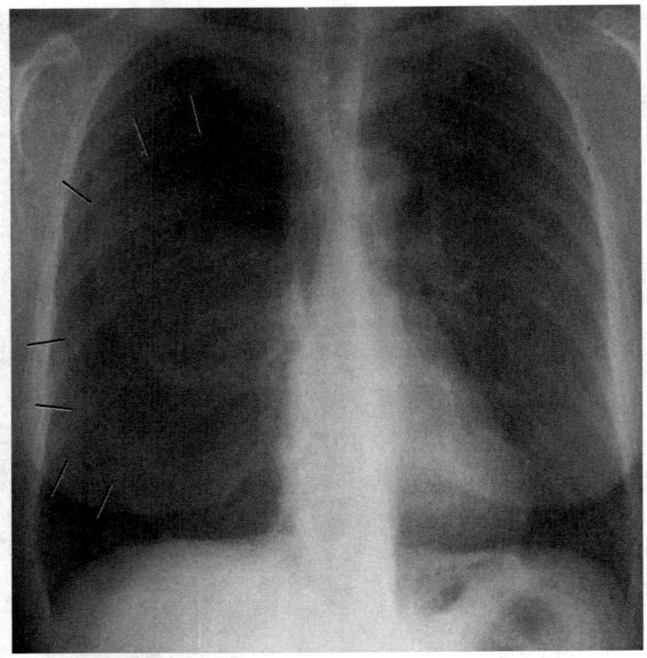

Fig. 19.12 Shortness of breath.

- **Discussion** This patient has a large **right pneumothorax.** Symptomatic pneumothoraces are treated with chest tube drainage. The causes for pneumothorax are myriad. Primary pneumothoraces have no definable cause. Secondary pneumothoraces are due to underlying pathology. This includes infection, asthma, and interstitial lung diseases. Barotrauma in patients on positive pressure ventilation and invasive procedures account for most iatrogenic pneumothoraces. Traumatic pneumothoraces are due to pulmonary parenchymal damage or central airway damage.

  A tension pneumothorax (not shown) compresses the systemic venous return to the heart and results in rapid death without treatment. The radiologic correlate is a shift of the mediastinum away from the pneumothorax. This latter sign is rarely seen, as these patients are diagnosed and treated on clinical grounds.

■ Chest Radiograph, Fig. 19.13 **Trauma.**

- **Description** Chest radiograph shows a displaced **mid-shaft fracture of the clavicle.**
- **Discussion** The clavicle is very commonly fractured and heals without significant intervention in over 95% of cases. Fractures are described by the portion of the bone in which they occur. Those at the extreme ends may need surgical repair because of associated ligamentous injury.

■ Chest Radiograph, Fig. 19.14 **Shortness of breath.**

- **Description** A single frontal radiograph of the chest demonstrates a globally **enlarged heart, engorged pulmonary vasculature** with predominance in the upper lobes (redistribution or cephalization), perihilar haze, blunt right costophrenic

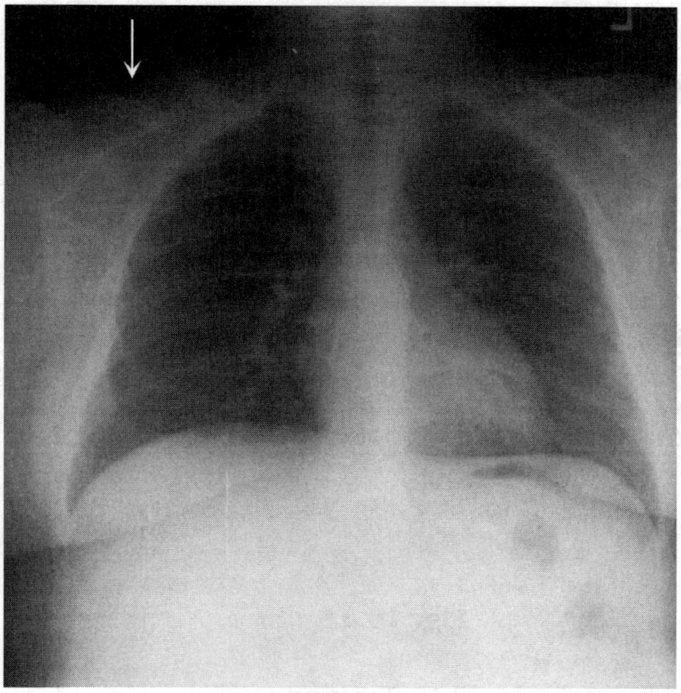

Fig. 19.13 Trauma.

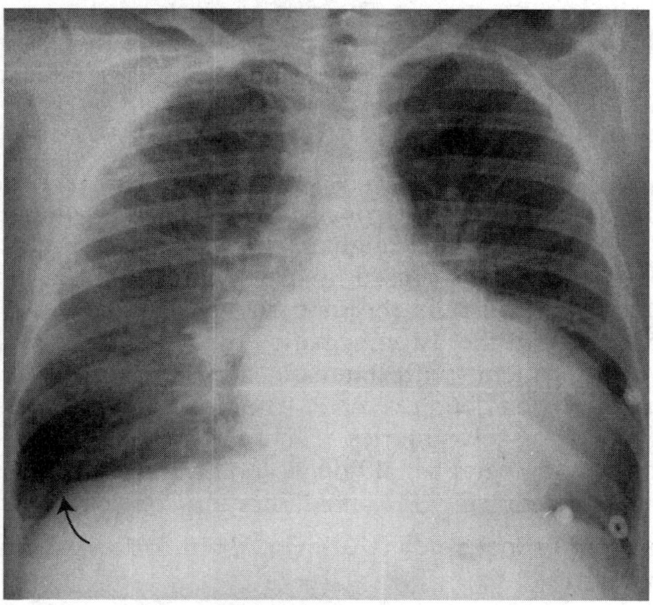

Fig. 19.14 Shortness of breath.

angle, and **Kerley's B lines.** The azygous vein is engorged as well.

- **Discussion** This patient is in **congestive heart failure.** The majority of the radiographic signs just mentioned indicate left-sided failure. Two important signs of right-sided failure are noted: distended systemic veins and an enlarged right heart.

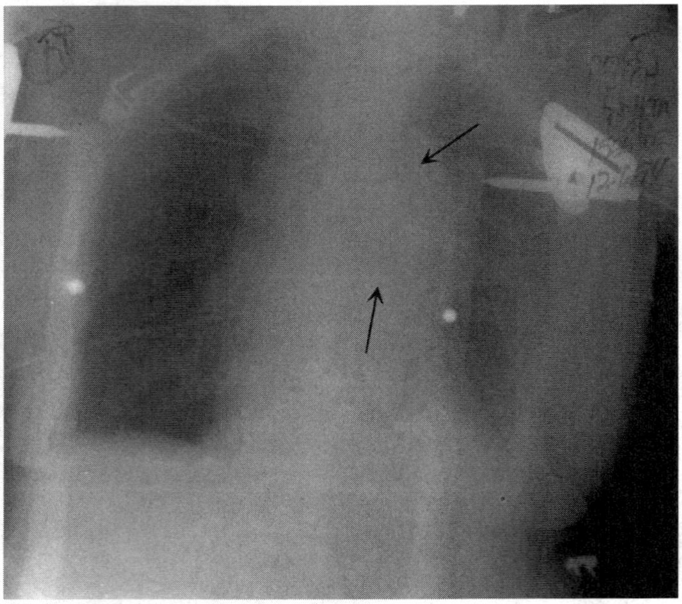

Fig. 19.15 Trauma.

- Chest Radiograph, Fig. 19.15 **Trauma.**
  - **Description** A portable radiograph of the chest obtained with the patient on a backboard. There is a **widened mediastinum** with abnormal contours. The aortic arch is too large and a mural calcification does not project at the periphery. A **left apical cap** is present as well as depression of the left mainstem bronchus. The right paratracheal stripe is widened. Bilateral rib fractures, including the left first rib, are noted.
  - **Discussion** This patient has many of the signs of a **mediastinal hematoma.** The largest concern would then be an aortic transection, which was proven angiographically in this case (not shown). This sort of injury is caused by deceleration and tension on the aorta at its fixed points: the root, the great vessels, the ligamentum arteriosum, and the diaphragmatic crura. It is estimated that only 10% of the patients with traumatic aortic transections survive long enough to reach definitive care. The differential diagnosis includes other major vascular injuries. The appropriate radiologic study in all cases is angiography, although certain trauma centers are beginning to use thoracic CT scanning first.
- Upper Gastrointestinal Study (UGI), Fig. 19.16 **Infant with projectile vomiting.**
  - **Description** Image from an upper gastrointestinal series shows barium in the stomach and proximal duodenum. The pyloric channel is long and narrow with shoulders of mucosa at each end.
  - **Discussion** This patient has hypertrophic **pyloric stenosis.** It is most commonly seen in firstborn males, often with a prior family history. Siblings may be affected too. The clinical and radiologic manifestations are due to hypertrophy of the circular muscle layers surrounding the pylorus. Ineffective hyperperistalsis in the proximal stomach causes **nonbilious projectile vomiting.** (Bile

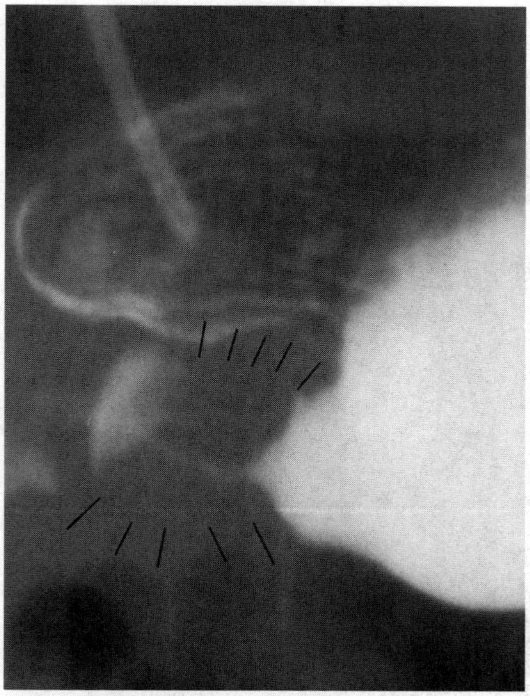

Fig. 19.16 Infant with projectile vomiting.

in the vomitus would indicate obstruction beyond the ampulla of vater.) Often the pylorus is palpable as an upper abdominal "olive." Radiologic diagnosis is by barium study or ultrasound. Treatment consists of rehydration and pyloromyotomy.

- Voiding Cystourethrogram (VCUG), Fig. 19.17 **Urinary tract infections.**
  - **Description** Image from a retrograde cystogram prior to voiding shows a normal bladder. The right ureter is filled with contrast and is markedly dilated and tortuous. The right collecting system is dilated as well and its papillae are effaced. The left side is less severely involved.
  - **Discussion** This child has **chronic severe vesicoureteral reflux** (VUR). VUR is graded on a scale of 1 to 5 based on the level reflux reaches and anatomic changes in the ureter and collecting system. This patient has grade 5 reflux on the right and grade 4-5 on the left.

    VUR is thought to be caused by improper insertion of the ureter into the bladder. Sterile, colonized, or infected urine ascends to the upper tracts, causing a focal tubulointerstitial nephritis most prominent in the polar regions. Patients may or may not be symptomatic. Mild cases are treated with antibiotics and severe cases with ureteral reimplantation. The feared sequelae are renal failure and hypertension.
- Abdominal Radiograph, Fig. 19.18 **Vomiting and abdominal pain.**
  - **Description** Supine radiograph of the abdomen shows dilated loops of small bowel with a small amount of air in the stomach and proximal colon. Surgical clips are present as well.

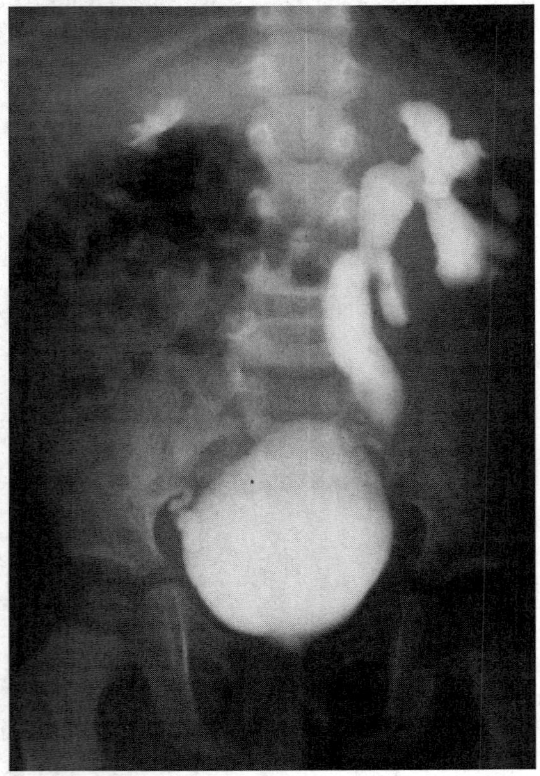

Fig. 19.17 Urinary tract infections.

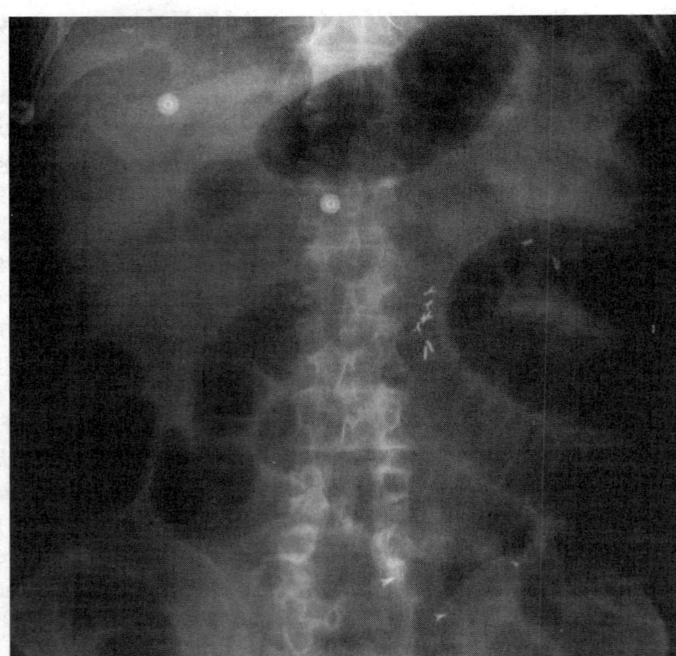

Fig. 19.18 Vomiting and abdominal pain.

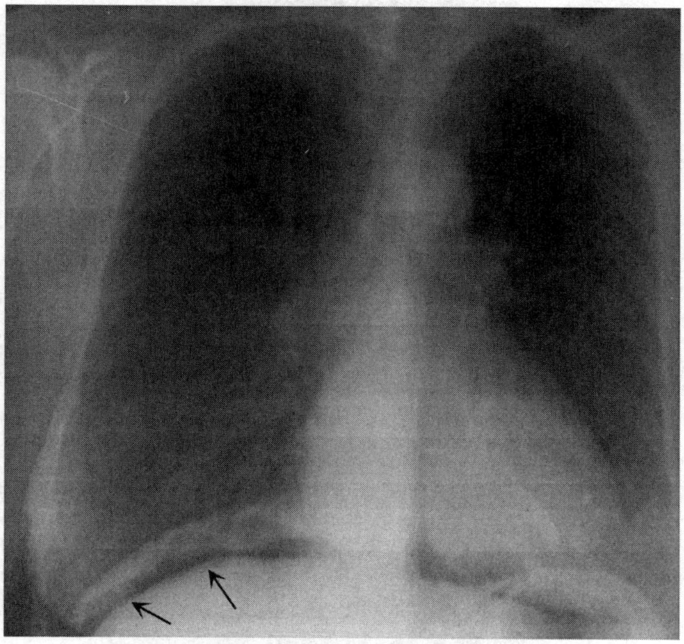

Fig. 19.19 Acute onset abdominal pain.

- **Discussion** This patient has a **small bowel obstruction.** The majority of air in the bowel is swallowed, and the small bowel normally has no visible air. Therefore dilated bowel containing easily seen air is abnormal. Dilated small bowel has the appearance of a "stack of coins." Thickening of the small bowel folds or wall may be seen as well. A horizontal beam radiograph of the abdomen (not shown) would show air-fluid levels in the bowel, most likely with differential levels in the same loop. The presence of air in the colon in this case indicates either early or partial obstruction.

  Small bowel obstruction in adults is usually caused by adhesions from prior surgery. Other causes include incarcerated hernias, intraluminal mass/foreign bodies, inflammatory processes, and intussusception. Treatment consists of nasogastric tube decompression and rehydration. Protracted cases may require surgery.

■ Chest Radiograph, Fig. 19.19 Acute onset **abdominal pain.**

- **Description Air** is present **between the right hemidiaphragm and the liver.** A smaller amount of gas lies between the left hemidiaphragm and the air-filled stomach.
- **Discussion** This patient has **free intraperitoneal air, indicating a perforated hollow viscus.** Common **causes include peptic ulcer disease, carcinoma, penetrating trauma, and post-laparotomy state.** Treatment is surgical repair of the perforation and debridement of the peritoneum. Antibiotics are given for peritonitis.

■ Upper Gastrointestinal Study (UGI), Fig. 19.20 **Dysphagia.**

- **Description** Air and barium contrast esophagram demonstrates a well-marginated, **irregularly contoured and ulcerated mass in the mid-esophagus.**

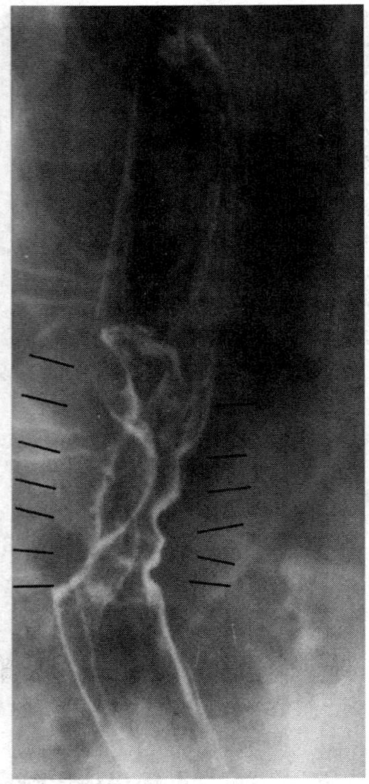

Fig. 19.20 Dysphagia.

- **Discussion** A mass with this appearance is almost pathognomonic for **esophageal carcinoma.** The majority are of squamous cell type. The esophagus has no serosa to limit the spread of tumor; lymphatic dissemination occurs early. The prognosis is very poor. Surgical treatment consists of resection with gastric mobilization or bypass. Chemotherapy and radiation are used as well.

■ Waters' View of Facial Radiograph, Fig. 19.21 **Facial pain and fever.**

- **Description** Waters' radiograph of the paranasal sinuses shows **air-fluid levels in both maxillary and frontal sinuses.**
- **Discussion** This patient has **acute sinusitis.** Treatment is with antibiotics. Differential diagnosis of this pattern would include facial fractures with bloody effusions and barotrauma. Totally opacified sinuses have a wider differential diagnosis. Involvement of the frontal and sphenoid sinuses is a concern because of the proximity of the brain and meninges and preformed pathways that may allow spread of infection intracranially. Intraorbital spread may occur from ethmoid sinusitis.

■ Elbow Radiograph, Fig. 19.22 **Fell on outstretched arm. Pain on pronation and supination.**

- **Description** Anteroposterior and lateral radiographs of the elbow demonstrate prominent **anterior and posterior fat pads ("sail sign").**

  A **nondisplaced fracture of the radial head is present.**
- **Discussion Radial head fractures** are common and are usually treated conservatively. Occasionally the fracture fragment is

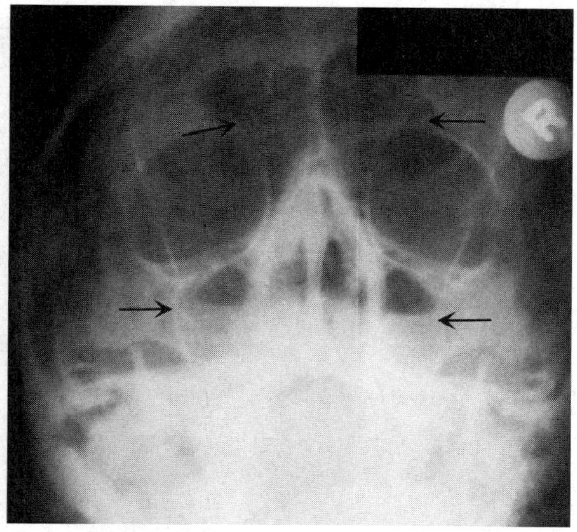

Fig. 19.21 Facial pain and fever.

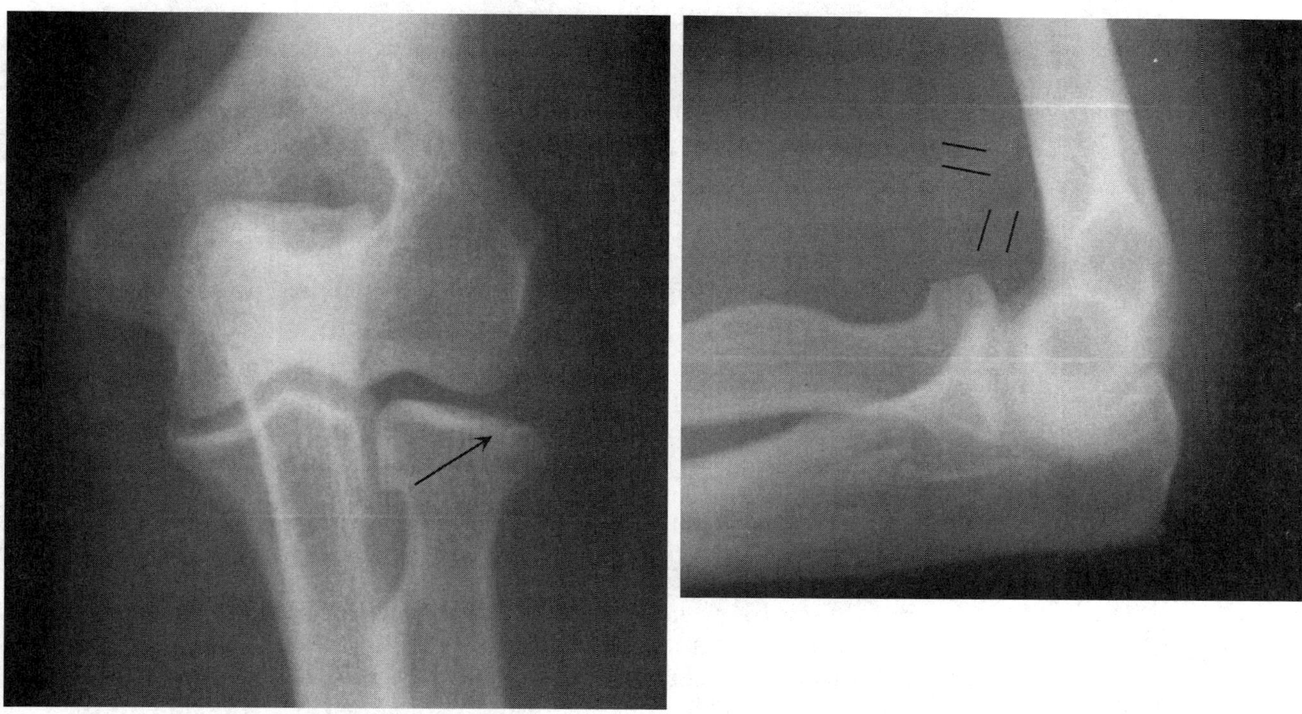

Fig. 19.22 Fell on outstretched arm; pain on pronation and supination.

excised. At times the fracture may not be seen on the initial radiographs. In this situation a nuclear medicine bone scan may be helpful.

- Transvaginal Ultrasound (US), Fig. 19.23 **Vaginal bleeding.**
  - **Description** Sagittal transvaginal ultrasound image shows a 1.5-cm complex cystic structure with a hyperechoic rim in the uterine fundus. An echogenic region is present eccentrically in the **cystic structure.**

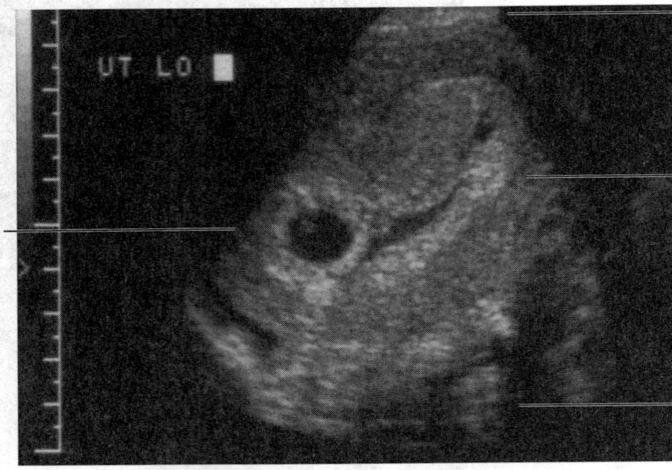

Fig. 19.23 Vaginal bleeding.

- **Discussion** This patient has an **intrauterine pregnancy of early gestational age.** One fourth of all pregnant women will have bleeding in the first trimester. Evaluation is best done by correlation of quantitative βhCG and ultrasound. Differential diagnosis would include ectopic pregnancy, threatened abortion, gestational trophoblastic disease, and non–pregnancy-related processes.

■ Upper Gastrointestinal Study (UGI), Fig. 19.24 **Abdominal pain.**

- **Description** Single radiograph from an upper gastrointestinal series shows a collection of **barium projecting beyond the radiologic confines of the lesser curvature of the stomach.**
- **Discussion** This patient has a **gastric ulcer.** Signs to suggest benignity include protrusion beyond the expected gastric lumen, folds radiating from the ulcer (not shown), a smooth edema mound, and healing in an appropriate time course. Currently, *Helicobacter pylori* is considered a major causative factor in peptic ulcer disease; however, its association with gastric ulcers is slightly lower than with duodenal ulcers.

■ Ankle Radiograph, Fig. 19.25 **Trauma.**

- **Description** A single radiograph of the **ankle** shows a **fracture line** extending from the lateral portion of the tibial metaphysis through the central portion of the epiphysis. A fracture through the fibular physis has occurred with associated widening of the mortise.
- **Discussion** This skeletally immature (pediatric) patient has a **Salter-Harris IV fracture of the tibia and a Salter-Harris I fracture of the fibula.** The ligaments stabilizing the ankle and the interosseous membrane are disrupted, too. Reduction may be by closed or open (i.e., operative) means. However, the patient must be monitored for abnormal growth patterns after clinical union (adequate weight bearing) has been achieved. Premature physeal closure may cause limb shortening. Partial closure will lead to angulation deformity. Hyperemia to the entire limb may cause overgrowth at the other physes, leading to overall lengthening.

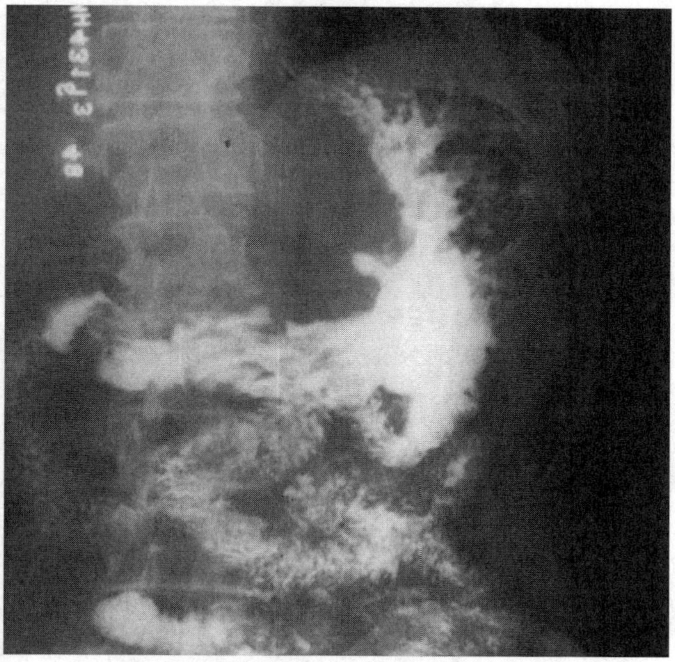

Fig. 19.24 Abdominal pain.

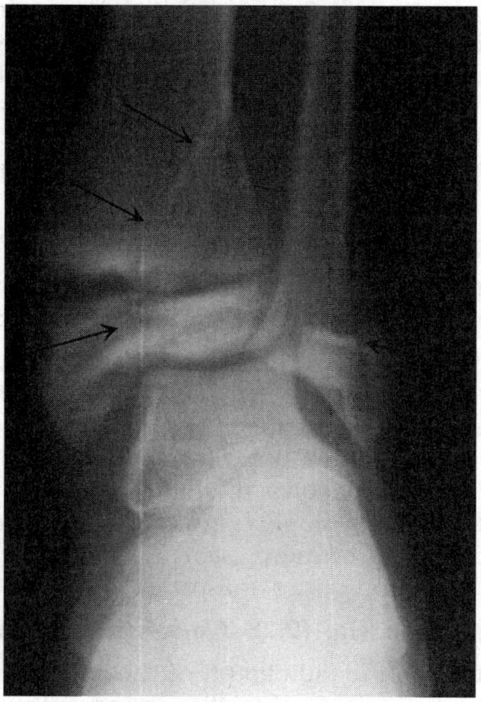

Fig. 19.25 Trauma.

■ Cervical Spine Radiograph, Fig. 19.26 **Trauma.**

- **Description** Cross-table lateral radiograph of the cervical spine shows a **displaced and angulated fracture** through the pars interarticularis **of C_2** bilaterally. Retropharyngeal soft tissue swelling is noted.

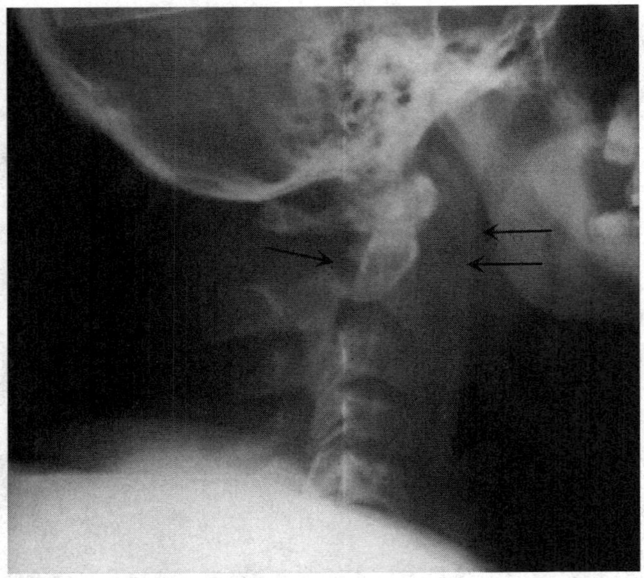

Fig. 19.26 Trauma.

- **Discussion** This patient has a **"hangman's" fracture.** This fracture is unstable and requires immobilization (a halo device). Immediate concerns include protection from further neurologic injury and diagnosis and treatment of other injuries. Evaluation of the amount and level of neurologic injury is also important.

  Note that the entire cervical spine is not visualized on this radiograph. If the cervical spine from the skull base through top of T1 cannot be seen on cross-table lateral radiographs, CT scan is the study of choice.

■ Leg Radiograph, Fig. 19.27 **Trauma.**

- **Description** Anteroposterior radiograph of the lower leg of a child shows **buckling of the distal tibial cortex.**
- **Discussion** This patient has a **torus fracture.** These occur only in young children and are due to relative plasticity of their bony structure. The mechanism is axial loading. Characteristic locations include the radius and tibia. Treatment is conservative.

  Another fracture peculiar to children is the greenstick fracture (not shown), where a single bony cortex fractures with bowing of the opposite side. Treatment involves completion of the fracture and closed reduction under anesthesia.

■ Abdominal Radiograph, Fig. 19.28 **Abdominal pain and fever.**

- **Description** Supine radiograph of the abdomen shows air throughout a redundant colon. Projecting over the right greater sciatic foramen is a well-marginated calcific density.
- **Discussion** This patient has **acute appendicitis.** Diagnosis is usually made purely on the basis of history and physical exam findings. Difficulty arises in the very young, women of childbearing age, and older patients. Plain films may be helpful, as in this case, but are often nonspecific. The presence of an appendicolith in a patient with appendicitis is indicative of more severe disease.

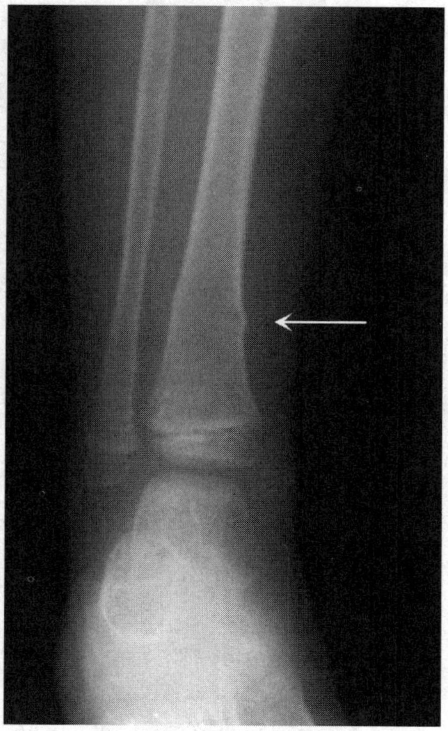

Fig. 19.27 Trauma.

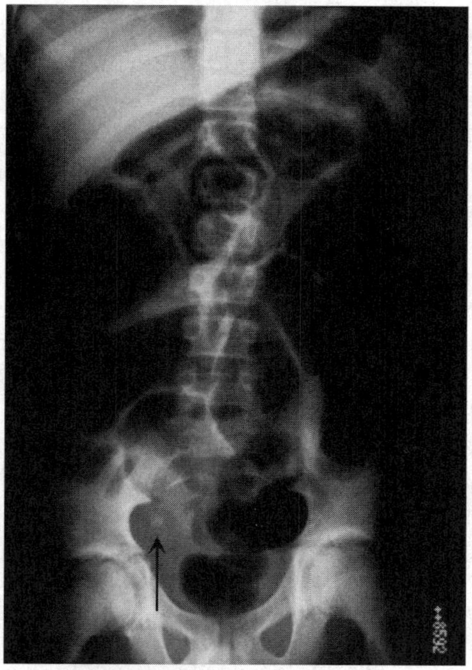

Fig. 19.28 Abdominal pain and fever.

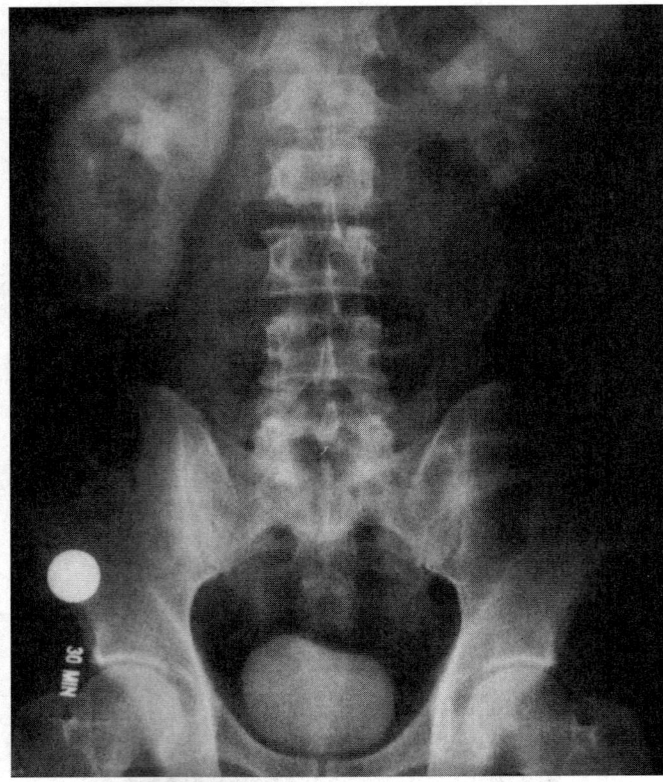

Fig. 19.29 Colicky right flank pain.

However, many normal appendices contain appendicoliths, making this sign less useful.

Ultrasound can be very specific if the appendix is visualized. Barium enema is also very good; unfortunately, 15% of normal appendixes do not fill with barium. CT scan has the highest sensitivity and specificity of all the possible radiologic tests.

- Excretory Urogram, Fig. 19.29 **Colicky right flank pain.**
  - **Description** Supine radiograph of the abdomen from an **excretory urogram obtained 30 minutes after the administration of intravenous contrast.** A dense nephrogram with a partial pyelogram is present on the right. The left nephrogram has already faded, although the kidney is small with diffuse cortical thinning. The left pyelogram is normal. The bladder is partially filled and unremarkable.
  - **Discussion** This patient has a functionally solitary **obstructed right kidney.** Obstruction produces an immediate, increasingly dense nephrogram. The pyelogram is variable as to its time of appearance based on the level and degree of obstruction. The left kidney shows signs of global atrophy due to long-standing unilateral disease.

    Obstruction is usually caused by mechanical means, such as a ureteral calculus, sloughed papilla, blood clot, or ureteral trauma. The underlying mechanism dictates the method of relieving the obstruction. Many stones will pass spontaneously. However, those that do not may be removed cystoscopically or be crushed by ex-

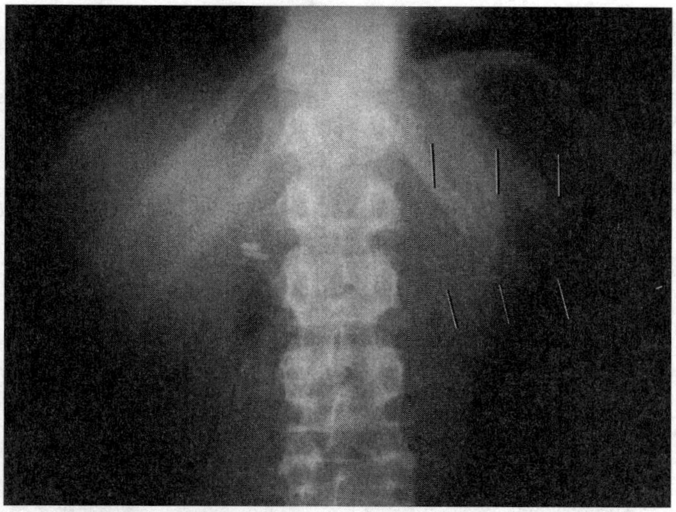

Fig. 19.30 Epigastric pain and vomiting.

tracorporeal shock wave lithotripsy. Other causes of obstruction may be relieved by creating a percutaneous nephrostomy. Infection behind an obstruction represents an emergency.

- Abdominal Radiograph, Fig. 19.30 **Epigastric pain and vomiting.**
  - **Description** Supine radiograph of the abdomen demonstrates **punctate calcifications** projecting in the expected position **of the pancreas.** Surgical clips are in the right upper quadrant. A nonspecific bowel gas pattern is present.
  - **Discussion** This patient has **acute pancreatitis superimposed on chronic pancreatitis.** No plain radiographic findings are specific for pancreatitis. A "colon cut-off" is only suggestive. The calcifications, however, are very specific for chronic pancreatitis. CT scan is the modality of choice for assessing the extent of disease and to detect complications (e.g., abscess, hemorrhage, or necrosis). Common etiologies of pancreatitis include alcoholism, cholelithiasis, hypercalcemia, certain medications, and trauma.
- Chest Radiograph, Fig. 19.31 **Shortness of breath.**
  - **Description** Frontal radiograph of the chest demonstrates diffuse **bilateral interstitial disease with** foci of air space **consolidation.** Apical cystic changes are present.
  - **Discussion** This patient has ***Pneumocystis carinii*** **pneumonia.** The radiograph shows all the classic findings for AIDS-associated PCP except one, a pneumothorax. Diagnosis is made by finding the organism in sputum or bronchial washings. Treatment is with pentamidine and respiratory support.

    Other pulmonary diseases seen in AIDS patients include tuberculosis, atypical mycobacterial infections, fungal pneumonia, and viral pneumonia. Kaposi's sarcoma and lymphoma may involve the lung as well.

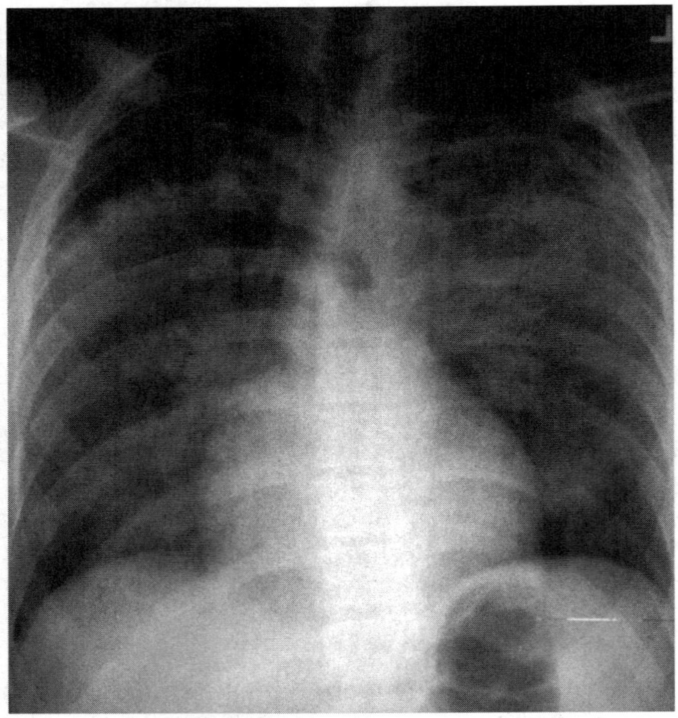

Fig. 19.31 Shortness of breath.

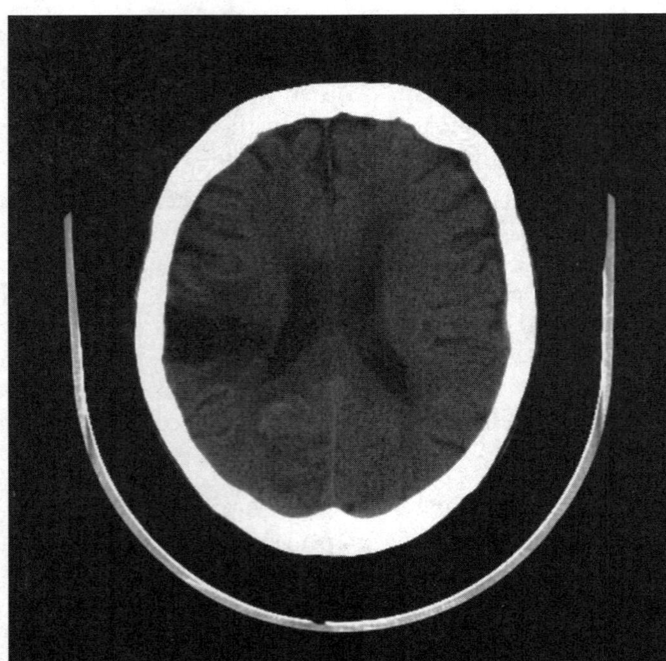

Fig. 19.32 Left-sided weakness.

- Head CT, Fig. 19.32 **Left-sided weakness.**
  - **Description** Axial CT image without contrast at the level of the cerebral hemispheres demonstrates **low attenuation in the right frontoparietal region** involving the cortex and subcortical white matter.

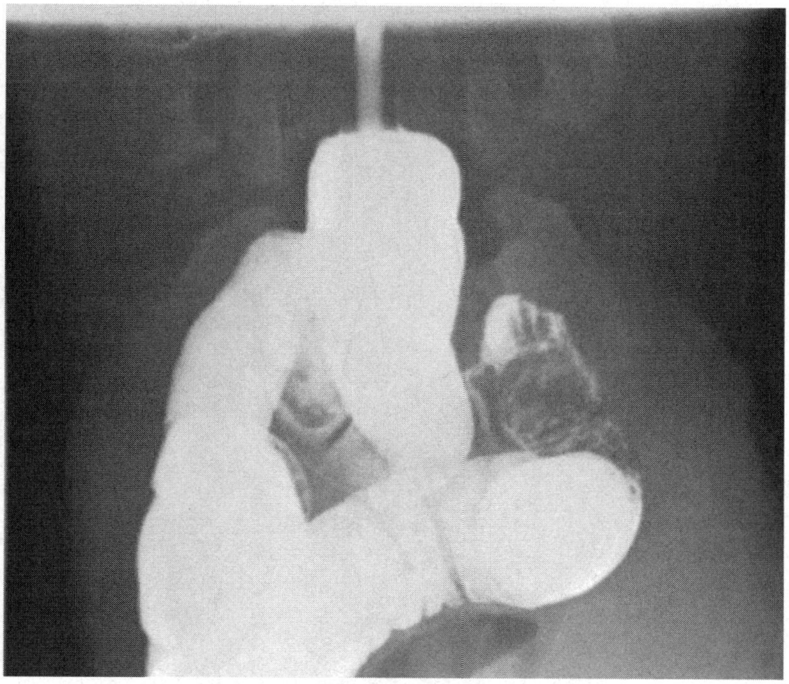

Fig. 19.33 Abdominal pain.

- **Discussion** This patient has a **right hemispheric stroke in the distribution of the middle cerebral artery.** Stroke syndromes are named for the vascular territory they involve. Each has a characteristic constellation of clinical findings. Strokes are further classified as bland (i.e., ischemic), such as this one, or hemorrhagic. Secondary hemorrhage into a bland stroke may occur. Risk factors include all those for atherosclerotic vascular disease, nonvalvular atrial fibrillation, and collagen-vascular diseases. Treatment consists of supportive care as well as anticoagulation for bland strokes.

■ Barium Enema, Fig. 19.33 **Child with intermittent abdominal pain.**

- **Description** A single radiograph from a barium enema shows near complete filling of the colon. **A filling defect is in the proximal ascending colon** and has a characteristic surrounding "coiled spring."
- **Discussion** This child has an **intussusception.** This happens when one segment of bowel telescopes into another, causing intermittent obstruction. Eventually, vascular compromise of the intussusceptum may occur. Therefore expeditious reduction is carried out. Most pediatric cases may be reduced by barium enema. No underlying lead point is found in the majority of cases. The age group most often affected is between 3 months and 3 years. In adults a pathologic lead point (usually a tumor) is often found, and reduction requires surgery.

Answers and Explanations to Multiple Choice Review Questions

Chapter 1: Neurology

1. Answer a:

The body temperature must be greater than or equal to 32° C to meet the criteria for brain death. A person cannot be declared dead by brain criteria without a documented body temperature.

b: Heart rate is not a required part of the examination for death by brain criteria though it is always a good idea to check heart rate. Blood pressure however is required, and systolic BP must be greater than or equal to 90 mm Hg for criteria to be met.

c: An ECG is not required to make the determination of death by brain criteria.

d: A chest x-ray is not required to make the determination of death by brain criteria.

e: A head CT is not required for a clinical determination of death by brain criteria though in head injury cases patients may get a head CT prior to the determination. A head CT is not a recommended confirmatory test in cases of brain death.

2. Answer e:

Spinal reflexes such as deep tendon reflexes and withdrawal response may be preserved in cases of brain death because these are mediated purely through the spinal cord and do not require any brain function.

a: Decorticate posturing is suggestive of some brain function below the level of the cortex. Because the red nucleus has a flexor bias, many think that decorticate posturing represents a loss of cortical input with an intact red nucleus and other structures below. Decerebrate posturing would then be below the level of the red nucleus and the lateral vestibulospinal tracts are felt to be the primary extensor bias in these cases. In patients with any type of such posturing, part of the brain is functional and thus these postures are never consistent with the diagnosis of death by brain criteria.

b: The pupillary light response, cough reflex, jaw jerk as well as the corneal reflexes, oculocephalics, oculovestibular, and gag reflexes are all cranial nerve tests and must always be absent in cases of death by brain criteria.

c: The pupillary light response, cough reflex, jaw jerk as well as the corneal reflexes, oculocephalics, oculovestibular, and gag reflexes are all cranial nerve tests and must always be absent in cases of death by brain criteria.

d: The pupillary light response, cough reflex, jaw jerk as well as the corneal reflexes, oculocephalics, oculovestibular, and gag reflexes are all cranial nerve tests and must always be absent in cases of death by brain criteria.

3. Answer e:

Because the patient is having a seizure you know that her cortex is partly functional. A seizure is never consistent with the diagnosis of death by brain criteria.

a: An EEG should not be done until the clinical picture is entirely consistent with the diagnosis of death by brain criteria.
b: An apnea test should not be done until the clinical picture is entirely consistent with the diagnosis of death by brain criteria.
c: An angiogram should not be done until the clinical picture is entirely consistent with the diagnosis of death by brain criteria.
d: Seizure activity is not consistent with brain death. The family should be informed of this.

4. Answer b:

Migraines are common in young women, are typically described as throbbing, may be unilateral or bilateral, and usually remit with sleep over several hours. Because there were no preceding visual or other neurologic symptoms in this patient, this is a migraine without aura.

a: Because there were no preceding visual or other neurologic symptoms in this patient, this is a migraine without aura.
c: Cluster headaches have associated unilateral tearing, rhinorrhea and flushing, and usually occur in men in their 40s to 50s.
d: The headache of subarachnoid hemorrhage is usually diffuse and maximal at onset. They are classically described as "the worst headache of my life."
e: Giant cell arteritis typically produces less severe unilateral headache and visual disturbances in patients over 60 years old.

5. Answer e:

Migraine without aura is not associated with focal abnormalities on neurologic examination.

a: Migraines without aura are not associated with a Horner's syndrome.
b: Migraines without aura are not associated with focal abnormalities.
c: Migraines are not usually associated with meningismus though subarachnoid hemorrhage and meningitis are.
d: Mass lesions may produce papilledema secondary to increases intracranial pressure, but migraines do not.

6. Answer d:

Nonsteroidals may be useful for aborting the migraine if taken immediately at the onset of headache.

a: Tricyclic antidepressants are useful as prophylaxis but not as abortive agents for migraines.
b: Beta-blockers are useful as prophylaxis but not as abortive agents for migraines.
c: Calcium channel blockers are useful for prophylaxis but not as abortive agents for migraines.
e: Narcotic analgesics are addictive and should be reserved for headaches not relieved by other means.

Chapter 2: Ophthalmology

1. Answer c:

Cataracts often present as painless progressive visual loss in the elderly population. Symptoms may include decreased contrast sensitivity and glare in brightly lit areas. Slit lamp examination and direct ophthalmoscopy will aid in this diagnosis.

a: Corneal abrasions are usually painful and cause acute visual change if located along the visual axis.

b: Retinal detachment often causes acute visual loss, the appearance of floaters, flashing lights, and sometimes a marked visual field loss in the affected eye. If not attended to immediately, complete and irreversible vision loss may occur.

d: In acute angle closure glaucoma, patients often present with marked eye pain relatively early in the course of the disease. The eye is usually very red and the pupil mid-dilated with a sluggish reaction to light relative to the fellow eye. It is important to obtain the intraocular pressure of any glaucoma suspect.

e: Central retinal artery occlusion usually causes *acute* painless loss of vision in the affected eye. A pale fundus with a characteristic cherry red spot of the macula is characteristic of this affliction.

2. Answer a:

Astigmatism is a refractive error usually caused by an aspherical shape to the cornea. This refractive error often changes as a result of the surgical procedure and many times is easily eliminated with spectacle correction.

b: Amblyopia, or lazy eye, is a condition due to ocular problems as a child. This usually is due to one eye being turned in or out at an age less than 7. For these children, surgery (or glasses in some cases), combined with patching therapy often results in good vision. An eye with severe amblyopia, however, will not see well.

c: Retinal disease, including diabetic retinopathy and macular degeneration, may limit the visual potential of a cataractous eye. It is therefore important to perform a retinal examination in order to assess the potential benefit of cataract surgery.

d: Corneal opacification will not be eliminated by cataract surgery. If a large corneal scar exists along the visual axis (in front of the pupil) of a cataractous eye, vision is unlikely to reach 20/20 following a successful cataract extraction. Corneal opacities are easily visualized with slit lamp biomicroscopy.

e: Optic nerve atrophy often causes marked irreversible loss of vision. This condition is often seen in multiple sclerosis and in patients with intracranial masses. Damaged optic nerve axons severely limit the potential for visual acuity in an affected eye.

3. Answer d:

Although many patients presenting with a red eye have simple viral or bacterial conjunctivitis, the decreased vision and photophobia strongly suggest more serious eye disease. The corneal infection can often be diagnosed in the primary care setting with the use of topical flourescein dye. A cobalt blue light source, which is built in to many ophthalmoscopes and available as a penlight attachment, often reveals dye uptake on the ocular surface. This usually takes the form of a "dendrite," or tree branch-shaped lesion. Occurrence of ocular herpes often follows fever and can be more frequent or more severe in atopic individuals. Referral for confirmation and antiviral treatment would be appropriate.

a: Bacterial conjunctivitis does not usually decrease vision significantly or cause photophobia.

b: Viral conjunctivitis can decrease vision but it usually occurs over several weeks. Both eyes are usually affected, although often asymmetrically.

c: Angle closure glaucoma usually occurs in older adults and is not associated with a mucus discharge. Nausea, vomiting, and periocular pain may be present.

e: Although allergic or atopic conjunctivitis may cause a red eye, it is usually bilateral and does not decrease vision or cause photophobia. Itching is usually prominent.

4. Answer a:

Bacterial conjunctivitis often presents with a unilateral or bilateral purulent discharge, red eye, and no significant visual loss. It often responds quickly to topical antibiotics, and it is moderately contagious. Good hand washing is prudent.

b: Viral epidemic conjunctivitis is extremely contagious but usually presents bilaterally with more watery or a mucus-type of discharge. Symptoms take somewhat longer to resolve, and there is no specific treatment for the infectious component. Cool compresses often provide some symptomatic relief. A history of a viral prodrome or preauricular lymph node may be present.
c: Herpes simplex virus ocular disease usually does not present as a contagious agent but as a reactivation of quiescent latent herpes resident in the trigeminal ganglion. The discharge is usually not mucopurulent and visual acuity can be decreased.
d: Fungal conjunctivitis is very uncommon.
e: Allergic conjunctivitis does not spread from person to person, is usually bilateral, and itching and watery discharge are often prominent.

5. Answer d:

Blepharitis is one of the most common eyelid conditions affecting adults. It is caused by a blockage of the meibomian glands in the lids. The meibomian glands produce lipid and this blockage results in a tear film devoid of lipid. This causes changes on the ocular surface resulting in the symptoms of red eyes, foreign body sensation, "heaviness" about the eyes, and pain. Bacterial infection within the glands may occur. Further, lipid may "back up" within the lid, causing a foreign body reaction or lipogranuloma. This may result in the formation of a sty, which in the acute phase, is swollen and tender.

a: Although the lids may become infected, the resulting conjunctivitis is usually toxic due to bacterial toxins and not primarily infection. Primary bacterial conjunctivitis usually presents with a purulent discharge and is more acute in nature.
b: Patients with angle closure glaucoma are generally quite ill with severe eye pain, precipitous drop in visual acuity, and nausea and vomiting without associated lid involvement.
c: Although the eyelids may be erythematous and irritated, patients with herpes simplex have lesions confined to the epidermis. Onset is typically unilateral. The corneas may become involved with the classic arborizing dendritic corneal epithelial defect.
e: Corneal abrasions usually have a history of some type of eye trauma. Onset is abrupt with significant eye pain, foreign body sensation, and tearing, prompting the patient to seek immediate medical attention.

6. Answer b:

Treatment is aimed at increasing the oil's transit time from the gland by the use of hot soaks applied to both eyes, usually twice a day for 5 to 10 minutes. Antibiotic ointment q HS is applied to the lids to treat the secondary bacterial infection.

a: In poorly responsive cases, lid margin cultures and oral doxycycline antibiotic may be of benefit but most do well without such measures. Lid margins cultures may disclose resistant organisms. Doxycycline is secreted in glandular secretions and acts as both an antibiotic as well as an agent that thins the secretions. The standard dose is 100 mg a day for 6 weeks.

c: Viral cultures would not be of help because the condition is of bacterial origin.
d: Herpes simplex cultures would not be of help because the condition is of bacterial origin.
e: Topical anesthetics such as Tetracaine are toxic to the cornea and severe corneal damage may result from their use. Topical anesthetics should never, under any circumstances, be prescribed for a patient.

CHAPTER 3: OTORHINOLARYNGOLOGY

1. Answer b:

A sore throat will be present in both tonsillitis and peritonsillar abscess.

a: Trismus is the inability to fully open the mouth. Pus adjacent to the pterygoid muscles induces spasm. Trismus is one of the hallmarks of a peritonsillar abscess.
c: Bulging of the anterior tonsillar pillar is caused by pus in the potential space around the tonsil capsule. It is one of the signs of a peritonsillar abscess.
d: A muffled voice or "hot potato" voice is caused by the medial displacement of the tonsil by the abscess.
e: In a peritonsillar abscess the bulging of the inferior tonsillar pillar on the side with the abscess often obscures the visualization of that tonsil.

2. Answer b:

Teenagers or young adults who do not respond to treatment for bacterial tonsillitis may have infectious mononucleosis. The appearance of the tonsils may not differentiate bacterial tonsillitis from mono tonsillitis. The monospot test with positive results will avoid further unncessary antibiotics and direct you toward offering symptomatic relief.

a: *Streptococcus* causes bacterial tonsillitis, but almost invariably responds to penicillin. Therefore, results of a throat culture, either positive or negative, would not be very helpful.
c: As long as the patient was taking his medication it should make no difference whether the antibiotic was given orally or parenterally.
d: Clindamycin may be useful in patients with chronic tonsillitis who do not respond to penicillin. Remember, this was this boy's first episode of tonsillitis.
e: A biopsy of the tonsil is useful for unilateral painless enlargement. Nothing about this case suggests malignancy.

3. Answer a:

History and physical examination is the most useful in differentiating acute sinusitis from viral rhinosinusitis or allergic rhinitis. Adequate physical examination requires decongestion of the nose with phenylephrine to visualize the middle meatus.

b: Plain x-rays rarely add additional useful information in evaluation of acute sinusitis.
c: A CT scan in the coronal plane is an essential road map for undertaking sinus surgery, but is unnecessary for initial evaluation of acute sinusitis.
d: CT scans in the axial plane are unnecessary for the evaluation of uncomplicated acute sinusitis.
e: A swab culture of nasal discharge only adds confusing information as it is usually contaminated by normal nasal flora. Selection of antibiotics should be based upon knowledge of the prevalence of pathogenic organisms.

4. Answer c:

Conjunctival edema (chemosis), proptosis, and restricted range of movement of extraocular muscles are all signs of orbital complications of sinusitis.

a: Excessive tearing (epiphora) may be caused by many ophthalmologic problems and is not specifically associated with orbital complications of sinusitis.
b: As sinusitis responds to treatment the nasal discharge clears, but may have flecks of blood or old purulent material that are emptying from the previously inflamed sinuses. This is not a sign of an orbital complication.
d: Conjunctival erythema and drainage on the lashes may be seen in several ophthalmologic infections and are not specific for orbital complication of sinusitis.
e: Enophthalmos, or a sunken in eye, is a complication of an unrepaired orbital floor fracture and not of sinusitis.

5. Answer d:

Remember back to basic life support and CPR training. The first step is inspection of the mouth for sources of obstruction, then a jaw thrust, which will bring the tongue forward.

a: This is a good treatment for most respiratory distress, but it is not the first step.
b: This is a good choice for patients who cannot be intubated, but it is not the first or even second management step.
c: Tracheotomy is not the preferred technique for management of acute upper airway obstruction. The thyroid gland will be in the way and have to be mobilized or divided.
e: A nasopharyngeal airway can relieve upper airway obstruction caused by a swollen tongue or large tonsils. It is not, however, the first step in management.

Chapter 4: Pulmonology

1. Answer c:

Inhaled corticosteroids are not indicated in the acute management of asthma because not enough medication can be delivered to the bronchospastic patient to be efficacious.

a: Nebulized beta-2 agonists are standard therapy for acute exacerbations of asthma because they promote bronchodilation.
b: Intravenous methylprednisolone is indicated in the acute management of asthma because it is antiinflammatory and thus treats the underlying pathology of asthma. Although steroids are given, they do not have a significant effect until several hours have passed.
d: Intravenous aminophylline is indicated in the acute management of asthma and may accentuate bronchodilation.
e: Subcutaneous epinephrine can be used in selected patients (usually younger patients) in the acute management of asthma.

2. Answer d:

This patient is more likely to have extrapulmonary tuberculosis secondary to his HIV disease.

a: HIV-positive patients are at higher risk of being infected with drug-resistant organisms, and a 4-drug regimen should be instituted.
b: HIV-positive patients who have pulmonary tuberculosis are more likely to present with "atypical" chest x-rays.
c: Tuberculosis has increased with the increased incidence of HIV disease.
e: The patient is more likely to be skin test–negative as a result of his HIV disease (secondary to immunosuppression).

3. Answer c:

Although the patient has no indication for systemic or local steroids at present, she should be followed serially to assess for disease activity and need for treatment.

a: A CT scan of the chest is not mandatory to evaluate sarcoidosis or to monitor the progression of the disease.

b: Because the patient is asymptomatic and there are no pulmonary function test abnormalities, there is no indication for prednisone for this patient's pulmonary involvement. There is also no evidence to suggest that prednisone will prevent the progression of this disease.

d: There is no well-defined role for gallium scanning in sarcoidosis.

e: In general, a neurologist is not required to make the diagnosis of neurosarcoidosis.

4. Answer a:

The findings of interstitial lung disease with bilateral hilar adenopathy and right paratracheal adenopathy is characteristic of sarcoidosis. The incidence of sarcoidosis is increased in blacks, especially females. In addition to lung disease there can be multisystem involvement, including the eyes, parotid glands, lymph nodes, heart, liver, spleen, skin, and neurologic system. It is characterized by noncaseating granulomas found at the sites of organ involvement.

b: Lymphoma is often the major diagnosis in the differential with sarcoidosis. It, however, typically presents with asymmetric adenopathy, and if lung involvement is present, it typically is more focal. This diagnosis, however, cannot be excluded completely without obtaining tissue for pathologic review.

c: Both TB and fungal infections can present with adenopathy and interstitial infiltrates. TB, however, is usually more focal and typically involves predominantly the upper lobes, especially if it is caused by reactivation of old disease. These infections, however, must always be considered when evaluating a patient with interstitial lung disease.

d: Subacute hypersensitivity pneumonitis could present with a nonproductive cough, dyspnea on exertion, and low-grade fevers. However, mediastinal adenopathy is not seen in hypersensitivity pneumonitis.

5. Answer b:

A nocturnal polysomnogram is indicated in the workup of excessive daytime sleepiness that is not simply secondary to chronic sleep deprivation.

a: Morning headaches are an indicator of severe obstructive sleep apnea and only occur in about 10% to 15% of patients. Often symptoms of excessive sleepiness only occur when sedentary, and so his presentation is quite typical for a sleep disorder.

c: Many patients with symptomatic obstructive sleep apnea may not desaturate below 90%, making overnight oximetry a poor screening procedure.

d: Pulmonary hypertension only occurs in approximately 15% to 20% of patients and, without signs or symptoms of pulmonary hypertension, there is no reason to evaluate for it.

Chapter 5: Cardiology

1. Answer a:

You must determine whether the patient has a history of underlying cardiac disease or whether she is taking any medications that might predispose to bradycardia and PACs.

b: There is no role for quinidine therapy to suppress PACs that are likely benign. Quinidine has a relatively high potential for proarrhythmic adverse effects.

c: From the information provided, this patient has no indication that her bradycardia or PACs are threatening to her.

d: It is reasonable to check serum electrolytes, especially if the patient is on medications that might alter chemistries (i.e., diuretics); however, potassium should never be given empirically without knowing the serum concentration. Hyperkalemia is considerably more dangerous than hypokalemia.
e: There is no indication for EP testing in this patient. You might want to consider Holter monitoring to look for arrhythmias.

2. Answer b:
Verapamil should be avoided because of its negative inotropic effect.

a: Digoxin is useful as a positive inotropic agent.
c: ACE inhibitor therapy is a mainstay in treatment of heart failure.
d: Diuretics are useful in relieving congestive symptoms and edema.
e: Alcohol must be discontinued to eliminate the causative agent for the cardiomyopathy.

3. Answer b:
Development of CHF after a probable viral illness raises a high suspicion for viral cardiomyopathy.

a: This patient is at low risk for MI, and there is nothing in the history to support this diagnosis.
c: Amyloidosis is relatively uncommon, especially in younger persons.
d: The history is not suggestive of pericarditis, and the patient has no other predispositions to developing pericardial constriction. The viral cardiomyopathy with associated pericarditis could ultimately lead to constrictive pericarditis.
e: There is no mention of a new murmur to support the diagnosis of papillary muscle dysfunction with mitral insufficiency.

4. Answer b:
The pain of pericarditis often lessens with sitting forward. The rub of pericarditis is a sandpaper sound that can be multicomponent during both systole and diastole.

a: Papillary muscle rupture would cause acute mitral regurgitation with a holosystolic murmur.
c: There is no evidence for cardiac dysfunction to suggest cardiomyopathy. Viral infections can be responsible for both pericarditis and cardiomyopathy.
d: The character of the pain is not ischemic, and the ECG changes are consistent with pericarditis, not diffuse myocardial ischemia.
e: Esophagitis would not be associated with acute ECG changes, especially ST segment elevation.

5. Answer d:
The pericarditis has resulted in a pericardial effusion and development of tamponade. The effusion limits the opposing surfaces of the visceral and parietal pericardium that was responsible for the rub.

a: There is no evidence for acute MI. This history and progression of disease are most consistent with pericarditis.
b: Acute bacterial endocarditis is unlikely given the lack of fever or other stigmata of endocarditis.
c: Acute tubular necrosis usually follows an episode of hypotension with poor renal perfusion.
e: The murmur is gone because of the pericardial effusion.

Chapter 6: Gastroenterology

1. Answer e:
The symptoms that the patient is presenting with is most consistent with an acute infectious gastroenteritis. Giardia can cause acute symptoms from either an enterotoxic or invasive process. In this particular senario the patient is presenting with symptoms most consistent with acute enteroinvasive process. Giardia is a parasite found

in brackish waters and is usually ingested with seafood. Symptoms of crampy abdominal pain, nausea, vomiting, and diarrhea occur within 12 to 24 hours of presentation.

a: Gastric ulcers usually present with upper GI bleeding, with hematemesis and coffee-ground emesis. Stool sampling is usually positive for blood, usually melenic, and the patient's symptoms of fever and chills are not commonly seen with upper GI bleeding.
b: Ulcerative colitis usually presents with lower GI bleeding. Abdominal cramping is common; fever can occasionally occur. Ulcerative colitis is an inflammatory process that starts from the rectum, and disease proximal to the ileo-cecal valve is uncommon. Vomiting is not commonly seen with ulcerative colitis.
c: Viral gastroenteritis can cause many of this patient's symptoms. However, the duration of the illness and the blood in the stool is not commonly seen with viral gastroenteritis.
d: *Staphylococcus* usually causes GI symptoms by producing an enterotoxin and usually presents within 4 to 6 hours after ingestion of contaminated food products, particularly dairy products.

2. Answer a:

The mainstay of treatment for acute infectious gastroenteritis is supportive care and correction of hemodynamic instability. The patient definitely requires intravenous fluids; although orthostatic blood pressures were not checked and the blood pressure is within normal limits, the patient has a "resting" tachycardia, out of proportion to his fever, a sign of significant volume depletion. Normal saline or lactated Ringer's should be given briskly until tachycardia and orthostasis is normalized and the patient has shown the ability to maintain fluid status.

b, e: Stool for fecal leukocytes and stool cultures help in determining the organism and, if necessary, appropriate antibiotic therapy.
c: Lomotil and other anticholinergics are contraindicated in patients with enteroinvasive gastroenteritis. These medications work by decreasing GI motility and can worsen symptoms, lengthen illness, and even contribute to toxic megacolon.
d: Although blood tests are not necessary for minor cases of gastroenteritis, these labs should be checked on any patient who is exhibiting significant volume loss and complications.

3. Answer e:

Bacillus cereus produces an enterotoxin that causes a diarrheal illness, usually within 12 hours. *B. cereus* is usually found in rice products.

a: *Trichonella spiralis* is a parasitic organism found in uncooked pork. Symptoms usually take years to develop, and include muscle weakness caused by infiltration of the parasite into the muscle. Acute gastroenteritis is not commonly seen with *Trichonella.*
b: Cytomegalovirus may cause diarrhea, but usually in immunocompromised patients, such as those with AIDS. Diarrhea is more commonly chronic.
c: *Helicobactor pylori* has been implicated in formation of gastric ulcers. Symptoms of infection include heartburn and upper GI bleeding.
d: *Clostridium botulinum* causes symptoms as a result of a severe neurotoxin that causes muscle tetany. Gastroenteritis symptoms are not the predominant clinical presentation.

4. Answer c:

Most gastroenteritides can and are managed appropriately as an outpatient. If the patient is hemodynamically stable, reliable, and does not have evidence of enteroinvasive etiology, outpatient management with increased oral hydration is appropriate.

a, d: Upper endoscopy and lower colonscopy are not indicated initially. Endoscopy may be indicated if the patient does not improve with conservative therapy within a week.
b: Incorrect. See C.
e: A tagged red cell scan is used to determine sites of lower GI bleeding, if the bleeding is greater than 0.3 ml/min.

5. Answer d:
Symptoms of dysphagia should not be discounted as psychiatric. This man may be presenting with gradual esophageal narrowing resulting an esophageal malignancy. Symptoms of the malignancy include dysphagia, initially with solids, then liquids. Weight loss, chest pain, bleeding, and aspiration all may occur. Other unusual presentations involve obstruction of adjacent organs.

a: See d.
b, c, e: A CT scan may be needed, but the barium swallow is the most sensitive and best initial first step of those listed. Upper endoscopy with biopsy should be completed after barium swallow has been performed. If these tests are normal, then an esophageal dysmotility study should be considered.

Chapter 7: Psychiatry

1. Answer e:
Schizoid personality disorder shares some features of schizophrenia but is not a psychotic illness with delusions and hallucinations. Individuals with personality disorders have enduring problems with features of their temperament and character. The schizoid individual tends to be aloof, friendless, odd, and lacks depth and warmth.

a: Amphetamine intoxication can create acute psychosis, complete with bizarre delusions and hallucinations, that is indistinguishable from schizophrenia.
b: Acute schizophrenic psychosis typically involves complex, bizarre delusions and auditory hallucinations.
c: Malingering may occur when there is some secondary gain from appearing to be ill—for example, to explain failing grades or get out of examinations.
d: Brief reactive psychosis may occur in response to serious stressors, such as in a student with high parental expectations who is failing courses in college. Delusions and hallucinations may adopt a theme related to the conflict.

2. Answer a:
The dexamethasone suppression test is not regarded as a diagnostic tool. It may be useful in some cases of major depression in monitoring response to antidepressant treatment.

b: History from the family is likely to be critical in illuminating past psychiatric and drug use history, family history, potential medical conditions, and developmental history that will aid in defining the diagnosis.
c: Toxicology screening will be critical for determining whether an intoxication is potentially responsible for the presentation.
d: A careful neurologic examination is indicated to rule out potential CNS abnormalities (e.g. tumor, bleeding, head injury) that may account for this change in mental state.
e: HIV testing is an important part of the examination of individuals presenting with new mental status changes. The first notable symp-

toms of HIV infection may be psychiatric.

3. Answer d:

Nortriptyline is a tricyclic antidepressant that has no utility in the management of acute agitation or psychosis.

a: Halperidol in small intramuscular doses is an excellent choice for the management of acute psychotic agitation. Watch carefully for acute dystonic reaction.
b: Lorazepam in small intramuscular doses is another excellent choice for the management of acute agitation. It is well absorbed intramuscularly and provides sedation and relief from anxiety. It will not have any antipsychotic efficacy.
c: Chlorpromazine given orally is a good choice to provide both sedation and antipsychotic efficacy. It is a low-potency and antipsychotic and must be monitored for creating hypotension and anticholinergic toxicities. Small initial test doses are indicated. Acute dystonia is less likely. Used intramuscularly, it is very painful on injection.
e: Trifluoperazine is a medium-potency antipsychotic that is generally well tolerated with a side effect profile that makes hypotension and anticholinergic toxicities unlikely. It may cause acute dystonia.

Chapter 8: Rheumatology

1. Answer a:

CPPD crystals of pseudogout are positively birefringent. As pseudogout is an inflammatory arthritis, synovial fluid would show low viscosity and 2000 to 75,000 WBCs. No organisms are involved, so gram stain is negative.

b: Negative birefringence is seen in urate crystals, and this synovial fluid is typical in a patient with gout.
c: Although the crystals of pseudogout are positively birefringent, the synovial fluid is inflammatory and thus would be translucent rather than transparent; 2000-75,000 WBS's are typically seen.
d: Urate, not CPPD crystals, are negatively birefringent, and the remainder of the synovial fluid analysis is that of normal synovial fluid.

2. Answer d:

Adequate calcium intake and weight-bearing exercise are important in obtaining and maintaining bone mass, even after adolescence, and are all that is necessary in an otherwise healthy young woman. As this patient reaches menopause, estrogen replacement may be appropriate.

a: Although estrogen replacement at menopause can be very effective in preventing osteoporosis, a young woman can build up bone mass now with adequate calcium intake and weight-bearing exercise.
b: Bone densometry measurements can help assess risk of osteoporosis-related fractures and are indicated for patients with estrogen-depleted states (including menopause), chronic corticosteroid use, and in patients with fractures or x-ray abnormalities. They are not indicated in a healthy young woman.
c: Calcitonin is sometimes used to treat osteoporosis but does not have a role in the prevention of osteoporosis.

3. Answer b:

In this patient with a probable *S. aureus* septic joint, daily (or more frequently if necessary) joint aspiration and culture is necessary until the fluid is sterile to prevent destruction of the joint by the infected synovial fluid. IV antibiotics produce adequate antibody levels in the synovial fluid to fight infection.

a: Daily aspiration and IV antibiotics are necessary, and IV antibiotics produce high enough levels of synovial fluid antibiotic levels that intraarticular injection is not necessary. Intraarticular antibiotics also can cause chemical synovitis.

c: IV antibiotics alone are inadequate to treat septic arthritis. Daily joint aspiration is necessary to prevent joint damage. Splinting of a joint will not affect spread of infection.

d: Although high-dose antibiotics are used after an adequate course of IV antibiotics, they are initially insufficient to treat a septic joint. Joint aspiration needs to be done daily until the fluid is sterile to prevent joint damage.

4. Answer d:

This patient is presenting with an acute onset of SLE with evidence of lupus nephritis and possible CNS involvement. Normal creatinine 1 week previous suggest that this is an acute process. Corticosteroid therapy in conjunction with immunosuppressives are the first-line agents in patients with severe flares, especially with lupus nephritis.

a: Corticosteroids and NSAIDs are probably inadequate to treat this severe flare of SLE.

b: Hemodialysis is not indicated at this time, although if renal function continues to worsen, it may be needed. Antiinflammatory/immunosuppressive therapy is necessary to try to halt progression of the disease flare.

c: Low-dose heparin is used in SLE patients who have evidence of hypercoagulability. Corticosteroids alone are probably inadequate to treat this severe flare of SLE.

5. Answer a:

A high percentage of PMNs in the synovial fluid supports a diagnosis of septic arthritis.

b: Fluid from a septic joint is usually cloudy and opaque.

c: WBC count in synovial fluid from a septic joint is usually >100,000.

d: Crystals are generally not seen in a septic joint unless a crystalline arthritis is also present.

Chapter 9: Orthopedics

1. Answer e:

Obese, adolescent, black males are the typical patient profile. The radiographic finding of the posterior-inferior displaced epiphysis is diagnostic of this condition.

a: Mixed radiodense and radiolucent findings are typical in Perthes disease. No slip of the epiphysis is found in this condition.

b: Articular space narrowing is common in this condition. Juxtaarticular erosions may also be found.

c: Septic arthritis without fever is unlikely.

d: Transient synovitis also produces no radiographic evidence of a slip.

2. Answer d:

The S1 nerve root provides sensation to the lateral aspect of the foot, motor to the gastroc/soleus complex, and reflex to the ankle.

a: The L3 nerve root provides sensation to the anterior thigh and motor to the quadriceps.

b: The L4 nerve root provides sensation to the anterior leg and motor to the anterior tibialis.

c: The L5 nerve root provides sensation to the medial aspect of the foot and motor to the extensor hallicus longus.

e: The S2 nerve root provides sensation to the bottom of the foot.

3. Answer b:

If a specific identifiable source is not available and urgent antibiotic administration is needed, blood cultures performed before antibiotic administration are required. This may be the only chance for organism identification once antibiotics are given, and would not delay the urgent administration.

a: Antibiotics should be administered after blood cultures are obtained. Broad-spectrum antibiotics can be continued until organism identification is achieved.
c: Aspiration without a definable source on x-ray is difficult, and the yield is low.
d: A bone scan should be obtained after antibiotic administration, not before.
e: A delay in treatment of this septic child should not await a consult service that is not immediately available.

4. Answer c:
The physical examination raises the possibility of a scaphoid fracture, even though no fracture is identified on radiographs. Long arm splinting, with immobilization of the thumb, should be performed until repeat radiographs can be obtained. Generally 10 days are required for bone resorption to reveal nondisplaced fractures.

a: Early restoration of motion is not advised before repeat evaluation secondary to the risk of fracture displacement.
b: Casting is never recommended in cases of acute fractures with significant swelling. The circumferential nature of the cast may produce tissue ischemia if elevated pressures are present.
d: Although a CT scan would likely reveal the presence of a fracture, the cost prohibits its use in the acute setting. If a significant possibility still exists despite normal radiographs in 10 days, a CT scan may then be considered.
e: The suspicion here is that of an occult scaphoid fracture. An MRI is not indicated.

5. Answer c:
New-onset carpal tunnel syndrome with relatively mild symptoms is best managed conservatively with neutral splinting and NSAIDs.

a: Surgical release is not indicated for mild symptoms, especially without a trial of conservative treatment.
b: Some sort of intervention should be rendered to prevent further compression to the nerve.
d: Systemic steroids are not indicated. Local steroid injection may be utilized if the patient's symptoms do not improve.
e: The symptoms and signs seem well localized to the median nerve. Cervical etiology would be rare.

Chapter 10: Dermatology

1. Answer c:
Specific antitreponemal antibodies are not routinely checked in secondary syphilis, as nontreponemal antibodies (RPR or VDRL) are almost always positive and their titer may be used to monitor response to treatment.

a: As above, a positive RPR or VDRL is used to confirm the diagnosis when there is a clinical suspicion of syphilis.
b: A potassium hydroxide preparation should be checked to rule out tinea corporis.
d: Patients with one sexually transmitted disease are at increased risk for others, including herpes genitalis, HIV, chlamydia, and others. They should be questioned about symptoms and counselled about HIV testing.
e: See answer D; although the primary chancre most commonly has resolved by the time the rash of secondary syphilis presents (2 to 12 weeks after contact), there may be some overlap. Patients should be examined for coexisting genital lesions

and possible sexually transmitted diseases.

2. Answer b:

Benzathine penicillin G 2×10^6 units intramuscularly $\times$ 1 is the treatment of choice for early syphilis, including secondary syphilis.

a: Erythromycin is a second-line therapy, as relapses have been noted.
c: Benzathine penicillin G 2×10^6 units $\times$ 3 is the recommended treatment for late syphilis, including late latent and tertiary syphilis.
d: Although the rash of secondary syphilis should resolve on its own, the patient is not "cured," and there is risk of development of tertiary syphilis.
e: Oral penicillins are not recommended treatment for syphilis.

3. Answer b:

The diagnosis of genital warts is chiefly clinical; a biopsy is a potentially traumatic procedure for a child and should be avoided unless there is nonresponsiveness to treatment, uncertainty of the diagnosis, or extenuating legal considerations.

a: A 10-year-old child is unlikely to have acquired perirectal warts at childbirth without signs before this age; autoinoculation from nongenital skin may rarely occur, but in this scenario sexual abuse must be considered and fully evaluated. Health care providers are usually mandated reporters of suspected sexual or physical abuse.
c: Anoscopy must be performed to rule out anal canal lesions.
d: Full skin and mucous membrane evaluation must be performed to rule out other signs of abuse, other STDs, and nongenital warts.
e: An RPR to rule out syphilis, either causing the skin changes (condylomata lata) or coexisting, is crucial.

4. Answer a:

Local destruction on a 3-week basis until clinical clearing is noted is necessary to adequately control genital warts. Children, especially if previously traumatized, may require topical anesthetics (using caution on mucosal surfaces).

b: Podophyllin is never dispensed for home application nor applied daily.
c: Surgical excision is preferred for recalcitrant warts or large exophytic lesions; it is not a first-line treatment for small, flat lesions.
d: Systemic penicillins are treatment for syphilis, not for genital warts.
e: A single liquid nitrogen treatment is unlikely to eradicate genital warts. Salicylic acid is not a recommended treatment for genital warts.

Chapter 11: Endocrinology

1. Answer a:

This patient presents with the characteristic findings of adrenocortical hypofunction, including asthenia, orthostatic hypotension, and hyperpigmentation. The ACTH stimulation test measures the reserve capacity of adrenal steroid production.

b: Measurement of serum aldosterone may be within normal limits in patients with mild adrenal insufficiency.
c: Measurement of urine cortisol may be within normal limits in patients with mild adrenal insufficiency.
d: Imaging studies will not be helpful in assessing the function of the adrenal glands.
e: Toxicology screening will not be helpful in assessing the function of the adrenal glands.

2. Answer b:

A high-dose dexamethasone test is useful in differentiating patients who fail to adequately suppress cortisol secretion after the low dose of dexamethasone. A pituitary source of

ACTH is generally suppressible, whereas an ectopic source of ACTH is not. Once hypercortisolism has been identified, its source must be found. Assuming that this patient has not received exogenous steroid, the most likely source is her pituitary. Much less likely are an adrenal source or an ACTH-producing nonpituitary neoplasm. Further localization may occur with an imaging study (e.g., abdominal CT or pituitary MRI) or petrosal sinus sampling.

a: A high-dose dexamethasone suppression test is the next most appropriate test in identifying the source of hypercotisolism.
c: Measurement of serum ACTH is the next most appropriate test in identifying the source of hypercortisolism.
d: The dexamethasone suppression test has established that this patient has Cushing's syndrome (hypercortisolism), therefore measurement of urinary free cortisol is unnecessary.
e: Empiric pharmacologic treatment is not indicated; furthermore, it is usually reserved for patients deemed to be nonsurgical candidates.

3. Answer d:

The lightheadedness in the afternoon is explained by hypoglycemia secondary to an increased activity level without changes in diet or insulin dosage. The peak effect of his NPH given before breakfast is during the afternoon, concomitant with his symptoms. Therefore a reduced dose of NPH should be given before breakfast.

a: Changes in his bedtime dose of insulin will not affect his afternoon symptoms.
b: A decrease in the regular insulin dose before breakfast will affect his morning rather than afternoon glucose levels.
c: Changes in his bedtime dose of insulin will not affect his afternoon symptoms.
e: Changes in his dinnertime dose of insulin will not affect his afternoon symptoms.

4. Answer d:

Severe diabetic nephropathy commonly results in oliguria rather than polyuria.

a: Neurogenic diabetes insipidus represents an ADH deficiency resulting in polydipsia and polyuria.
b: Nephrogenic diabetes insipidus represents a failure of the kidneys to respond to ADH, resulting in polydipsia and polyuria.
c: Psychogenic polydipsia is a condition found predominantly in patients with psychiatric disorders and is characterized by polydipsia and polyuria.
e: Diabetes mellitus may result in an osmotic diuresis when plasma glucose exceeds the reabsorptive capability of the kidney.

5. Answer d:

The half-life of thyroxine is seven days; therefore hypothyroidism would be unexpected this soon.

a: Thyroid storm is a complication following thyroid surgery in patients who are not adequately treated with antithyroid drugs preoperatively.
b: Wound hemorrhage may complicate up to 1% of thyroid operations, resulting in airway compromise.
c: Recurrent laryngeal nerve injury is a known complication following thyroid surgery, resulting in a change in voice.
e: Hypoparathyroidism is a complication during thyroid surgery. Care must be taken to leave at least one parathyroid gland to prevent postoperative hypocalcemia.

Chapter 12: Urology

1. Answer e:

This patient has history of ESRD and, therefore, difficulty excreting potassium normally. The indication for acute dialysis in this patient is hyperkalemia *refractory* to medical management. Her ECG has already shown the effects of elevated potassium, with rising levels despite medical therapy.

a: This will take at least 30 minutes to act and, with rising serum levels, will not be enough to lower serum levels effectively.
b: These will temporarily force potassium intracellularly, but will not effectively treat refractory hyperkalemia.
c: Although a repeat ECG is helpful in monitoring progression of hyperkalemia, medial intervention at this time is critical. Hemodialysis is an emergent therapy needed in this patient.
d: Bicarbonate also may help temporarily treat hyperkalemia but will be ineffective in reducing this level of serum potassium quickly and effectively enough.

2. Answer b:
Most patients suffer from either a microcytic hypochromic anemia primarily as a result of iron deficiency anemia or a normochromic, normocytic anemia from chronic disease.

a: Patients with renal failure are subject to coagulopathies primarily as a function of platelet dysfunction. The etiology is multifactorial. It may be exacerbated as well because of chronic anemia.
c: An inability to excrete phosphorous on a chronic basis is the result of the kidney's inability to excrete phosphorous because of a reduced GFR or increased tubular resorption.
d: Metabolic acidosis develops in renal failure because the kidney is unable to excrete H^+ ions, related to its inability to excrete ammonia.
e: As evidenced in the patient's first scenario; hyperkalemia is a chronic and sometimes life-threatening problem. Potassium retention's etiology is also multifactorial—decreased GFR, renal tubular dysfunction, oliguria, acidosis, and other factors.

3. Answer e:
Staphylococcus aureus is not a typical organism that causes urethritis.

a: *Neisseria gonorrhea* is a gram-negative diplococci often cultured. It occurs through sexual contact.
b: Chlamydia may cause up to one half of cases of infection.
c: *Ureaplasma urealyticum* may cause 30% of cases of infection.
d: Syphilis may cause penile discharge. It is often associated with sexual contact.

4. Answer d:
N. gonorrhea does not usually grow in the vagina.

a: Infertility may be a complication caused by obstruction and scarring of the fallopian tubes.
b: Urethral structure may be a result of infection.
c: Tubal pregnancy may be a complication of infection leading to obstruction and scarring of the fallopian tubes.
e: PID may occur in up to 20% of patients with a gonococcal infection.

5. Answer e:
The use of tampons has not been found to be associated with cystitis.

a: Pregnancy has been associated with increased risk for UTI.
b: Bacterial colonization during intercourse may predispose to increased risk for cystitis.
c: Use of a diaphragm has been associated with increased risk of cystitis.
d: Delayed micturition after intercourse may predispose to increased risk for cystitis because of increased bacterial colonization.

6. Answer e:

Vancomycin is not a first-line antibiotic for cystitis.

a: Often Bactrim is first-line therapy, specifically aimed at typical gram-negative organisms.
b: Amoxicillin is often a first-line drug for cystitis.
c: Nitrofurantoin may be used for cystitis therapy.
d: Phenazopyridine (Pyridium) is a urinary analgesic that may be used 3 times a day to help with dysuria.

Chapter 13: Obstetrics and Gynecology

1. Answer c:

Fibroadenomas are freely mobile, have distinct borders, and are smooth in consistency.

a: Pain is not associated with a fibroadenoma.
b: Indistinct borders are usually associated with malignant masses.
d: Bloody discharge is usually associated with intraducal papilloma.

2. Answer b:

Tissue diagnosis is mandatory in any solid breast mass. Fine-needle aspiration is the best route.

a: Following the lesion is not adequate. Tissue diagnosis is mandatory.
c: Danazol would treat fibrocystic breast disease.
d: Wide excision of the lesion is not indicated unless fine-needle aspiration cannot be done.

3. Answer b:

Hysteroscopy would be indicated for bleeding secondary to a cervical lesion.

a: Cultures would be appropriate in any sexually active woman with a discharge.
c: Pap smear is indicated in the presence of a uterine lesion.
d: When a gross lesion is present, biopsy is indicated.

4. Answer b:

90% of cervical cancers are squamous cell cancer.

a: HPV is probably the most common risk factor for the development of cervical cancer.
c: Cone biopsy (cold, laser, LLETZ) gives cure rates of 95% for carcinoma in situ.
d: 45 is the median age for invasive cancer.

5. Answer c:

The diagnosis of preeclampsia cannot be made by the clinical facts presented, but red flags are raised. Further investigation is warranted to clarify the diagnosis and assure maternal and fetal well-being.

a: There are no indications for hospitalization and hydration.
b: The criteria for diagnosis of preeclampsia are not met in the presentation.
d: Further evaluation of the patient is indicated before sending the patient home.

6. Answer b:

Laboratory evaluation supports the diagnosis of mild preeclampsia with the level of hypertension, proteinuria, generalized edema, and elevated uric acid. Her epigastric tenderness is likely secondary to emesis, as the liver function tests were normal. She has mild hyperreflexia but no other acute CNS manifestations. Viral gastroenteritis would not account for the proteinuria or elevated uric acid. Hyperemesis gravidarum is a disease of the first trimester.

a: Hyperemesis gravidarum is only in the first trimester of pregnancy.
c: Severe preeclampsia would not be supported with these elevations of

blood pressure and amount of proteinuria.

d: Based on blood pressure and proteinuria, mild preeclampsia must be diagnosed.

Chapter 14: Hematology and Oncology

1. Answer b:

This severe hemophiliac with arm swelling is having bleeding in the soft tissues until proven otherwise and must be treated promptly with factor VIII. He is also developing signs of compartment syndrome (abnormal sensation) and should evaluated by a vascular or plastic surgeon.

a: Red blood cell transfusion will not correct the underlying coagulation disorder and, unless the patient is volume-depleted from significant blood loss, is not necessary.

c: This patient's presentation is not consistent with venous occlusion. Venogram will not help in the diagnosis or management of his case and will delay appropriate management with factor replacement.

d: This patient's symptoms are not the result of a bone fracture. Soft tissue hemorrhage is the cause of the swelling. Radiograph will demonstrate soft tissue swelling only and will delay appropriate management with factor replacement.

2. Answer c:

Because of the abnormal or decreased factor VIII, the PTT is prolonged.

a: The PT is normal in patients with hemophilia unless the patient has another reason for coagulation defects, such as liver disease.

b: A bleeding time in this patient should not be performed.

d: The PTT is prolonged in hemophiliac patients.

3. Answer b:

This patient appears to be having a vasocoocclusion crisis and is clearly dehydrated. Given that she is afebrile and most likely had a viral syndrome, broad-spectrum antibiotics are not necessary in this situation.

a: Women of childbearing age with nausea should have a pregnancy test.

c: This patient is clearly dehydrated and should be aggressively hydrated.

d: It is likely that the patient's hematocrit will drop even further with hydration, and transfusion is indicated.

e: This patient's pain should be managed with parenteral narcotic analgesics, particularly because her oral medication has been ineffective. It is crucial to achieve adequate pain control for patients with sickle cell crisis.

4. Answer a:

Liver biopsy is not indicated in this patient without first evaluating the gastrointestinal bleeding. Although the hepatomegaly and elevated alkaline phosphatase should not be ignored, the most straightforward approach would be to evaluate the lower GI tract first.

b, c: One common cause for slow gastrointestinal bleeding in a patient of this age group is colon carcinoma. Both colonoscopy and sigmoidoscopy with barium enema can be used in the workup for suspected colon carcinoma and would be appropriate in this patient.

d: Iron studies would be helpful in confirming iron deficiency anemia, which, if present, would confirm that the blood loss is chronic. (Iron studies would not be necessary in this patient.)

e: A CT scan of the abdomen would be appropriate in this patient to evaluate the size and texture of the liver. It may also show abnormalities of the bowel but should not be relied on in any patient suspected of having colon carcinoma.

Chapter 15: Internal Medicine

1. Answer d:

The patient is competent to make the "do not resuscitate" decision and it should not be changed without his consent.

a: The patient's wishes must be respected despite the wishes of the cardiologist.
b: Despite noble intentions, a competent patient's wishes must be respected by the treating physician(s).
c: The administration of appropriate care should occur if a patient is likely to benefit and agrees to the treatment plan.
e: The cost of a procedure currently has no place in matters of ethics and patient autonomy.

2. Answer d:

Although hemodialysis patients can tolerate slightly higher potassium levels, a measurement of 6.9 is dangerous for anyone. This needs to be treated as an emergency. The goal is to prevent cardiac arrhythmias. This plan correctly gives the sequence necessary and stresses the urgency of the situation.

a: Hyperkalemia is an expected lab derangement in a hemodialysis patient who has missed his treatments. Furthermore, it requires immediate aggressive treatment while awaiting hemodialysis.
b: Using diuretics in this patient is a waste of precious time and risks the development of cardiac arrhythmias.
c: Sending a patient with this degree of hyperkalemia to a regular medical floor is dangerous, especially without first checking for the ECG manifestations of hyperkalemia.
e: This plan does not address the urgent nature of the patient's condition. The nurse should be called and the patient should be placed on telemetry, but these are not the first interventions.

3. Answer c:

T wave peaking followed by flattening are commonly documented ECG manifestations in patients with hyperkalemia.

a: T wave inversion is a finding associated with myocardial ischemia, not hyperkalemia.
b: Although arrhythmias occur, ventricular fibrillation is the most common, and atrial flutter is not necessarily associated with hyperkalemia.
d: Left bundle branch block may be present, but it is not causally related to this electrolyte abnormality.
e: Localized inferior ST elevations should suggest myocardial infarction rather than hyperkalemia.

4. Answer b:

The most frequent cause of treatment failure in Lyme disease is lack of true disease being present. This patient treated himself without many symptoms indicating Lyme disease. Before committing him to more antibiotics it is important to confirm or refute the diagnosis.

a: Before initiating further treatment the diagnosis must be confirmed.
c: Intravenous penicillin is reserved for patients with advanced disease (e.g., involving the central nervous system); before initiating another therapeutic trial of antibiotics a serologic diagnosis should be made.
d: Avoiding wooded areas may lessen his risk of ultimately contracting Lyme disease but is unnecessary.
e: Before initiating another therapeutic trial of antibiotics a serologic diagnosis should be made.

5. Answer b:

The p value represents the chance in percentages that the difference noted was due to chance alone.

a: A p value of <0.01 indicates a very definite difference between the two groups.
c: The p value of <0.01 is the same as saying "less than 1%."
d: Although the p value suggests that the difference between the groups is significant, it does not suggest what accounts for the difference.
e: A p value of <0.01 suggests that the difference between the groups be accounted for by chance alone less than 1% of the time.

Chapter 16: Emergency Medicine

1. Answer d:

Cold dressings may relieve pain but are contraindicated in large burns because the resultant vasoconstriction can deprive vulnerable tissue of blood flow. An antibiotic ointment followed by a sterile dry dressing is indicated.

a: Adequate analgesia is important not only for comfort, but also to make possible a through débridement. Copious fluids are needed to replace fluids lost from burn sites.
b: Débridement of blisters, necrotic skin, and foreign bodies is indicated for large burns.
c: Tetanus immunization should be addressed in any burn.
e: Copious fluids are needed to replace fluids lost from burn sites.

2. Answer b:

Coins lodged in the trachea lie in the sagittal plain. (The cartilaginous rings are C-shaped, with soft tissue along the posterior side). This child, though stable at the moment, will require constant monitoring until surgical removal can be arranged. No attempt to remove the coin in the ED should be made for fear that a shift in its position could compromise the airway.

a: The sagittal orientation of the coin implies it is lodged in the trachea. Answer A (water sips, midazolam and or glucagon) describes methods to enhance passage through the lower esophagus.
c: Some coins can be removed from the esophagus via an inflated catheter, but this method is far too risky to use for coins lodged in the trachea.
d: Otolaryngologic help is warranted, but the patient has the potential to obstruct his airway at any time. Because of the potential airway compromise, this child should not leave the emergency department unless he is closely monitored.
e: The Heimlich maneuver is reserved for complete obstruction of the airway.

3. Answer d:

As with any emergency, the first priority is to secure and protect the airway and breathing. This may include a chest x-ray, pulse oximetry, arterial blood gas, and oxygen. If the patient is asymptomatic, has a clear chest x-ray, and good follow-up can be assured, he can be discharged within 6 hours.

a: Pulmonary irritation is the most frequent serious effect of hydrocarbon ingestion. Ipecac is contraindicated because resultant emesis could increase the chance of aspiration.
b: Gasoline and other hydrocarbons have the potential to be very dangerous, but overall morbidity is low. Most problems arise in the pulmonary system, not the GI tract. GI decontamination is indicated only if there has been a toxic additive.
c: Charcoal is ineffective at absorbing gasoline.
e: Gasoline ingestion is often benign, leaving the GI tract uninjured, but it has the potential to cause seri-

ous respiratory damage, even in small amounts.

4. Answer c:

Cutting the umbilical cord is not needed and can wait until the child arrives at the ED. There is always the risk that the volunteer may use nonsterile trauma shears rather than a sterile blade. Nonsterile blades have been linked to neonatal tetanus.

a: The ABCs of delivery start with suctioning the airway.
b: The ABCs of delivery continue with tying the umbilical cord to prevent excessive transfusion to or from the child.
d: A newborn infant is wet and unable to shiver, thus at risk of hypothermia. Even in warm conditions, drying and wrapping the child is needed.
e: Removing the placenta is not an emergency and can safely be delayed until arrival at the hospital.

5. Answer c:

A very brief examination is indicated; splinting is often wise to minimize pain and to protect the joint from further injury, and pain control can be started. An x-ray should be performed before any flexion or extension of the elbow is performed. Nursemaid's elbow is caused by a pulling on a child's relaxed outstretched arm. This child is at high risk of a much different injury, posterior dislocation and possible fracture of her elbow.

a: The description of the treatment of nursemaid's elbow is correct; the history, however, is inconsistent with nursemaid's elbow.
b: It is wise to rule out snuff box tenderness and neurovascular compromise in any upper extremity injury. The chief complaint, however, is elbow tenderness, and a fracture must be ruled out.
d: A brief examination is indicated, but if a fracture is suspected, a motor examination (i.e., flexion of the joint) should be deferred until after a fracture is ruled out by examination.
e: This limb will receive a temporary splint. Before the fractures have been identified, it is best to splint the limb in the position in which it is found, rather than risk injury by manipulating a broken limb.

Chapter 17: Pediatrics

1. Answer c:

This is a typical presentation of colic; however, other causes of irritability must be ruled out. The first step in this process is a thorough physical examination, including vital signs. Additional tests may be necessary based on the presence or absence of physical findings.

a: Without fever or other signs or symptoms of illness, these tests are unnecessary.
b: Although this is a classic history for colic, a careful, complete physical examination is required to rule out other causes of irritibility, such as hair tourniquet, shaken baby syndrome, or even bacterial illness.
d: Without a suspicious story or suggestive physical findings, this action would not be appropriate.

2. Answer d:

If the infant had a history of seizure within 3 days of receiving the DTP vaccine, further immunization with pertussis vaccine would be contraindicated; however, family history would not place this infant at any increased risk of reaction to vaccination.

a, b, c: This family history does not fit the criteria for altering the infant's immunizations.

3. Answer b:

Live-virus vaccines are contraindicated in immune-compromised patients (OPV, MMR, varicella vaccine). Live-virus vaccines with shedding (OPV) are contraindicated in housemates of immune-compromised individuals, which is the case in this situation. Therefore OPV is contraindicated. When she is 12 to 15 months of age, it will be safe for this infant to receive MMR.

a, c: Each has OPV, a live-virus vaccine that is contraindicated in housemates of immunosuppressed patients.
d: MMR is also a live-virus vaccine and, therefore, contraindicated in housemates of immunosuppressed patients.

4. Answer c:
This is a typical presentation of measles. Most otherwise healthy children require only rest and antipyretics; however, adults on board may be susceptible to atypical measles. Thus respiratory isolation would be appropriate.

a: Unless the child were to develop a petichial rash or become toxic-appearing, this would not be necessary.
b: Because measles is a viral illness, this action would be tempting but inappropriate, and would probably confuse the picture.
d: The oral findings are Koplik's spots, not the result of Group A strep. In the absence of other throat findings, this would not be necessary.

5. Answer c:
This is a classic presentation of gastroesophageal reflux. Key points are that the infant appears well, with appropriate weight gain and appetite. With an ill or dehydrated-appearing infant, the presence of fever or poor weight gain would require further workup, but in this case a trial of presumptive therapy for reflux with frequent reassessment would be appropriate.

a: Although the history concerns the appearance of the infant, particularly the weight, the appetite and the small amount of emesis on observation would not warrent a pediatric emergency center visit at this point.
b: Unless the emesis were to increase to include most of the amount ingested, this would not be necessary.
d: It is very tempting to switch formulas, but rarely appropriate. Evidence of formula intolerance usually includes fussiness, diarrhea, and poor weight gain. In the absence of these, it is better to use a diagnostic and therapeutic trial that better fits the most likely diagnosis, i.e., GE reflux.

Chapter 18: Surgery

1. Answer c:
Murphy's sign is the abrupt interruption in inspiration while palpating the RUQ and is indicative of biliary pathology.

a: A positive Rovsing's sign is a helpful physical sign of local peritoneal irritation, secondary to acute appendicitis.
b: A positive Psoas sign is also a physical sign of local retroperitoneal irritation, secondary to acute appendicitis.
d: A rectal examination is mandatory and can often be the only positive physical finding in acute appendicitis.
e: A gas-filled appendix on plain abdominal x-ray is suggestive of acute appendicitis.

2. Answer b:
A patient with a known AAA complaining of lower abdominal and flank pain is at high risk for further dissection and rupture. This possibility must be immediately ruled out in a setting where if rupture is

found emergent operation can be performed. CT scan would be the diagnostic modality of choice.

a: Symptomatic abdominal aortic aneurysms demand urgent evaluation.
c: Newly symptomatic AAAs cannot be managed electively.
d: A CT scan is the diagnostic test of choice.
e: Although a CT scan is indicated, it should not be done on an outpatient basis.

3. Answer a:

A pulsating abdominal mass can be palpated in greater than 90% of patients presenting with a symptomatic AAA.

b: Femoral and distal pulses may be lost secondary to distal occlusion.
c: Severe abdominal pain is most often seen in ruptured AAA.
d: Altered mental status may indicate hypotension and shock.
e: Fever may be present in mycotic aneurysms.

4. Answer b:

Internal hemorrhoids are covered with columnar epithelium proximal to the dentate line and most often associated with bleeding upon defecation.

a: External hemorrhoids are covered with innervated stratified squamous epithelium distal to the dentate line.
c: Pain is most often associated with external hemorrhoids, especially if thrombosed.
d: Hemorrhoids are not premalignant.
e: Painless rectal bleeding is most consistent with internal, rather than thrombosed, hemorrhoids.

5. Answer c:

Asymptomatic hemorrhoids require no specific therapy.

a, e: Injection sclerotherapy and hemorrhoidectomy are possible therapies in symptomatic external hemorrhoids.
b: Rubber band ligation should not be routinely used to treat external hemorrhoids due to the pain induced.
d: Asymptomatic hemorrhoids will receive no benefit from the topical application of steroids.

INDEX

Pages in italics indicate figures; pages with *t* indicate tables.

E

H

J

Q

R

Mosby's Reviews Series

How to install this program—Windows users

1. Place the disk in Drive A: (or B:)
2. From Program Manager, select File, then Run, then enter:
   A:SETUP (or B:SETUP if your disk drive is B:)
3. Follow the instructions on screen.

How to run this program—Windows users

Open the MOSBY Program Group and select the Mosby's Reviews Series program.

How to install this program—Macintosh users

1. Insert the disk into the disk drive. Double-click on the disk icon.
2. Double-click on the Install icon.

How to run this program—Macintosh users

Open the MOSBY folder and select the Mosby's Reviews Series program.

The software enclosed, Version 2.0, represents an upgrade in the FIND function. Now you can search not only by Question Number, but also by Level 1, 2, or 3 descriptions! (See the Help menu for details.) And, Version 2.0 also upgrades all your previously loaded testbanks, so that *all* your testbanks will now have this improved feature.

If later on you install a *Mosby's Reviews Series* diskette that does not have this upgraded function, you will need to reinstall this diskette, so that Version 2.0 remains the most currently installed version on your system (check the "About" screen in the Help menu to verify your version).

Mosby's toll-free Technical Support number is 1-800-692-9010.

For complete instructions on using the program, please read the Help Index.